The Integrative Functions of the Basal Ganglia

This volume is the first comprehensive and single-authored book on the functions of the basal ganglia. The goal is to provide a new synthesis of diverse areas of research on the basal ganglia, from cellular mechanisms of synaptic transmission and plasticity to neural circuit mechanisms underlying behavior. A global theory of basal ganglia function incorporating research from the last 40 years is presented. I hope to explain for the first time how the basal ganglia generate behavior, how they contribute to learning and memory, and how impairments in basal ganglia function can lead to neurological and psychiatric disorders.

Features

- The only single-authored book on the basal ganglia with coverage of the latest literature.
- Spans multiple levels of analysis, from cellular physiology to behavior.
- Includes coverage of clinical symptoms, encompassing neuropsychology, movement disorders, and psychiatric disorders.
- Discusses the role of the basal ganglia in learning and memory.

The Integrative Functions of the Basal Ganglia

Henry H. Yin

CRC Press is an imprint of the
Taylor & Francis Group, an **informa** business

Cover Image: Image from Anders Nelson. Spiny projection neurons in the dorsolateral striatum (blue), surrounded by axons from motor cortex (yellow).

First edition published 2024
by CRC Press
6000 Broken Sound Parkway NW, Suite 300, Boca Raton, FL 33487-2742

and by CRC Press
4 Park Square, Milton Park, Abingdon, Oxon, OX14 4RN

CRC Press is an imprint of Taylor & Francis Group, LLC

© 2024 Taylor & Francis Group, LLC

Reasonable efforts have been made to publish reliable data and information, but the author and publisher cannot assume responsibility for the validity of all materials or the consequences of their use. The authors and publishers have attempted to trace the copyright holders of all material reproduced in this publication and apologize to copyright holders if permission to publish in this form has not been obtained. If any copyright material has not been acknowledged please write and let us know so we may rectify in any future reprint.

Except as permitted under U.S. Copyright Law, no part of this book may be reprinted, reproduced, transmitted, or utilized in any form by any electronic, mechanical, or other means, now known or hereafter invented, including photocopying, microfilming, and recording, or in any information storage or retrieval system, without written permission from the publishers.

For permission to photocopy or use material electronically from this work, access www.copyright.com or contact the Copyright Clearance Center, Inc. (CCC), 222 Rosewood Drive, Danvers, MA 01923, 978-750-8400. For works that are not available on CCC please contact mpkbookspermissions@tandf.co.uk

Trademark notice: Product or corporate names may be trademarks or registered trademarks and are used only for identification and explanation without intent to infringe.

ISBN: 9781498768696 (hbk)
ISBN: 9781032580333 (pbk)
ISBN: 9780429154461 (ebk)

DOI: 10.1201/9780429154461

Typeset in Times
by codeMantra

To my parents, who taught me more than I can remember.

Contents

Preface xiii
About the Author xv

Chapter 1 Introduction 1
          1.1 Discovering the Basal Ganglia 1
          1.2 Fortunes of the BG 2
          1.3 Extrapyramidal System 3
          1.4 Basic Motif of Cerebral Organization 4
          1.5 Action Selection 6
          1.6 Reinforcement Learning 7
          1.7 Behavior: The Neglected Problem 8
          1.8 Explaining BG Function 9
          1.9 Summary 11
          References 11

Chapter 2 Anatomical Organization of the Basal Ganglia 15
          2.1 What's in a Name? 15
          2.2 Topographical Inputs to the BG 17
                  2.2.1 Limbic Striatum 18
                  2.2.2 Associative Striatum 18
                  2.2.3 Sensorimotor Striatum 18
          2.3 Chemical Compartments 19
          2.4 Sources of Corticostriatal Projections 20
          2.5 Convergence and Divergence 21
          2.6 Thalamostriatal Projections 22
          2.7 Dopaminergic Projections to the Striatum 23
          2.8 Direct and Indirect Pathways 23
          2.9 BG Output Nuclei 25
          2.10 Globus Pallidus External Segment (GPe) 27
          2.11 Globus Pallidus Internal Segment (GPi) 28
          2.12 Ventral Pallidum 28
          2.13 Ventral Tegmental Area (VTA) 29
          2.14 Substantia Nigra 30
                  2.14.1 Nigral Outputs 31
                  2.14.2 Substantia Nigra Pars Lateralis 32
          2.15 Subthalamic Nucleus (STN) 32
          2.16 Summary 33
          References 34

Chapter 3 Synaptic Transmission and Plasticity in the Basal Ganglia 43
          3.1 Striatal Neurons 43
                  3.1.1 Spiny Projection Neurons (SPNs) 44
                  3.1.2 Fast-Spiking Interneurons (FSIs) 45
                  3.1.3 Low-Threshold Spiking Interneurons (LTSIs) 45

		3.1.4 Cholinergic Interneurons (CINs)...46

- 3.2 Pallidal Neurons 47
  - 3.2.1 GPe Neurons 47
  - 3.2.2 GPi Neurons 48
  - 3.2.3 Nigral Neurons 48
- 3.3 Glutamate 49
  - 3.3.1 Corticostriatal Transmission..50
  - 3.3.2 Thalamostriatal Transmission 50
- 3.4 Gamma-Aminobutyric Acid (GABA) 50
  - 3.4.1 Feedforward Inhibition in the Striatum...51
  - 3.4.2 GABAergic Transmission in the Pallidum 52
  - 3.4.3 Lateral Inhibition...52
- 3.5 Dopamine 53
  - 3.5.1 How DA Influences Striatal Outputs...54
  - 3.5.2 Dopaminergic Modulation of GABAergic Transmission 55
- 3.6 Acetylcholine (ACh)..55
- 3.7 Endocannabinoid (eCB) 57
- 3.8 Adenosine 58
- 3.9 Synaptic Plasticity in the BG 59
  - 3.9.1 Induction of Plasticity 59
- 3.10 Striatal LTP..60
- 3.11 Striatal LTD 60
- 3.12 LTP or LTD?..62
- 3.13 Functional Implications of Striatal Synaptic Plasticity 62
- 3.14 Summary 63
- References 63

Chapter 4 Current Ideas on BG Function 75

- 4.1 Parallel Loops..75
- 4.2 Convergence and Divergence 78
- 4.3 Interaction between Loops 79
- 4.4 Disinhibition..81
- 4.5 Rate Model 83
- 4.6 Action Selection Models 85
  - 4.6.1 Focused Selection..85
  - 4.6.2 Central Selection 86
  - 4.6.3 Problems with Action Selection 87
- 4.7 Reinforcement Learning..87
- 4.8 The Challenge of Behavioral Analysis 89
- 4.9 Summary 90
- References 90

Chapter 5 Behavior and Control 97

- 5.1 Insufficiency Principle...97
- 5.2 Solutions to the Calculation Problem 98
- 5.3 The Definition of Control 99
- 5.4 Computing in a Control System 100
- 5.5 Misunderstanding Control 101
- 5.6 Hierarchical Control 103

	5.7	Beyond Sensorimotor Transformations 104
	5.8	Reinforcement and Teleology 104
	5.9	Neural Signaling and Control Systems 105
	5.10	Information and Coding 107
	5.11	Summary 108
	References 109

Chapter 6 The Place of the BG in the Hierarchy 111

  6.1 Posture and Movement 111
      6.1.1 Parkinsonian Rigidity 112
  6.2 Control of Muscle Tension and Length 112
      6.2.1 Muscle Length Control and the Fusimotor System 113
  6.3 Bidirectional Control 115
      6.3.1 Stiffness Control 115
      6.3.2 Bandwidth Limitations 116
  6.4 Reticulospinal Pathway 116
  6.5 Position Controllers for Orientation 116
  6.6 Midbrain Contributions to Orienting 118
  6.7 Nigrocollicular Pathway and Eye Movements 119
  6.8 BG Regulation of Steering and Orienting 120
  6.9 SNr and Postural Control 122
  6.10 BG and Locomotion 123
  6.11 SNr and Position Coordinates 123
      6.11.1 SNr and Representation of Position Vectors 124
  6.12 VTA Output and Head Position 126
  6.13 Functional Significance of BG Outputs 127
      6.13.1 Two-Way Comparison Functions 129
  6.14 Summary 129
  References 130

Chapter 7 Transition Control 135

  7.1 Velocity Control 135
  7.2 DA and Kinematics 137
  7.3 Position Control versus Velocity Control 138
  7.4 Vector Integration to Endpoint (VITE) Model 139
  7.5 A Neural Integrator in the BG 140
      7.5.1 Integrator Dynamics 140
  7.6 Direct and Indirect Pathways 142
  7.7 Neurobiological Implementation of Integration 143
  7.8 Dopamine and Gain Control 145
      7.8.1 Bradykinesia and Akinesia 146
      7.8.2 Adaptive Gain 146
  7.9 Adaptive Gain and Reinforcement of Action Parameters 147
  7.10 Corticostriatal Circuit and Transition Control 148
  7.11 Compared with Other Models of BG Function 150
  7.12 Limitations in Previous Experimental Designs 151
  7.13 Summary 152
  References 153

| **Chapter 8** | Higher-Order Transitions and Cognition | 157 |

8.1 Event Repetition and Control of Tempo ... 157
8.2 Regulation of Rhythmic Behavior ... 158
    8.2.1 Licking ... 158
8.3 Related Rates and Gear Coupling ... 161
8.4 Interval Timing ... 163
    8.4.1 BG and Timing ... 164
    8.4.2 DA Modulation of Timing ... 165
    8.4.3 Direct and Indirect Pathways ... 166
8.5 Serial Order ... 167
8.6 Dorsolateral Striatum and Grooming ... 168
8.7 Sequence Learning ... 168
8.8 Imagination ... 171
    8.8.1 BG and the Imagination Mode ... 172
    8.8.2 Mental Rotation ... 173
8.9 Working Memory ... 173
8.10 Summary ... 174
References ... 175

Chapter 9 Motivation ... 181

9.1 Aspects of Motivation ... 181
9.2 Limbic BG and Reward ... 182
9.3 Valence and Bidirectional Control ... 184
9.4 Distinct Accumbens Outputs Regulate Reparatory and Consummatory Behaviors ... 185
9.5 Reinforcement ... 186
9.6 Self-Stimulation ... 186
9.7 Effort Exertion ... 188
    9.7.1 Limbic BG and Effort Regulation ... 190
    9.7.2 DA Contribution to Effort ... 190
    9.7.3 Conflating Reward Rate and Performance ... 191
9.8 DA and Force Generation ... 191
9.9 A Labile Motivational Hierarchy ... 193
9.10 Parallel BG Networks and Motivated Behaviors ... 194
9.11 Summary ... 196
References ... 196

Chapter 10 Actions and Goals ... 201

10.1 Approaching a Goal ... 201
10.2 Compulsory Approach ... 201
10.3 A Striatal Circuit for Relationship Control and Continuous Pursuit ... 202
10.4 Approach Behavior and Feedback ... 205
10.5 Learning to Approach ... 206
10.6 Contingency, Associative Structures, and Analysis of Conditioning Experiments ... 208
10.7 Goal-Directed Actions ... 210
10.8 Neural Basis of Action–Outcome Learning ... 211
    10.8.1 Striatal Activity Modulated by Reward Expectancy ... 212
    10.8.2 Associative Cortico-BG Network and A–O Learning ... 213

		10.8.3	Posterior Dorsomedial Striatum	213
		10.8.4	Direct and Indirect Pathways	215
		10.8.5	Synaptic Plasticity in pDMS	217
		10.8.6	The Role of DA in Goal-Directed Actions	217
	10.9	BG Contributions to Neuroprosthetic Control		218
	10.10	Summary		219
	References			219

Chapter 11 Corticostriatal Contributions to Habits and Behavioral Automaticity ... 225

	11.1	Multiple Memory Systems	225
	11.2	Place and Response	226
	11.3	BG and Procedural Learning	227
	11.4	Limitations of Multiple Memory Systems Framework	228
	11.5	From Actions to Habits	229
	11.6	Sensorimotor Striatum and Habit Formation	230
		11.6.1 Plasticity Mechanisms Underlying Habits	232
	11.7	Development of Automaticity	232
	11.8	Habits and Skills	234
		11.8.1 Reduced Attentional Demand	234
		11.8.2 Effector Specificity	234
	11.9	Nature of Feedback	235
	11.10	Compulsive Behavior	236
	11.11	Summary	237
	References		238

Chapter 12 Dopamine and Reinforcement Learning ... 243

	12.1	Prediction Errors	243
		12.1.1 Rescorla–Wagner Model	243
		12.1.2 Temporal Difference Algorithm	244
	12.2	Principles of RL	244
	12.3	Phasic DA and RPE	246
	12.4	Results that Challenge the RPE Hypothesis	247
	12.5	Value, Performance, and RPE	248
	12.6	Phasic DA and Performance	248
	12.7	Is DA Necessary and Sufficient for Learning?	251
	12.8	Learning and Performance	253
	12.9	Adaptive Gain and Reinforcement	255
	12.10	Reinforcement Learning versus Control	257
	12.11	Summary	259
	References		259

Chapter 13 Reorganization, Exploration, and Plasticity ... 265

	13.1	Learning from a Control Perspective	265
	13.2	Exploration and Reorganization	267
	13.3	BG and Exploration	267
	13.4	Bird Song Learning	268
	13.5	BG Contributions to Bird Song	270
	13.6	DA and Bird Song	271

	13.7	Lessons from Bird Song	272
	13.8	Neural Plasticity and Reorganization	273
		13.8.1 Adaptive Gain and Induction of Long-Term Plasticity	273
		13.8.2 Changes in DA Dependence with Learning	275
	13.9	Summary	276
		References	276

Chapter 14 Interpretation of Clinical Symptoms 281

    14.1 Analysis of Symptoms 281
        14.1.1 Impaired Input Function 281
        14.1.2 Impaired Output Function 282
        14.1.3 Change in Gain 282
        14.1.4 Oscillations 282
    14.2 Postural Control Deficits 282
    14.3 Bradykinesia, Akinesia, and Paradoxical Kinesia 284
    14.4 Deficits in Locomotion 285
    14.5 Deep Brain Stimulation (DBS) 287
    14.6 Hyperkinetic Symptoms 288
        14.6.1 Chorea 288
        14.6.2 L-DOPA–Induced Dyskinesia (LID) 289
        14.6.3 Reduced Damping 289
        14.6.4 Loss of Selectivity 290
        14.6.5 Loss of Feedback 290
    14.7 Perseveration, Stereotypy, and Compulsion 291
    14.8 Attentional Deficits 292
    14.9 Psychosis and Schizophrenia 293
        14.9.1 Positive and Negative Symptoms 293
        14.9.2 Hallucinations 293
        14.9.3 Delusions 294
    14.10 Summary 295
    References 296

Chapter 15 Synthesis 301

    15.1 Behavior and Control 301
    15.2 Kinematics as a Gateway to Understanding BG Function 302
    15.3 Cortex versus BG 304
    15.4 BG Outputs and the Coordination Problem 304
    15.5 Direct and Indirect Pathways 305
    15.6 DA and Adaptive Gain 306
    15.7 Motivational Hierarchy 307
    15.8 Goal Seeking and Control of Relationship 308
    15.9 Higher Functions 308
    15.10 Learning and Reorganization 309
    15.11 Gaps in Understanding 310
    15.12 A New Vista and the Way Forward 310
    References 311

Index 313

Preface

This book is an attempt to explain how the basal ganglia work, with a focus on how they contribute to behavior. It aims to develop a new model of basal ganglia function based on recent experimental findings and the principles of hierarchical feedback control. While I harbor the hope that this book may prove useful to those doing research on the basal ganglia, I have not written for specialists only. My imaginary reader is anyone with an interest in the relationship between the brain and behavior.

I knew that writing such a book would be a formidable challenge. While writing it, I was constantly reminded of my own ignorance on a wide range of topics. But it was also an opportunity to study the work of many scientists, past and present, whose admirable work is covered in this book. Despite my struggles, the long process of writing this book has proved to be a rewarding voyage of discovery.

It was Sidney Simon who first invited me to write this book. Without his persistence, I would not have agreed to such a major undertaking. I am particularly indebted to the following individuals for their advice, conversation, and feedback: Konstantin Bakhurin, Joseph Barter, Hongwei Dong, Isabella Fallon, Alexander Friedman, Charles Gerfen, Frank Hirth, Ryan Hughes, Xin Jin, Grace Lee, Richard Mooney, Peter Redgrave, Charles Wilson, and Fu-Ming Zhou. I also gratefully acknowledge the steady support of my own research over the years by the National Institute on Alcohol Abuse and Alcoholism, the National Institute of Drug Abuse, the National Institute of Mental Health, and the National Institute of Neurological Disorders and Stroke. Finally, I would like to express my gratitude to my wife, Cara, without whose support it would have been impossible to complete this book.

About the Author

Henry H. Yin is currently a professor of Psychology and Neuroscience at Duke University. Yin received his B.A. from Washington University in Saint Louis and his Ph.D. from UCLA. His research has, from the very start, focused on the function of the basal ganglia, combining different levels of analysis, from cellular to behavioral mechanisms. He has published over 80 papers on various aspects of basal ganglia function.

1 Introduction

So long as one clings to the study of elements, one is dealing with well circumscribed units, a well-defined subject, presenting clear-cut problems, and one can call on familiar and approved methods of analysis. As soon as one raises the eye from the unit to the whole system, the subject becomes fuzzy, the problems ill-descript, and the prospect of fruitful attack discouraging in its indefiniteness.

Paul Weiss

The basal ganglia (BG) are a collection of cell groups beneath the cerebral cortex. Known for centuries, they have increasingly attracted attention in recent years. Although the BG have long been implicated in many disorders, from Parkinson's disease to schizophrenia, their function remains obscure despite decades of research. As Konorski (1967, p. 267) noted: "Confusions in the analysis of the function of the basal ganglia lies not in the fact that it is difficult to discover any symptoms after lesions, but, on the contrary, that the symptoms are so manifold and variable". Lesions can produce deficits in reaching, grooming, and orofacial movements (Pisa, 1988; Pisa & Schranz, 1988); abnormal postures and rigidity (Martin, 1967); loss of speech (aphasia) (Jackson, 1875); learning and memory deficits (Packard & Knowlton, 2002; Saint-Cyr, Taylor, & Lang, 1988); and even hallucinations and delusions (Cummings, 1985; Middleton & Strick, 2000).

Reading the vast literature on the BG, which is full of contradictions and inconsistencies, the story of blind men describing the elephant comes to mind. The list of labels attached to the BG is long and ever-expanding, including motor control, language, attention, working memory, and procedural learning. Although the accumulation of facts has added new pieces to the puzzle, it also makes it harder to see how the pieces fit together.

Here we shall begin with a historical sketch of how the BG were first discovered and how models of BG function have evolved over time. We shall then consider the limitations of current models and the obstacles, both conceptual and technical, that have prevented progress.

1.1 DISCOVERING THE BASAL GANGLIA

The BG were first clearly described by Thomas Willis in his *Cerebri Anatome* published in 1664, the first modern neuroanatomy book based on dissections from multiple animal species. Willis was assisted by a talented team he assembled at Oxford, including the polymath Christopher Wren, the architect of St. Paul's Cathedral and the Sheldonian Theatre. According to Willis, Wren not only finished most of the drawings in the book with "his most skillful hand," but was also present at the dissections "to confer and reason about the uses of the parts" (Willis, 1664).

In Chapter 13 of *Cerebri Anatome*, we find Wren's famous drawing of the corpus striatum, which roughly corresponds to what is now called the striatum, the largest of the BG nuclei (Figure 1.1). In this drawing, subcortical structures in a sheep brain are exposed by removing the overlying cerebral cortex, and sectioning of the corpus striatum reveals a highly striated appearance. According to Willis, these striated bodies, which join the brain and the medulla oblongata, "in man and four-footed beasts are constantly found of the same species or form, and in every one of them, figured after the same manner" (Willis, 1664).

A gifted clinician, who coined "neurology," Willis examined the brains of patients who had died of various neurological diseases and attempted to relate the location and extent of tissue damage to their symptoms. He observed that in patients who died of palsy, the striated bodies were less firm, discolored, and less striated in appearance. According to him, the striated appearance is due to both

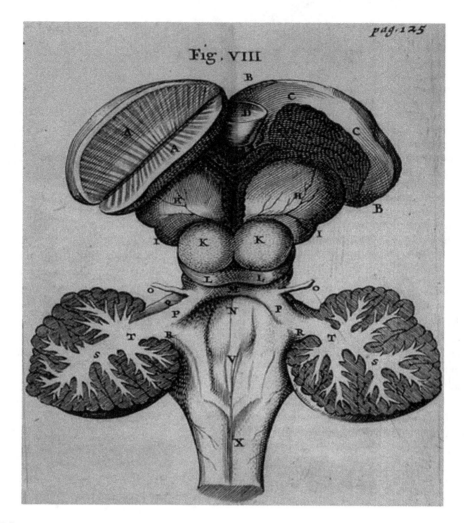

FIGURE 1.1 Willis, Wren, and the corpus striatum. Wren's drawing of the corpus striatum (striated bodies, labeled A) from the *Cerebri Anatome*.

descending and ascending pathways that connect the brain and the brainstem: the ascending pathways provide multimodal sensory inputs, which eventually could lead to the faculty of imagination, whereas the descending projections to the brainstem and spinal cord are responsible for generating spontaneous movements. Thus, largely based on anatomical examination, Willis offered the initial speculations on BG function, but nowhere in his book did he define sensation, movement, or imagination; nor did he develop a model for how these faculties may be implemented by the corpus striatum.

1.2 FORTUNES OF THE BG

Following Willis' discovery, the other major parts of the BG were identified by a succession of distinguished neuroanatomists (Parent, 2012). For example, Vicq-d'Azyr characterized the substantia nigra (black substance), which contains dopaminergic neurons darkened by high levels of neuromelanin. Burdach divided the striatum into the caudate and putamen and identified the internal and external segments of the globus pallidus.

Despite these discoveries, there was no consensus on the function of the BG, and estimation of their importance fluctuated in subsequent history. To Reil, who coined "psychiatry", the BG were the core nuclei of the cerebral hemispheres. In his enthusiasm, he even went so far as to compare the relationship between the BG and the rest of the brain to that between the sun and other planets in the solar system (Schiller, 1967). Being described as the sun of the nervous system, however, was perhaps the zenith in the fortunes of the BG, for soon they were eclipsed by the cerebral cortex.

The golden age of clinical neurology in the 19th century was defined by reports of dramatic symptoms following cortical damage, starting with Broca and Wernicke's observations on aphasia. These sensational discoveries led to efforts to localize psychological functions in the cerebral cortex. Lacking modern histological techniques, however, in many early clinical reports the extent of brain damage was not always well-characterized. Since the cerebral cortex is far larger than the underlying BG in the human brain, cortical damage is relatively easy to identify, whereas BG damage, which frequently accompanied cortical damage, was not always recognized. Perhaps for this reason, as the cerebral cortex took center stage, the contributions of the BG to behavior became neglected. In Cajal's magisterial *Histology of the Nervous System* (Cajal, 1909), for example, the corpus striatum was the subject of the shortest and least inspired chapter: "Endowed early on with very important and complex functions, it has come by way of a series of reductions simply to help coordinate reflex movements." Instead, Cajal devoted far more space to describing the cerebral cortex.

Cajal is only one of many distinguished investigators to conclude that the BG merely play an accessory role, one that is phylogenetically primitive and later replaced by the cerebral cortex. According to Herrick, the striatum is "the highest center of dominance in the control of the skeletal musculature, a role which is enormously enlarged in reptiles and birds. In mammals, parallel with the elaboration of cortex, the part which the striatum plays in the patterning of the behavior is progressively reduced, but it retains important functions of coordinating and stabilizing motor systems for facilitation of muscular coordination, so in mammals the striatal complex is interpolated in the efferent cortical systems as an accessory facilitating mechanism" (Herrick, 1948). Such views contributed to the neglect of the BG for much of the 20th century.

It was not until the last few decades that the study of the BG once again gathered momentum, partly due to the development of techniques for accurate anatomical tracing and physiological recording, which led to reliable knowledge of cellular elements in the BG and their connectivity (Wilson, 2016).

1.3 EXTRAPYRAMIDAL SYSTEM

Historically, there has been a persistent tendency to contrast the functions of the BG and the cerebral cortex. It is often assumed that the cortex is the center for voluntary actions, whereas subcortical structures like the BG subserve involuntary, reflexive, or postural mechanisms that are considered more primitive and accessory. An example of such thinking is the influential distinction between "pyramidal" and "extrapyramidal" systems, a distinction that is common in the clinical literature.

Wilson first reported a rare degenerative disease (Wilson's disease) with extensive BG damage that results in postural deficits, rigidity, and akinesia. Since these symptoms are distinct from those from pyramidal tract damage, Wilson attributed them to a separate, phylogenetically older, "extrapyramidal" motor system (Wilson, 1914, 1925). This phylogenetically older motor system, which has become nearly synonymous with the BG, is thought to be responsible for postural reflexes and innate automatic behaviors (Denny-Brown, 1962; Martin, 1967). For example, striatal lesions often result in abnormal postures (Richter & Klüver, 1944), and globus pallidus lesions can impair the ability to position one's head while eating or drinking (Labuszewski, Lockwood, McManus, Edelstein, & Lidsky, 1981). Such deficits are often contrasted with those resulting from damage to the pyramidal tract, such as paresis and inability to initiate volitional movements (Jung & Hassler, 1960).

Yet the pyramidal/extrapyramidal distinction neglects the strong connections between cortex and BG. For example, corticostriatal projections often stem from axon collaterals of cortical pyramidal neurons projecting to the brainstem and spinal cord (Cajal, 1909; Reiner, Hart, Lei, & Deng, 2010). BG outputs also reach the reticulospinal pathway, either directly via the projections to the reticular formation or indirectly via projections to the mesencephalic locomotor region (Takakusaki, Habaguchi, Ohtinata-Sugimoto, Saitoh, & Sakamoto, 2003). The anatomical connectivity therefore suggests that the BG are both extrapyramidal (reticulospinal) and pyramidal (corticospinal).

Merely a generation after Wilson proposed the concept of the extrapyramidal system, Denny-Brown argued that "the pyramidal system is useless to the organism without the extrapyramidal system. Nor is the pyramidal system alone employed for movements projected into the environment" (Derek Denny-Brown, 1962, p. 130). Nevertheless, the concept of the extrapyramidal system has had an enduring influence on the field of BG research. For example, it is common to contrast action and posture, implicitly assuming that they are generated by distinct neural substrates (Mink, 1996).

According to another influential view, the role of the BG is mainly to facilitate cortical function. For example, the motor cortex issues movement commands for volitional actions, while the BG merely assists the motor cortex (Anderson & Horak, 1985; Marsden, 1982; Mink, 1996; Mitchell, Richardson, Baker, & DeLong, 1987). Unlike the motor cortex, the BG are not considered essential for movement because BG damage does not produce unconditional deficits in movement. In Parkinson's disease, dopamine (DA) depletion can lead to the inability to initiate movements (akinesia), yet patients are able to move under certain conditions, such as emotional excitement or the presence of very salient sensory inputs. Such "paradoxical" movements, however, do not invalidate the key role of the BG in generating volitional movements. After all, only motor neurons are unconditionally necessary for movement. Motor cortical or pyramidal tract lesions also fail to disrupt movement unconditionally. For example, large corticospinal lesions mainly affect finger and manipulative movements, leaving other types of movements relatively intact, whereas reticulospinal lesions produce more severe deficits in torso movement and posture control (Lawrence & Kuypers, 1968a, b).

Modern studies do not support Cajal and Herrick's dismissal of the BG as a primitive relic replaced by the more modern cerebral cortex. Comparative studies have suggested that the cortex and BG evolved together during vertebrate evolution (Grillner & Robertson, 2016; Reiner, 2010). Extensive damage to the BG can often produce more dramatic deficits than cortical damage: while most voluntary movements are intact in animals with complete removal of the cortex (Bjursten et al, 1976), they are abolished after extensive damage to the BG (Sorenson & Ellison, 1970). Animals without the BG, being unable to feed or drink by themselves, could not survive on their own.

Surprisingly, the cortex-centric view of brain function has also been influential in BG research. Many behavioral tasks, especially those used in primate studies, were designed for the purpose of investigating "pyramidal" function. For example, many studies used head-fixed monkeys performing discrete arm or finger movements, which are thought to be initiated by the motor cortical areas. Some argue that the BG are not necessary for movement per se but facilitate the movements generated by the motor cortex by lifting tonic suppression of behavior (Mink, 1996). Yet decorticate animals at birth still show an extensive behavioral repertoire. Cats that had complete cortical lesions neonatally could still eat, drink, and groom adequately, display maternal and sexual behavior, and even perform visual discrimination in a T-maze (Bjursten, 1976).

The view that the cerebral cortex is uniquely important for volitional behavior is therefore largely a bias inherited from 19th century neurology. It has led to misguided attempts to contrast the roles of the cortex and BG in behavior, and to the neglect of the integrative function of cortico-BG circuits.

1.4 BASIC MOTIF OF CEREBRAL ORGANIZATION

As Willis noted first, the BG appear to connect the cortex above with the brainstem and diencephalon below. This anatomical motif is evolutionarily conserved (Figure 1.2). This motif, which can be discerned in all vertebrates, includes not just the BG but also other components, including the cortex and thalamus. Excitatory cortical and thalamic inputs reach the BG, which in turn send inhibitory projections to the brainstem and thalamus.

The striatum, being the input nucleus, can inhibit output nuclei, including the globus pallidus and substantia nigra, which are also inhibitory and often fire at high rates. Thus, the basic circuit involves two inhibitory synapses with a net excitatory effect. When the striatum is activated, it inhibits the output nucleus (e.g. substantia nigra pars reticulata), and the downstream targets of nigral outputs are then disinhibited (Chevalier & Deniau, 1990). There are also pathways, most notably the indirect pathway, that exert the opposite effect on downstream targets of BG output

Introduction

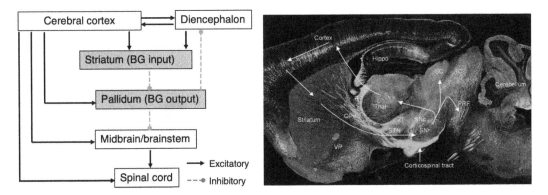

FIGURE 1.2 The basic motif of cerebral organization. On the left is a schematic summary of the basic components of the central nervous system, indicating the place of the cerebral cortex and basal ganglia (BG, shaded). On the right is a sagittal section of the rat brain showing most of the components (From Zhou, 2016 with permission). GPe, external segment of globus pallidus (pallidum); VP, ventral pallidum; SNr, substantia nigra pars reticulata (pallidum); SNc, substantia nigra pars compacta; SC, superior colliculus (midbrain); IC, inferior colliculus (midbrain); STN, subthalamic nucleus (diencephalon); ZI, zona incerta (diencephalon), Thal, thalamus (diencephalon); PRF, pontine reticular formation (brainstem).

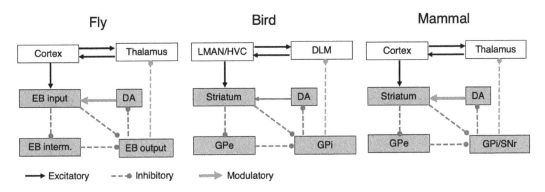

FIGURE 1.3 Evolutionary conservation of the basic motif. Comparison of the basic cerebral organization in the fly brain, bird brain, and mammalian brain. EB, ellipsoid body, which contains input layers, intermediate layers, and output layers. LMAN and HVC are analogs of the frontal cortex in the bird brain. DLM, dorsolateral nucleus of the anterior thalamus; GPe, external globus pallidus; GPi, internal globus pallidus; DA, dopamine. See the text for a more detailed explanation.

projections, with a net inhibitory effect. Finally, the striatum also receives strong modulatory inputs, the most famous of which are the projections from midbrain DA neurons.

Both the cerebral cortex and the BG have been present since the very beginning of vertebrate evolution (Butler & Hodos, 2005; Grillner & Robertson, 2016). In the most primitive vertebrates like the lamprey, which diverged from the main vertebrate evolution roughly 560 million years ago, we can still discern the basic cortico-BG circuit (Grillner & Robertson, 2015; Swanson, 2012). The lamprey BG also contain areas comparable to the striatum and pallidum and receive dopaminergic inputs as well as glutamatergic inputs from a layered cortex or pallium (Robertson et al., 2012; Suryanarayana, Robertson, Wallén, & Grillner, 2017).

The BG circuits in songbirds, whose common ancestor with humans lived at least 300 million years ago, are remarkably similar to those in mammals (Jarvis et al., 2005; Reiner et al., 2004). The bird BG also contain the analogs of the striatum and pallidum, though they are not spatially segregated as in the mammalian BG and use similar transmitters like γ-Aminobutyric acid (GABA) and DA (Luo, Ding, & Perkel, 2001; Mooney, 2009). The songbird BG are a part of an anterior forebrain circuit that is critical for the learning and performance of bird song (Figure 1.3).

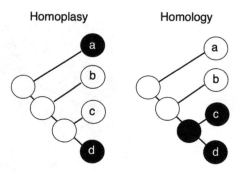

FIGURE 1.4 Homology and homoplasy. On the left, a and d are independently evolved traits. On the right, c and d are homologous traits that are similar due to a shared ancestor.

Perhaps most remarkable is the existence of a proto-BG circuit in invertebrates. Hirth and colleagues have argued that the central complex in arthropods is analogous to the vertebrate BG (Kottler, Faville, Bridi, & Hirth, 2019; Strausfeld & Hirth, 2013a). The central complex contains a striatum-like area called the fan-shaped body as well as the analog of the direct (striatonigral) pathway (protocerebral bridge). As in the mammalian BG, this pathway is characterized by the expression of D1 DA receptors and the neuropeptide substance P. Outputs from the fly "striatum" project to the ellipsoid body, which also contains GABAergic projection neurons much like the vertebrate pallidum. There is also extensive dopaminergic innervation of the central complex, and DA depletion produces Parkinsonian symptoms such as akinesia, as also found in humans (Hirth, 2010a; Strausfeld & Hirth, 2013b).

There are two possible interpretations of the similar anatomical organization found in vertebrates and invertebrates. Homoplasy attributes similar features in different species to independent development (convergent evolution), whereas homology attributes such similarities to common descent (Figure 1.4). Although similarities between the invertebrate central complex and the vertebrate BG could conceivably be examples of homoplasy, as some have argued (Murray, Wise, & Rhodes, 2011), there is evidence indicating that they are due to common descent. In addition to morphological resemblances, there are also similarities in embryological derivation, anatomical topography, neurochemical makeup, behavioral function, and gene regulatory networks that orchestrate neural development (Buhl, Kottler, Hodge, & Hirth, 2021; Green et al., 2017; Guo & Ritzmann, 2013; Mehlman, Winter, Valerio, & Taube, 2019; Strausfeld & Hirth, 2013b; Turner-Evans et al., 2017). As convergent evolution cannot fully explain such detailed correspondence both in structure and in function, the most parsimonious explanation is a common evolutionary origin. Hirth speculates that the common ancestor of vertebrates and invertebrates that first developed the basic BG circuit is a bilaterian from ~600 million years ago, before the split into Protostomia and Deuterostomia (Hirth, 2010b).

The chief components of the cerebrum, cortex and BG, evolved together during vertebrate evolution, and both have increased in size since the mammalian radiation over 65 million years ago (Reiner, 2010). Since the BG cannot function normally without their cortical afferents, it would be misleading to consider their integrative function independently of the cortex. Rather, specific regions of the cortex and their connected BG areas function as a larger unit. In the last few decades, the idea that cortico-BG-thalamocortical networks are the basic units of behavioral function has become widely accepted (Alexander, DeLong, & Strick, 1986), though it remains unclear exactly how each component of the network contributes to the overall output.

1.5 ACTION SELECTION

The evolutionary conservation of the BG suggests a unifying function. What do all animals with the prototypical BG circuit have in common? The most popular answer to this question is action selection (Alexander et al., 1986; Mink, 1996). According to this idea, the BG output selects a motor program

Introduction

while inhibiting competing programs. Because BG projection neurons are inhibitory, BG output is assumed to provide tonic suppression of behavior. The basic circuit, known as the direct pathway, contains two successive inhibitory synapses, thus allowing disinhibition of the target region. Strong inputs to the BG can produce a transient pause in the BG output, enabling the selection of a particular action (Hikosaka, Takikawa, & Kawagoe, 2000; Mink, 1996). At the same time, competing actions are presumably inhibited by increased BG output in other channels. The BG output is assumed to play a permissive role, allowing cortical input to activate the relevant channel for action selection.

The action selection model assumes that actions are all or none, varying only in the probability of being selected. The detailed parameters of action, such as velocity, direction, or amplitude, are not a part of the model and are therefore relegated to other systems, presumably outside of the BG. As we shall see in later chapters, the basic assumptions of this model are contradicted by more recent experimental results.

1.6 REINFORCEMENT LEARNING

The BG have also been implicated in reinforcement learning (RL) (Barto, 1995; Miller, 1981; Sutton & Barto, 2018). RL is a popular branch of machine learning inspired by animal learning theory. In RL, the agent is not given the "correct" answer but must improve its behavior through feedback. This is considered "unsupervised" learning, which allows the agent to maximize rewards over time by adjusting its behavioral policy.

The key idea in RL is based on Thorndike's law of effect: Any behavior followed by a satisfactory effect is most likely to be repeated in the future, whereas behavior followed by a bad effect is less likely to be repeated. Thorndike assumed the biological substrate was the strengthening of neural connections from stimulus to response: "The connections formed between situation and responses are represented by connections between neurones and neurones, whereby the disturbance or neural current arising in the former is conducted to the latter across their synapses. The strength or weakness of a connection means the greater or less likelihood that the same current will be conducted from the former to the latter rather than to some other place" (p. 246).

The law of effect was developed into a model of behavior by Hull (Hull, 1943). In Hull's model, the stimulus-response (S-R) association is known as habit strength (sHr), which is modified whenever the organism receives reinforcement. Performance is determined by multiplying incentive motivation and habit strength.

$$_sE_R = {_sH_R} \times D \times V \times K$$

where:

$_sE_R$ is excitatory potential, the likelihood that the organism would produce response r to stimulus s.
$_sH_R$ is the current habit strength.
D is drive strength, which could be quantified by hours of deprivation.
V is stimulus intensity.
K is incentive or how attractive the reward is.

Based on this model, learning algorithms were developed that consisted of a subtraction between maximum associative strength and current associative strength, resulting in an effective reinforcement signal that is used to adjust the state-action association. A change in the association also changes the probability of action in a specific state (Bush & Mosteller, 1951). This school of thought later blossomed into a popular field in computer science and machine learning, now known simply as RL.

The BG are often considered the major neural substrates for RL (Miller, 1981). In particular, the corticostriatal pathway is often viewed as the biological implementation of state-action association (Barto, 1995). The key teaching signal that updates this value is a reward prediction error, the difference

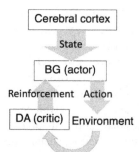

FIGURE 1.5 Action selection and reinforcement learning. The basal ganglia (BG) are thought to be the actor that selects actions based on state information from the cerebral cortex. The state-action activation function can be modified by the effective reinforcement signal, thought to be carried by dopamine (DA).

between actual and predicted rewards and implemented by DA signaling to the BG (Schultz, Dayan, & Montague, 1997). In the most influential variant, the BG and DA together implement the actor-critic model in RL, with the BG serving as the actor that selects actions and the DA as the critic that provides evaluative feedback (good or bad) to adjust the behavioral policy (Figure 1.5).

1.7 BEHAVIOR: THE NEGLECTED PROBLEM

The action selection model provides a theory of performance, while the RL model provides a theory of learning. Since performance is equivalent to the selection of a categorical action, it is assumed that reinforcement alters the probability of action selection. But this raises the question of how the action is defined in the first place. What exactly is reinforced or repeated? When does it start, and when does it end? For example, in instrumental or operant conditioning, a rat may learn to press a lever for a food reward, so that lever press is the action that is being reinforced. But if we examine this action closely, there appears to be tremendous variability. The rat might approach the lever from the left side or from the right. It might press it down using its paw or its snout. Early in the session, it might stuff the food in its mouth, consuming it as soon as it is collected, chewing and pressing the lever at the same time. Later, as it becomes sated, it might slow down and consume the food while in a more comfortable sitting posture before returning to the lever. Which part of this variable cycle of behaviors is the relevant action? The one that is closest in time to the reward or the one that is causally responsible for reward delivery? How can these be demarcated? These questions are challenging for current RL models.

To appreciate the challenges in behavioral analysis and their implications for studying brain function, we must remove the veils of conventional thinking and pay attention to details. Consider the final common path from motor neurons to muscles. This path causes muscle contraction, which is assumed to move the body parts and create the externally observable patterns that we recognize as behavior. Yet the relationship between muscle output and externally observable behavior is more complex than is commonly realized. Although muscle contraction can cause acceleration of the load, say an arm, at the same time the arm is also affected by external forces like wind and gravity, as well as the movement of the rest of the body. The selection of specific muscle tensions cannot guarantee specific behaviors. The contribution of the motor neuron output varies depending on the load, fatigue, blood circulation, angular velocity, and mechanical advantage depending on the joint angles. Hundreds of muscle "effectors" are found in the body, anchored essentially by elastic bands. Only the *net* acceleration is integrated to yield velocity, which is integrated to yield position.

A closer look at behavior, then, reveals the inadequacies of common descriptions like action selection. Such descriptions are not only vague but more importantly also constrain the type of questions asked since they limit measures or analysis to detecting the presence or absence of the discrete category. The decision on what to measure reflects underlying assumptions about behavior.

For example, if we assume that the organism generates some action in response to incoming sensory input, then it would be reasonable to ask how the probability of action given some sensory input may be altered. Following this assumption, typical *in vivo* experiments are performed on restrained animals using cue-based tasks and analyze the neural activity in relation to discrete events, which are recorded as time stamps. As we shall see, this approach has failed to shed light on BG function.

More recently, a process-based approach has been introduced. This approach treats both neural activity and behavior as continuous processes that change over time. By minimizing the use of physical restraint on the animal while simultaneously recording neural activity and behavior with high temporal and spatial resolution, it is possible to elucidate the relationship between the firing rates of specific neuronal populations in the BG and specific aspects of behavior. Using this approach, studies have found that BG activity represents continuous kinematic variables (Barter et al., 2015a, b; Kim, Barter, Sukharnikova, & Yin, 2014; Park, Coddington, & Dudman, 2020; Yin, 2017). These results suggest that, in studying BG function, it is not enough to ask whether some action is selected, but also how far, how fast, and in which direction. In later chapters, we will review these results in detail and explore their implications for understanding how the BG contribute to action generation.

1.8 EXPLAINING BG FUNCTION

In neuroscience, there is a persistent tendency to look for *where* some function is to be found instead of asking what it is and how it works (Von Holst, 1954). Explanations of brain function have been dominated by labels, e.g. hippocampus for memory, amygdala for emotion, prefrontal cortex for cognitive control, etc. The implicit goal is to create a giant lookup table with each brain area (or more microscopically cell types and molecules) listed in one column and a corresponding list of associated functions in the next column. While this product would be convenient for multiple-choice tests, it is unlikely to resemble a viable model of brain function.

Attaching psychological categories to different brain areas is also a legacy of 19th-century neurology. The "localizationists" argued that specific brain areas perform distinct functions, and that lesioning one cortical area produce different effects from lesioning another. The influence of the localizationist school is clear in current thinking on BG function. Distinct cortical functional territories are assumed to have distinct functions, which are also reflected in their target regions in the BG (Alexander et al., 1986). But the list of functions associated with different networks seems arbitrary: some refer to movements of discrete body parts (eyes versus arms), but other labels such as "cognitive" refer to more abstract psychological categories, such as working memory and attention.

Interpretation of the effects of discrete brain lesions is difficult due to the ambiguity in the description of behavior. Hughlings Jackson analyzed common symptoms following damage to the cortex and striatum, such as weakness or paresis on the side contralateral to the side of lesion in patients with hemiplegia (Jackson, 1873). He concluded that symptoms that appear similar could be generated in different ways. According to Jackson, the nervous system is hierarchically organized, and the ambiguity in intepreting behavior stems from the hiearchical oganization. The removal of the higher volitional level may spare lower-level function, though a similar set of effectors is used. What is lost after brain damage is not movement per se, but a particular way of generating the movement. For example, a patient may struggle to perform voluntary movements while still being capable of performing more automatic movements, which can be distinguished from volitional action by the level at which the action is initiated. Yet at first glance, this difference may not be obvious because similar effectors may be used. According to Jackson, the higher levels of the hierarchy, found in the cortex and BG, are characterized by more global and diverse representations of movements and sensations. Instead of specifying muscle contractions, they specify higher parameters like the serial order of actions.

Although Jackson suggested that a hierarchical organization is needed to explain behavior, he did not provide a working model relating the higher levels to the lower levels. But his observations suggest that the debate on the localization of function hinges on how "function" is defined. The ambiguity in such definitions is a major source of confusion, as functional labels often refer to processes at different levels. The discovery of functional specialization, e.g. the somatotopic organization of primary sensory or motor cortices, does not solve the problem of how the brain implements functions at the behavioral level, which is more relevant when considering common psychological functions. For example, on a keyboard, each key is associated with one letter. Typing the letter A is one function, which only requires one key, but writing a message or a novel would require the whole keyboard. Localization of function depends on the nature of the function in question—whether it is like typing the letter A or writing a novel.

That the cortex excites the striatum is an accurate functional description, but the function in this case is local. A far more difficult challenge is to understand how the corticostriatal projection may contribute to more global functions at the level of behavior. To do so, we must incorporate the local function of excitation into a circuit model that performs specific computational operations. This working model might exhibit emergent properties which may include selecting a goal, initiating an action, obtaining a reward, and so on.

The traditional focus on the *where* question has led to the neglect of the question of *how* different elements are coordinated as well as *when* different behavioral elements are expressed. Consider the following observation: a rat walks toward a piece of food, picks it up, and eats it. This sequence, which lasts perhaps a few seconds, involves hundreds of muscles contracting and relaxing at different times. Locomotor regions would be activated first, followed by sniffing and whisking, and a posture change shifting to a sitting position with forepaws now reaching for and grabbing the food. While sitting, the rat holds the food pellet in its paws and nibble on it, before swallowing. Throughout, neither sensory input nor effector output stops—rather, there is continuous modulation. As some smells increase in intensity, others perhaps become weaker. There are precise targeting movements of the forepaws and placement of the food, and then coordination between forepaws and mouth opening, and between chewing and swallowing. All the while, orientation, body posture, and balance are still controlled.

Thus "getting food" is not a localizable function, as it can be achieved in many different ways. Damage to some brain area responsible for the movement of a given body part would have limited effects on achieving this goal, the state of having obtained the food. Many actions are thus functionally equivalent, and the only thing they have in common is the goal. This raises the question of how distinct and variable actions can be recruited to achieve some goal and how one can learn about the causal relationship between the action and the goal. As we shall see, these are among the key functions of the cortico-BG networks.

Instead of adding conventional psychological categories, such as volition, memory, attention, habit, and so on, to the list of BG "functions", what is needed is a more precise description of what these nuclei do, beyond a list of arbitrary psychological categories based on superficial observation, and a model of how the circuits implement the function. A model makes predictions about observable variables, but the predictions can be deduced based on the functional relationships specified within the model, independent of empirical observations. With respect to a particular brain region, two major types of predictions are possible. First, by describing the nature of neural signals and how they are transformed within a circuit, a model may generate predictions about the pattern of neural activity to be found and its relationship with behavioral variables. Secondly, by describing the functional role of specific components, a model can generate predictions about what happens when certain components are damaged or stimulated. Manipulations of specific populations or components should also yield predictable patterns both at the neural activity level and at the behavioral level. Ultimately, an explanation of BG function should lead to a working model that can actually run—that is, it can be used, either in simulation or in a real robot, to generate the relevant behaviors, including malfunctioning states when specific parts are damaged. This is the criterion by which the model of BG function advanced in this book, or indeed any other model of BG function, should be judged.

1.9 SUMMARY

Willis, the first person to describe the BG clearly, believes that these nuclei contribute to sensation, movement, and imagination. Others argue that the BG are merely a relic in the mammalian brain, its function largely replaced by the cerebral cortex. More recent work suggests that BG evolved together with the cerebral cortex, together forming a basic cerebral anatomical motif that has been conserved for over 500 million years. This motif is essential for volitional behavior, as extensive damage to the BG eliminates volitional behavior while preserving reflexes. Studies of brain evolution do not support the idea that there is an evolutionarily old subcortical system whose function was replaced by a newer cortical system.

Damage to BG can affect many functions such as learning, memory, movement, and attention, but such observations do not shed light on the actual computations performed by BG circuits. The standard model of BG function explains performance as action selection and learning as RL. In both cases, the underlying assumption is that behavior is categorical. However, recent work has begun to question this assumption, as many BG neurons appear to represent continuous variables like movement kinematics. What is missing in current models is an adequate functional description of what the BG might do. At the level of integrative function, the challenge is to determine how different cortico-BG networks are coordinated to generate observable behavior.

REFERENCES

Alexander, G. E., DeLong, M. R., & Strick, P. L. (1986). Parallel organization of functionally segregated circuits linking basal ganglia and cortex. *Annual Review of Neuroscience*, 9, 357–381.

Anderson, M., & Horak, F. (1985). Influence of the globus pallidus on arm movements in monkeys. Iii. Timing of movement-related information. *Journal of Neurophysiology*, 54(2), 433–448.

Barter, J. W., Li, S., Lu, D., Rossi, M., Bartholomew, R., Shoemaker, C. T., ... Yin, H. H. (2015a). Beyond reward prediction errors: The role of dopamine in movement kinematics. *Frontiers in Integrative Neuroscience*, 9, 39.

Barter, J. W., Li, S., Sukharnikova, T., Rossi, M. A., Bartholomew, R. A., & Yin, H. H. (2015b). Basal ganglia outputs map instantaneous position coordinates during behavior. *Journal of Neuroscience*, 35(6), 2703–2716.

Barto, A. G. (1995). Adaptive critics and the basal ganglia. *Models of Information Processing in the Basal Ganglia*, 215–232.

Bjursten, L.-M., Norrsell, K., & Norrsell, U. (1976). Behavioural repertory of cats without cerebral cortex from infancy. *Experimental Brain Research*, 25(2), 115–130.

Buhl, E., Kottler, B., Hodge, J. J. L., & Hirth, F. (2021). Thermoresponsive motor behavior is mediated by ring neuron circuits in the central complex of drosophila. *Scientific Reports*, 11(1), 155. doi: 10.1038/s41598-020-80103-9.

Bush, R. R., & Mosteller, F. (1951). A mathematical model for simple learning. *Psychological Review*, 58(5), 313.

Butler, A. B., & Hodos, W. (2005). *Comparative Vertebrate Neuroanatomy: Evolution and Adaptation*: New Jersey: John Wiley & Sons.

Cajal, S. R. (1909). *Histology of the Nervous System* (Swanson, N., & Swanson, L. W., Trans.): New York: Oxford University Press.

Chevalier, G., & Deniau, J. M. (1990). Disinhibition as a basic process in the expression of striatal functions. *Trends in Neurosciences*, 13(7), 277–280.

Cummings, J. L. (1985). Organic delusions: Phenomenology, anatomical correlations, and review. *The British Journal of Psychiatry*, 146(2), 184–197.

Denny-Brown, D. (1962). *The Basal Ganglia and their Relation to Disorders of Movement*: Oxford: Oxford University Press.

Green, J., Adachi, A., Shah, K. K., Hirokawa, J. D., Magani, P. S., & Maimon, G. (2017). A neural circuit architecture for angular integration in drosophila. *Nature*, 546(7656), 101.

Grillner, S., & Robertson, B. (2015). The basal ganglia control of downstream brainstem motor centres—an evolutionarily conserved strategy. *Current Opinion in Neurobiology*, 33, 47–52.

Grillner, S., & Robertson, B. (2016). The basal ganglia over 500 million years. *Current Biology*, 26(20), R1088–R1100.

Guo, P., & Ritzmann, R. E. (2013). Neural activity in the central complex of the cockroach brain is linked to turning behaviors. *Journal of Experimental Biology, 216*(6), 992–1002.

Herrick, C. J. (1948). *The Brain of the Tiger Salamander, Ambystoma Tigrinum*: Chicago, Univ. of Chicago Press.

Hikosaka, O., Takikawa, Y., & Kawagoe, R. (2000). Role of the basal ganglia in the control of purposive saccadic eye movements. *Physiological Reviews, 80*(3), 953–978.

Hirth, F. (2010a). Drosophila melanogaster in the study of human neurodegeneration. *CNS & Neurological Disorders-Drug Targets (Formerly Current Drug Targets-CNS & Neurological Disorders), 9*(4), 504–523.

Hirth, F. (2010b). On the origin and evolution of the tripartite brain. *Brain, Behavior and Evolution, 76*(1), 3–10.

Hull, C. (1943). *Principles of Behavior*: New York: Appleton-Century-Crofts.

Jackson, J. H. (1873). On the Localisation of Movements in the Brain. *The Lancet 101.2581*(1873), 232–235.

Jackson, J. H. (1875). A lecture on softening of the brain. *The Lancet, 106*(2714), 335–339.

Jarvis, E. D., Gunturkun, O., Bruce, L., Csillag, A., Karten, H., Kuenzel, W., ... Butler, A. B. (2005). Avian brains and a new understanding of vertebrate brain evolution. *Nature Reviews Neuroscience, 6*(2), 151–159. doi: 10.1038/nrn1606.

Jung, R., & Hassler, R. (1960). The extrapyramidal motor system. *Handbook of Physiology, 2*, 863–927.

Kim, N., Barter, J. W., Sukharnikova, T., & Yin, H. H. (2014). Striatal firing rate reflects head movement velocity. *European Journal of Neuroscience, 40*(10), 3481–3490. doi: 10.1111/ejn.12722.

Konorski, J. (1967). *Integrative Activity of the Brain*: Chicago: University of Chicago Press.

Kottler, B., Faville, R., Bridi, J. C., & Hirth, F. (2019). Inverse control of turning behavior by dopamine d1 receptor signaling in columnar and ring neurons of the central complex in drosophila. *Current Biology, 29*(4), 567–577 e566. doi: 10.1016/j.cub.2019.01.017.

Labuszewski, T., Lockwood, R., McManus, F., Edelstein, L., & Lidsky, T. (1981). Role of postural deficits in oro-ingestive problems caused by globus pallidus lesions. *Experimental Neurology, 74*(1), 93–110.

Lawrence, D. G., & Kuypers, H. G. (1968a). The functional organization of the motor system in the monkey i. The effects of bilateral pyramidal lesions. *Brain, 91*(1), 1–14.

Lawrence, D. G., & Kuypers, H. G. (1968b). The functional organization of the motor system in the monkey ii. The effects of lesions of the descending brain-stem pathways. *Brain, 91*(1), 15–36.

Luo, M., Ding, L., & Perkel, D. J. (2001). An avian basal ganglia pathway essential for vocal learning forms a closed topographic loop. *Journal of Neuroscience, 21*(17), 6836–6845.

Marsden, C. (1982). The mysterious motor function of the basal ganglia: The robert wartenberg lecture. *Neurology, 32*(5), 514–539.

Martin, J. P. (1967). *The Basal Ganglia and Posture*: Philadelphia, PA: Lippincott.

Mehlman, M. L., Winter, S. S., Valerio, S., & Taube, J. S. (2019). Functional and anatomical relationships between the medial precentral cortex, dorsal striatum, and head direction cell circuitry. I. Recording studies. *Journal of Neurophysiology, 121*(2), 350–370. doi: 10.1152/jn.00143.2018.

Middleton, F. A., & Strick, P. L. (2000). Basal ganglia and cerebellar loops: Motor and cognitive circuits. *Brain Research Reviews, 31*(2–3), 236–250.

Miller, R. (1981). *Meaning and Purpose in the Intact Brain*: New York: Oxford University Press.

Mink, J. W. (1996). The basal ganglia: Focused selection and inhibition of competing motor programs. *Progress in Neurobiology, 50*(4), 381–425.

Mitchell, S., Richardson, R., Baker, F., & DeLong, M. (1987). The primate globus pallidus: Neuronal activity related to direction of movement. *Experimental Brain Research, 68*(3), 491–505.

Mooney, R. (2009). Neural mechanisms for learned birdsong. *Learning & Memory, 16*(11), 655–669.

Murray, E. A., Wise, S. P., & Rhodes, S. E. (2011). What can different brains do with reward. In: Gottfried, J. A. (Ed.), *Neurobiology of Sensation and Reward* (pp. 61–98). New York: CRC Press.

Packard, M. G., & Knowlton, B. J. (2002). Learning and memory functions of the basal ganglia. *Annual Review of Neuroscience, 25*, 563–593.

Parent, A. (2012). The history of the basal ganglia: The contribution of karl friedrich burdach. *Neuroscience & Medicine, 3*(4), 374–379.

Park, J., Coddington, L. T., & Dudman, J. T. (2020). Basal ganglia circuits for action specification. *Annual Review of Neuroscience, 43*(2020), 485–507.

Pisa, M. (1988). Motor functions of the striatum in the rat: Critical role of the lateral region in tongue and forelimb reaching. *Neuroscience, 24*(2), 453–463.

Pisa, M., & Schranz, J. A. (1988). Dissociable motor roles of the rat's striatum conform to a somatotopic model. *Behavioral Neuroscience, 102*(3), 429.

Reiner, A. (2010). The conservative evolution of the vertebrate basal ganglia. In: Steiner, H. (Ed.), *Handbook of Behavioral Neuroscience* (Vol. 20, pp. 29–62). New York: Elsevier.

Reiner, A., Perkel, D. J., Bruce, L. L., Butler, A. B., Csillag, A., Kuenzel, W., ... Striedter, G. (2004). Revised nomenclature for avian telencephalon and some related brainstem nuclei. *Journal of Comparative Neurology*, *473*(3), 377–414.

Richter, R., & Klüver, H. (1944). Spontaneous striatal degeneration in a monkey. *Journal of Neuropathology & Experimental Neurology*, *3*(1), 49–62.

Robertson, B., Huerta-Ocampo, I., Ericsson, J., Stephenson-Jones, M., Pérez-Fernández, J., Bolam, J. P., ... Grillner, S. (2012). The dopamine d2 receptor gene in lamprey, its expression in the striatum and cellular effects of d2 receptor activation. *PLoS One*, *7*(4), e35642.

Saint-Cyr, J., Taylor, A. E., & Lang, A. (1988). Procedural learning and neostriatal dysfunction in man. *Brain*, *111*(4), 941–960.

Schiller, F. (1967). The vicissitudes of the basal ganglia (further landmarks in cerebral nomenclature). *Bulletin of the History of Medicine*, *41*(6), 515–538.

Schultz, W., Dayan, P., & Montague, P. R. (1997). A neural substrate of prediction and reward. *Science*, *275*(5306), 1593–1599.

Sorenson, C. A., & Ellison, G. D. (1970). Striatal organization of feeding behavior in the decorticate rat. *Experimental Neurology*, *29*(1), 162–174.

Strausfeld, N. J., & Hirth, F. (2013a). Deep homology of arthropod central complex and vertebrate basal ganglia. *Science*, *340*(6129), 157–161.

Strausfeld, N. J., & Hirth, F. (2013b). Homology versus convergence in resolving transphyletic correspondences of brain organization. *Brain, Behavior and Evolution*, *82*(4), 215–219.

Suryanarayana, S. M., Robertson, B., Wallén, P., & Grillner, S. (2017). The lamprey pallium provides a blueprint of the mammalian layered cortex. *Current Biology*, *27*(21), 3264–3277. e3265.

Sutton, R. S., & Barto, A. G. (2018). *Reinforcement Learning: An Introduction*: Cambridge, MA: MIT Press.

Swanson, L. W. (2012). *Brain Architecture: Understanding the Basic Plan*: New York: Oxford University Press.

Takakusaki, K., Habaguchi, T., Ohtinata-Sugimoto, J., Saitoh, K., & Sakamoto, T. (2003). Basal ganglia efferents to the brainstem centers controlling postural muscle tone and locomotion: A new concept for understanding motor disorders in basal ganglia dysfunction. *Neuroscience*, *119*(1), 293–308.

Turner-Evans, D., Wegener, S., Rouault, H., Franconville, R., Wolff, T., Seelig, J. D., ... Jayaraman, V. (2017). Angular velocity integration in a fly heading circuit. *Elife*, *6*, e23496.

Von Holst, E. (1954). Relations between the central nervous system and the peripheral organs. *British Journal of Animal Behaviour*, *2*, 89–94.

Willis, T. (1664). *Cerebral Anatomy*: Amsterdam: Joannes de Someren.

Wilson, C. (2016). The history of the basal ganglia: Cells and circuits. In: Steiner, H. (Ed.), *Handbook of Behavioral Neuroscience* (vol. 24, pp. 45–62). New York: Elsevier.

Wilson, K. (1925). Croonian lectures. *Lancet*, *2*(1), 53.

Wilson, S. K. (1914). An experimental research into the anatomy and physiology of the corpus striatum. *Brain*, *36*(3–4), 427–492.

Yin, H. H. (2017). The basal ganglia in action. *Neuroscientist*, *23*(3), 299–313. doi: 10.1177/1073858416654115.

Zhou, F. (2016). The substantia nigra pars reticulata. In: Steiner, H. (Ed.), *Handbook of Behavioral Neuroscience, Second Edition* (pp. 293–316). New York: Elsevier.

2 Anatomical Organization of the Basal Ganglia

In navigating the immense literature on the basal ganglia (BG), a major obstacle is the confusing anatomical terminology. Although this is a general problem in neuroscience, nowhere else is it as acute as in the study of the BG. As conventional terminology was developed without accurate knowledge of cell types and connectivity, it is unnecessarily complex. Consequently, BG anatomy is notoriously difficult to master.

This chapter presents an overview of BG anatomy. The goal is to describe the basic components as well as general principles of organization, in order to make later chapters comprehensible to those unfamiliar with the BG. Readers who wish to find more comprehensive treatments may consult other sources (Gerfen & Wilson, 1996; Parent, 1986; Steiner & Tseng, 2016).

2.1 WHAT'S IN A NAME?

When brain structures were first named, the names were based on their visual appearance. In neuroanatomy, as in everything else, judging things by their labels is a questionable practice. Arbitrary anatomical labels often prevent us from discerning similarities in physiological properties and connectivity. Early anatomists like Willis and Burdach relied on visual inspection of the brain with naked eyes—hence terms like globus pallidus (pale globes) or corpus striatum (striated bodies). Yet these terms only reflect gross anatomical features, without shedding light on connectivity or function. The use of Latin and Greek, while providing an aura of prestige, did not add much scientific value, and what little descriptive value they once possessed all but disappeared with the decline of classical learning.

Often multiple terms are used to describe the same structure. For example, Burdach was responsible for introducing 'putamen' and 'caudate nucleus,' two names that are commonly used in the literature to refer to different parts of the dorsal striatum. This division is based on bundles of ascending and descending axons that form the internal capsule in primates, separating the caudate on the medial side and the putamen on the lateral side. The distinction between caudate and putamen is therefore largely arbitrary. Moreover, although the olfactory tubercle and nucleus accumbens are now both considered ventral extensions of the striatum (De Olmos & Heimer, 1999; Swanson, 2000), it is not clear from the names that they could be closely related, or that all striatal regions have similar projection neurons and interneurons.

Swanson proposed a classification scheme that simplifies the anatomy. According to him, the brain (cerebrum) can be divided into the cerebral cortex and BG (cerebral nuclei) (Swanson, 2000). The cerebral cortex is a layered structure that can be distinguished from the BG based on the type of projection neurons used. Projection neurons have axons that leave the structure of origin to synapse onto neurons in other brain regions. While the cortex contains glutamatergic projection neurons, the BG use GABAergic projection neurons. Since glutamate is the major excitatory transmitter and GABA is the major inhibitory transmitter, they have opposite effects on target neurons. According to this criterion, many heterogeneous regions are classified as cortical, such as the basolateral amygdala, hippocampus, and visual cortex. On the other hand, the BG contain two parts, the striatum and the pallidum, both with projection neurons that use GABA as the primary transmitter. As the input nucleus, the striatum mainly receives the excitatory projections from cortical pyramidal neurons and sends inhibitory projections to the pallidum. As the output nucleus, the pallidum mainly receives inhibitory projections from the striatum and sends inhibitory projections to nearly all regions in the diencephalon, midbrain, and brainstem. Both the striatum and the pallidum are heterogeneous, and traditionally many names are given to different regions with distinct visual appearances (Table 2.1).

Instead of simply memorizing arbitrary names, it helps to understand how different structures are related to each other. For example, the caudate, putamen, nucleus accumbens, and olfactory tubercle all have similar spiny projection neurons (SPNs). Even though they were initially thought to be unrelated regions based on their distinct visual appearances and locations, their projection neurons and connectivity suggests that they may perform analogous operations on their respective inputs. In contrast, by Swanson's criterion, areas like the subthalamic nucleus (STN) and intralaminar thalamus would not be classified as parts of the BG, as is often done. Although they are strongly connected with the BG, their projection neurons are glutamatergic and excitatory. Moreover, it is also inconsistent to classify these structures as parts of the BG, but not the zona incerta or mediodorsal nucleus of the thalamus, which are also structures with glutamatergic projection neurons that are similarly connected with the BG.

Developmental studies have provided support for Swanson's proposed classification. In the neural tube, the diencephalic neuroepithelium differentiates into the hypothalamus, thalamus, subthalamus, and epithalamus (habenula), all of which are major targets of BG outputs. The dorsolateral walls of the neuroepithelium around the lateral ventricle give rise to the cerebral cortex, and the ventrolateral wall to the BG. Along the ventrolateral wall of the neural tube, there are two longitudinal ridges: progenitor domains

TABLE 2.1
Anatomical Terminology

	Dorsal	Ventral	Amygdaloid
Striatum	Caudate, putamen	Nucleus accumbens (shell and core) Olfactory tubercle Island of Calleja	Central nucleus of the amygdala
Pallidum	Globus pallidus internus (GPi), globus pallidus externus (GPe), substantia nigra pars reticulata (SNr), substantia nigra pars lateralis (SNl)	Ventral tegmental area (VTA), ventral pallidum (VP)	Bed nucleus of the stria terminalis (BNST), substantia innominata

Swanson's classification of the basal ganglia

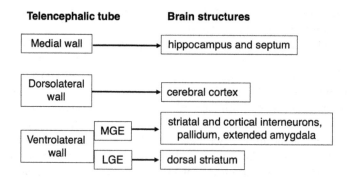

FIGURE 2.1 Development of the basal ganglia (BG). Three distinct neuroepithelium regions within each telencephalic tube give rise to different brain regions during development. The BG (striatum and pallidum) originate from two ridges in the ventrolateral wall. MGE, medial ganglionic eminence; LGE, lateral ganglionic eminence.

Anatomical Organization of the Basal Ganglia

from the lateral ganglionic eminence, which develop into the striatum, and progenitor domains from the medial ganglionic eminence, which develop into the pallidum, including parts of what is sometimes called the extended amygdala (Deacon, Pakzaban, & Isacson, 1994; Flandin, Kimura, & Rubenstein, 2010; Heimer et al., 1997; Stenman, Toresson, & Campbell, 2003). According to Swanson's classification, the extended amygdala actually includes both striatal and pallidal components. Thus, pallidal structures seem to share a common developmental origin. These observations suggest that the distinction between striatum and pallidum is supported by their distinct developmental trajectories (Figure 2.1).

2.2 TOPOGRAPHICAL INPUTS TO THE BG

The striatum can be divided into subregions based on the pattern of the corticostriatal projections. Three large regions are usually recognized: limbic, associative, and sensorimotor (McGeorge & Faull, 1989; Parent, 1990). Although recent work has shown that finer divisions are possible (Foster et al., 2021), this tripartite division provides a convenient starting point for a discussion of the major inputs to the BG. The three divisions receive largely distinct classes of sensory inputs: the limbic striatum receives interoceptive inputs monitoring the internal state of the body, the associative striatum receives exteroceptive inputs monitoring the external environment, and the sensorimotor striatum receives proprioceptive and somatosensory inputs monitoring the muscles and body surface. This division is at least partly preserved in the pallidum (Parent & Hazrati, 1995a) (Figure 2.2).

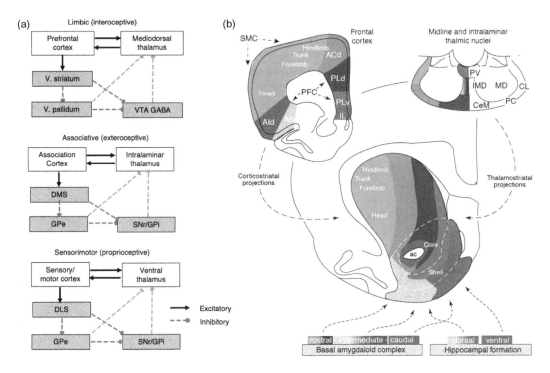

FIGURE 2.2 Tripartite division of the BG. (a) Summary of connectivity of three major divisions. (b) Illustration of cortical and thalamic projections to different striatal regions, organized in ventromedial to dorsolateral zones. ac, anterior commissure; ACd, dorsal anterior cingulate cortex; AId, dorsal agranular insular cortex; AIv, ventral agranular insular cortex; BG, basal ganglia; CeM, central medial thalamic nucleus; CL, central lateral thalamic nucleus; GPe, globus pallidus externus; GPi, globus pallidus internus; IL, infralimbic cortex; IMD, intermediodorsal thalamic nucleus; MD, mediodorsal thalamic nucleus; PC, paracentral thalamic nucleus; PFC, prefrontal cortex; PLd, dorsal prelimbic cortex; PLv, ventral prelimbic cortex; PV, paraventricular thalamic nucleus; SMC, sensorimotor cortex; SNr, substantia nigra pars reticulata; VTA, ventral tegmental area. From Voorn, Vanderschuren, Groenewegen, Robbins, and Pennartz (2004) with permission.

2.2.1 LIMBIC STRIATUM

The limbic striatum includes most ventral striatal regions, like the nucleus accumbens and the olfactory tubercle. Heimer and colleagues demonstrated the similarities in connectivity patterns between the ventral striatopallidal circuit and its dorsal counterparts (Groenewegen, Wright, Beijer, & Voorn, 1999; Heimer, Switzer, & Van Hoesen, 1982). Like the dorsal circuits, there are strong projections from cortical regions to the ventral striatum, which projects to the ventral pallidum and the ventral tegmental area (VTA), which in turn project to the brainstem and thalamic regions. Some of the thalamic target regions, such as the mediodorsal thalamus, send projections back to the prefrontal cortex, where many of the limbic corticostriatal projections originate.

The limbic striatum receives inputs from virtually all allocortical areas, including the hippocampus and cortical amygdala. Its largest and most well-known region, the nucleus accumbens, is usually divided into two parts, core and shell, based on immunohistochemical staining for calbindin. The core is continuous with and similar to the dorsal striatum, whereas the shell resembles some of the amygdalar nuclei. Core outputs primarily target the ventral pallidum (VP) and VTA, which can influence areas that are critical for locomotion and approach behavior. The outputs of the shell mainly target the VP, bed nuclei of stria terminalis (BNST), and the lateral hypothalamus. Shell outputs can ultimately influence a number of neuroendocrine and autonomic effectors.

Sometimes called the 'olfacto-striatum' in the older literature, the limbic striatum has long been implicated in locomotion and orofacial behaviors (Herrick, 1926), as well as reward and motivation (Ikemoto & Panksepp, 1999; Salamone, Correa, Farrar, Nunes, & Pardo, 2009). It is the major target of the mesolimbic dopamine (DA) pathway. The olfactory tubercle, situated in the ventral striatum, receives heavy projections from the ventral olfactory bulb and the vomeronasal amygdala, including the medial and cortical posteromedial nuclei of the amygdala (Lanuza et al., 2008). Through these projections, neural signals representing sexual pheromonal inputs can reach the ventral striatum, suggesting a role in reproductive behavior (Zhang et al., 2017).

2.2.2 ASSOCIATIVE STRIATUM

The associative striatum, occupying the dorsomedial region in rodents, is the main target of projections from association cortices, including the prefrontal cortex and many multimodal association cortical areas (Hintiryan et al., 2016). Behavioral observations suggest a key role in orienting behavior, including eye movements (Hikosaka, Sakamoto, & Usui, 1989; Mettler & Mettler, 1942). In humans, damage to the associative striatum (caudate) is also associated with a variety of deficits related to motivation, including loss of drive (apathy), compulsive behavior, and hyperactivity (Villablanca & Salinas-Zeballos, 1972). Studies have also shown that the associative striatum is critical for learning the instrumental contingency—the relationship between actions and their outcomes (Balleine, Peak, Matamales, Bertran-Gonzalez, & Hart, 2021; Yin, Ostlund, Knowlton, & Balleine, 2005).

2.2.3 SENSORIMOTOR STRIATUM

The sensorimotor striatum comprises mainly the lateral or dorsolateral striatum (putamen in primates), which receives very extensive projections from primary somatosensory and motor cortical areas (Künzle, 1975, 1977). Corticostriatal projections to the sensorimotor striatum are somatotopically organized (Hintiryan et al., 2016; McGeorge & Faull, 1989; Webster, 1961, 1965). For example, in primates, the rostral sensorimotor cortex (head areas) projects to the central and ventral regions, whereas the more caudal sensorimotor cortex (limb areas) projects to dorsal regions of the dorsolateral striatum. Neurons in the sensorimotor striatum fire during the movement of specific body parts (Alexander & DeLong, 1985a, b; Carelli & West, 1991; Cho & West, 1997). Electrical stimulation in a cluster of putative SPNs can evoke movement in a single body part. In primates, neurons related to

Anatomical Organization of the Basal Ganglia

movements of the lower extremity are found in the dorsolateral putamen; neurons related to orofacial movements are located in a more ventromedial region; and neurons related to movements of the upper extremity are located in an intermediate position (Alexander & DeLong, 1985b).

In addition, the sensorimotor striatum has also been implicated in procedural learning (Saint-Cyr, Taylor, & Lang, 1988). Lesions in this area can impair the learning of arbitrary serial order of actions as well as habit formation when behavior becomes more stimulus-driven and automatic following extensive training (Graybiel, 2008; Yin, 2010; Yin & Knowlton, 2006).

2.3 CHEMICAL COMPARTMENTS

The striatum can also be divided into two major chemical compartments—striosome (also known as patch) and matrix (Gerfen, 1992; Graybiel, 1990). These compartments are characterized by their expression of chemical markers and connectivity. The markers commonly used to delineate these compartments include neuromodulators, receptors, or enzymes related to their production and reuptake (Crittenden & Graybiel, 2011). For example, cholinesterase, the enzyme that breaks down acetylcholine, is highly expressed in the striatum. Striosomes were initially described as small islands with poor cholinesterase staining embedded in the much larger matrix compartment, which is intensely stained with cholinesterase (Graybiel & Ragsdale, 1978).

Developmentally, striosome neurons are born first, from fate-restricted apical intermediate progenitors, whereas matrix neurons are born later, from basal intermediate progenitors (Kelly et al., 2018). Striosomal neurons usually migrate out of the lateral ganglionic eminence earlier than matrix neurons. Matrix neurons are born later and migrate out into the striatum. Early on, nigrostriatal DA neurons innervate small islands in the striatum, which later become striosomes (Fishell & van der Kooy, 1987; Graybiel, 1984; van der Kooy & Fishell, 1987). Later, the DA neurons also send projections to the matrix.

As summarized in Figure 2.3, these compartments are characterized by distinct patterns of connectivity. Striosomes receive inputs primarily from limbic cortical regions, whereas matrix neurons receive inputs from associative and sensorimotor cortical regions (Eblen & Graybiel, 1995; Gerfen, 1989; Jimenez-Castellanos & Graybiel, 1987). Layer 5b pyramidal neurons preferentially target the

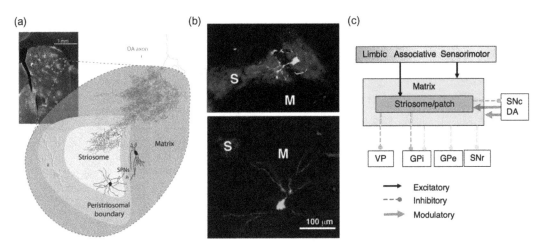

FIGURE 2.3 Chemical compartments in the striatum. (a) Illustration of chemical compartments in relation to major striatal cell types. From Brimblecomb and Cragg (2016) with permission. (b) Labeling for mu opioid receptors (in red), a marker for the striosome compartment. In green is green fluorescent protein from spiny projection neuron (SPN) in the two compartments. The striosome SPN (top) has smaller dendritic arborizations than the matrix SPN (bottom). S, striosome; M, matrix. From Fujiyama et al. (2011) with permission. (c) Summary of connectivity for the two compartments.

striosome, whereas layers 3–5a target the matrix. The striosome and matrix also have distinct output channels. For example, striosome neurons project to the globus pallidus internus (GPi) region that projects to the habenula (part of the epithalamus), whereas the matrix projects to GPi regions that project to the thalamus (Rajakumar, Elisevich, & Flumerfelt, 1993). Activation of matrix neurons usually inhibits other matrix neurons but not striosome neurons, suggesting that they do not project strongly to striosome neurons (Lopez-Huerta et al., 2016).

2.4 SOURCES OF CORTICOSTRIATAL PROJECTIONS

The cerebral cortex is the largest source of projections to the striatum, providing about 70% of all inputs (Doig, Moss, & Bolam, 2010). Based on his own work, Cajal argued that corticostriatal projections are axon collaterals of pyramidal neurons that reach the brainstem and spinal cord, but he also acknowledged previous work by Meynert showing corticostriatal projections that only innervate the striatum (Cajal, 1909). Projections arising from axon collaterals of cortical pyramidal neurons are now called the "pyramidal tract" (PT) pathway, and cortical projections that mainly target the striatum are called the intratelencephalic (IT) pathway (Donoghue & Kitai, 1981; Jinnai & Matsuda, 1979; Levesque, Charara, Gagnon, Parent, & Deschenes, 1996; Reiner, Hart, Lei, & Deng, 2010; Shepherd, 2013).

The PT pathway, which originates mainly from deep layer 5 neurons, sends mainly ipsilateral projections with axon collaterals synapsing in many areas on their way to the spinal cord. This pathway is characterized by thick and well-myelinated axons that allow fast conduction of signals over long distances. In contrast, the IT pathway has bilateral intracortical projections and intrastriatal axon collaterals, often with callosal axonal branches that cross to the contralateral hemisphere (Cowan & Wilson, 1994; Kincaid & Wilson, 1996). In addition to the striatum, the IT pathway also targets other brain areas, though the axons do not usually extend beyond the diencephalic or midbrain level. Compared to the PT pathway, the IT has thinner, slower conducting axons. Figure 2.4 summarizes the major differences between these two pathways.

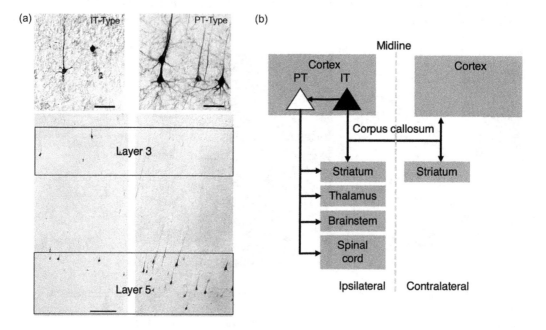

FIGURE 2.4 Pyramidal type (PT) and intratelencephalic type (IT) neurons. (a) Top, images of representative IT-type and PT-type neurons in rat cortex. Scale bar=50 μm. IT neurons are more commonly found in layer 3 and upper layer 5, whereas PT neurons are more commonly found in deep layer 5. Scale bars: 200 μm. Modified with permission from Reiner, Jiao, Del Mar, Laverghetta, and Lei (2003). (b) Summary of IT and PT projections.

Anatomical Organization of the Basal Ganglia

Early studies of corticostriatal projections from the rat somatosensory cortex to the dorsolateral striatum found two distinct patterns: a localized dense projection and a more diffuse projection (Burton, Alloway, & Rosenthal, 1988). The denser projection is topographically organized, and largely maintains the whisker barrel pattern of the somatosensory cortex, where adjacent whiskers are represented in adjoining cortical barrels. In the striatum, these whisker barrels are represented in overlapping regions. This pattern is characteristic of PT projections. The PT axonal arborization has focal clusters of terminals spread over a large striatal region, estimated to be 1–2 mm in the rat striatum (Cowan & Wilson, 1994). In contrast to the highly discrete PT terminal endings, IT projections are more diffuse, with sparse *en passant* terminals over a very large area.

2.5 CONVERGENCE AND DIVERGENCE

Instead of a precise point-to-point topographical relationship, the corticostriatal projections are characterized by a complex pattern of convergence and divergence, as shown in Figure 2.5 (Parent & Hazrati, 1995a). The convergent pattern is not surprising since cortical pyramidal neurons far outnumber SPNs. For example, it has been estimated that one hemisphere in a rat brain contains about 15 million cortical neurons and only about 2.8 million striatal neurons, with a roughly 6:1 ratio. In the mouse brain, nearly all striatal regions receive diffuse projections from at least five cortical subregions (Hunnicutt et al., 2016). These inputs usually arise from cortical subregions that are interconnected. For example, primary and secondary somatosensory and motor cortices contain recurrent networks, and they also send convergent corticostriatal projections to subregions in the sensorimotor striatum. Likewise, in the prelimbic and anterior cingulate prefrontal cortical projection fields in the striatum, axon terminals overlap with those from medial, orbital, and lateral prefrontal cortical areas (Mailly, Aliane, Groenewegen, Haber, & Deniau, 2013).

Corticostriatal projections can also be divergent. From most cortical areas, projections can innervate widely distributed sites in the striatum (Alloway, Mutic, & Hoover, 1998; Berendse, Galis-de Graaf, & Groenewegen, 1992; Ebrahimi, Pochet, & Roger, 1992; Malach & Graybiel, 1986). In the

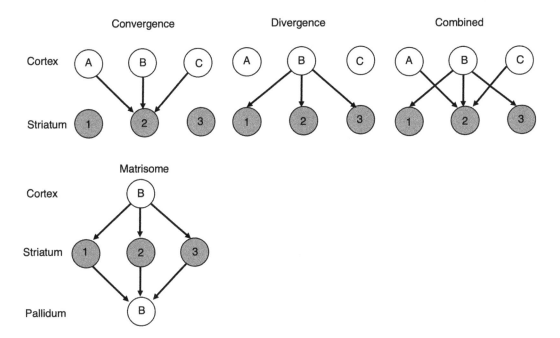

FIGURE 2.5 Patterns of corticostriatal inputs. A combination of convergence and divergence is observed in corticostriatal projections. In the matrisome organization, a single cortical area projects to multiple dispersed striatal regions, which in turn converge upon a single pallidal region.

somatosensory barrel cortex of the rat, for example, projections from a single cortical column can reach multiple dorsolateral striatal regions (Burton, Alloway, & Rosenthal, 1988). Primary motor cortical afferents innervate not only the dorsolateral striatum but also the dorsomedial striatum (Johansson & Silberberg, 2020). Even a single cortical neuron can innervate multiple striatal regions (Parent & Hazrati, 1995a, b).

Graybiel and colleagues identified discrete areas in the matrix that are sites of corticostriatal divergence. These areas, which they called matrisomes, appear to be dispersed striatal modules that receive inputs from body-part representations in the primary somatosensory and motor cortices. But in turn, they send convergent outputs to discrete locations in both GPi and globus pallidus externus (GPe; Flaherty & Graybiel, 1991, 1993, 1994). Multiple copies of the same cortical signals can be sent to widely dispersed striatal modules, which then send convergent inputs to the pallidum. A similar pattern is also found in the limbic striatum (Groenewegen et al., 1999).

Selemon and Goldman-Rakic showed that there is a mixture of intermixed and interdigitation patterns in corticostriatal projections (Selemon & Goldman-Rakic, 1985). With an interdigitation pattern, projections from two distinct cortical regions remain segregated in the striatum: even if they converge in the same striatal region, the axon terminals might target distinct neurons. Input specificity is thus maintained. On the other hand, the intermixing pattern suggests recombination of cortical inputs by striatal neurons. Projections from the frontal and temporal areas have interdigitating patterns where the two fields overlap, but the prefrontal and parietal terminals appear to be intermixed rather than interdigitated.

2.6 THALAMOSTRIATAL PROJECTIONS

The second major source of excitatory inputs to the striatum is the thalamus (Smith et al., 2014; Vogt & Vogt, 1941). Thalamostriatal projections, first described by the Vogts, can be divided into two major classes (Figure 2.6). One class originates from the intralaminar group (midline, parafascicular, and centromedial), which is often considered to be a part of the reticular activating system for arousal (Jones, 2012; Krauthamer & Dalsass, 1978; Sherman & Guillery, 2006; Sidibé, Bevan, Bolam, & Smith, 1997). Unlike many other thalamic nuclei, the intralaminar nuclei project strongly and topographically to the striatum, especially the matrix compartment, but weakly and diffusely to the cortex (Dubé, Smith, & Bolam, 1988; Herkenham & Pert, 1981). The second class of thalamostriatal projections originate from most of the other major thalamic nuclei, such as the ventral

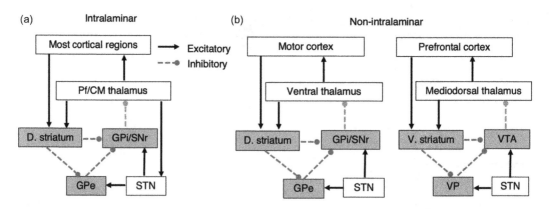

FIGURE 2.6 Thalamostriatal projections. (a) Schematic illustration of intralaminar thalamostriatal projections, and reentrant projections back to the intralaminar thalamus. Pf, parafascicular nucleus; CM, centromedial nucleus. (b) Two examples of nonintralaminar thalamostriatal projections and associated circuitry. GPi, globus pallidus internus; SNr, substantia nigra pars reticulata; GPe, globus pallidus externus; STN, subthalamic nucleus; VTA, ventral tegmental area.

motor nuclei (McFarland & Haber, 2001). Unlike the intralaminar nuclei, which mainly target the striatum, these nonintralaminar nuclei send major projections to the cerebral cortex with modest innervation of the striatum (Takada, Itoh, Yasui, Sugimoto, & Mizuno, 1985). Moreover, whereas intralaminar projections more commonly synapse on dendritic shafts of SPNs (Raju, Shah, Wright, Hall, & Smith, 2006), most nonintralaminar thalamostriatal projections, like corticostriatal projections, target dendritic spines of SPNs (Dubé et al., 1988; Lacey, Bolam, & Magill, 2007).

Cortical and intralaminar thalamic axon terminals can be distinguished based on their expression of glutamate transporters. Corticostriatal terminals are vesicular glutamate transporter 1 (VGLUT1)-positive, whereas intralaminar thalamic terminals are VGLUT2-positive (Fremeau et al., 2001). On the other hand, ventral thalamic inputs to the striatum can express both VGLUT1 and VGLUT2 (Barroso-Chinea, Castle, Aymerich, & Lanciego, 2008).

Most thalamic nuclei that project to the striatum also receive projections from the BG output nuclei. There are thalamo-BG-thalamic loops, analogous to the cortico-BG-thalamo-cortical loop. For example, the medial posterior thalamus receives projections from the substantia nigra lateralis. As this region is also a source of projection to the striatum, it appears to be a key component in a reentrant neural circuit for regulating orienting behavior (Takada et al., 1985). Likewise, the ventral motor thalamic nuclei are a source of thalamostriatal projections as well as a target of BG output projections (McFarland & Haber, 2001).

2.7 DOPAMINERGIC PROJECTIONS TO THE STRIATUM

Receptors for neuromodulators are usually G-protein–coupled and initiate second messenger signaling to affect synaptic transmission and neuronal excitability. The BG receive many neuromodulatory inputs, such as DA, serotonin, and norepinephrine. Of these, the best-known example is DA, which is affected in disorders as diverse as Parkinson's, schizophrenia, and addiction.

As shown in Figure 2.7, dopaminergic projections to the striatum arise primarily from two major pathways, the nigrostriatal pathway from the A9 DA cell group in the substantia nigra pars compacta (SNc) and the mesolimbic pathway from the A10 DA cell group in the VTA (Björklund & Dunnett, 2007). The nigrostriatal pathway, the largest DA pathway in the brain, is also the pathway most affected by Parkinson's disease (Fallon & Loughlin, 1995). The nigrostriatal pathway mainly targets the striatum but also projects strongly to the cortex, especially frontal cortical regions. The mesolimbic pathway originates from VTA DA neurons and mainly targets the ventromedial striatum and the prefrontal cortex.

DA projections to the striosome and matrix compartments also come from distinct sets of DA neurons, which can be distinguished on the basis of calbindin expression (Gerfen, Herkenham, & Thibault, 1987; Jimenez-Castellanos & Graybiel, 1987). DA neurons from the dorsal tier are calbindin-positive and project mainly to the matrix, whereas those from the ventral tier are calbindin-negative and project to the striosome.

2.8 DIRECT AND INDIRECT PATHWAYS

DA receptor expression is commonly used to define direct (striatonigral) and indirect (striatopallidal) pathways (Gerfen et al., 1990). Direct pathway spiny projection neurons (dSPNs) express D1-like DA receptors and substance P, whereas indirect pathway spiny projection neurons (iSPNs) express D2-like receptors and adenosine 2A receptors (Gerfen & Keefe, 1994; Gerfen & Surmeier, 2011). The striatonigral pathway projects directly to the substantia nigra pars reticulata (SNr) and GPi, while the striatopallidal pathway projects to the GPe, which then projects to the SNr and GPi.

Direct and indirect pathway neurons, which receive partly overlapping inputs from many different brain areas, are roughly equal in number. They differ in their expression of a number of receptors and chemicals (Gerfen et al., 1990; Lei, Jiao, Del Mar, & Reiner, 2004; Wall, De La Parra, Callaway, & Kreitzer, 2013). The two populations are intermixed in most of the striatum

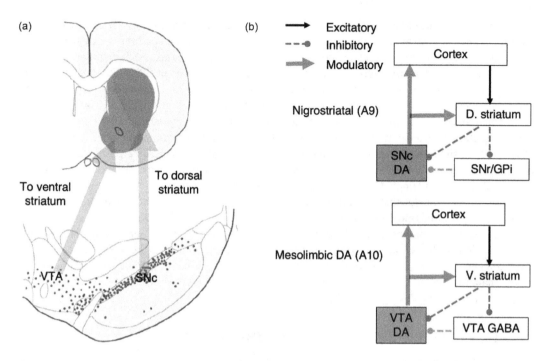

FIGURE 2.7 Dopaminergic projections. (a) Coronal sections of the rat brain showing different regions targeted by dopamine (DA) projections from the more medial ventral tegmental area (VTA) and the more lateral substantia nigra pars compacta (SNc). Dots represent the cell bodies of individual DA neurons. Modified from Bjorklund and Dunnett (2007) with permission. (b) Summary of major connections of nigrostriatal and mesolimbic DA pathways. GPi, globus pallidus internus; SNr, substantia nigra pars reticulata; GPe, globus pallidus externus.

except in the most caudal area, which contains predominantly dSPNs and receives inputs from the primary auditory cortex and the medial geniculate thalamic nucleus (Gangarossa et al., 2013; Hunnicutt et al., 2016).

Within each chemical compartment, there can be both direct and indirect pathway neurons. Developmentally, the segregation into direct and indirect pathways takes place mostly after the striosome/matrix division. The transcription factor Nolz1 is critical for the specification of dSPNs (Soleilhavoup et al., 2020). Without Nolz1, expression of dSPN-specific genes is reduced, resulting in the induction of genes that are specific to the iSPNs, and the generation of iSPNs. While the specification of dSPNs and iSPNs is largely independent, the intermingling of these two populations in the striatum is a result of active migration of iSPNs during compartment formation (Tinterri et al., 2018).

While the direct pathway is relatively well understood, the indirect pathway is more controversial (Wilson, 2016). As illustrated in Figure 2.8, there are multiple 'indirect' pathways originating in the D2-positive SPNs. Compared with the direct pathway from dSPN to SNr, this circuit is indirect because it has an additional relay in the GPe. Consequently, it is expected to have a net inhibitory effect on the targets of the BG by increasing SNr output. In another common version of the indirect pathway, the STN is included (iSPN to GPe to STN to SNr). Since STN neurons are excitatory, this pathway also produces the same net effect of enhancing BG output. However, there is also a prominent projection from the STN to the GPe that is often neglected (Wilson, 2016).

Whether direct and indirect pathways are truly segregated has also been a point of contention (Chang, Wilson, & Kitai, 1981; Hazrati & Parent, 1992b; Parent & Hazrati, 1995b; Wilson, 2016). To begin with, many striatonigral neurons send axon collaterals to the GPe (Fujiyama et al., 2011; Loopuijt & Van der Kooy, 1985; Parent & Hazrati, 1995a; Wu, Richard, & Parent, 2000), whereas striatopallidal neurons do not send comparable collaterals to the SNr (Beckstead & Cruz, 1986). The collaterals from dSPNs to GPe are also known as 'bridging collaterals' (Cazorla et al., 2014).

Anatomical Organization of the Basal Ganglia

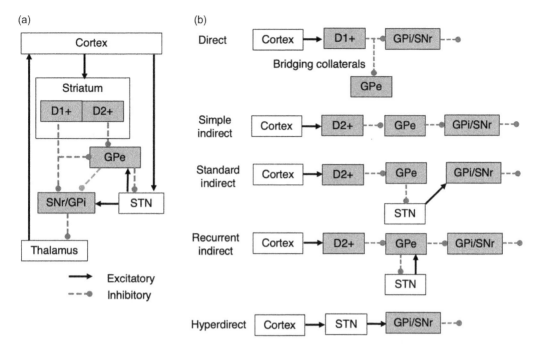

FIGURE 2.8 Direct, indirect, and hyperdirect pathways. (a) Illustration of the overall cortico-basal gannglia circuit. (b) Illustration of the direct pathway, including bridging collaterals, three different versions of the indirect pathway, and the hyperdirect pathway. GPi, globus pallidus internus; SNr, substantia nigra pars reticulata; SNc, substantia nigra pars compacta; GPe, globus pallidus externus.

Kellendonk and colleagues found that these collateral connections are highly plastic and regulated by the activation of D2 receptors. In the ventral striatum, there is an analogous direct pathway from ventral striatal output to VTA GABA neurons and an indirect pathway to the ventral pallidum, which in turn projects to the VTA. Both dSPNs and iSPNs in the ventral striatum project to the ventral pallidum, but iSPNs do not directly project to the VTA (Kupchik et al., 2015).

Even allowing for some overlap, there is now consensus that these pathways are indeed segregated and characterized by distinct patterns of gene expression (Gerfen & Surmeier, 2011; Gokce et al., 2016; Gong et al., 2007; Parent, Bouchard, & Smith, 1984; Shuen, Chen, Gloss, & Calakos, 2008). Single-cell RNA sequencing studies have revealed strong differences in the transcriptional profiles of these two major populations (Chen et al., 2021).

Axons of SPNs from both direct and indirect pathways have collaterals within the striatum. These collaterals target other SPNs, forming a recurrent inhibition network. Compared with the striatonigral pathway, the striatopallidal pathway shows less convergence of inputs from different regions (Foster et al., 2021). Different striatal domains project to distinct areas in the GPe, with little overlap in their adjacent terminal fields.

2.9 BG OUTPUT NUCLEI

The pallidum is here used as an umbrella term for all BG output nuclei. In this sense, it comprises the globus pallidus (internal and external segments), the ventral pallidum, and the GABAergic projection neurons in the VTA, SNr, and substantia nigra pars lateralis (SNl). Traditionally, these dispersed nuclei have been studied as separate entities, but they can all be classified as pallidal on account of their similarities in intrinsic cellular elements and overall connectivity. They are situated to process the convergent striatal inputs and, in turn, project to the diencephalon, midbrain, and brainstem. They are capable of influencing a wide range of effectors, including skeletomotor, autonomic, and neuroendocrine (Figure 2.9).

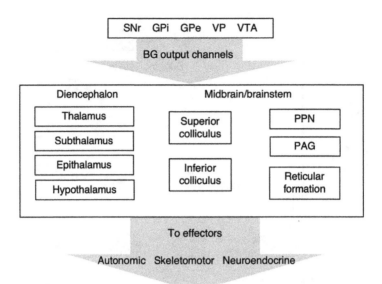

FIGURE 2.9 Summary of basal gannglia (BG) outputs. Pallidal output is directed at the diencephalon (dorsal and ventral thalamus, subthalamus, and epithalamus/habenula), midbrain (superior and inferior colliculi, PPN), and brainstem (pontine and medullary reticular formation). These projections to midbrain and brainstem nuclei can influence a variety of effectors, including skeletomotor, autonomic, and neuroendocrine outputs. GPi, globus pallidus internus; SNr, substantia nigra pars reticulata; GPe, globus pallidus externus; PPN, pedunculopontine nucleus; PAG, periaqueductal gray; VP, ventral pallidum; VTA, ventral tegmental area.

Like the striatum, the pallidum also contains GABAergic projection neurons. But unlike the SPNs, pallidal neurons have far fewer spines, as they do not receive massive glutamatergic projections. Also distinct from SPNs, which show sparse firing, pallidal output neurons often show high tonic firing rates. The combination of a very quiet input nucleus (striatum) and a continuously high-firing-output nucleus (pallidum) is a key feature of the BG, with major functional implications, as we shall see in later chapters.

Because the pallidal components are dispersed and often separated by fiber tracts, traditionally they were considered separate nuclei. Because the GPi and SNr are similar, both in intrinsic cellular elements and connectivity, some have argued that they are one continuous structure divided by the internal capsule (Parent & Hazrati, 1995a). The SNr has a greater proportion of associative striatal inputs, whereas the GPi receives more sensorimotor striatal inputs (Yoshida, Nakajima, & Niijima, 1981; Yoshida & Precht, 1971). These projections are highly convergent, with the SPNs far outnumbering their pallidal target neurons (Oorschot, 1996). Finally, the VP can be considered a ventral extension of the GPe. Both dSPN and iSPN axons from the nucleus accumbens core reach the VP (Haber, Groenewegen, Grove, & Nauta, 1985; Kalivas, Churchill, & Klitenick, 1993).

The GPe, along with its ventral extension the VP, is often considered a relay station in the indirect pathway. But both the GPe and VP in fact project to many areas outside of the BG, so they should be recognized as BG output nuclei. The GPe sends direct projections to many of the same areas that also receive GPi projections.

The striatopallidal projections are also topographically organized, largely reflecting the pattern imposed by corticostriatal projections. Different pallidal structures receive preferential projections from different striatal regions: limbic (VTA and VP), associative (SN), and sensorimotor (GPe and GPi). Even within a particular region, there are also limbic, associative, and sensorimotor divisions (Parent & Hazrati, 1995a). Again, finer divisions are possible. Recent work in mice has revealed at least 14 distinct domains within the SNr and 36 domains in the GPe, but the functional significance of these finer divisions remains unclear (Foster et al., 2021).

Anatomical Organization of the Basal Ganglia

Often, pallidal axons provide collaterals that reach multiple structures. For example, in the SNr, the same neuron can send multiple axon collaterals that target the tectum, thalamus, and medullary reticular formation (Beckstead, 1983; Gandia, De Las Heras, García, & Giménez-Amaya, 1993; Niijima & Yoshida, 1982; Pare, Hazrati, Parent, & Steriade, 1990; Schneider, Manetto, & Lidsky, 1985).

2.10 GLOBUS PALLIDUS EXTERNAL SEGMENT (GPe)

In rodents, the GPe is simply known as globus pallidus, and the equivalent of GPi is called the entopeduncular nucleus, on account of its location next to the cerebral peduncle. For the sake of consistency, we shall use "GPi" throughout this book. The GPe is similar to the GPi but receives inputs mainly from iSPNs, and projects mostly to the GPi, SNr, and STN (Sato, Lavallée, Lévesque, & Parent, 2000).

There is considerable convergence of striatal and STN inputs upon GPe neurons (Hazrati & Parent, 1992a, b). The intralaminar nucleus, particularly the parafascicular nucleus (Pf), also projects to the GPe, as well as striatum, GPi, and STN through collateral axons (Deschênes, Bourassa, Doan, & Parent, 1996).

The GPe is usually considered an intrinsic nucleus, a relay station for the indirect pathway, on its way to the GPi and SNr, but it also projects to many areas outside of the BG, including the same thalamic nuclei targeted by the GPi and SNr (Carter & Fibiger, 1978; Foster et al., 2021; Mastro, Bouchard, Holt, & Gittis, 2014). For example, parvalbumin-positive GPe neurons project to the PF of the thalamus (Mastro et al., 2014).

GPe projections to SNr converge with striatonigral axons. The same SNr neurons can receive inputs from both the striatum and the GPe. Recent work also showed that the GPe more faithfully reflected the striatal topography compared to the SNr (Foster et al., 2021). In particular, compared with the SNr, there is less convergence of inputs from different striatal regions.

The GPe also projects to targets that do not usually receive SNr or GPi projections (Figure 2.10). For example, whereas the SNr sends a massive projection to the superior colliculus (visual tectum), the GPe projects ipsilaterally to the inferior colliculus (auditory tectum) (Yasui, Kayahara, Kuga, & Nakano, 1990). Interestingly, the same caudal GPe region that projects to the inferior colliculus also projects to the auditory cortex, suggesting that it may play a role in auditory processing (Moriizumi & Hattori, 1991). At least some of these projections are cholinergic (Parent, Boucher, & O'Reilly-Fromentin, 1981).

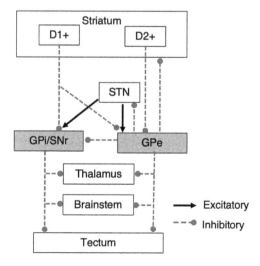

FIGURE 2.10 Globus pallidus externus (GPe) and globus pallidus internus (GPi) connectivity. STN, subthalamic nucleus.

Traditionally, the emphasis is on the forward direction of signaling in the BG, from the input nucleus to the output nucleus. While projections from the striatum to the pallidum are dominant, there are also significant projections in the opposite direction, especially from the GPe to the striatum. About a quarter of GPe neurons, known as the arkypallidal neurons, provide most of the GABAergic projections to the striatum (Kita & Kita, 1994; Kita & Kita, 2001; Mallet et al., 2012). Arkypallidal axons are oriented toward the striatum, with long and thin collaterals and many *en passant* varicosities. Despite the relatively small number of neurons in the pallidostriatal pathway, they synapse on nearly all striatal cell types, and exert a major influence on striatal activity and behavior (Bevan, Booth, Eaton, & Bolam, 1998; Mallet et al., 2016). The other major class of GPe neurons, known as the prototypic neurons, also send some projections to the striatum (Fujiyama et al., 2016), but their major targets are the STN, GPi, and SNr (Bevan et al., 1998; Kita & Kita, 1994).

2.11 GLOBUS PALLIDUS INTERNAL SEGMENT (GPi)

The GPi (entopeduncular nucleus in rodents) is also a major output nucleus of the BG. Like the SNr, it is part of the direct pathway. It receives topographically organized projections from the striatum (dSPNs primarily), STN, and GPe (Bolam & Smith, 1992; Van Der Kooy & Carter, 1981).

The connectivity of the GPi is similar to that of the SNr. Striatal and GPe projections converge in both structures (Bolam, Smith, Ingham, von Krosigk, & Smith, 1993). The GABAergic GPe inputs also converge with glutamatergic STN inputs. This organization suggests that the direct and indirect pathways have parallel but opposite influences on the BG output nuclei.

In the rat, the rostral GPi projects to the lateral habenula, whereas the caudal GPi projects to the ventral anterior-ventral lateral thalamus, intralaminar thalamic nuclei, and pedunculopontine nucleus (PPN). The strong projection to the lateral habenula distinguishes the GPi from the SNr. Through this output projection, the limbic GPi can ultimately affect many behaviors, including ingestion, mating, avoidance, and neuroendocrine function. Single-cell transcriptional analysis revealed distinct classes of GPi/entopeduncular neurons with distinct projections to the lateral habenula and thalamus (Wallace et al., 2017). The parvalbumin-positive GABAergic neurons project widely to the thalamus and brainstem, often through extensive axon collaterals (Rajakumar, Elisevich, & Flumerfelt, 1994). The somatostatin-positive neurons, however, exclusively project to the habenula. The glutamatergic parvalbumin-positive neuronal population, which also projects to the lateral habenula, presents an exception to the rule that BG output is inhibitory. These neuronal populations also receive largely distinct projections from the striatum and GPe. Somatostatin and parvalbumin-positive neurons that project to the lateral habenula receive mostly projections from striosome neurons, whereas parvalbumin-positive neurons that project to thalamus receive projections from matrix neurons.

In primates, work has focused on GPi projections to the thalamus. Strick and colleagues showed that GPi regions are organized into discrete output channels that target thalamic regions that in turn reach various motor cortical regions (Hoover & Strick, 1993). At least five parallel BG thalamocortical loops have been identified: skeletomotor, oculomotor, dorsolateral prefrontal, lateral orbitofrontal, and anterior cingulate (Alexander, DeLong, & Strick, 1986; Middleton & Strick, 1994).

2.12 VENTRAL PALLIDUM

Whereas the GPe receives projections mostly from the dorsal striatum, the VP receives projections mostly from the ventral striatum (Heimer & Wilson, 1975; Root, Melendez, Zaborszky, & Napier, 2015; Zahm, 1989; Zahm & Heimer, 1988). The ventromedial VP is strongly innervated by neurotensin-positive inputs and receives projections from the nucleus accumbens shell, while the

Anatomical Organization of the Basal Ganglia

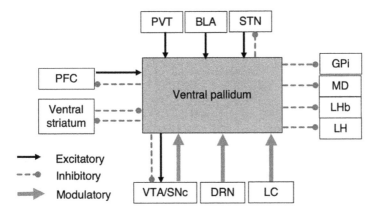

FIGURE 2.11 Ventral pallidum connectivity. MD, mediodorsal thalamus; BLA, basolateral amygdala; PVT, paraventricular thalamus; LH, lateral hypothalamus; LHb, lateral habenula; DRN, dorsal raphe nucleus; LC, locus coeruleus. STN, subthalamic nucleus; PFC, prefrontal cortex; GPi, globus pallidus internus.

dorsolateral VP contains calbindin-positive fibers, and receives mainly accumbens core projections (Zahm & Heimer, 1988). There is also a ventrolateral region that receives inputs from the shell and olfactory tubercle (Figure 2.11).

Direct and indirect pathways are also found in the ventral striatopallidal system (Sesack & Grace, 2010; Tripathi, Prensa, & Mengual, 2013). Like the GPe, the VP receives massive projections from ventral striatal iSPNs (Alheid, 2003), but there are also significant projections from dSPNs (Bengtson & Osborne, 2000; Haber et al., 1985). Some argue that, in the ventral striatum, segregation of direct and indirect pathways is not as clear-cut as in the dorsal striatum (Kupchik et al., 2015). For example, co-expression of D1 and D2 appears to be more common in the accumbens shell (Bertran-Gonzalez et al., 2008). But it should be noted that this difference is a matter of degree, as in the dorsal striatum the bridging collaterals from dSPNs also target the GPe neurons.

The GABAergic projections from the accumbens to VP often co-release neuropeptides such as enkephalin, dynorphin, or substance P (Moskowitz & Goodman, 1984). Enkephalin is a ligand for mu opioid receptors, whereas dynorphin activates kappa opioid receptors. Glutamatergic inputs come primarily from the medial prefrontal cortex and the medial STN. The prefrontal inputs originate mostly from the infralimbic region (Sesack, Deutch, Roth, & Bunney, 1989). In addition, there is also significant input from VTA DA neurons as well as serotonergic input from the dorsal raphe (Vertes, 1991).

VP outputs target the STN, SNr, and frontal cortex, the amygdala, lateral habenular and mediodorsal thalamus, hypothalamus, VTA, and other regions in the midbrain tegmentum (Haber, Groenewegen, Grove, & Nauta, 1985). Most of these targets also receive projections from the GPe. Just like the GPe, there is also a significant arkypallidal neuronal population in the VP that projects to the ventral striatum, specifically targeting the accumbens shell (Vachez et al., 2021).

2.13 VENTRAL TEGMENTAL AREA (VTA)

First described as a distinct brain region by Tsai (1925), the VTA is the source of the mesolimbic DA pathway, but it also contains other cell types, especially GABAergic projection neurons that are similar to those found in the SNr (Beckstead, Domesick, & Nauta, 1979; Morales & Margolis, 2017). The VTA is often considered to be critical for reward and motivation. It is a very effective site for intracranial self-stimulation: rats will press a lever just to deliver electrical stimulation to this area (Olds, 1977; Wise, 2004). VTA lesions usually result in loss of DA innervation of the ventral striatum and prefrontal cortex, and produce profound deficits in motivation (Mogenson, Jones, & Yim, 1980; Salamone et al., 2009).

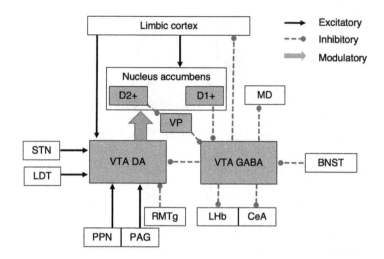

FIGURE 2.12 Ventral tegmental area (VTA) connectivity. CeA, central amygdala; BNST, bed nucleus of stria terminalis; LDT, laterodorsal tegmentum; LHb, lateral habenula; MD, mediodorsal nucleus of the thalamus; PAG, periaqueductal gray; RMTg, rostromedial tegmental nucleus. STN, subthalamic nucleus; DA, dopamine; PPN, pedunculopontine nucleus.

VTA connectivity is summarized in Figure 2.12. The major inputs to the VTA come from the ipsilateral ventral striatum, especially from the dSPNs (Zhou, Furuta, & Kaneko, 2003). The dSPN–VTA projection is the analog of the striatonigral pathway, whereas D2+ SPNs in the ventral striatum mostly project to the VP. In addition, the VTA also receives glutamatergic inputs from many regions, especially the prefrontal cortex (Beier et al., 2015; Carr & Sesack, 2000; Faget et al., 2016).

VTA GABAergic neurons serve as a major output for the limbic BG circuits, analogous to the SNr and GPi. Projections from the VTA GABA neurons reach many target regions, including the VP, lateral and magnocellular preoptic nuclei, lateral hypothalamus, and lateral habenula (Root et al., 2018; Taylor et al., 2014). These regions are largely distinct from the targets of VTA DA projections. There are also neurons that can release both DA and glutamate and target the ventral striatum (Stuber, Hnasko, Britt, Edwards, & Bonci, 2010; Tecuapetla et al., 2010) (Figure 2.12).

Different parts of the VTA are also connected with each other and with the neighboring SNc and retrorubral nucleus, mainly via axon collaterals of GABAergic projections (Ferreira, Del-Fava, Hasue, & Shammah-Lagnado, 2008; Phillipson, 1979). The dorsolateral VTA projects to itself, as well as to the SNc and retrorubral nucleus. The ventrolateral VTA mainly innervates the interfascicular nucleus.

2.14 SUBSTANTIA NIGRA

The substantia nigra (black substance), first described by Vicq d'Azyr, is among the oldest BG structures identified due to its conspicuous coloration (Parent, 2016). The dark appearance is due to the high level of neuromelanin in DA neurons (Rabey & Hefti, 1990). The substantia nigra is similar to and continuous with the more medially located VTA. Both contain mostly GABA and DA neurons. The substantia nigra is usually divided into three parts: SNc, pars reticulata (SNr), and pars lateralis (SNl) (Fallon & Loughlin, 1995).

In rodents, the SNr is the major source of BG output (Deniau, Mailly, Maurice, & Charpier, 2007). The SNc, which contains the largest DA cell population in the brain, is the source of the nigrostriatal pathway. In Parkinson's disease, the death of DA neurons contribute to a variety of symptoms, including bradykinesia and rigidity (Hornykiewicz, 2006). The SNl is the smallest and most obscure nigral component. As its name suggests, it is lateral to the SNr, but the exact boundary is difficult to define (Vankova, Arluison, Leviel, & Tramu, 1992).

Most inputs to the SNr are GABAergic projections from the GPe and the striatum (Deniau, Kitai, Donoghue, & Grofova, 1982; Ribak, Vaughn, & Roberts, 1980). There are fewer glutamatergic inputs, which mostly come from the cortex and STN (Cacciola et al., 2016; Kitano, Tanibuchi, & Jinnai, 1998; Milardi et al., 2015; Rinvik & Ottersen, 1993; Yoland Smith & Wichmann, 2015). There is also a substantial projection from the PPN (Lavoie & Parent, 1994).

Striatal terminals synapse on the distal dendrites of SNr cells, whereas GPe terminals synapse on the perikarya and proximal dendrites (Smith & Bolam, 1991). A given striatal domain may project to multiple SNr domains, and axons from a single SPN can extend throughout the rostrocaudal length of SNr. Dendrites of SNr neurons near the margins of a domain often extend into adjacent domains (Foster et al., 2021). There is an onion-like pattern in the striatal axonal arbors in the SNr (Deniau, Menetrey, & Charpier, 1996; Deniau & Chevalier, 1992; Mailly, Charpier, Menetrey, & Deniau, 2003). Striatal axonal arbors and SNr neurons that receive their projections are organized along longitudinal and curved laminas enveloping a core. Each lamina contains a group of SNr neurons innervating a specific thalamic and collicular region. Striatal projections related to auditory and visual inputs target the most ventral lamina, and somatosensory projections are found more dorsally.

2.14.1 Nigral Outputs

SNr outputs reach multiple targets, especially the tectum (superior and inferior colliculi), thalamus, and brainstem (Afifi & Kaelber, 1965; Beckstead, 1983; Beckstead et al., 1979; Faull & Mehler, 1978; Redgrave, Marrow, & Dean, 1992; Rinvik, Grofová, & Ottersen, 1976; Von Krosigk & Smith, 1991; Yasui et al., 1992). The pattern of termination of nigral axons vary depending on the target, with dense and focal innervation in the brainstem, but more widespread arborization in the thalamus and colliculus (summarized in Figure 2.13).

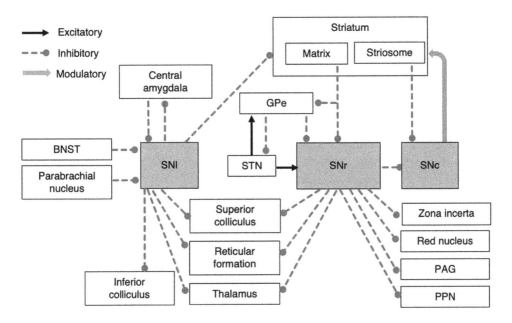

FIGURE 2.13 Substantia nigra connectivity. BNST, bed nuclei of stria terminalis; GPe, globus pallidus externus; STN, subthalamic nucleus; SNr, substantia nigra pars reticulata; SNc, substantia nigra pars compacta; PAG, periaqueductal gray; PPN, pedunculopontine nucleus.

SNr projections to the superior colliculus, also known as the nigrotectal pathway, are especially prominent. They target the deep/intermediate layers, synapsing on both glutamatergic projection neurons and local GABAergic interneurons (Harting, Huerta, Hashikawa, Weber, & Van Lieshout, 1988; Kaneda, Isa, Yanagawa, & Isa, 2008).

2.14.2 Substantia Nigra Pars Lateralis

The SNl receives a strong projection from central amygdaloid nucleus, which is striatum-like (Swanson & Petrovich, 1998). This amygdalo-nigral pathway uses neuropeptides like substance P, dynorphin, and neurotensin. SNl also receives inputs from a number of other limbic areas, including the parabrachial complex, hypothalamus, and the BNST (Vankova et al., 1992).

SNl outputs are similar to those from the SNr. For example, it also projects to the superior colliculus (May & Hall, 1986). There is also a parallel pathway originating in the SNl that targets the inferior colliculus (Yasui, Nakano, Kayahara, & Mizuno, 1991). Whereas the SNr appears to be involved in orienting guided mostly by visual inputs, the SNl may contribute to orienting guided by auditory inputs (Watanabe & Kawana, 1979). Another difference that distinguishes the SNl from the SNr is the lack of recurrent inhibition in the SNl. Whereas virtually all SNr output neurons have axons with local collaterals arborizing within the SNr, SNl axons seem to lack this pattern of collateralization.

2.15 SUBTHALAMIC NUCLEUS (STN)

A crucial nucleus that is intimately connected with the pallidum is the STN, first described by Luys (also called corpus Luysii in the older literature). Although the STN is not strictly speaking a part of the BG, it plays a critical role in determining BG output. It has also attracted much attention for its role in movement disorders (Hamani, Saint-Cyr, Fraser, Kaplitt, & Lozano, 2004). Unilateral STN lesions result in hemiballismus, or violent movements of the extremities, especially on the side contralateral to lesion (Carpenter, Whittier, & Mettler, 1950). Deep brain stimulation of the STN is among the most popular treatments for Parkinson's disease.

Reciprocally connected regions of the STN and GPe reach the same output neurons in GPi and SNr (Figure 2.14). The STN sends glutamatergic projections to the GPe, GPi, SNr, ventral pallidum, and VTA (Groenewegen & Berendse, 1990; Kita & Kitai, 1987; Koshimizu, Fujiyama, Nakamura, Furuta, & Kaneko, 2013).

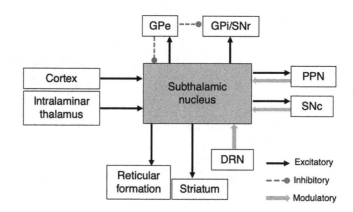

FIGURE 2.14 Subthalamic nucleus (STN) connectivity. DRN, dorsal raphe nucleus; PPN, pedunculopontine nucleus; GPe, globus pallidus externus; GPi, globus pallidus internus; SNr, substantia nigra pars reticulata.

Traditionally, the STN is viewed as a relay nucleus in the indirect pathway (Parent & Hazrati, 1995b). But the STN is also strongly connected with many other brain regions. For example, on account of extensive motor cortical inputs, Nambu and colleagues proposed that the cortico-STN-pallidum circuit forms a 'hyperdirect pathway' (Nambu, Tokuno, & Takada, 2002). As shown in Figure 2.8, this pathway can bypass the striatum, thus allowing the cortex direct control over BG output nuclei. Moreover, the STN also projects to a variety of midbrain and brainstem regions, often with multiple targets via axon collaterals (Groenewegen & Berendse, 1990; Koshimizu et al., 2013; Robledo & Féger, 1990). It is therefore in a position to influence motor pathways more directly, for example, via the PPN (Watson et al., 2021).

The STN receives ipsilateral cortical inputs, in particular from the motor cortex (Afsharpour, 1985). This projection is part of the hyperdirect circuit, through which cortical output is thought to excite pallidal output directly. On the other hand, the GABAergic afferents come mainly from the GPe (Kita, Chang, & Kitai, 1983; Kita & Kita, 2012; Kitai & Deniau, 1981; Künzle, 1977). This projection is often neglected in standard illustrations of BG anatomy. By reciprocating the excitatory STN projection with an inhibitory projection, the GPe–STN projection may play a role in generating and shaping specific patterns of neural activity.

The tripartite division of the BG into limbic, associative, and sensorimotor regions has also been applied to the STN (Parent & Hazrati, 1995b). The dorsolateral sensorimotor region of the STN is strongly and reciprocally connected with the GPe. Reciprocally connected regions of the STN and GPe also innervate common BG output neurons (Bevan, Bolam, & Crossman, 1994; Bevan, Crossman, & Bolam, 1994; Shink, Bevan, Bolam, & Smith, 1996). In addition to pallidal structures, the STN also sends projections to many other areas, especially the PPN and the cerebral cortex.

2.16 SUMMARY

The BG contain two major components, the striatum and pallidum. The striatum receives excitatory inputs from the cortex and thalamus, and sends outputs to the pallidum. Although it appears quite homogeneous to the naked eye, there are significant regional differences in connectivity and expression of key receptors and chemicals. Three major divisions have been proposed for the striatum, though finer divisions based on connectivity analysis are possible. The limbic striatum receives largely interoceptive inputs from allocortical areas; the associative division receives largely exteroceptive inputs from association cortices; and the sensorimotor striatum receives largely proprioceptive and somatosensory inputs from the primary somatosensory and motor cortices.

The striatum can also be divided into two major chemical compartments: striosome (patch) and matrix. These compartments are characterized by distinct expression of enzymes, receptors, and neuromodulators, as well as distinct anatomical connectivity. The striosome provides direct inhibition of DA neurons in the midbrain, whereas the matrix provides the major projections to BG output neurons.

The pallidum is also a heterogeneous structure with many spatially dispersed components, including the GPe, GPi, SN, VP, and VTA. The major inhibitory inputs to the pallidum are from the striatum, and the major excitatory inputs from the STN. Pallidal outputs target a wide variety of areas, including brainstem, tectum, and diencephalon (thalamus, subthalamus, and epithalamus).

The basic BG pathway involves two consecutive inhibitory synapses, also known as the direct pathway. The net effect of activating this pathway is selective activation of the target neuron receiving pallial innervation. In addition, there are also other pathways, most notably the indirect pathway, that can exert the opposite effect on the targets of the BG. Some of these areas, in turn, send descending projections to motor neurons, thus influencing the final common path from motor neurons to muscles. BG projections to the thalamic nuclei can also influence the thalamocortical network, often via reentrant projections back to the cortical areas whence corticostriatal projections originate.

The BG output nuclei also provide feedback, either directly via pallidostriatal projections or more indirectly via neuromodulator systems. For example, both VTA and SNr projection neurons have collateral branches that target neighboring DA neurons, which in turn project back to the striatum. The most well-known neuromodulatory inputs to the BG are the nigrostriatal and mesolimbic DA pathways. These pathways originate from DA cell groups in the midbrain, near the projection neurons from VTA and SNr, and target not only the striatum but also cortical regions that project to the striatum.

An examination of the anatomy therefore shows that the BG receive inputs from virtually all regions of the cerebral cortex, including cortical areas associated with all sensory modalities as well as multimodal inputs from association cortical areas. In turn, the BG outputs are in a position to influence, albeit indirectly, all types of effectors—skeletomotor, autonomic, and neuroendocrine.

REFERENCES

Afifi, A., & Kaelber, W. W. (1965). Efferent connections of the substantia nigra in the cat. *Experimental Neurology*, *11*(4), 474–482.

Afsharpour, S. (1985). Topographical projections of the cerebral cortex to the subthalamic nucleus. *Journal of Comparative Neurology*, *236*(1), 14–28.

Alexander, G. E., & DeLong, M. R. (1985a). Microstimulation of the primate neostriatum. I. Physiological properties of striatal microexcitable zones. *Journal of Neurophysiology*, *53*(6), 1401–1416.

Alexander, G. E., & DeLong, M. R. (1985b). Microstimulation of the primate neostriatum. II. Somatotopic organization of striatal microexcitable zones and their relation to neuronal response properties. *Journal of Neurophysiology*, *53*(6), 1417–1430.

Alexander, G. E., DeLong, M. R., & Strick, P. L. (1986). Parallel organization of functionally segregated circuits linking basal ganglia and cortex. *Annual Reviews of Neuroscience*, *9*, 357–381.

Alheid, G. F. (2003). Extended amygdala and basal forebrain. *Annals of the New York Academy of Sciences*, *985*(1), 185–205.

Alloway, K. D., Mutic, J. J., & Hoover, J. E. (1998). Divergent corticostriatal projections from a single cortical column in the somatosensory cortex of rats. *Brain Research*, *785*(2), 341–346.

Balleine, B. W., Peak, J., Matamales, M., Bertran-Gonzalez, J., & Hart, G. (2021). The dorsomedial striatum: An optimal cellular environment for encoding and updating goal-directed learning. *Current Opinion in Behavioral Sciences*, *41*, 38–44.

Barroso-Chinea, P., Castle, M., Aymerich, M. S., & Lanciego, J. L. (2008). Expression of vesicular glutamate transporters 1 and 2 in the cells of origin of the rat thalamostriatal pathway. *Journal of Chemical Neuroanatomy*, *35*(1), 101–107. doi: 10.1016/j.jchemneu.2007.08.001.

Beckstead, R. (1983). Long collateral branches of substantia nigra pars reticulata axons to thalamus, superior colliculus and reticular formation in monkey and cat. Multiple retrograde neuronal labeling with fluorescent dyes. *Neuroscience*, *10*(3), 767–779.

Beckstead, R., & Cruz, C. (1986). Striatal axons to the globus pallidus, entopeduncular nucleus and substantia nigra come mainly from separate cell populations in cat. *Neuroscience*, *19*(1), 147–158.

Beckstead, R. M., Domesick, V. B., & Nauta, W. J. (1979). Efferent connections of the substantia nigra and ventral tegmental area in the rat. *Brain Research*, *175*(2), 191–217.

Beier, K. T., Steinberg, E. E., DeLoach, K. E., Xie, S., Miyamichi, K., Schwarz, L., . . . Luo, L. (2015). Circuit architecture of vta dopamine neurons revealed by systematic input-output mapping. *Cell*, *162*(3), 622–634. doi: 10.1016/j.cell.2015.07.015.

Bengtson, C. P., & Osborne, P. B. (2000). Electrophysiological properties of cholinergic and noncholinergic neurons in the ventral pallidal region of the nucleus basalis in rat brain slices. *Journal of Neurophysiology*, *83*(5), 2649–2660.

Berendse, H. W., Galis-de Graaf, Y., & Groenewegen, H. J. (1992). Topographical organization and relationship with ventral striatal compartments of prefrontal corticostriatal projections in the rat. *Journal of Computer Neurology*, *316*(3), 314–347.

Bertran-Gonzalez, J., Bosch, C., Maroteaux, M., Matamales, M., Hervé, D., Valjent, E., & Girault, J.-A. (2008). Opposing patterns of signaling activation in dopamine d1 and d2 receptor-expressing striatal neurons in response to cocaine and haloperidol. *Journal of Neuroscience*, *28*(22), 5671–5685.

Bevan, M. D., Bolam, J., & Crossman, A. (1994). Convergent synaptic input from the neostriatum and the subthalamus onto identified nigrothalamic neurons in the rat. *European Journal of Neuroscience*, *6*(3), 320–334.

Bevan, M. D., Booth, P. A., Eaton, S. A., & Bolam, J. P. (1998). Selective innervation of neostriatal interneurons by a subclass of neuron in the globus pallidus of the rat. *Journal of Neuroscience, 18*(22), 9438–9452.

Bevan, M. D., Crossman, A., & Bolam, J. (1994). Neurons projecting from the entopeduncular nucleus to the thalamus receive convergent synaptic inputs from the subthalamic nucleus and the neostriatum in the rat. *Brain Research, 659*(1–2), 99–109.

Björklund, A., & Dunnett, S. B. (2007). Dopamine neuron systems in the brain: An update. *Trends in Neurosciences, 30*(5), 194–202.

Bolam, J. P., & Smith, Y. (1992). The striatum and the globus pallidus send convergent synaptic inputs onto single cells in the entopeduncular nucleus of the rat: A double anterograde labelling study combined with postembedding immunocytochemistry for gaba. *Journal of Comparative Neurology, 321*(3), 456–476.

Bolam, J. P., Smith, Y., Ingham, C. A., von Krosigk, M., & Smith, A. D. (1993). Convergence of synaptic terminals from the striatum and the globus pallidus onto single neurones in the substantia nigra and the entopeduncular nucleus. *Progress in Brain Research, 99*, 73–88.

Brimblecombe, K. R., & Cragg, S. J. (2016). The striosome and matrix compartments of the striatum: A path through the labyrinth from neurochemistry toward function. *ACS Chemical Neuroscience, 8*(2), 235–242.

Burton, H., Alloway, K., & Rosenthal, P. (1988). Somatotopic organization of the second somatosensory cortical area after lesions of the primary somatosensory area in infant and adult cats. *Brain Research, 448*(2), 397–402. doi: 10.1016/0006-8993(88)91285-1.

Cacciola, A., Milardi, D., Anastasi, G. P., Basile, G. A., Ciolli, P., Irrera, M., . . . Mondello, S. (2016). A direct cortico-nigral pathway as revealed by constrained spherical deconvolution tractography in humans. *Frontiers in Human Neuroscience, 10*, 374.

Cajal, S. R. (1909). *Histology of the Nervous System* (Swanson, N. & Swanson, L. W., Trans.): New York: Oxford University Press.

Carelli, R. M., & West, M. O. (1991). Representation of the body by single neurons in the dorsolateral striatum of the awake, unrestrained rat. *Journal of Comparative Neurology, 309*(2), 231–249.

Carpenter, M. B., Whittier, J. R., & Mettler, F. A. (1950). Analysis of choreoid hyperkinesia in the rhesus monkey. Surgical and pharmacological analysis of hyperkinesia resulting from lesions in the subthalamic nucleus ol luys. *Journal of Comparative Neurology, 92*(3), 293–331.

Carr, D. B., & Sesack, S. R. (2000). Projections from the rat prefrontal cortex to the ventral tegmental area: Target specificity in the synaptic associations with mesoaccumbens and mesocortical neurons. *Journal of Neuroscience, 20*(10), 3864–3873.

Carter, D., & Fibiger, H. (1978). The projections of the entopeduncular nucleus and globus pallidus in rat as demonstrated by autoradiography and horseradish peroxidase histochemistry. *Journal of Comparative Neurology, 177*(1), 113–123.

Cazorla, M., de Carvalho, F. D., Chohan, M. O., Shegda, M., Chuhma, N., Rayport, S., . . . Kellendonk, C. (2014). Dopamine d2 receptors regulate the anatomical and functional balance of basal ganglia circuitry. *Neuron, 81*(1), 153–164.

Chang, H. T., Wilson, C. J., & Kitai, S. T. (1981). Single neostriatal efferent axons in the globus pallidus: A light and electron microscopic study. *Science, 213*(4510), 915–918.

Chen, R., Blosser, T. R., Djekidel, M. N., Hao, J., Bhattacherjee, A., Chen, W., . . . Zhang, Y. (2021). Decoding molecular and cellular heterogeneity of mouse nucleus accumbens. *Nature Neuroscience, 24*(12), 1757–1771.

Cho, J., & West, M. O. (1997). Distributions of single neurons related to body parts in the lateral striatum of the rat. *Brain Reearchs, 756*(1–2), 241–246.

Cowan, R. L., & Wilson, C. J. (1994). Spontaneous firing patterns and axonal projections of single corticostriatal neurons in the rat medial agranular cortex. *Journal of Neurophysiology, 71*(1), 17–32.

Crittenden, J. R., & Graybiel, A. M. (2011). Basal ganglia disorders associated with imbalances in the striatal striosome and matrix compartments. *Frontiers in Neuroanatomy, 5*, 59. doi: 10.3389/fnana.2011.00059.

De Olmos, J. S., & Heimer, L. (1999). The concepts of the ventral striatopallidal system and extended amygdala. *Annals of the New York Academy of Sciences, 877*(1), 1–32.

Deacon, T., Pakzaban, P., & Isacson, O. (1994). The lateral ganglionic eminence is the origin of cells committed to striatal phenotypes: Neural transplantation and developmental evidence. *Brain Research, 668*(1–2), 211–219.

Deniau, J. M., & Chevalier, G. (1992). The lamellar organization of the rat substantia nigra pars reticulata: Distribution of projection neurons. *Neuroscience, 46*(2), 361–377.

Deniau, J. M., Kitai, S. T., Donoghue, J. P., & Grofova, I. (1982). Neuronal interactions in the substantia nigra pars reticulata through axon collaterals of the projection neurons. An electrophysiological and morphological study. *Experimental Brain Research, 47*(1), 105–113.

Deniau, J. M., Mailly, P., Maurice, N., & Charpier, S. (2007). The pars reticulata of the substantia nigra: A window to basal ganglia output. *Progress in Brain Research, 160*, 151–172.

Deniau, J. M., Menetrey, A., & Charpier, S. (1996). The lamellar organization of the rat substantia nigra pars reticulata: Segregated patterns of striatal afferents and relationship to the topography of corticostriatal projections. *Neuroscience, 73*(3), 761–781.

Deschênes, M., Bourassa, J., Doan, V. D., & Parent, A. (1996). A single-cell study of the axonal projections arising from the posterior intralaminar thalamic nuclei in the rat. *European Journal of Neuroscience, 8*(2), 329–343.

Doig, N. M., Moss, J., & Bolam, J. P. (2010). Cortical and thalamic innervation of direct and indirect pathway medium-sized spiny neurons in mouse striatum. *The Journal of Neuroscience, 30*(44), 14610–14618.

Donoghue, J. P., & Kitai, S. T. (1981). A collateral pathway to the neostriatum from corticofugal neurons of the rat sensory-motor cortex: An intracellular hrp study. *Journal of Computer Neurology, 201*(1), 1–13.

Dubé, L., Smith, A. D., & Bolam, J. P. (1988). Identification of synaptic terminals of thalamic or cortical origin in contact with distinct medium-size spiny neurons in the rat neostriatum. *Journal of Comparative Neurology, 267*(4), 455–471.

Eblen, F., & Graybiel, A. M. (1995). Highly restricted origin of prefrontal cortical inputs to striosomes in the macaque monkey. *Journal of Neuroscience, 15*(9), 5999–6013.

Ebrahimi, A., Pochet, R., & Roger, M. (1992). Topographical organization of the projections from physiologically identified areas of the motor cortex to the striatum in the rat. *Neuroscience Research, 14*(1), 39–60.

Faget, L., Osakada, F., Duan, J., Ressler, R., Johnson, A. B., Proudfoot, J. A., . . . Hnasko, T. S. (2016). Afferent inputs to neurotransmitter-defined cell types in the ventral tegmental area. *Cell Reports*, 15(12), 2796–2808. doi: 10.1016/j.celrep.2016.05.057.

Fallon, J., & Loughlin, S. (1995). Substantia Nigra. *The Rat Nervous System* (2nd edition), In Paxinos, G. (Ed.) (pp. 215–237). Sydney: Academic Press.

Faull, R., & Mehler, W. (1978). The cells of origin of nigrotectal, nigrothalamic and nigrostriatal projections in the rat. *Neuroscience, 3*(11), 989–1002.

Ferreira, J., Del-Fava, F., Hasue, R., & Shammah-Lagnado, S. (2008). Organization of ventral tegmental area projections to the ventral tegmental area–nigral complex in the rat. *Neuroscience, 153*(1), 196–213.

Fishell, G., & van der Kooy, D. (1987). Pattern formation in the striatum: Developmental changes in the distribution of striatonigral neurons. *Journal of Neuroscience, 7*(7), 1969–1978.

Flaherty, A. W., & Graybiel, A. M. (1991). Corticostriatal transformations in the primate somatosensory system. Projections from physiologically mapped body-part representations. *Journal of Neurophysiology, 66*(4), 1249–1263.

Flaherty, A. W., & Graybiel, A. M. (1993). Output architecture of the primate putamen. *Journal of Neuroscience, 13*(8), 3222–3237.

Flaherty, A. W., & Graybiel, A. M. (1994). Input-output organization of the sensorimotor striatum in the squirrel monkey. *Journal of Neuroscience, 14*(2), 599–610.

Flandin, P., Kimura, S., & Rubenstein, J. L. (2010). The progenitor zone of the ventral medial ganglionic eminence requires nkx2-1 to generate most of the globus pallidus but few neocortical interneurons. *Journal of Neuroscience, 30*(8), 2812–2823.

Foster, N. N., Barry, J., Korobkova, L., Garcia, L., Gao, L., Becerra, M., . . . Dong, H. W. (2021). The mouse cortico-basal ganglia-thalamic network. *Nature, 598*(7879), 188–194. doi: 10.1038/s41586-021-03993-3.

Fremeau Jr, R. T., Troyer, M. D., Pahner, I., Nygaard, G. O., Tran, C. H., Reimer, R. J., . . . Edwards, R. H. (2001). The expression of vesicular glutamate transporters defines two classes of excitatory synapse. *Neuron, 31*(2), 247–260.

Fujiyama, F., Nakano, T., Matsuda, W., Furuta, T., Udagawa, J., & Kaneko, T. (2016). A single-neuron tracing study of arkypallidal and prototypic neurons in healthy rats. *Brain Structure and Function, 221*(9), 4733–4740.

Fujiyama, F., Sohn, J., Nakano, T., Furuta, T., Nakamura, K. C., Matsuda, W., & Kaneko, T. (2011). Exclusive and common targets of neostriatofugal projections of rat striosome neurons: A single neuron-tracing study using a viral vector. *European Journal of Neuroscience, 33*(4), 668–677.

Gandia, J., De Las Heras, S., García, M., & Giménez-Amaya, J. M. (1993). Afferent projections to the reticular thalamic nucleus from the globus pallidus and the substantia nigra in the rat. *Brain Research Bulletin, 32*(4), 351–358.

Gangarossa, G., Espallergues, J., Mailly, P., De Bundel, D., De Kerchove, D. E., Hervé, D., . . . Krieger, P. (2013). Spatial distribution of d1r-and d2r-expressing medium-sized spiny neurons differs along the rostro-caudal axis of the mouse dorsal striatum. *Frontiers in Neural Circuits, 7*, 124.

Gerfen, C. R. (1989). The neostriatal mosaic: Striatal patch-matrix organization is related to cortical lamination. *Science, 246*(4928), 385–388.

Gerfen, C. R. (1992). The neostriatal mosaic: Multiple levels of compartmental organization in the basal ganglia. *Annual Review of Neuroscience, 15*, 285–320.

Gerfen, C. R., Engber, T. M., Mahan, L. C., Susel, Z., Chase, T. N., Monsma, F. J., Jr., & Sibley, D. R. (1990). D1 and d2 dopamine receptor-regulated gene expression of striatonigral and striatopallidal neurons. *Science, 250*(4986), 1429–1432.

Gerfen, C. R., Herkenham, M., & Thibault, J. (1987). The neostriatal mosaic: II. Patch-and matrix-directed mesostriatal dopaminergic and non-dopaminergic systems. *Journal of Neuroscience, 7*(12), 3915–3934.

Gerfen, C. R., & Keefe, K. A. (1994). Neostriatal dopamine receptors. *Trends in Neuroscience, 17*(1), 2–3; discussion 4–5.

Gerfen, C. R., & Surmeier, D. J. (2011). Modulation of striatal projection systems by dopamine. *Annual Review of Neuroscience, 34*, 441–466. doi: 10.1146/annurev-neuro-061010-113641.

Gerfen, C. R., & Wilson, C. J. (1996). The basal ganglia. In Swanson, L. W., Bjorklund, A., & Hokfelt, T. (Eds.), *Handbook of Chemical Neuroanatomy* (vol. 12, pp. 371–468). Amsterdam: Elsevier.

Gokce, O., Stanley, G. M., Treutlein, B., Neff, N. F., Camp, J. G., Malenka, R. C., . . . Quake, S. R. (2016). Cellular taxonomy of the mouse striatum as revealed by single-cell rna-seq. *Cell Reports, 16*(4), 1126–1137. doi: 10.1016/j.celrep.2016.06.059.

Gong, S., Doughty, M., Harbaugh, C. R., Cummins, A., Hatten, M. E., Heintz, N., & Gerfen, C. R. (2007). Targeting cre recombinase to specific neuron populations with bacterial artificial chromosome constructs. *Journal of Neuroscience, 27*(37), 9817–9823. doi: 10.1523/JNEUROSCI.2707-07.2007.

Graybiel, A. M. (1984). Correspondence between the dopamine islands and striosomes of the mammalian striatum. *Neuroscience, 13*(4), 1157–1187.

Graybiel, A. M. (1990). Neurotransmitters and neuromodulators in the basal ganglia. *Trends in Neurosciences, 13*(7), 244–254.

Graybiel, A. M. (2008). Habits, rituals, and the evaluative brain. *Annual Review of Neuroscience, 31*, 359–387. doi: 10.1146/annurev.neuro.29.051605.112851.

Graybiel, A. M., & Ragsdale, C. W., Jr. (1978). Histochemically distinct compartments in the striatum of human, monkeys, and cat demonstrated by acetylthiocholinesterase staining. *Proceedings of the National Academy of Sciences, 75*(11), 5723–5726.

Groenewegen, H. J., & Berendse, H. W. (1990). Connections of the subthalamic nucleus with ventral striatopallidal parts of the basal ganglia in the rat. *Journal of Comparative Neurology, 294*(4), 607–622.

Groenewegen, H. J., Wright, C. I., Beijer, A. V., & Voorn, P. (1999). Convergence and segregation of ventral striatal inputs and outputs. *Annals of the New York Academy of Sciences, 877*(1), 49–63.

Haber, S. N., Groenewegen, H. J., Grove, E. A., & Nauta, W. J. (1985). Efferent connections of the ventral pallidum: Evidence of a dual striato pallidofugal pathway. *Journal of Comparative Neurology, 235*(3), 322–335.

Hamani, C., Saint-Cyr, J. A., Fraser, J., Kaplitt, M., & Lozano, A. M. (2004). The subthalamic nucleus in the context of movement disorders. *Brain, 127*(1), 4–20.

Harting, J., Huerta, M., Hashikawa, T., Weber, J., & Van Lieshout, D. (1988). Neuroanatomical studies of the nigrotectal projection in the cat. *Journal of Comparative Neurology, 278*(4), 615–631.

Hazrati, L. N., & Parent, A. (1992a). Convergence of subthalamic and striatal efferents at pallidal level in primates: An anterograde double-labeling study with biocytin and pha-l. *Brain Research, 569*(2), 336–340. doi: 10.1016/0006-8993(92)90648-S.

Hazrati, L. N., & Parent, A. (1992b). Differential patterns of arborization of striatal and subthalamic fibers in the two pallidal segments in primates. *Brain Research, 598*(1–2), 311–315. doi: 10.1016/0006-8993(92)90199-J.

Heimer, L., Alheid, G. F., de Olmos, J. S., Groenewegen, H. J., Haber, S. N., Harlan, R. E., & Zahm, D. S. (1997). The accumbens: Beyond the core-shell dichotomy. *Journal of Neuropsychiatry, 9*(3), 354–381.

Heimer, L., Switzer, R. D., & Van Hoesen, G. W. (1982). Ventral striatum and ventral pallidum: Components of the motor system? *Trends in Neuroscience, 5*, 83–87.

Heimer, L., Wilson, R. D. (1975). *The Subcortical Projections of the Allocortex: Similarities in the Neural Associations of the Hippocampus, the Piriform Cortex, and the Neocortex*: New York: Raven.

Herkenham, M., & Pert, C. B. (1981). Mosaic distribution of opiate receptors, parafascicular projections and acetylcholinesterase in rat striatum. *Nature, 291*(5814), 415.

Herrick, C. J. (1926). *Brains of Rats and Men*: Chicago: University of Chicago.

Hikosaka, O., Sakamoto, M., & Usui, S. (1989). Functional properties of monkey caudate neurons. I. Activities related to saccadic eye movements. *Journal of Neurophysiology, 61*(4), 780–798.

Hintiryan, H., Foster, N. N., Bowman, I., Bay, M., Song, M. Y., Gou, L., . . . Dong, H. W. (2016). The mouse cortico-striatal projectome. *Nature Neuroscience*. doi: 10.1038/nn.4332.

Hoover, J. E., & Strick, P. L. (1993). Multiple output channels in the basal ganglia. *Science, 259*(5096), 819–821.

Hornykiewicz, O. (2006). The discovery of dopamine deficiency in the parkinsonian brain. In Riederer, P. (Ed.), *Parkinson's Disease and Related Disorders* (pp. 9–15). New York: Springer.

Hunnicutt, B. J., Jongbloets, B. C., Birdsong, W. T., Gertz, K. J., Zhong, H., & Mao, T. (2016). A comprehensive excitatory input map of the striatum reveals novel functional organization. *Elife*, 5. doi: 10.7554/eLife.19103.

Ikemoto, S., & Panksepp, J. (1999). The role of nucleus accumbens dopamine in motivated behavior: A unifying interpretation with special reference to reward-seeking. *Brain Research Reviews*, *31*(1), 6–41.

Jimenez-Castellanos, J., & Graybiel, A. M. (1987). Subdivisions of the dopamine-containing a8-a9-a10 complex identified by their differential mesostriatal innervation of striosomes and extrastriosomal matrix. *Neuroscience*, *23*(1), 223–242.

Jinnai, K., & Matsuda, Y. (1979). Neurons of the motor cortex projecting commonly on the caudate nucleus and the lower brain stem in the cat. *Neuroscience Letters*, *13*(2), 121–126.

Johansson, Y., & Silberberg, G. (2020). The functional organization of cortical and thalamic inputs onto five types of striatal neurons is determined by source and target cell identities. *Cell Reports*, *30*(4), 1178–1194. e1173.

Jones, E. G. (2012). *The Thalamus*: Cambridge: Cambridge University Press.

Kalivas, P., Churchill, L., & Klitenick, M. (1993). Gaba and enkephalin projection from the nucleus accumbens and ventral pallidum to the ventral tegmental area. *Neuroscience*, *57*(4), 1047–1060.

Kaneda, K., Isa, K., Yanagawa, Y., & Isa, T. (2008). Nigral inhibition of gabaergic neurons in mouse superior colliculus. *Journal of Neuroscience*, *28*(43), 11071–11078.

Kelly, S. M., Raudales, R., He, M., Lee, J. H., Kim, Y., Gibb, L. G., . . . Graybiel, A. M. (2018). Radial glial lineage progression and differential intermediate progenitor amplification underlie striatal compartments and circuit organization. *Neuron*, *99*(2), 345–361. e344.

Kincaid, A. E., & Wilson, C. J. (1996). Corticostriatal innervation of the patch and matrix in the rat neostriatum. *Journal of Comparative Neurology*, *374*(4), 578–592.

Kita, H., Chang, H., & Kitai, S. (1983). The morphology of intracellularly labeled rat subthalamic neurons: A light microscopic analysis. *Journal of Comparative Neurology*, *215*(3), 245–257.

Kita, H., & Kita, S. T. (1994). The morphology of globus pallidus projection neurons in the rat: An intracellular staining study. *Brain Research*, *636*(2), 308–319.

Kita, H., & Kita, T. (2001). Number, origins, and chemical types of rat pallidostriatal projection neurons. *Journal of Comparative Neurology*, *437*(4), 438–448.

Kita, H., & Kitai, S. T. (1987). Efferent projections of the subthalamic nucleus in the rat: Light and electron microscopic analysis with the pha-1 method. *Journal of Comparative Neurology*, *260*(3), 435–452.

Kita, T., & Kita, H. (2012). The subthalamic nucleus is one of multiple innervation sites for long-range corticofugal axons: A single-axon tracing study in the rat. *Journal of Neuroscience*, *32*(17), 5990–5999.

Kitai, S. T., & Deniau, J. M. (1981). Cortical inputs to the subthalamus: Intracellular analysis. *Brain Research*, *214*(2), 411–415.

Kitano, H., Tanibuchi, I., & Jinnai, K. (1998). The distribution of neurons in the substantia nigra pars reticulata with input from the motor, premotor and prefrontal areas of the cerebral cortex in monkeys. *Brain Research*, *784*(1–2), 228–238.

Koshimizu, Y., Fujiyama, F., Nakamura, K. C., Furuta, T., & Kaneko, T. (2013). Quantitative analysis of axon bouton distribution of subthalamic nucleus neurons in the rat by single neuron visualization with a viral vector. *Journal of Comparative Neurology*, *521*(9), 2125–2146.

Krauthamer, G., & Dalsass, M. (1978). Differential synaptic modulation of polysensory neurons of the intralaminar thalamus by medial and lateral caudate nucleus and substantia nigra. *Brain Research*, *154*(1), 137–143.

Künzle, H. (1975). Bilateral projections from precentral motor cortex to the putamen and other parts of the basal ganglia. An autoradiographic study in macaca fascicularis. *Brain Research*, *88*(2), 195–209.

Künzle, H. (1977). Projections from the primary somatosensory cortex to basal ganglia and thalamus in the monkey. *Experimental Brain Research*, *30*(4), 481–492.

Kupchik, Y. M., Brown, R. M., Heinsbroek, J. A., Lobo, M. K., Schwartz, D. J., & Kalivas, P. W. (2015). Coding the direct/indirect pathways by d1 and d2 receptors is not valid for accumbens projections. *Nature Neuroscience*, *18*(9), 1230.

Lacey, C. J., Bolam, J. P., & Magill, P. J. (2007). Novel and distinct operational principles of intralaminar thalamic neurons and their striatal projections. *Journal of Neuroscience*, *27*(16), 4374–4384.

Lanuza, E., Novejarque, A., Martínez-Ricós, J., Martínez-Hernández, J., Agustín-Pavón, C., & Martínez-García, F. (2008). Sexual pheromones and the evolution of the reward system of the brain: The chemosensory function of the amygdala. *Brain Research Bulletin*, *75*(2–4), 460–466.

Lavoie, B., & Parent, A. (1994). Pedunculopontine nucleus in the squirrel monkey: Cholinergic and glutamatergic projections to the substantia nigra. *Journal of Comparative Neurology, 344*(2), 232–241. doi: 10.1002/cne.903440205.

Lei, W., Jiao, Y., Del Mar, N., & Reiner, A. (2004). Evidence for differential cortical input to direct pathway versus indirect pathway striatal projection neurons in rats. *Journal of Neuroscience, 24*(38), 8289–8299.

Levesque, M., Charara, A., Gagnon, S., Parent, A., & Deschenes, M. (1996). Corticostriatal projections from layer v cells in rat are collaterals of long-range corticofugal axons. *Brain Research, 709*(2), 311–315.

Loopuijt, L. D., & Van der Kooy, D. (1985). Organization of the striatum: Collateralization of its efferent axons. *Brain Research, 348*(1), 86–99.

Lopez-Huerta, V. G., Nakano, Y., Bausenwein, J., Jaidar, O., Lazarus, M., Cherassse, Y., . . . Arbuthnott, G. (2016). The neostriatum: Two entities, one structure? *Brain Structure and Function, 221*(3), 1737–1749.

Mailly, P., Aliane, V., Groenewegen, H. J., Haber, S. N., & Deniau, J. M. (2013). The rat prefrontostriatal system analyzed in 3d: Evidence for multiple interacting functional units. *Journal of Neuroscience, 33*(13), 5718–5727. doi: 10.1523/JNEUROSCI.5248-12.2013.

Mailly, P., Charpier, S., Menetrey, A., & Deniau, J.-M. (2003). Three-dimensional organization of the recurrent axon collateral network of the substantia nigra pars reticulata neurons in the rat. *Journal of Neuroscience, 23*(12), 5247–5257.

Malach, R., & Graybiel, A. M. (1986). Mosaic architecture of the somatic sensory-recipient sector of the cat's striatum. *Journal of Neuroscience, 6*(12), 3436–3458.

Mallet, N., Micklem, B. R., Henny, P., Brown, M. T., Williams, C., Bolam, J. P., . . . Magill, P. J. (2012). Dichotomous organization of the external globus pallidus. *Neuron, 74*(6), 1075–1086.

Mallet, N., Schmidt, R., Leventhal, D., Chen, F., Amer, N., Boraud, T., & Berke, J. D. (2016). Arkypallidal cells send a stop signal to striatum. *Neuron, 89*(2), 308–316. doi: 10.1016/j.neuron.2015.12.017.

Mastro, K. J., Bouchard, R. S., Holt, H. A., & Gittis, A. H. (2014). Transgenic mouse lines subdivide external segment of the globus pallidus (gpe) neurons and reveal distinct gpe output pathways. *The Journal of Neuroscience, 34*(6), 2087–2099.

May, P., & Hall, W. (1986). The sources of the nigrotectal pathway. *Neuroscience, 19*(1), 159–180.

McFarland, N. R., & Haber, S. N. (2001). Organization of thalamostriatal terminals from the ventral motor nuclei in the macaque. *Journal of Comparative Neurology, 429*(2), 321–336.

McGeorge, A. J., & Faull, R. L. (1989). The organization of the projection from the cerebral cortex to the striatum in the rat. *Neuroscience, 29*(3), 503–537.

Mettler, F. A., & Mettler, C. C. (1942). The effects of striatal injury. *Brain: A Journal of Neurology, 65*, 242–255.

Middleton, F. A., & Strick, P. L. (1994). Anatomical evidence for cerebellar and basal ganglia involvement in higher cognitive function. *Science, 266*(5184), 458–461.

Milardi, D., Gaeta, M., Marino, S., Arrigo, A., Vaccarino, G., Mormina, E., . . . Baglieri, A. (2015). Basal ganglia network by constrained spherical deconvolution: A possible cortico-pallidal pathway? *Movement Disorders, 30*(3), 342–349.

Mogenson, G. J., Jones, D. L., & Yim, C. Y. (1980). From motivation to action: Functional interface between the limbic system and the motor system. *Progress in Neurobiology, 14*(2–3), 69–97.

Morales, M., & Margolis, E. B. (2017). Ventral tegmental area: Cellular heterogeneity, connectivity and behaviour. *Nature Reviews Neuroscience, 18*(2), 73–85. doi: 10.1038/nrn.2016.165.

Moriizumi, T., & Hattori, T. (1991). Pallidotectal projection to the inferior colliculus of the rat. *Experimental Brain Research, 87*(1), 223–226.

Moskowitz, A. S., & Goodman, R. (1984). Light microscopic autoradiographic localization of mu and delta opioid binding sites in the mouse central nervous system. *Journal of Neuroscience, 4*(5), 1331–1342.

Nambu, A., Hironobu T., and Masahiko T. (2002). Functional significance of the cortico–subthalamo–pallidal 'hyperdirect'pathway. *Neuroscience Research, 43*(2), 111–117.

Niijima, K., & Yoshida, M. (1982). Electrophysiological evidence for branching nigral projections to pontine reticular formation, superior colliculus and thalamus. *Brain Research, 239*(1), 279–282.

Olds, J. (1977). *Drives and Reinforcements: Behavioral Studies of Hypothalamic Functions*: New York: Raven Press.

Oorschot, D. E. (1996). Total number of neurons in the neostriatal, pallidal, subthalamic, and substantia nigral nuclei of the rat basal ganglia: A stereological study using the cavalieri and optical disector methods. *Journal of Comparative Neurology, 366*(4), 580–599.

Pare, D., Hazrati, L. N., Parent, A., & Steriade, M. (1990). Substantia nigra pars reticulata projects to the reticular thalamic nucleus of the cat: A morphological and electrophysiological study. *Brain Research, 535*(1), 139–146. doi: 10.1016/0006-8993(90)91832-2.

Parent, A. (1986). *Comparative Neurobiology of the Basal Ganglia*: Hoboken, NJ: J. Wiley.
Parent, A. (1990). Extrinsic connections of the basal ganglia. *Trends in Neurosciences, 13*(7), 254–258.
Parent, A. (2016). The history of the basal ganglia: The nuclei. In: Steiner, H. (Ed.), *Handbook of Behavioral Neuroscience* (vol. 24, pp. 33–44). New York: Elsevier.
Parent, A., Bouchard, C., & Smith, Y. (1984). The striatopallidal and striatonigral projections: Two distinct fiber systems in primate. *Brain Research, 303*(2), 385–390.
Parent, A., Boucher, R., & O'Reilly-Fromentin, J. (1981). Acetylcholinesterase-containing neurons in cat pallidal complex: Morphological characteristics and projection towards the neocortex. *Brain Research, 230*(1–2), 356–361.
Parent, A., & Hazrati, L. N. (1995a). Functional anatomy of the basal ganglia. I. The cortico-basal ganglia-thalamo-cortical loop. *Brain Research Reviews, 20*(1), 91–127. doi: 10.1016/0165-0173(94)00007-c.
Parent, A., & Hazrati, L. N. (1995b). Functional anatomy of the basal ganglia. Ii. The place of subthalamic nucleus and external pallidum in basal ganglia circuitry. *Brain Research Reviews, 20*(1), 128–154. doi: 10.1016/0165-0173(94)00008-D.
Phillipson, O. (1979). A golgi study of the ventral tegmental area of tsai and interfascicular nucleus in the rat. *Journal of Comparative Neurology, 187*(1), 99–115.
Rabey, J., & Hefti, F. (1990). Neuromelanin synthesis in rat and human substantia nigra. *Journal of Neural Transmission-Parkinson's Disease and Dementia Section, 2*(1), 1–14.
Rajakumar, N., Elisevich, K., & Flumerfelt, B. (1993). Compartmental origin of the striato-entopeduncular projection in the rat. *Journal of Comparative Neurology, 331*(2), 286–296.
Rajakumar, N., Elisevich, K., & Flumerfelt, B. (1994). Parvalbumin-containing gabaergic neurons in the basal ganglia output system of the rat. *Journal of Comparative Neurology, 350*(2), 324–336.
Raju, D. V., Shah, D. J., Wright, T. M., Hall, R. A., & Smith, Y. (2006). Differential synaptology of vglut2-containing thalamostriatal afferents between the patch and matrix compartments in rats. *Journal of Comparative Neurology, 499*(2), 231–243.
Redgrave, P., Marrow, L., & Dean, P. (1992). Topographical organization of the nigrotectal projection in rat: Evidence for segregated channels. *Neuroscience, 50*(3), 571–595.
Reiner, A., Hart, N. M., Lei, W., & Deng, Y. (2010). Corticostriatal projection neurons–dichotomous types and dichotomous functions. *Frontiers in Neuroanatomy, 4*, 142.
Reiner, A., Jiao, Y., Del Mar, N., Laverghetta, A. V., & Lei, W. L. (2003). Differential morphology of pyramidal tract-type and intratelencephalically projecting-type corticostriatal neurons and their intrastriatal terminals in rats. *Journal of Comparative Neurology, 457*(4), 420–440.
Ribak, C. E., Vaughn, J. E., & Roberts, E. (1980). Gabaergic nerve terminals decrease in the substantia nigra following hemitransections of the striatonigral and pallidonigral pathways. *Brain Research, 192*(2), 413–420.
Rinvik, E., Grofová, I., & Ottersen, O. P. (1976). Demonstration of nigrotectal and nigroreticular projections in the cat by axonal transport of proteins. *Brain Research, 112*(2), 388–394.
Rinvik, E., & Ottersen, O. P. (1993). Terminals of subthalamonigral fibres are enriched with glutamate-like immunoreactivity: An electron microscopic, immunogold analysis in the cat. *Journal of Chemical Neuroanatomy, 6*(1), 19–30.
Robledo, P., & Féger, J. (1990). Excitatory influence of rat subthalamic nucleus to substantia nigra pars reticulata and the pallidal complex: Electrophysiological data. *Brain Research, 518*(1–2), 47–54.
Root, D. H., Melendez, R. I., Zaborszky, L., & Napier, T. C. (2015). The ventral pallidum: Subregion-specific functional anatomy and roles in motivated behaviors. *Progress in Neurobiology, 130*, 29–70.
Root, D. H., Zhang, S., Barker, D. J., Miranda-Barrientos, J., Liu, B., Wang, H.-L., & Morales, M. (2018). Selective brain distribution and distinctive synaptic architecture of dual glutamatergic-gabaergic neurons. *Cell Reports, 23*(12), 3465–3479.
Saint-Cyr, J., Taylor, A. E., & Lang, A. (1988). Procedural learning and neostriatal dysfunction in man. *Brain, 111*(4), 941–960.
Salamone, J. D., Correa, M., Farrar, A. M., Nunes, E. J., & Pardo, M. (2009). Dopamine, behavioral economics, and effort. *Frontiers in Behavioral Neuroscience, 3*, 13. doi: 10.3389/neuro.08.013.2009.
Sato, F., Lavallée, P., Lévesque, M., & Parent, A. (2000). Single-axon tracing study of neurons of the external segment of the globus pallidus in primate. *Journal of Comparative Neurology, 417*(1), 17–31.
Schneider, J., Manetto, C., & Lidsky, T. (1985). Substantia nigra projection to medullary reticular formation: Relevance to oculomotor and related motor functions in the cat. *Neuroscience Letters, 62*(1), 1–6.
Selemon, L. D., & Goldman-Rakic, P. S. (1985). Longitudinal topography and interdigitation of corticostriatal projections in the rhesus monkey. *Journal of Neuroscience, 5*(3), 776–794.
Sesack, S. R., Deutch, A. Y., Roth, R. H., & Bunney, B. S. (1989). Topographical organization of the efferent projections of the medial prefrontal cortex in the rat: An anterograde tract-tracing study with phaseolus vulgaris leucoagglutinin. *Journal of Comparative Neurology, 290*(2), 213–242.

Sesack, S. R., & Grace, A. A. (2010). Cortico-basal ganglia reward network: Microcircuitry. *Neuropsychopharmacology, 35*(1), 27–47.
Shepherd, G. M. (2013). Corticostriatal connectivity and its role in disease. *Nature Reviews Neuroscience, 14*(4), 278–291.
Sherman, S. M., & Guillery, R. W. (2006). *Exploring the Thalamus and its Role in Cortical Function*: Cambridge, MA: MIT Press.
Shink, E., Bevan, M., Bolam, J., & Smith, Y. (1996). The subthalamic nucleus and the external pallidum: Two tightly interconnected structures that control the output of the basal ganglia in the monkey. *Neuroscience, 73*(2), 335–357.
Shuen, J. A., Chen, M., Gloss, B., & Calakos, N. (2008). Drd1a-tdtomato bac transgenic mice for simultaneous visualization of medium spiny neurons in the direct and indirect pathways of the basal ganglia. *Journal of Neuroscience, 28*(11), 2681–2685.
Sidibé, M., Bevan, M. D., Bolam, J. P., & Smith, Y. (1997). Efferent connections of the internal globus pallidus in the squirrel monkey: I. Topography and synaptic organization of the pallidothalamic projection. *Journal of Comparative Neurology, 382*(3), 323–347.
Smith, Y., & Bolam, J. P. (1991). Convergence of synaptic inputs from the striatum and the globus pallidus onto identified nigrocollicular cells in the rat: A double anterograde labelling study. *Neuroscience, 44*(1), 45–73. doi: 10.1016/0306-4522(91)90250-r.
Smith, Y., Galvan, A., Ellender, T. J., Doig, N., Villalba, R. M., Ocampo, I. H., . . . Bolam, P. (2014). The thalamostriatal system in normal and diseased states. *Frontiers in Systems Neuroscience, 8*, 5.
Smith, Y., & Wichmann, T. (2015). The cortico-pallidal projection: An additional route for cortical regulation of the basal ganglia circuitry. *Movement Disorders: Official Journal of the Movement Disorder Society, 30*(3), 293.
Soleilhavoup, C., Travaglio, M., Patrick, K., Garção, P., Boobalan, E., Adolfs, Y., . . . Oosterveen, T. (2020). Nolz1 expression is required in dopaminergic axon guidance and striatal innervation. *Nature Communications, 11*(1), 1–17.
Steiner, H., & Tseng, K. Y. (2016). *Handbook of Basal Ganglia Structure and Function* (vol. 24): New York: Academic Press.
Stenman, J., Toresson, H., & Campbell, K. (2003). Identification of two distinct progenitor populations in the lateral ganglionic eminence: Implications for striatal and olfactory bulb neurogenesis. *Journal of Neuroscience, 23*(1), 167–174.
Stuber, G. D., Hnasko, T. S., Britt, J. P., Edwards, R. H., & Bonci, A. (2010). Dopaminergic terminals in the nucleus accumbens but not the dorsal striatum corelease glutamate. *Journal of Neuroscience, 30*(24), 8229–8233. doi: 10.1523/JNEUROSCI.1754-10.2010.
Swanson, L. W. (2000). Cerebral hemisphere regulation of motivated behavior. *Brain Research, 886*(1–2), 113–164.
Swanson, L. W., & Petrovich, G. D. (1998). What is the amygdala? *Trends in Neurosciences, 21*(8), 323–331.
Takada, M., Itoh, K., Yasui, Y., Sugimoto, T., & Mizuno, N. (1985). Topographical projections from the posterior thalamic regions to the striatum in the cat, with reference to possible tecto-thalamo-striatal connections. *Experimental Brain Research, 60*(2), 385–396.
Taylor, S. R., Badurek, S., Dileone, R. J., Nashmi, R., Minichiello, L., & Picciotto, M. R. (2014). Gabaergic and glutamatergic efferents of the mouse ventral tegmental area. *Journal of Comparative Neurology, 522*(14), 3308–3334.
Tecuapetla, F., Patel, J. C., Xenias, H., English, D., Tadros, I., Shah, F., . . . Tepper, J. M. (2010). Glutamatergic signaling by mesolimbic dopamine neurons in the nucleus accumbens. *Journal of Neuroscience, 30*(20), 7105–7110.
Tinterri, A., Menardy, F., Diana, M. A., Lokmane, L., Keita, M., Coulpier, F., . . . Merchan-Sala, P. (2018). Active intermixing of indirect and direct neurons builds the striatal mosaic. *Nature Communications, 9*(1), 4725.
Tripathi, A., Prensa, L., & Mengual, E. (2013). Axonal branching patterns of ventral pallidal neurons in the rat. *Brain Structure and Function, 218*(5), 1133–1157.
Tsai, C. (1925). The optic tracts and centers of the opossum. Didelphis virginiana. *Journal of Comparative Neurology, 39*(2), 173–216.
Vachez, Y. M., Tooley, J. R., Abiraman, K., Matikainen-Ankney, B., Casey, E., Earnest, T., . . . Uddin, O. (2021). Ventral arkypallidal neurons inhibit accumbal firing to promote reward consumption. *Nature Neuroscience, 24*(3), 379–390.
Van Der Kooy, D., & Carter, D. A. (1981). The organization of the efferent projections and striatal afferents of the entopeduncular nucleus and adjacent areas in the rat. *Brain Research, 211*(1), 15–36.
van der Kooy, D., & Fishell, G. (1987). Neuronal birthdate underlies the development of striatal compartments. *Brain Research, 401*(1), 155–161.

Vankova, M., Arluison, M., Leviel, V., & Tramu, G. (1992). Afferent connections of the rat substantia nigra pars lateralis with special reference to peptide-containing neurons of the amygdalo-nigral pathway. *Journal of Chemical Neuroanatomy, 5*(1), 39–50.

Vertes, R. P. (1991). A pha-l analysis of ascending projections of the dorsal raphe nucleus in the rat. *Journal of Comparative Neurology, 313*(4), 643–668.

Villablanca, J., & Salinas-Zeballos, M. E. (1972). Sleep-wakefulness, eeg and behavioral studies of chronic cats without the thalamus: The" athalamic" cat. *Archives italiennes de biologie, 110*(3), 383–411.

Vogt, C., & Vogt, O. (1941). Thalamusstudien i–iii. *Journal Fur Psychologie Und Neurologie, 50*, 32–154.

Von Krosigk, M., & Smith, A. (1991). Descending projections from the substantia nigra and retrorubral field to the medullary and pontomedullary reticular formation. *European Journal of Neuroscience, 3*(3), 260–273.

Voorn, P., Vanderschuren, L. J., Groenewegen, H. J., Robbins, T. W., & Pennartz, C. M. (2004). Putting a spin on the dorsal-ventral divide of the striatum. *Trends in Neurosciences, 27*(8), 468–474.

Wall, N. R., De La Parra, M., Callaway, E. M., & Kreitzer, A. C. (2013). Differential innervation of direct-and indirect-pathway striatal projection neurons. *Neuron, 79*(2), 347–360.

Wallace, M. L., Saunders, A., Huang, K. W., Philson, A. C., Goldman, M., Macosko, E. Z., . . . Sabatini, B. L. (2017). Genetically distinct parallel pathways in the entopeduncular nucleus for limbic and sensorimotor output of the basal ganglia. *Neuron, 94*(1), 138–152. e135.

Watanabe, K., & Kawana, E. (1979). Nigral projections to the inferior and the superior colliculus in the rat: A horseradish peroxidase study. *Okajimas Folia Anatomica Japonica, 56*(5), 289–295.

Watson, Glenn D.R., et al. (2021). Thalamic projections to the subthalamic nucleus contribute to movement initiation and rescue of parkinsonian symptoms. *Science Advances, 7*(6), eabe9192.

Webster, K. (1961). Cortico-striate interrelations in the albino rat. *Journal of Anatomy, 95*(Pt 4), 532.

Webster, K. (1965). The cortico-striatal projection in the cat. *Journal of Anatomy, 99*(Pt 2), 329.

Wilson, C. (2016). The history of the basal ganglia: Cells and circuits. In: Steiner, H. (Ed.), *Handbook of Behavioral Neuroscience* (vol. 24, pp. 45–62). Cambridge, MA: Academic Press.

Wise, R. A. (2004). Dopamine, learning and motivation. *Nature Reviews Neuroscience, 5*(6), 483–494. doi: 10.1038/nrn1406.

Wu, Y., Richard, S., & Parent, A. (2000). The organization of the striatal output system: A single-cell juxtacellular labeling study in the rat. *Neuroscience Research, 38*(1), 49–62.

Yasui, Y., Kayahara, T., Kuga, Y., & Nakano, K. (1990). Direct projections from the globus pallidus to the inferior colliculus in the rat. *Neuroscience Letters, 115*(2–3), 121–125.

Yasui, Y., Nakano, K., Kayahara, T., & Mizuno, N. (1991). Non-dopaminergic projections from the substantia nigra pars lateralis to the inferior colliculus in the rat. *Brain Research, 559*(1), 139–144.

Yasui, Y., Nakano, K., Nakagawa, Y., Kayahara, T., Shiroyama, T., & Mizuno, N. (1992). Non-dopaminergic neurons in the substantia nigra project to the reticular formation around the trigeminal motor nucleus in the rat. *Brain Research, 585*(1–2), 361–366.

Yin, H. H. (2010). The sensorimotor striatum is necessary for serial order learning. *Journal of Neuroscience, 30*(44), 14719–14723. doi: 10.1523/JNEUROSCI.3989-10.2010.

Yin, H. H., & Knowlton, B. J. (2006). The role of the basal ganglia in habit formation. *Nature Reviews Neuroscience, 7*(6), 464–476.

Yin, H. H., Ostlund, S. B., Knowlton, B. J., & Balleine, B. W. (2005). The role of the dorsomedial striatum in instrumental conditioning. *European Journal of Neuroscience, 22*, 513–523.

Yoshida, M., Nakajima, N., & Niijima, K. (1981). Effect of stimulation of the putamen on the substantia nigra in the cat. *Brain Research, 217*(1), 169–174.

Yoshida, M., & Precht, W. (1971). Monosynaptic inhibition of neurons of the substantia nigra by caudato-nigral fibers. *Brain Research, 32*(1), 225–228.

Zahm, D. (1989). The ventral striatopallidal parts of the basal ganglia in the rat—ii. Compartmentation of ventral pallidal efferents. *Neuroscience, 30*(1), 33–50.

Zahm, D., & Heimer, L. (1988). Ventral striatopallidal parts of the basal ganglia in the rat: I. Neurochemical compartmentation as reflected by the distributions of neurotensin and substance p immunoreactivity. *Journal of Comparative Neurology, 272*(4), 516–535.

Zhang, Z., Zhang, H., Wen, P., Zhu, X., Wang, L., Liu, Q., . . . Xu, F. (2017). Whole-brain mapping of the inputs and outputs of the medial part of the olfactory tubercle. *Frontiers in Neural Circuits, 11*, 52.

Zhou, L., Furuta, T., & Kaneko, T. (2003). Chemical organization of projection neurons in the rat accumbens nucleus and olfactory tubercle. *Neuroscience, 120*(3), 783–798.

3 Synaptic Transmission and Plasticity in the Basal Ganglia

Having described the anatomy, we now turn to the cellular elements and transmission of neural signals in the basal ganglia (BG). Neurotransmitters released by presynaptic neurons diffuse across the synaptic cleft and produce specific effects on postsynaptic neurons by binding to dedicated receptors. Excitatory transmitters depolarize the neuron, bringing the membrane potential closer to the threshold for spike generation, whereas inhibitory transmitters usually hyperpolarize it, making it less likely to fire. As the same transmitter can act on multiple types of receptors, the impact of a transmitter depends on the spatial distribution and density of receptors expressed by postsynaptic neurons. When the neuron reaches spike threshold, voltage-gated sodium channels may be activated transiently to trigger action potentials, which are rapidly propagated along the length of the axon and ultimately result in transmitter release at the axon terminal.

As the major excitatory transmitter, glutamate binds to receptors that allow the influx of cations like sodium. As the major inhibitory transmitter, gamma-Aminobutyric acid (GABA) binds to receptors that allow the influx of the anion chloride. In addition, there are other transmitters that exert modulatory effects by binding to G-protein–coupled receptors (GPCRs). Because these receptors are not ion channels, their activation does not directly produce currents but instead engages intracellular pathways that can alter cellular excitability as well as synaptic transmission. In addition, synapses can change, resulting in long-lasting changes in transmission. Such 'synaptic plasticity' is often considered a key biological basis for learning and memory.

Instead of providing a comprehensive review of synaptic transmission and plasticity in the BG, here we shall focus on a few well-studied examples to illustrate key principles. We shall first review the major cell types and major forms of synaptic transmission in the BG. We shall then discuss long-term synaptic plasticity, with a focus on long-term potentiation (LTP) and long-term depression (LTD) in the striatum.

3.1 STRIATAL NEURONS

Although the spiny projection neuron (SPNs) constitute over 90% of striatal neurons, they did not attract much attention at first, perhaps due to their modest size (~10–15 μm in diameter for a rat SPN). Cajal thought they were interneurons and considered the much larger cholinergic neurons (~20–50 μm in diameter) to be the projection neurons. It was not until the 1970s that SPNs were recognized as projection neurons, when DiFiglia and colleagues showed that the axons of SPNs leave the striatum and reach the globus pallidus and beyond (DiFiglia, Pasik, & Pasik, 1976; Preston, Bishop, & Kitai, 1980).

Numerous interneuron classes have been identified in the striatum, though together they make up less than 10% of the total striatal neuronal population (Table 3.1). Like SPNs, striatal interneurons also receive glutamatergic afferents from the cortex and thalamus, but their output is directed primarily to SPNs and other interneurons (Kawaguchi, Wilson, Augood, & Emson, 1995; Tepper, Tecuapetla, Koos, & Ibanez-Sandoval, 2010). Unlike SPNs, most striatal interneurons are pacemakers that fire tonically even in the absence of synaptic inputs.

Different populations of striatal neurons have very different firing rates and intrinsic membrane properties. For example, although the SPNs are capable of sustained firing, they are normally quiet, and fire in short bursts, whereas most of the interneurons have higher firing rates and often show

TABLE 3.1
Interneurons in the Striatum

Name	Marker	Transmitter	Characteristic	Percentage of Total Cells
Cholinergic interneuron (CIN)	ChAT	Acetylcholine (ACh)	Hyperpolarization-activated currents and spontaneous pacemaking. Receives strong intralaminar thalamic input.	~1–2
Fast-spiking interneuron (FSI)	PV	Gamma-Aminobutyric acid (GABA)	Receives cortical inputs and project to SPNs. Capable of sustained firing with little spike frequency adaptation.	~3–5
Low-threshold spiking interneuron (LTSI)	Somatostain (SOM), Neuropeptide Y(NPY), Nitric oxide (NOS)	SOM, NPY, NOS, GABA	Low-threshold Ca^{2+} spike, high input resistance, depolarized resting membrane potential. Innervate matrix, but cell body is found in both compartments.	~1
Tyrosine hydroxylase-positive (TH+)	TH+	GABA	Often receive projections from cortex and SPNs. Synapse on SPNs.	<0.1
Calretinin (CR+)	Calretinin	GABA?	Physiological properties largely unknown	~0.5

sustained firing, as they are equipped with membrane properties that enable spontaneous pacemaking, i.e. spiking in the absence of synaptic inputs. They can also be distinguished by their connectivity (Johansson & Silberberg, 2020; Lee, Yonk et al., 2019). Table 3.1 lists common striatal interneurons. In addition to SPNs, here we will only consider three classes of striatal interneurons that are well characterized: low-threshold spiking interneurons (LTSIs), fast-spiking interneurons (FSIs), and cholinergic interneurons (CINs).

3.1.1 Spiny Projection Neurons (SPNs)

SPNs, also known as medium spiny neurons, are characterized by thousands of dendritic spines, sites of glutamatergic synapses from the cortex and thalamus (Figure 3.1). The spines are absent from the cell body, but very dense about 50–60 μm from the soma and less dense at the dendritic tips (Kita & Kitai, 1988). Although receiving massive excitatory inputs, the mean firing rate of the SPN is quite low, about 1–2 Hz *in vivo* in awake animals.

In the awake animal, SPNs show rapid fluctuations in membrane potential (Mahon et al., 2006). During sleep or under anesthesia, SPN membrane potentials shift between two states: a hyperpolarized 'down state' potentials (−90 to −70 mV) and a more depolarized 'up state,' during which spiking can occur (−60 to −40 mV) (Wilson & Groves, 1981; Wilson & Kawaguchi, 1996). These states reveal intrinsic membrane properties that contribute to the low firing rates of SPNs. The down state is mainly due to a potassium conductance (Kir2) that maintains the membrane potential close to the potassium equilibrium potential (close to −90 mv) (Jiang & North, 1991; Kita, Kita, & Kitai, 1984). This inwardly rectifying conductance can reduce input resistance and shunt excitatory inputs even without hyperpolarizing the neuron.

Since potassium conductance reduces the impact of excitatory inputs, convergent excitatory inputs are needed to activate the SPN (Wilson & Kawaguchi, 1996). To reach the threshold for spiking, presumably an SPN must be activated by inputs from an ensemble of cortical neurons.

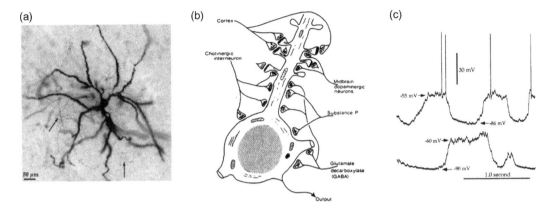

FIGURE 3.1 Striatal projection neuron (SPN). (a) A picture of an SPN from Wu and Parent (2000) with permission. With thousands of dendritic spines, they are also known as medium spiny neurons. (b) Schematic drawing showing the major synaptic inputs to the SPN. From Smith and Bolam (1990) with permission. (c) Sample traces showing up and down states in the membrane potential of an SPN. From Wilson and Kawaguchi (1996) with permission.

3.1.2 Fast-Spiking Interneurons (FSIs)

FSIs, which are also GABAergic, express parvalbumin (PV), a protein that chelates intracellular calcium (Bolam, Clarke, Smith, & Somogyi, 1983; Kita, Kosaka, & Heizmann, 1990). They are capable of sustained high firing rates with little frequency adaptation (Kawaguchi, 1993; Kawaguchi et al., 1995). As high firing rates can introduce sharp increases in intracellular calcium concentration, PV may be needed to buffer intracellular calcium.

Like SPNs, FSIs receive excitatory input from both cortex and thalamus, but spiking can be triggered more easily than SPNs (Ramanathan, Hanley, Deniau, & Bolam, 2002; Kita, 1993; Sidibe & Smith, 1999). FSIs also receive inhibitory inputs from other interneurons (Chang & Kita, 1992), pallidostriatal projections (Bevan, Booth, Eaton, & Bolam, 1998), as well as the reticular nucleus of the thalamus (Klug et al., 2018). Although they express gap junctions, which allow them to be electrotonically coupled (Kita et al., 1990), FSIs are not usually synchronized in behaving animals (Berke, 2008).

3.1.3 Low-Threshold Spiking Interneurons (LTSIs)

LTSIs are another type of GABAergic interneuron in the striatum, known for their low-threshold calcium spikes. They express somatostatin, neuropeptide Y, and nitric oxide, though not all three are necessarily expressed in the same neuron. With high input resistance and depolarized membrane potential, LTSIs show tonic activity, though usually at a lower firing rate than FSIs (Beatty, Sullivan, Morikawa, & Wilson, 2012; Kawaguchi, 1993; Sharott, Doig, Mallet, & Magill, 2012).

LTSIs have sparsely branching axons that are much longer than those of FSIs (Figure 3.2). Unlike FSIs, which target proximal dendrites of SPNs, LTSIs synapse on distal dendrites and evoke weak inhibitory postsynaptic currents (IPSCs) in SPNs (Gittis, Nelson, Thwin, Palop, & Kreitzer, 2010; Straub et al., 2016). Given their lower levels of GABA compared to FSIs, it is possible that they influence target neurons not mainly through GABA release, but rather through co-release of somatostain, nitric oxide synthase, or neuropeptide Y. However, recent work has shown that LTSI neurons synapse near dopamine (DA) terminals, where they can reduce striatal DA via GABA-B signaling (Holly, Davatolhagh, España, & Fuccillo, 2021).

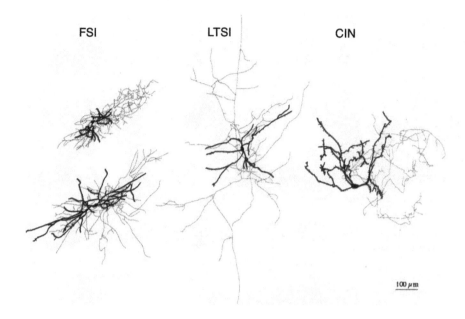

FIGURE 3.2 Striatal interneurons. Dendrites are solid black lines, and axons are light black lines. From Kawaguchi, Wilson, Augood, and Emson (1995) with permission. FSI, fast-spiking interneuron; LTSI, low-threshold spiking interneuron; CIN, cholinergic interneuron.

3.1.4 Cholinergic Interneurons (CINs)

Although few in number, CINs are among the largest neurons in the brain, with a soma of 20–50 μm in diameter and a very large axonal arborization within the striatum, spanning as much as 400 μm in the rat striatum (Difiglia, 1987; Kemp & Powell, 1971). CINs receive inputs from intralaminar nuclei and mediodorsal thalamic nuclei, association cortices, and the brainstem (Butcher & Hodge, 1976; Klug et al., 2018; Lapper & Bolam, 1992).

CINs are also pacemakers, firing tonically at 4–5 Hz. For this reason, they are also known as tonically active neurons (TANs). Pacemaking is enabled by sodium currents and hyperpolarization-activated cation currents (Bennett, Callaway, & Wilson, 2000). CINs express the hyperpolarization and cyclic adenosine monophosphate-dependent cation (HCN) channels responsible for the I_h current, which depolarizes the cell when it is hyperpolarized, a common membrane mechanism for pacemaking. Calcium influx through voltage-dependent calcium channels leads to the opening of small conductance potassium (SK) channels that contribute to long-lasting afterhyperpolarization, which limits spiking (Wilson & Goldberg, 2006). As intracellular calcium levels fall, SK channel currents also decrease. Consequently, neurons can once again be depolarized by the opening of HCN channels.

The cell bodies of CINs are found in both striosome and matrix compartments. While their dendrites often cross compartment boundaries, their axons usually arborize in the matrix. Consequently, acetylcholine (ACh) is mainly released in the matrix compartment, where high levels of acetylcholinesterase can rapidly terminate ACh signaling, avoiding desensitization of nicotinic acetylcholine receptors (nAChRs) (Crittenden & Graybiel, 2011). With tonic ACh pulses from tonic CIN pacemaking, nAChRs can be continuously active, at least in the matrix compartment (Aosaki et al., 1995; Bennett & Wilson, 1999).

In the striatum, ACh signaling is closely associated with dopamine (DA) signaling. There is extensive overlap of DA and ACh axon fibers, which are as little as 1 μm apart (DeBoer & Abercrombie, 1996). Both axon fibers synapse on distal dendrites and spine necks of SPNs

(Bolam, Wainer, & Smith, 1984). ACh can modulate presynaptic DA release from axon terminals, and DA can also regulate the firing rate of CINs. ACh and DA appear to have antagonistic roles. After DA depletion in Parkinson's disease, there is excessive ACh in the striatum, which is thought to be responsible for some of the motor symptoms (Pisani, Bernardi, Ding, & Surmeier, 2007).

Phasic DA activity promotes a pause in the tonic firing of CINs, by activating D2 receptors (Aosaki, Graybiel, & Kimura, 1994; Straub, Tritsch, Hagan, Gu, & Sabatini, 2014). Cragg and colleagues showed that the pause in CINs is largely due to a decay of excitatory input, which activates a potassium current (Zhang, Reynolds, & Cragg, 2018). The Kv7.2/3 potassium channels provide the major current (I_{kr}) responsible for the pause. This delayed rectifier current (traditionally called the M current) is slow, noninactivating, and activated by the termination of excitatory synaptic inputs to CINs. DA facilitates the pause of CINs through indirect potentiation of synaptic input and the M current. In order to produce the pause, there is no need for a large input to generate an action potential and a prolonged afterhyperpolarization. Rather all that is needed is a relatively small change in the membrane potential that is sufficient to activate I_{Kr}. The activation of D2 receptors is not necessary for pause generation, but can promote the pause by indirectly potentiating I_{Kr}.

3.2 PALLIDAL NEURONS

As mentioned in the last chapter, a key reason for classifying areas like VP, globus pallidus externus (GPe), globus pallidus internus (GPi), substantia nigra pars reticulata (SNr), and ventral tegmental area (VTA) as pallidal is the similarity in their projection neurons. Pallidal neurons have features that clearly distinguish them from striatal neurons. The most striking feature of pallidal neurons is their tonic spiking, as most of them are equipped with mechanisms for spontaneous pacemaking. Many are capable of sustained high firing rates, as high as 100 Hz or more. Spontaneous firing is due to a persistent sodium current activated at hyperpolarized membrane potentials (Do & Bean, 2003, 2004).

3.2.1 GPe Neurons

Most inputs to the GPe come from the striatum and STN. Striatal fibers course in a rostrocaudal direction and parallel to each other. Each GPe neuron can receive as many as 10,000 striatal boutons (Difiglia, Pasik, & Pasik, 1982; Parent & Hazrati, 1995).

Despite considerable heterogeneity, most GPe neurons are tonically active in the absence of synaptic inputs (Bevan, Magill, Terman, Bolam, & Wilson, 2002; Hegeman, Hong, Hernández, & Chan, 2016; Kita & Kita, 1994). Two major classes of GPe neurons, prototypic and arkypallidal neurons, can be discerned (Abdi et al., 2015; Dodson et al., 2015; Mallet et al., 2016). They are shown in Figure 3.3.

The more common prototypic GPe neurons express PV. Usually found in the lateral GPe, these neurons have narrower action potentials and high tonic firing rates. They receive indirect pathway spiny projection neuron (iSPN) projections as well as some axon collaterals from direct pathway spiny projection neurons (dSPNs), and they send strong projections to the STN and parafascicular nucleus of the thalamus. In contrast, arkypallidal neurons express preproenkephalin and often also the transcription factors Npas1 and Foxp2. They project to the striatum, where they synapse on LTSIs and FSIs but rarely on SPN (Dodson et al., 2015; Hernández et al., 2015; Mallet et al., 2012; Saunders, Huang, & Sabatini, 2016). In addition, some GPe neurons express the transcription factor Lhx6, which overlaps with the Npas1+ neurons (Mastro, Bouchard, Holt, & Gittis, 2014). Lhx6+ neurons, found mostly in the medial GPe, are distinct from PV+ neurons in connectivity and intrinsic properties. They have lower spontaneous firing rates and project strongly to the striatum and pars compacta (SNc).

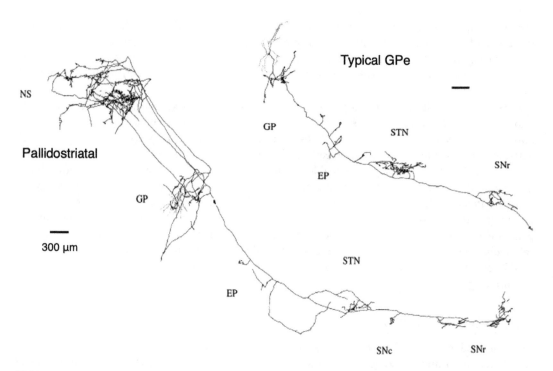

FIGURE 3.3 Common cell types in the globus pallidus externus (GPe). On the left is a rat pallidostriatal neuron, with axon collaterals to most other basal ganglia structures. Right, a more typical GPe neuron that does not project to the striatum but instead sends axons to multiple BG output nuclei. NS, neostriatum; EP, endopeduncular nucleus; . Modified from Bevan, Booth, Eaton, and Bolam (1998) with permission.

3.2.2 GPi Neurons

The main inputs to GPi neurons come from dSPNs, sometimes from collaterals of axons that also reach the SNr (Parent, Charara, & Pinault, 1995). These projections can produce powerful inhibition of GPi neurons. In addition, the GPi is strongly and reciprocally connected with the STN, which provides the major source of glutamatergic inputs (Shink, Bevan, Bolam, & Smith, 1996). The dendrites of many GPi cells are oriented perpendicular to incoming striatal axons (Percheron, Yelnik, & François, 1984). Each pallidal neuron can receive axons from many SPNs in a large area.

There are at least three classes of GPi neurons (Nakanishi, Kita, & Kitai, 1991): somatostatin-positive neurons that co-release glutamate/GABA, GABAergic PV-positive neurons, and glutamatergic PV-positive neurons. More recent work using single-cell mRNA sequencing reveals transcriptionally distinct neuronal populations with distinct connectivity and physiological function (Wallace et al., 2017). For example, GPi neurons that project to the lateral habenula (epithalamus) receive input from striosomes and the limbic striatum, whereas PV+ neurons receive input exclusively from the matrix and in turn project to the ventral thalamus. Somatostain-positive neurons co-release GABA and glutamate in the lateral habenula, whereas PV+ neurons release only glutamate. In addition, there is also a less common class of glutamatergic neurons that project to the lateral habenula (Shabel, Proulx, Trias, Murphy, & Malinow, 2012; Stephenson-Jones et al., 2016).

3.2.3 Nigral Neurons

SNr neurons are mostly GABAergic, while SNc neurons are mostly dopaminergic (Figure 3.4). Most SNr GABAergic neurons express parvalbumin (PV) (Rajakumar, Elisevich, & Flumerfelt, 1994). They are tonically active, largely due to a persistent sodium conductance with slow inactivation

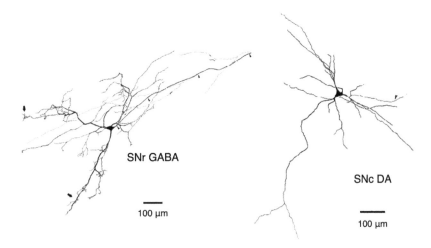

FIGURE 3.4 Common cell types in the substantia nigra. Left, a GABAergic substantia nigra pars reticulata (SNr) neuron. From Grofova, Deniau, and Kitai (1982). Right, a dopaminergic pars compacta (SNc) neuron. From Tepper, Martin, and Anderson (1995) with permission.

(Atherton & Bevan, 2005). Voltage-gated potassium currents in SNr GABA neurons can repolarize the neuron (Puopolo, Raviola, & Bean, 2007). These neurons also have a fast delayed rectifier current that activates quickly but inactivates slowly, contributing to sustained high firing rates (Ding, Wei, & Zhou, 2011). In nigral DA neurons, on the other hand, this current decreases when the firing rate is high (above 10 Hz). Consequently, DA neurons are less capable of sustained high firing rates but typically show brief and high-frequency bursts (Grace & Bunney, 1980, 1983, 1984).

3.3 GLUTAMATE

Glutamate receptors are either ionotropic (*N*-methyl-D-aspartate or NMDA, kainate, AMPA), or metabotropic (mGluR). The ionotropic glutamate receptors are permeable to a number of positively charged ions (cations). The binding of glutamate to AMPA receptors results in depolarization. While AMPA conductance is relatively linear, NMDA conductance is nonlinear, being lower at hyperpolarized membrane potentials due to magnesium block, which is removed at more depolarized membrane potentials. Unlike AMPA receptors, which are usually not permeable to calcium, NMDARs allow substantial calcium entry. With depolarization, the voltage-dependent NMDA receptors and L-type calcium channels begin to contribute to further depolarization (Carter & Sabatini, 2004).

NMDARs play a major role in long-term synaptic plasticity, e.g. in the well-known example of Schaeffer collateral LTP in the hippocampus (Malenka & Bear, 2004). They are often considered coincidence detectors because their activation requires depolarization and presynaptic input at the same time. The depolarization requirement is due to the magnesium block at hyperpolarized potentials. But this view is an oversimplification, as NMDARs are certainly not the only coincidence detectors in this sense; nor are they exclusively involved in plasticity, as they have a well-established role in synaptic transmission. Since the NMDA current has a slow rise time and a more prolonged delay, it can contribute to the temporal summation of excitatory synaptic inputs (Collingridge, Herron, & Lester, 1988). In the visual cortex, NMDA activation can amplify the postsynaptic response in a multiplicative fashion (Daw, Stein, & Fox, 1993).

In the BG, there is also a high expression of metabotropic glutamate receptors (mGluRs), which are G-protein–coupled. They are activated by strong and sustained input by detecting extra glutamate spillover outside of the dendritic spine. They can activate signaling pathways that often reduce glutamatergic transmission (Conn, Battaglia, Marino, & Nicoletti, 2005). Group I mGluRs are usually found postsynaptically; they are Gs-coupled, increase intracellular calcium release and adenylyl cyclase, and activate phospholipase C. Group II and III mGluRs are found at both presynaptic and postsynaptic sites, where their activation decreases adenylyl cyclase. When postsynaptic neurons receive strong excitatory input, activation of group I mGluRs can release endocannabinoids (eCBs, see below) that travel back to the presynaptic terminal to reduce glutamate release.

3.3.1 CORTICOSTRIATAL TRANSMISSION

In the striatum, most glutamatergic synaptic sites are found on the head and neck of the spine, as well as distal dendritic shafts of SPNs (Gerfen, 1988). But all known striatal interneurons also express glutamate receptors and typically receive glutamatergic inputs from both cortex and thalamus.

Cortical axons can synapse on many SPNs, but they do not form many synapses on any one neuron (Jones, Coulter, Burton, & Porter, 1977; Zheng & Wilson, 2002). Because individual corticostriatal axons synapse sparsely, neighboring striatal neurons do not necessarily receive input from any corticostriatal axon. A given SPN is expected to receive many different inputs, with relatively few synapses from any one source (Wilson, 2004). As each cortical axon only makes a few synapses with each SPN, neighboring SPNs are assumed to have few cortical inputs in common (Wilson, 2004; Zheng & Wilson, 2002).

Brief glutamate uncaging at distal dendrites evokes up-states lasting hundreds of milliseconds in both dSPNs and iSPNs, even though as few as 10 spines were affected. Stimulation of distal dendrites appears to be more effective in evoking dendritic plateau potentials and up-states in the cell body than stimulation of proximal dendrites (Plotkin, Day, & Surmeier, 2011). These up-states depend on both NMDA receptors and voltage-dependent calcium channels.

3.3.2 THALAMOSTRIATAL TRANSMISSION

Thalamostriatal projections synapse on virtually all striatal cell types (Smith et al., 2014). There is considerable variability in the probability of presynaptic glutamate release among different thalamostriatal inputs (Ding, Peterson, & Surmeier, 2008; Hunnicutt et al., 2016). For example, anteromedial thalamic terminals in the striatum show low release probability (synaptic facilitation), whereas axons from centrolateral and mediodorsal thalamic nuclei show high release probability (synaptic depression). Activation of thalamostriatal axons typically generates bursting in CINs, followed by a pause (Ding, Guzman, Peterson, Goldberg, & Surmeier, 2010).

Functionally related thalamic and cortex areas can target the same striatal region. For example, inputs from the ventral thalamic motor nuclei and frontal cortex converge in the same striatal region (McFarland & Haber, 2000). It is possible that they synergistically regulate striatal output. Thalamic input may alter the activation function of SPNs, effectively serving as a multiplicative gain control mechanism. For example, intralaminar thalamic input may bring the SPNs to a depolarized state, and other additional cortical inputs may drive firing. Inputs from the secondary motor cortex and parafascicular thalamic nucleus can act in a multiplicative fashion in activating SPNs (Lee, Bakhurin et al., 2019).

3.4 GAMMA-AMINOBUTYRIC ACID (GABA)

Most cell types in the BG are GABAergic, which explains the intense expression of glutamic acid decarboxylase, an enzyme that converts the metabolic precursor glutamate to GABA. There are two major types of GABA receptors. GABA-A receptors are usually found on the postsynaptic soma (Sivilotti & Nistri, 1991), whereas GABA-B receptors are either postsynaptic receptors activating G-protein–coupled inwardly rectifying potassium (GIRK) channels and causing hyperpolarization or autoreceptors involved in the regulation of GABA release from presynaptic terminals (Dutar & Nicoll, 1988; Nicoll, 2004). GABA-A receptors are permeable to chloride and bicarbonate, thereby capable of hyperpolarizing the cell (Hevers & Lüddens, 1998). When GABA binds to GABA-A receptors, the opening of the channels allows an influx of chloride, which hyperpolarizes the neuron. For this mechanism to be effective, however, intracellular chloride concentration must be kept at a low level (e.g. ~5–10 mM vs. ~120–130 mM outside) to maintain the concentration gradient needed for the inwardly directed electrochemical driving force. If chloride is not removed from the cytoplasm, the concentration gradient will be lost, and GABA will cease to be hyperpolarizing and may even become depolarizing. Neurons that express GABA-A receptors typically have mechanisms for chloride extrusion, such as K+/Cl− cotransporter type 2 (KCC2) (Li, Tornberg, Kaila, Airaksinen, & Rivera, 2002).

The reversal potential of GABA-A receptors is usually close to the resting membrane potential (around −70 mV). At this membrane potential, GABA is not hyperpolarizing as there is virtually no net driving force, but it can still reduce neural activity through shunting. In shunting inhibition, chloride conductance does not change the membrane potential but instead reduces postsynaptic input resistance (Fatt & Katz, 1953). Thus GABAergic inputs can still prevent depolarization and action potential generation by clamping the membrane close to the chloride reversal potential and reducing the impact of concurrent excitatory inputs on membrane potential (Alvarez-Leefmans & Delpire, 2009).

3.4.1 Feedforward Inhibition in the Striatum

Each FSI projects to hundreds of SPNs, including both dSPNs and iSPNs (Koós & Tepper, 1999; Planert, Szydlowski, Hjorth, Grillner, & Silberberg, 2010), though some have reported a preference for striatonigral neurons (Gittis et al., 2010). This organization is known as feedforward inhibition, which is distinct from lateral inhibition among SPNs (Figure 3.5). Due to their potent GABAergic synapses with high release probability, FSIs are often thought to inhibit SPNs (Plenz & Kitai, 1998). While inhibition has indeed been observed, the effect of GABA could be more complex (Bracci & Panzeri, 2006; O'Hare et al., 2017). For example, Bracci and colleagues found that the effect of GABA on SPN activity depends on the timing of the GABA input relative to excitatory inputs (Ayling, Panzeri, & Bracci, 2007; Bracci & Panzeri, 2006). When weak excitatory input immediately follows the GABAergic input, its depolarizing effect can be amplified by GABA.

In addition to the dependence on the timing of other inputs, recent work has suggested that the effect of FSI GABA on SPN activity depends on the chloride reversal potential. When the GABA reversal potential is just below the firing threshold (Zadeh, Turner, Calakos, & Brunel, 2021), there is a nonmonotonic regime in which the effect of GABA depends on input strength: small GABAergic currents have an excitatory effect, while large currents are inhibitory. This can explain why both excitatory and inhibitory effects of FSI on SPN firing have been observed in vivo (O'Hare et al., 2017).

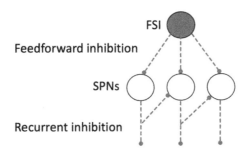

FIGURE 3.5 Local inhibitory circuits within the striatum. Top, feedforward inhibition circuit formed by fast-spiking interneuron (FSIs) and spiny projection neurons (SPNs). Each FSI synapses on many SPNs (both direct pathway spiny projection neurons and indirect pathway spiny projection neurons). Bottom, lateral inhibition circuit formed by axon collaterals of SPNs. Each SPN sends axon collaterals to nearby SPNs.

3.4.2 GABAergic Transmission in the Pallidum

The SNr receives convergent inputs from the dSPNs and the GPe (Ding, Li, & Zhou, 2015; Hikosaka, Sakamoto, & Miyashita, 1993). Pallidonigral projections from the GPe and striatonigral fibers converge onto SNr neurons, but they synapse onto distinct postsynaptic locations. Pallidonigral terminals synapse on the cell bodies and proximal dendrites of SNr neurons, whereas striatonigral synapses are found on more distal dendrites (von Krosigk, Smith, Bolam, & Smith, 1992).

Reynolds and colleagues compared striatonigral and pallidonigral transmission in rat brain slices (Connelly, Schulz, Lees, & Reynolds, 2010). Using voltage clamp recordings, they found that striatonigral IPSCs showed paired-pulse facilitation—in response to a pair of inputs separated by a very short time, the second evoked IPSC is higher than the first IPSC. Facilitation indicates a low baseline release probability of GABA at the striatonigral axon terminal. In contrast, pallidonigral synapses show paired-pulse depression, indicating high release probability of GABA at the axon terminal. Consequently, prolonged bursts of striatal IPSPs suppressed SNr activity, whereas the pallidonigral inputs became less potent over time with prolonged firing (Figure 3.6). In addition, pallidonigral IPSCs were significantly faster than striatonigral IPSCs when measured at the soma.

Pallidothalamic projections form large and densely spaced synapses on proximal dendrites and somata of thalamic neurons (Kaneda, Isa, Yanagawa, & Isa, 2008; Kultas-Ilinsky & Ilinsky, 1990). Activation of SNr nigrothalamic neurons reduces spontaneous firing in thalamic neurons (MacLeod, James, Kilpatrick, & Starr, 1980; Ueki, Uno, Anderson, & Yoshida, 1977). Nigrocollicular axons also synapse on deep/intermediate layer neurons in the superior colliculus, where they target not only excitatory projection neurons but also GABAergic interneurons (Bickford & Hall, 1992; Kaneda et al., 2008). The nigrocollicular synapse is neither facilitating nor depressing.

3.4.3 Lateral Inhibition

A prominent feature of the BG is the presence of lateral (recurrent) inhibition circuits. This type of organization is found among both striatal and pallidal projection neurons, but it has been more extensively studied in the striatum. In the striatal lateral inhibition network, axonal collaterals of

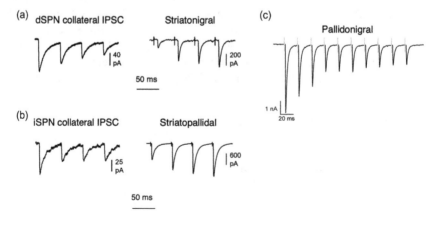

FIGURE 3.6 GABAergic transmission at striatonigral, striatopallidal, and pallidonigral synapses. (a) GABAergic collateral output from direct pathway spiny projection neurons (dSPNs) is depressing, i.e. inhibitory postsynaptic currents (IPSC) amplitude decreases progressively with additional pulses of input at high frequencies (20 Hz shown). But striatonigral synapses (from dSPN to substantia nigra pars reticulata [SNr]) are facilitating, i.e. the amplitude of the inhibitory postsynaptic current (IPSC) increases progressively. From Wei, Ding, and Zhou (2017) with permission. (b) Collateral output from iSPNs are depressing, but striatopallidal synapses (from iSPN to GPe) are facilitating. From Wei, Ding, and Zhou (2017) with permission. (c) Pallidonigral synapses are depressing. From Connelly, Schulz, Lees, and Reynolds (2010) with permission.

SPNs synapse on distal dendrites of other SPNs (Park, Lighthall, & Kitai, 1980; Wilson & Groves, 1980). Paired recordings in brain slices showed that dSPNs send axon collaterals preferentially to other dSPNs, and iSPNs to other iSPNs. These projections are usually one-way, with no reciprocal projections from the target neuron, suggesting some hierarchical organization among SPNs in both direct and indirect pathways (Wei et al., 2017). There is also some asymmetry in connectivity between dSPNs and iSPNs: dSPNs project to other dSPNs but rarely to iSPNs, whereas iSPNs project to both themselves and dSPNs (Taverna, Ilijic, & Surmeier, 2008; Tunstall, Oorschot, Kean, & Wickens, 2002; Wei et al., 2017).

Although individual release sites at SPN–SPN synapses and FSI–SPN synapses have similar quantal sizes and baseline release probabilities, unitary IPSCs at FSI–SPN synapses are larger due to more release sites and a more proximal location on the neuron (Koós & Tepper, 1999; Koos, Tepper, & Wilson, 2004). Most inhibitory synapses formed by collaterals are weak (Tunstall et al., 2002), but some SPNs synapse directly on nearby SPNs via multiple somatic innervations (Oorschot, 2013). Such somatic synapses can shunt dendritic input to the soma and provide effective inhibition. Moreover, although low release probability of GABA is unlikely to be effective in inhibiting target neurons with isolated inputs, there can be temporal summation given a train of inputs with high firing rate. Like the striatum, many pallidal projection neurons are also connected via axon collaterals. This has been studied most extensively in the SNr, though it is also a feature of other pallidal regions such as the VTA (Omelchenko & Sesack, 2009). SNr GABAergic neurons inhibit other SNr neurons with axon collaterals (Tepper & Lee, 2007; Zhou, 2016).

The SNr lateral inhibitory connections are also sparse, with low release probability. Although the SNr is much smaller than the striatum, the axonal arbor of each projection neuron is larger than that of the SPN and targets many other SNr neurons, even those at a distance. Despite tonically high firing rates, there are intrinsic conductances that reduce the impact of lateral inhibition on the baseline firing rates. Collateral inhibition in the SNr therefore has limited influence on tonic firing, yet there is potent inhibition if many SNr neurons are activated transiently (Brown, Pan, & Dudman, 2014).

3.5 DOPAMINE

Midbrain DA neurons can be divided into two major groups. The dorsal tier, which includes the dorsal SNc, VTA and SNl, contains calbindin-positive cells. The ventral tier, which includes cells in both the densocellular SNc and the ventral group in the SNr, does not express calbindin. SNc DA neurons project mostly to the ipsilateral dorsal striatum, forming the largest DA pathway in the brain, the nigrostriatal pathway (Björklund & Dunnett, 2007). Nigrostriatal axons also reach other BG areas as well as the cortex (Gauthier, Parent, Levesque, & Parent, 1999). VTA DA neurons target the ipsilateral ventromedial striatum as well as the prefrontal cortex, forming the mesolimbic pathway.

DA neurons can show regular firing even in the absence of synaptic inputs. In the VTA, such tonic activity depends primarily on a voltage-independent background sodium leak current, a voltage-dependent sodium current, and voltage-dependent calcium currents (Khaliq & Bean, 2010). Inputs can generate a random firing pattern in which inter-spike-intervals assume a Poisson-like distribution, and a phasic burst pattern, with firing rates as high as 50 Hz. Bursting is thought to be a result of disinhibition as well as afferent excitatory inputs (Paladini & Roeper, 2014; Tepper & Lee, 2007). *In vivo*, DA neurons normally show low tonic activity when the animal is at rest, but following salient stimuli or during actions, DA neurons usually display short bursts of activity, which produce widespread DA signaling in their targets in the cortex and striatum (Dommett et al., 2005; Horvitz, 2000; Schultz, 1998).

DA axons in the striatum are highly branched, forming very large axonal arbors (Prensa & Parent, 2001). A single DA neuron is estimated to have about 30 cm of axon branches and 1 million axonal varicosities (Andén et al., 1966). In some cases, the axonal arborization of single nigrostriatal neurons can reach multiple targets (e.g. pallidum and STN) beyond the striatum (Gauthier et al., 1999; Lindvall, Björklund, & Skagerberg, 1984).

DA receptors are expressed in most striatal neurons, including interneurons (Gerfen & Surmeier, 2011). Axons of DA neurons synapse on the neck of the dendritic spines in SPNs, in close proximity with the major glutamatergic inputs from the cortex, which target the heads of the spines (Freund, Powell, & Smith, 1984). They are thus in a position to modulate excitatory inputs to the striatum (Gerfen & Surmeier, 2011). Five subtypes of DA receptors have been identified (Missale, Nash, Robinson, Jaber, & Caron, 1998). D1 and D5 are classified as D1-like receptors (D1Rs). D2, D3, and D4 are classified as D2-like receptors (D2Rs). D1Rs are coupled to Gas/olf proteins, which stimulate adenylyl cyclase, elevate cyclic adenosine monophosphate (cAMP) levels, and activate protein kinase A (PKA). By contrast, D2-like receptors, being Gi/o-coupled, have the opposite effect. The D2 gene itself can be alternatively spliced: the short form often serves as the presynaptic autoreceptor that inhibits DA release, whereas the long form is more likely to be found postsynaptically (Khan et al., 1998). In addition, D2 activation can also activate the B-arrestin pathway (Urs, Peterson, & Caron, 2017).

3.5.1 How DA Influences Striatal Outputs

Nonselective DA agonists like amphetamine can produce a long-lasting increase in the responsiveness of SPNs to inputs, while DA depletion reduces the response to inputs (Schneider, Levine, Hull, & Buchwald, 1984). D1R activation is known to enhance L-type calcium currents (Carter & Sabatini, 2004) as well as NMDA currents (Cepeda, Buchwald, & Levine, 1993; Carlos Cepeda, Radisavljevic, Peacock, Levine, & Buchwald, 1992; Lee et al., 2002). Both allow calcium entry, which in turn initiates intracellular signaling (Figure 3.7). In contrast, D2R activation usually reduces SPN firing. Because D2 receptors exist primarily in a high-affinity state in the striatum, low nanomolar DA concentration is sufficient to suppress iSPNs via D2 activation. With burst firing, the concentration can increase rapidly, reaching micromolar ranges and activating D1 receptors (Bass et al., 2010).

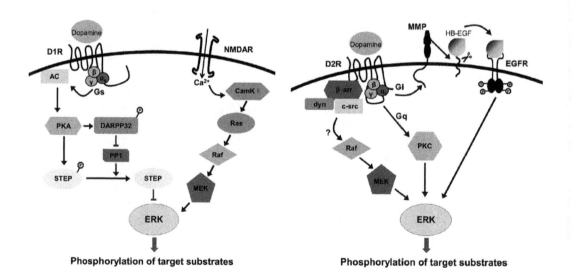

FIGURE 3.7 Dopaminergic signaling. Left, Signaling pathways following D1 receptor activation. Right, signaling pathways following D2 receptor activation. AC, adenylate cyclase; PKA, protein kinase A; PKC, protein kinase C; STEP, striatal enriched protein tyrosine phosphatase; ERK, extracellular signal-regulated kinase; MEK, MAPK/ERK Kinase; CamKII; DARPP32, dopamine- and cAMP-regulated neuronal phosphoprotein; PP1. From Baik (2013) with permission.

Recent work by Bevan and colleagues sheds light on how DA can rapidly modulate striatal output (Lahiri & Bevan, 2020). In brain slices, they used optogenetic stimulation to activate DA neurons, while measuring postsynaptic responses from dSPNs using perforated patch clamp recording, which preserves the intracellular mechanisms critical for GPCR signaling. They found that DA rapidly increased the excitability of dSPNs by activating D1 receptors, promoting transitions to the up state. The increased excitability was observed regardless of whether the neurons were in the up or down state. It also persisted for minutes following the DA release. Via a PKA pathway, D1 activation reduced potassium currents activated by depolarization and calcium influx. The overall effect on SPN activity is to reduce spike latency and afterhyperpolarization.

Synchronous activation of nigral DA neurons can result in DA release from both axon terminals in large target regions and from somatodendritic sites locally. The local release via vesicular exocytosis activates D2 autoreceptors and produces outward GIRK current or IPSCs (Beckstead, Grandy, Wickman, & Williams, 2004; Rice, Patel, & Cragg, 2011). In addition, DA can also exert an excitatory effect on SNr GABA neurons by activating D1-like receptors, an effect that requires the activation of TRPC3 channels (Zhou, Jin, Matta, Xu, & Zhou, 2009). Somatodendritic DA release may induce tonic activation of D1-like receptors and exert a tonic excitatory influence on SNr GABA neurons.

3.5.2 Dopaminergic Modulation of GABAergic Transmission

DA can also modulate inhibitory transmission in the striatum (Surmeier, Carrillo-Reid, & Bargas, 2011). For example, D1R activation potentiates GABA release at collaterals formed between dSPNs, whereas D2R activation reduces GABAergic transmission (Aceves et al., 2011; Chuhma, Tanaka, Hen, & Rayport, 2011; Guzman et al., 2003; Radnikow & Misgeld, 1998). In addition to D1R expression on the dSPN dendrites receiving glutamatergic inputs, the axon terminals of dSPNs in the SNr and GPi also express D1Rs (Levey et al., 1993; Yung et al., 1995). Striatal inputs to the GPi are also potentiated by D1R activation, whereas GPe inputs are depressed by D2R activation. DA modulation of inputs to the nigra contributes to bursting in nigral neurons (Aceves et al., 2011).

3.6 ACETYLCHOLINE (ACh)

Unlike the cerebral cortex, which mainly receives ACh from basal forebrain ACh neurons, the key source of striatal ACh is the CINs (Bolam et al., 1984). There are also extrinsic sources of ACh from the midbrain: the laterodorsal tegmental nucleus (LDT) contains ACh neurons that project mainly to the dorsomedial striatum, while cholinergic neurons in the PPN project to the dorsolateral striatum, where they excite CINs (Dautan et al., 2020; Dautan et al., 2014). By contrast, CINs inhibit other CINs as well as SPNs (Dautan et al., 2020; Witten et al., 2010). LDT and PPN also appear to have a net inhibitory effect on SPNs, though the underlying mechanism is unclear.

Intralaminar thalamostriatal inputs strongly drive CINs and increase ACh release (Consolo, Baldi, Giorgi, & Nannini, 1996). By contrast, cortical inputs are rare on the cell body and proximal dendrites, presumably exerting a much weaker influence on CINs. The outputs of CINs form synapses primarily on SPNs (Bolam et al., 1984), but also on FSIs (Chang & Kita, 1992; Koos & Tepper, 2002).

Given the large and extensive axonal arborizations of CINs in the striatum, ACh is released widely. Traditionally, it was believed that ACh signaling in the striatum is extrasynaptic and based on volume transmission. Recent work has suggested that the temporospatial distribution of ACh after release is more specific than previously believed and consistent with point-to-point transmission with a steep concentration gradient (Nosaka & Wickens, 2022).

ACh can serve as an excitatory transmitter via its action on nicotinic ACh receptors (nAChRs), but it can also act as a modulator via its action on muscarinic ACh receptors (mAChRs) (Role & Berg, 1996). In the striatum, nAChRs are located primarily on axon terminals, where they can enhance transmitter release (Zhou, Wilson, & Dani, 2002), whereas mAChRs are found both presynaptically and postsynaptically. Synchronized activation of CINs can activate nAChRs on the presynaptic axon terminals and increase presynaptic release of glutamate, DA, and GABA (Exley & Cragg, 2008; Koós & Tepper, 2002; Raz, Feingold, Zelanskaya, Vaadia, & Bergman, 1996; Zhou, Liang, & Dani, 2001).

Presynaptic mAChRs are primarily M2 type (M2 and M4), which are coupled to Gi/o proteins (Caulfield & Birdsall, 1998). M2/M4 activation can reduce ACh release by closing calcium channels and increasing the opening of Kir3 potassium channels (Goldberg, Ding, & Surmeier, 2012; Higley, Soler-Llavina, & Sabatini, 2009; Koós & Tepper, 2002; Yan & Surmeier, 1996). M4-receptors are also expressed postsynaptically in dSPNs, where their activation can reduce cAMP formation (Yan, Flores-Hernandez, & Surmeier, 2001).

Most postsynaptic mAChRs in the striatum are M1-like and coupled to Gq proteins that activate phospholipases and mobilize intracellular calcium. They are expressed on both dSPNs and iSPNs, where they can increase neuronal excitability (Akins, Surmeier, & Kitai, 1990; Pisani et al., 2007; Uchimura & North, 1990). M1 activation increases the dendritic excitability of iSPNs by reducing potassium conductance (Pisani et al., 2007; Shen et al., 2007) (Figure 3.8).

A striking anatomical feature of the striatum is the extensive overlap of striatal dopaminergic and cholinergic axon fibers, as little as 1 μm apart, suggesting strong interactions between these two transmitters (DeBoer & Abercrombie, 1996; Descarries, Gisiger, & Steriade, 1997). Both synapse on distal dendrites and dendritic spine necks of SPNs (Bolam et al., 1984). At the presynaptic release sites on DA terminals, the ACh concentration can reach very high levels upon release (up to 1 mM), leading to nAChR activation and potentiation of DA release (McGehee, Heath, Gelber, Devay, & Role, 1995; Zhou et al., 2001).

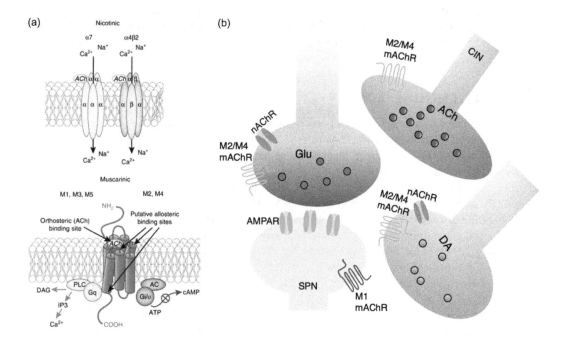

FIGURE 3.8 Cholinergic signaling. (a) Acetylcholine (ACh) effects on muscarinic and nicotinic receptors. From Jones, Byun, and Bubser (2012) with permission. (b) In the striatum, ACh can modulate presynaptic release from glutamatergic or dopaminergic axon terminals by activating nicotinic acetylcholine receptors (nAChRs). ACh activation of M2 class (M2 and M4) muscarinic ACh receptors can also suppress transmitter release from presynaptic terminals. ACh can modulate the activity of SPNs postsynaptically by activating M1-like receptors.

3.7 ENDOCANNABINOID (eCB)

eCBs, which include anandamide and 2-arachidonylglycerol (2-AG), are lipophilic molecules synthesized and released by neurons (Howlett et al., 2002; Piomelli, 2003). In the brain, the most common eCB receptor is the CB_1 receptor, which is highly expressed on corticostriatal terminals and on the axon terminals of striatonigral neurons (Herkenham, Lynn, de Costa, & Richfield, 1991; Hohmann & Herkenham, 2000; Lovinger & Mathur, 2016). CB1 receptors are coupled to the Gi/o protein, resulting in inhibition of adenylate cyclase. By closing calcium channels at the presynaptic terminal, they can suppress neurotransmitter release at both excitatory (e.g. corticostriatal) and inhibitory (e.g. striatonigral) synapses (Gerdeman & Lovinger, 2001; Wallmichrath & Szabo, 2002).

As summarized in Figure 3.9, endocannbinoid signaling is the best understood example of a retrograde messenger in synaptic transmission (Alger, 2002; Kreitzer & Regehr, 2001; Wilson, Kunos, & Nicoll, 2001). It can be triggered by multiple mechanisms that lead to the release of intracellular calcium. eCBs like 2-AG and anandamide then travel from the postsynaptic neuron to the presynaptic terminal, where they can reduce calcium influx and transmitter release (Adermark & Lovinger, 2007b; Alger, 2002; Szabo, Dörner, Pfreundtner, Nörenberg, & Starke, 1998).

The SNr has an extremely high concentration of CB1 receptors, which are expressed on the axon terminals of the many converging axonal fibers from dSPNs. As in the striatum, depolarization of postsynaptic SNr neurons can produce retrograde eCB signaling to reduce presynaptic GABA release from the striatonigral projections (Wallmichrath & Szabo, 2002).

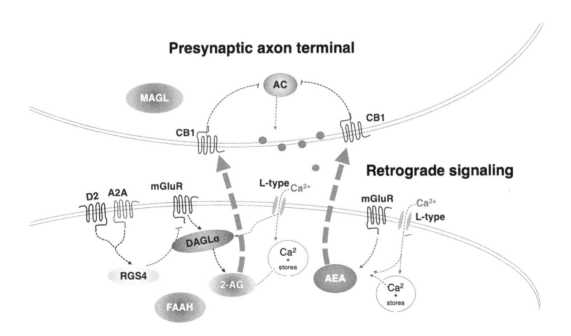

FIGURE 3.9 Endocannabinoid signaling in the BG. AEA, anandamide; 2-AG, 2-Arachidonoylglycerol; FAAH, fatty acid amide hydrolase; AC, adenylyl cyclase; DAGLα, diacylglycerol lipase alpha; mGluR, metabotropic glutamate receptor; RGS4, regulator of G-protein signaling 4. Modified from Lovinger and Mathur (2016) with permission.

3.8 ADENOSINE

Adenosine is a metabolite of adenosine triphosphate (ATP), which provides energy to living cells. Caffeine, the most common stimulant in the world, is an antagonist of A2A adenosine receptors, which are very common in the BG, especially in iSPNs, where they are co-expressed with D2 receptors. A2A receptors are therefore common markers for the indirect pathway.

Adenosine has four GPCRs: A1, A2A, A2B, and A3 (Fredholm, 1995). A1 and A3 are coupled to Gi/GO and decrease cAMP levels in the cell, whereas A2A and A2B are coupled to Gs/o and increase cAMP and PKA activity. A2B receptors are activated only by high concentrations of adenosine, and A3 receptors are sparsely expressed in the BG. The receptors that are most relevant for BG function are A1 and A2A receptors (Fredholm, Chen, Cunha, Svenningsson, & Vaugeois, 2005; Martinez-Mir, Probst, & Palacios, 1991; Schiffmann, Fisone, Moresco, Cunha, & Ferré, 2007).

Figure 3.10 summarizes adenosine signaling. A1 receptors are located presynaptically on corticostriatal terminals and postsynaptically on all SPNs. A1 activation is known to reduce transmitter

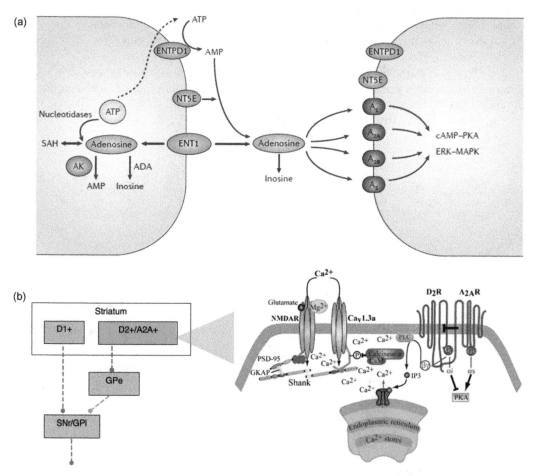

FIGURE 3.10 Adenosine signaling. (a) Adenosine release and activation of adenosine receptors. ENT1, equilibrative nucleoside transporter 1; ERK-MAPK, extracellular signal-regulated kinase, mitogen-activated protein kinase; PKA, protein kinase A ENTPD1, ectonucleoside triphosphate diphosphohydrolase 1. Modified from Chen, Eltzschig, and Fredholm (2013) with permission. (b) Proposed antagonistic actions of D2R and A2AR in iSPNs. D2R activation reduces N-methyl-D-aspartate (NMDA) signaling, but A2AR activation counteracts this effect at the membrane through heteromerization. Cav1.3a, Cav1.3a L-type calcium channel; PLC, phospholipase C; CaM, calmodulin; IP3, inositol 1,4,5-triphosphate; Shank, multiple ankyrin repeats-SH3 domain-PDZ domain-proline-rich region-sterile-α motif containing protein; PSD-95, postsynaptic density 95; GKAP, guanylate kinase-associated protein. From Azdad et al. (2009) with permission.

release. For example, it can reduce glutamate release from corticostriatal afferents and DA release from the dopaminergic terminals (Borycz et al., 2007; Lovinger & Choi, 1995). A2A receptors, on the other hand, are densely expressed on iSPNs, where they are colocalized with D2 receptors. Like D1 receptors, A2A receptors are coupled to Gs/Go pathways and facilitate excitatory synaptic transmission and excitability. A2A and D2R act on the same intracellular signaling pathways, cAMP–PKA and MAPK, but have opposite effects (Ferre, Von Euler, Johansson, Fredholm, & Fuxe, 1991; Fredholm et al., 2005). Activation of A2ARs can increase iSPN excitability (Ferré, Fuxe, Fredholm, Morelli, & Popoli, 1997).

3.9 SYNAPTIC PLASTICITY IN THE BG

Plasticity can involve changes in synaptic transmission and neuronal excitability (Malinow & Malenka, 2002; Zhang & Linden, 2003). Both can change a neuron's activation function, i.e. how its inputs are converted to outputs. Intrinsic excitability refers to changes in membrane potential in response to depolarization and hyperpolarization. Excitatory changes are usually attributed to changes in the expression of voltage-gated ion channels. Synaptic plasticity, on the other hand, refers to short-term and long-term changes in synaptic transmission. Here we will focus on synaptic plasticity, which has been extensively studied in the BG.

Normally, synaptic transmission is stable. A given excitatory input will consistently produce an output of a certain magnitude. Although the postsynaptic responses can change depending on many factors, such as neuromodulation and short-term changes in release probability, such changes are seldom long-lasting (Pitier & Alger, 1994; Zucker & Regehr, 2002). Under some conditions, however, long-lasting changes in the response of the postsynaptic neuron can be produced, including LTP and LTD. Because these changes can last weeks or longer and are often restricted to specific sets of synapses, they are often considered the biological substrate of learning and memory (Martin, Grimwood, & Morris, 2000).

In discussing mechanisms underlying synaptic plasticity, a useful distinction can be made between the site of induction, where the process leading to plasticity is initiated, and the site of expression, where the mechanism responsible for the change in synaptic transmission is found. Because the postsynaptic response is mediated by ligand-gated ion channels, removal and insertion of such channels can determine synaptic strength (Malinow & Malenka, 2002). With a postsynaptic site of expression, synaptic strength can be physically implemented by the number of ligand-gated receptors. On the other hand, with a presynaptic site of expression, the release probability can be regulated; long-term changes in release probability may be implemented by changes in voltage-gated calcium channels that depolarize the axon terminal (Gerdeman & Lovinger, 2003).

In the BG, plasticity is most commonly studied at the corticostriatal synapse and is often linked to reinforcement learning (Miller, 1981; Wickens & Kötter, 1995). Here we will focus on striatal LTP and LTD, the most extensively studied forms of synaptic plasticity in the BG, though the striatum is not the only site of synaptic plasticity.

3.9.1 Induction of Plasticity

High-frequency (tetanic) stimulation and spike-timing manipulations are the most common methods used for inducing synaptic plasticity. High-frequency stimulation protocols attempt to mimic a strong afferent train of activity. Spike-timing (spike-timing–dependent plasticity or STDP) protocols manipulate the relative timing of presynaptic input and postsynaptic depolarization without requiring high frequencies of stimulation. For example, in hippocampal neurons, presynaptic activation slightly preceding postsynaptic activation results in LTP, whereas the opposite order results in LTD (Bi & Poo, 1998).

In studies of synaptic plasticity, the term Hebbian generally refers to a requirement for both presynaptic input and postsynaptic depolarization (Hebb, 1949). Corticostriatal LTP is usually Hebbian in this sense. In studies of spike-timing–dependent plasticity, it is conventional to call the pre-post timing requirement 'Hebbian' and the post-pre 'anti-Hebbian.' The timing requirement can be explained by the overall correlation between presynaptic and postsynaptic activity (Widrow, Kim, & Park, 2015), rather than the original interpretation of a precise spike-timing requirement. There is no consensus on the spike-timing requirement for corticostriatal plasticity, with some reporting Hebbian results (Shen, Flajolet, Greengard, & Surmeier, 2008) and some reporting the opposite (anti-Hebbian) (Fino, Glowinski, & Venance, 2005). When GABAergic transmission is blocked, spike-timing–dependent plasticity (STDP) protocols generate Hebbian plasticity (Pawlak & Kerr, 2008). The molecular mechanisms underlying STDP, however, are similar to those found for conventional induction protocols (Fino & Venance, 2010; Shen et al., 2008). Here we will briefly review mechanisms underlying striatal LTP and LTD, focusing on mechanisms learned from high-frequency protocols.

3.10 STRIATAL LTP

As shown in Figure 3.11, striatal LTP usually requires the activation of NMDA receptors as well as D1 receptors in dSPNs (Calabresi, Pisani, Mercuri, & Bernardi, 1992; Dang et al., 2006). In iSPNs, A2A activation appears to contribute to LTP (Shen et al., 2008). An important feature of corticostriatal LTP is the contribution of DA as a third factor, in addition to presynaptic activation and postsynaptic depolarization. Corticostriatal and DA afferents can contact the same dendrites on the SPN, often in close proximity, with the DA terminals synapsing on the neck of the spines (Bouyer, Park, Joh, & Pickel, 1984; Freund et al., 1984). Even if no explicit manipulation of DA signaling is used, common LTP induction protocols using electrical stimulation can depolarize DA terminals, thus causing significant DA release (Calabresi, Picconi, Tozzi, & Di Filippo, 2007; Frémaux & Gerstner, 2016).

Heterosynaptic plasticity at the corticostriatal synapse resembles certain conditions for behavioral learning. In the most common interpretation, the activation of the presynaptic neuron is interpreted as a stimulus and postsynaptic activation as a response, and DA is interpreted as a reinforcement signal that can strengthen the association between the two (Bailey, Giustetto, Huang, Hawkins, & Kandel, 2000; Frémaux & Gerstner, 2016; Miller, 1981).

Depletion of DA abolishes LTD and LTP in brain slices (Calabresi, Maj, Pisani, Mercuri, & Bernardi, 1992). In dSPNs, LTP induction at the glutamatergic synapses depends on D1R activation (Kerr & Wickens, 2001; Pawlak & Kerr, 2008). On the other hand, in iSPNs, which lack D1Rs, LTP induction requires activation of A2ARs (Shen et al., 2008). This is not surprising since A2AR activation engages similar intracellular pathways as D1R activation: both A2A and D1 activation can increase cAMP, activate PKA, and increase phosphorylation of DARPP-32 (Calabresi et al., 2000; Svenningsson et al., 2004). Since DARPP-32 inhibits phosphatases, it could contribute to AMPAR trafficking, a widely established molecular mechanism underlying postsynaptic LTP expression (Malinow & Malenka, 2002). It should be noted, however, that DA is not always required for striatal plasticity. Striatal LTP can be induced even in DA-depleted mice (Kreitzer & Malenka, 2007; Shen et al., 2008).

3.11 STRIATAL LTD

Striatal LTD, first reported by Calabresi and colleagues, is usually induced by high-frequency stimulation of the corticostriatal pathway combined with postsynaptic SPN depolarization (Calabresi, Maj, Mercuri, & Bernardi, 1992; Calabresi, Maj, Pisani, et al., 1992; Lovinger, Tyler, & Merritt, 1993). It is especially common in the dorsolateral striatum but is also found in the nucleus accumbens (Robbe, Kopf, Remaury, Bockaert, & Manzoni, 2002).

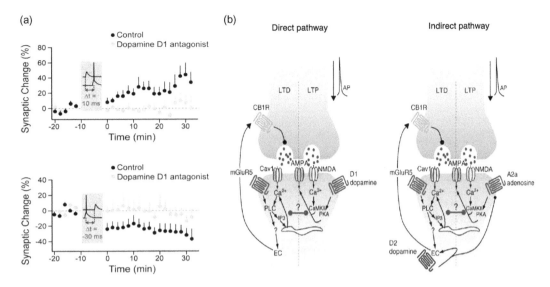

FIGURE 3.11 Striatal long-term potentiation (LTP). (a) Striatal LTP induced by spike-timing protocol in brain slices requires D1 receptor activation. Pre-post activation resulted in LTP, whereas post-pre activation resulted in long-term depression (LTD). LTP was blocked by bath application of a D1 antagonist. From Pawlak, Wickens, Kirkwood, and Kerr (2010) with permission. (b) Proposed mechanisms underlying LTP induction in the striatum. Modified from Surmeier, Plotkin, and Shen (2009) with permission.

Following the induction of striatal LTD, decreased mEPSC frequency (but not amplitude) and increased paired-pulse ratio suggest reduced release probability (Choi & Lovinger, 1997). LTD induction usually requires the activation of postsynaptic L-type calcium channels and mGluR5 receptors (Figure 3.12). In addition, Lovinger and colleagues discovered that LTD expression requires the postsynaptic generation of eCBs that diffuse retrogradely to activate presynaptic CB1 receptors (Gerdeman, Ronesi, & Lovinger, 2002). Depending on the stimulation protocol, anandamide or 2-AG could be mobilized to produce LTD (Lovinger & Mathur, 2012). CB1 receptors are common on the axon terminals of cortical pyramidal neurons and SPNs but rare on thalamic terminals. Consequently, eCB-dependent LTD of glutamatergic synapses is only expressed at corticostriatal inputs but not thalamostriatal inputs (Wu et al., 2015).

Because LTD involved D2 activation, it was initially thought to occur only in iSPNs (Kreitzer & Malenka, 2005). Later work, however, showed that both dSPNs and iSPNs can express this form of plasticity (Tozzi et al., 2011; Wang et al., 2006; Wu et al., 2015). Activation of D2Rs on CINs can promote a pause in CIN firing and briefly reduce ACh tone in the striatum. Normally, activation of M1Rs on both dSPNs and iSPNs suppresses L-type calcium channels, which are required for LTD induction (Adermark & Lovinger, 2007a). A brief reduction in ACh tone, however, is expected to reduce M1 activation and disinhibit L-type calcium channels, thus promoting LTD induction. In support of this account, genetic deletion of D2Rs on CINs, which is expected to reduce the pause in CIN activity, impaired LTD induction in both dSPNs and iSPNs. On the other hand, deletion of D2Rs in iSPNs did not abolish LTD (Augustin, Chancey, & Lovinger, 2018).

LTD can also be found at GABAergic synapses within the striatum (Adermark, Talani, & Lovinger, 2009). Whereas corticostriatal LTD requires DA and D2R activation, LTD at the inhibitory synapses does not. The involvement of CB1, D2, and group I mGluR receptors in LTD are all transient: their activation is only needed during the first few minutes after induction. The downstream signaling cascades are responsible for the long-lasting expression of LTD.

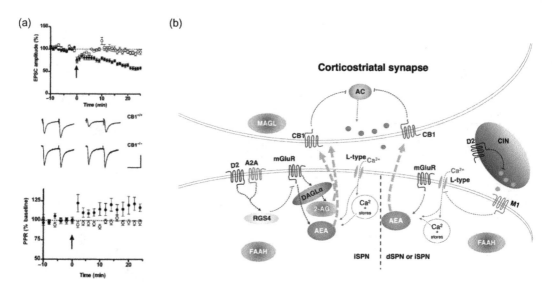

FIGURE 3.12 Striatal long-term depression (LTD). (a) Striatal LTD depends on endocannabinoid signaling. LTD was not found in CB1 knockout mice. Traces show changes in paired-pulse ratio (PPR), a measure that is inversely correlated with presynaptic release probability. LTD increases PPR, suggesting a reduction in release probability in corticostriatal axon terminals. From Gerdeman, Ronesi, and Lovinger (2002) with permission. (b) Molecular mechanisms underlying striatal LTD. See Figure 3.9 for abbreviations.

3.12 LTP OR LTD?

There appear to be significant regional differences in LTP and LTD, in part due to gradients of expression of various receptors (Gerdeman, Partridge, Lupica, & Lovinger, 2003; Partridge, Tang, & Lovinger, 2000; Smith, Musleh, Akopian, Buckwalter, & Walsh, 2001). For example, CB1 receptor expression is higher in the dorsolateral striatum, where LTD is commonly induced, and lower in the dorsomedial striatum, where LTP is more common.

Both NMDA-dependent LTP and eCB-dependent LTD can be induced simultaneously, but LTD is usually masked by LTP (Ma et al., 2018). Using the same induction protocol, blocking LTP could result in LTD and vice versa, suggesting that these two forms of plasticity may compete for expression (Yin, Park, Adermark, & Lovinger, 2007).

Some studies have suggested that when DA level is low, LTD is induced, but when there is a large increase in DA concentration, LTP will be produced (Reynolds & Wickens, 2002). It has been proposed that a single rule can explain the direction of plasticity at the corticostriatal synapses based on the dynamics of intracellular calcium concentration (Jędrzejewska-Szmek, Damodaran, Dorman, & Blackwell, 2017). Because the LTP threshold is higher than the LTD threshold, blocking NMDARs could result in LTD given calcium entry through L-type calcium channels, but the calcium concentration may not be sufficient for LTP induction. LTP is expected if the amplitude of the postsynaptic calcium transient is above some threshold, but LTD is expected if the postsynaptic calcium transient is below the potentiation threshold but above the depression threshold. With normal DA tone, conjunction of presynaptic activity and postsynaptic depolarization can result in LTD, but when coupled with a large phasic DA signal, it can produce LTP.

3.13 FUNCTIONAL IMPLICATIONS OF STRIATAL SYNAPTIC PLASTICITY

Synaptic strength is often assumed to be a fixed entity modified by synaptic plasticity and reflects associative strength in theories of learning or weights in artificial neural networks. However, the efficacy of a synapse can depend on many factors, such as the ongoing activity of other synapses and neuromodulation. It can fluctuate from moment to moment based on recent history (use dependence) (Zucker & Regehr, 2002).

Striatal LTP and LTD are not symmetric processes by which the strength of the corticostriatal synapse may be regulated bidirectionally. Once induced, LTP cannot be simply reversed by LTD, as the latter regulates presynaptic release rather than postsynaptic receptor expression. There is also saturation in synaptic strength when repeated high-frequency trains are delivered (Yin et al., 2009).

Postsynaptically expressed LTP is expected to increase response to excitatory input, but the effect of striatal LTD is not necessarily reduced activation of SPNs. LTD can alter the threshold and selectivity of striatal responses to the pattern of inputs. Reduced probability of glutamate release is expected to have the largest effect on responses to single afferent inputs and in the initial response to sustained presynaptic activity (Lovinger, 2010). But reduced release probability could also result in more synaptic facilitation, so responses to subsequent inputs in a spike train are not necessarily reduced. A combination of postsynaptically expressed LTP and presynaptically expressed LTD could reduce response to single synaptic inputs but potentiate response to burst inputs. It is therefore too simplistic to assume that LTP and LTD increase or decrease the postsynaptic response. Instead, they can shape specific parameters in the activation function. Without detailed modeling, the consequences of LTD on network output at both glutamatergic and GABA synapses are difficult to predict.

3.14 SUMMARY

While the projection neurons in both the striatum and pallidum are GABAergic, they have very different properties. SPNs are quiet at rest and occasionally activated by coherent excitatory inputs. Pallidal neurons are usually tonically active with high firing rates. These characteristics can be explained by their distinct membrane properties.

The cortex and thalamus send glutamatergic projections to the striatum. Within the BG, synaptic transmission is dominated by GABAergic transmission, both from inhibitory interneurons to SPNs (feedforward inhibition) and from lateral inhibition implemented by SPN axon collaterals on other SPNs. In addition, many neuromodulators rely on slower GPCR intracellular signaling that regulate neuronal excitability and synaptic transmission. Neuromodulators like DA, adenosine, and eCB can have different, even antagonistic, effects on target neurons depending on the receptor subtype.

Two forms of synaptic plasticity in the BG have been extensively studied. Corticostriatal LTP is postsynaptically expressed and usually requires NMDA and D1 activation. On the other hand, striatal LTD requires mobilization and release of eCBs, which travel back to the presynaptic terminal to reduce presynaptic transmitter release. A combination of LTP and LTD can shape the activation function of striatal neurons in response to synaptic inputs, especially from the cerebral cortex.

REFERENCES

Abdi, A., Mallet, N., Mohamed, F. Y., Sharott, A., Dodson, P. D., Nakamura, K. C., . . . Garas, F. N. (2015). Prototypic and arkypallidal neurons in the dopamine-intact external globus pallidus. *Journal of Neuroscience*, 35(17), 6667–6688.

Aceves, J. D. J., Rueda-Orozco, P. E., Hernández, R., Plata, V., Ibañez-Sandoval, O., Galarraga, E., & Bargas, J. (2011). Dopaminergic presynaptic modulation of nigral afferents: Its role in the generation of recurrent bursting in substantia nigra pars reticulata neurons. *Frontiers in Systems Neuroscience*, 5, 6.

Adermark, L., & Lovinger, D. M. (2007a). Combined activation of l-type ca2+ channels and synaptic transmission is sufficient to induce striatal long-term depression. *Journal of Neuroscience*, 27(25), 6781–6787. doi: 10.1523/JNEUROSCI.0280-07.2007.

Adermark, L., & Lovinger, D. M. (2007b). Retrograde endocannabinoid signaling at striatal synapses requires a regulated postsynaptic release step. *Proceedings of the National Academy of Sciences*, 104(51), 20564–20569. doi: 10.1073/pnas.0706873104.

Adermark, L., Talani, G., & Lovinger, D. M. (2009). Endocannabinoid-dependent plasticity at gabaergic and glutamatergic synapses in the striatum is regulated by synaptic activity. *European Journal of Neuroscience*, *29*(1), 32–41. doi: 10.1111/j.1460-9568.2008.06551.x.

Akins, P. T., Surmeier, D. J., & Kitai, S. (1990). Muscarinic modulation of a transient k+ conductance in rat neostriatal neurons. *Nature*, *344*(6263), 240–242.

Alger, B. E. (2002). Retrograde signaling in the regulation of synaptic transmission: Focus on endocannabinoids. *Progress in Neurobiology*, *68*(4), 247–286.

Alvarez-Leefmans, F. J., & Delpire, E. (2009). *Physiology and Pathology of Chloride Transporters and Channels in the Nervous System: From Molecules to Diseases*: Cambridge, MA: Academic Press.

Andén, N. E., Dahlström, A., Fuxe, K., Larsson, K., Olson, L., & Ungerstedt, U. (1966). Ascending monoamine neurons to the telencephalon and diencephalon. *Acta Physiologica Scandinavica*, *67*(3–4), 313–326.

Aosaki, T., Graybiel, A. M., & Kimura, M. (1994). Effect of the nigrostriatal dopamine system on acquired neural responses in the striatum of behaving monkeys. *Science*, *265*(5170), 412–415.

Aosaki, T., Kimura, M., & Graybiel, A. M. (1995). Temporal and spatial characteristics of tonically active neurons of the primate's striatum. *Journal of Neurophysiology*, *73*(3), 1234–1252.

Atherton, J. F., & Bevan, M. D. (2005). Ionic mechanisms underlying autonomous action potential generation in the somata and dendrites of gabaergic substantia nigra pars reticulata neurons in vitro. *The Journal of Neuroscience*, *25*(36), 8272–8281.

Augustin, S. M., Chancey, J. H., & Lovinger, D. M. (2018). Dual dopaminergic regulation of corticostriatal plasticity by cholinergic interneurons and indirect pathway medium spiny neurons. *Cell Reports*, *24*(11), 2883–2893.

Ayling, M., Panzeri, S., & Bracci, E. (2007). Gabaergic excitation in striatal projection neurons: Simulations and experiments. *Neurocomputing*, *70*(10–12), 1870–1876.

Azdad, K., Gall, D., Woods, A. S., Ledent, C., Ferre, S., & Schiffmann, S. N. (2009). Dopamine d2 and adenosine a2a receptors regulate nmda-mediated excitation in accumbens neurons through a2a-d2 receptor heteromerization. *Neuropsychopharmacology*, *34*(4), 972–986.

Baik, J.-H. (2013). Dopamine signaling in reward-related behaviors. *Frontiers in Neural Circuits*, *7*, 152.

Bailey, C. H., Giustetto, M., Huang, Y. Y., Hawkins, R. D., & Kandel, E. R. (2000). Is heterosynaptic modulation essential for stabilizing hebbian plasticity and memory? *Nature Reviews Neuroscience*, *1*(1), 11–20.

Bass, C. E., Grinevich, V. P., Vance, Z. B., Sullivan, R. P., Bonin, K. D., & Budygin, E. A. (2010). Optogenetic control of striatal dopamine release in rats. *Journal of Neurochemistry*, *114*(5), 1344–1352.

Beatty, J. A., Sullivan, M. A., Morikawa, H., & Wilson, C. J. (2012). Complex autonomous firing patterns of striatal low-threshold spike interneurons. *Journal of Neurophysiology*, *108*(3), 771–781.

Beckstead, M. J., Grandy, D. K., Wickman, K., & Williams, J. T. (2004). Vesicular dopamine release elicits an inhibitory postsynaptic current in midbrain dopamine neurons. *Neuron*, *42*(6), 939–946.

Bennett, B. D., Callaway, J. C., & Wilson, C. J. (2000). Intrinsic membrane properties underlying spontaneous tonic firing in neostriatal cholinergic interneurons. *Journal of Neuroscience*, *20*(22), 8493–8503.

Bennett, B. D., & Wilson, C. J. (1999). Spontaneous activity of neostriatal cholinergic interneurons in vitro. *Journal of Neuroscience*, *19*(13), 5586–5596.

Berke, J. D. (2008). Uncoordinated firing rate changes of striatal fast-spiking interneurons during behavioral task performance. *Journal of Neuroscience*, *28*(40), 10075–10080.

Bevan, M. D., Booth, P. A., Eaton, S. A., & Bolam, J. P. (1998). Selective innervation of neostriatal interneurons by a subclass of neuron in the globus pallidus of the rat. *Journal of Neuroscience*, *18*(22), 9438–9452.

Bevan, M. D., Magill, P. J., Terman, D., Bolam, J. P., & Wilson, C. J. (2002). Move to the rhythm: Oscillations in the subthalamic nucleus-external globus pallidus network. *Trends in Neurosciences*, *25*(10), 525–531.

Bi, G. Q., & Poo, M. M. (1998). Synaptic modifications in cultured hippocampal neurons: Dependence on spike timing, synaptic strength, and postsynaptic cell type. *Journal of Neuroscience*, *18*(24), 10464–10472.

Bickford, M., & Hall, W. (1992). The nigral projection to predorsal bundle cells in the superior colliculus of the rat. *Journal of Comparative Neurology*, *319*(1), 11–33.

Björklund, A., & Dunnett, S. B. (2007). Dopamine neuron systems in the brain: An update. *Trends in Neurosciences*, *30*(5), 194–202.

Bolam, J., Clarke, D., Smith, A., & Somogyi, P. (1983). A type of aspiny neuron in the rat neostriatum accumulates [3h] γ-aminobutyric acid: Combination of golgi-staining, autoradiography, and electron microscopy. *Journal of Comparative Neurology*, *213*(2), 121–134.

Bolam, J., Wainer, B., & Smith, A. (1984). Characterization of cholinergic neurons in the rat neostriatum. A combination of choline acetyltransferase immunocytochemistry, golgi-impregnation and electron microscopy. *Neuroscience*, *12*(3), 711–718.

Borycz, J., Pereira, M. F., Melani, A., Rodrigues, R. J., Köfalvi, A., Panlilio, L., ... Ferré, S. (2007). Differential glutamate-dependent and glutamate-independent adenosine a1 receptor-mediated modulation of dopamine release in different striatal compartments. *Journal of Neurochemistry*, *101*(2), 355–363.

Bouyer, J., Park, D., Joh, T., & Pickel, V. (1984). Chemical and structural analysis of the relation between cortical inputs and tyrosine hydroxylase-containing terminals in rat neostriatum. *Brain Research*, *302*(2), 267–275.

Bracci, E., & Panzeri, S. (2006). Excitatory gabaergic effects in striatal projection neurons. *Journal of Neurophysiology*, *95*(2), 1285–1290. doi: 10.1152/jn.00598.2005.

Brown, J., Pan, W.-X., & Dudman, J. T. (2014). The inhibitory microcircuit of the substantia nigra provides feedback gain control of the basal ganglia output. *eLife 3*, e02397.

Butcher, L. L., & Hodge, G. K. (1976). Postnatal development of acetylcholinesterase in the caudate-putamen nucleus and substantia nigra of rats. *Brain Research*, *106*(2), 223–240.

Calabresi, P., Gubellini, P., Centonze, D., Picconi, B., Bernardi, G., Chergui, K., ... Greengard, P. (2000). Dopamine and camp-regulated phosphoprotein 32 kda controls both striatal long-term depression and long-term potentiation, opposing forms of synaptic plasticity. *Journal of Neuroscience*, *20*(22), 8443–8451.

Calabresi, P., Maj, R., Mercuri, N. B., & Bernardi, G. (1992). Coactivation of d1 and d2 dopamine receptors is required for long-term synaptic depression in the striatum. *Neuroscience Letters*, *142*(1), 95–99.

Calabresi, P., Maj, R., Pisani, A., Mercuri, N. B., & Bernardi, G. (1992). Long-term synaptic depression in the striatum: Physiological and pharmacological characterization. *Journal of Neuroscience*, *12*(11), 4224–4233.

Calabresi, P., Picconi, B., Tozzi, A., & Di Filippo, M. (2007). Dopamine-mediated regulation of corticostriatal synaptic plasticity. *Trends in Neurosciences*, *30*(5), 211–219.

Calabresi, P., Pisani, A., Mercuri, N. B., & Bernardi, G. (1992). Long-term potentiation in the striatum is unmasked by removing the voltage-dependent magnesium block of nmda receptor channels. *European Journal of Neuroscience*, *4*(10), 929–935.

Carter, A. G., & Sabatini, B. L. (2004). State-dependent calcium signaling in dendritic spines of striatal medium spiny neurons. *Neuron*, *44*(3), 483–493.

Caulfield, M. P., & Birdsall, N. J. (1998). International union of pharmacology. Xvii. Classification of muscarinic acetylcholine receptors. *Pharmacological Reviews*, *50*(2), 279–290.

Cepeda, C., Buchwald, N. A., & Levine, M. S. (1993). Neuromodulatory actions of dopamine in the neostriatum are dependent upon the excitatory amino acid receptor subtypes activated. *Proceedings of the National Academy of Sciences*, *90*(20), 9576–9580.

Cepeda, C., Radisavljevic, Z., Peacock, W., Levine, M. S., & Buchwald, N. A. (1992). Differential modulation by dopamine of responses evoked by excitatory amino acids in human cortex. *Synapse*, *11*(4), 330–341.

Chang, H., & Kita, H. (1992). Interneurons in the rat striatum: Relationships between parvalbumin neurons and cholinergic neurons. *Brain Research*, *574*(1–2), 307–311.

Chen, J. F., Eltzschig, H. K., & Fredholm, B. B. (2013). Adenosine receptors as drug targets—what are the challenges?. *Nature Reviews Drug Discovery*, *12*(4), 265–286.

Choi, S., & Lovinger, D. M. (1997). Decreased frequency but not amplitude of quantal synaptic responses associated with expression of corticostriatal long-term depression. *Journal of Neuroscience*, *17*(21), 8613–8620.

Chuhma, N., Tanaka, K. F., Hen, R., & Rayport, S. (2011). Functional connectome of the striatal medium spiny neuron. *Journal of Neuroscience*, *31*(4), 1183–1192.

Collingridge, G. L., Herron, C. E., & Lester, R. A. (1988). Synaptic activation of n-methyl-d-aspartate receptors in the schaffer collateral-commissural pathway of rat hippocampus. *Journal of Physiology*, *399*, 283–300. doi: 10.1113/jphysiol.1988.sp017080.

Conn, P. J., Battaglia, G., Marino, M. J., & Nicoletti, F. (2005). Metabotropic glutamate receptors in the basal ganglia motor circuit. *Nature Reviews Neuroscience*, *6*(10), 787–798.

Connelly, W. M., Schulz, J. M., Lees, G., & Reynolds, J. N. (2010). Differential short-term plasticity at convergent inhibitory synapses to the substantia nigra pars reticulata. *Journal of Neuroscience*, *30*(44), 14854–14861.

Consolo, S., Baldi, G., Giorgi, S., & Nannini, L. (1996). The cerebral cortex and parafascicular thalamic nucleus facilitate in vivo acetylcholine release in the rat striatum through distinct glutamate receptor subtypes. *European Journal of Neuroscience*, *8*(12), 2702–2710.

Crittenden, J. R., & Graybiel, A. M. (2011). Basal ganglia disorders associated with imbalances in the striatal striosome and matrix compartments. *Frontiers in Neuroanatomy*, *5*, 59. doi: 10.3389/fnana.2011.00059.

Dang, M. T., Yokoi, F., Yin, H. H., Lovinger, D. M., Wang, Y., & Li, Y. (2006). Disrupted motor learning and long-term synaptic plasticity in mice lacking nmdar1 in the striatum. *Proceedings of the National Academy of Sciences*, *103*(41), 15254–15259.

Dautan, D., Huerta-Ocampo, I., Gut, N. K., Valencia, M., Kondabolu, K., Kim, Y., . . . Mena-Segovia, J. (2020). Cholinergic midbrain afferents modulate striatal circuits and shape encoding of action strategies. *Nature Communications, 11*(1), 1–19.

Dautan, D., Huerta-Ocampo, I., Witten, I. B., Deisseroth, K., Bolam, J. P., Gerdjikov, T., & Mena-Segovia, J. (2014). A major external source of cholinergic innervation of the striatum and nucleus accumbens originates in the brainstem. *Journal of Neuroscience, 34*(13), 4509–4518.

David Smith, A., & Paul Bolam, J. (1990). The neural network of the basal ganglia as revealed by the study of synaptic connections of identified neurones. *Trends in Neurosciences, 13*(7), 259–265.

Daw, N., Stein, P., & Fox, K. (1993). The role of nmda receptors in information processing. *Annual Review of Neuroscience, 16*(1), 207–222.

DeBoer, P., & Abercrombie, E. D. (1996). Physiological release of striatal acetylcholine in vivo: Modulation by d1 and d2 dopamine receptor subtypes. *Journal of Pharmacology and Experimental Therapeutics, 277*(2), 775–783.

Descarries, L., Gisiger, V., & Steriade, M. (1997). Diffuse transmission by acetylcholine in the cns. *Progress in Neurobiology, 53*(5), 603–625.

Difiglia, M. (1987). Synaptic organization of cholinergic neurons in the monkey neostriatum. *Journal of Comparative Neurology, 255*(2), 245–258.

Difiglia, M., Pasik, P., & Pasik, T. (1976). A golgi study of neuronal types in the neostriatum of monkeys. *Brain Research, 114*(2), 245–256.

Difiglia, M., Pasik, P., & Pasik, T. (1982). A golgi and ultrastructural study of the monkey globus pallidus. *Journal of Comparative Neurology, 212*(1), 53–75.

Ding, J. B., Guzman, J. N., Peterson, J. D., Goldberg, J. A., & Surmeier, D. J. (2010). Thalamic gating of corticostriatal signaling by cholinergic interneurons. *Neuron, 67*(2), 294–307.

Ding, J. B., Peterson, J. D., & Surmeier, D. J. (2008). Corticostriatal and thalamostriatal synapses have distinctive properties. *Journal of Neuroscience, 28*(25), 6483–6492.

Ding, S., Li, L., & Zhou, F. M. (2015). Nigral dopamine loss induces a global upregulation of presynaptic dopamine d1 receptor facilitation of the striatonigral gabaergic output. *Journal of Neurophysiology, 113*(6), 1697–1711. doi: 10.1152/jn.00752.2014.

Ding, S., Wei, W., & Zhou, F.-M. (2011). Molecular and functional differences in voltage-activated sodium currents between gaba projection neurons and dopamine neurons in the substantia nigra. *Journal of Neurophysiology, 106*(6), 3019–3034.

Do, M. T. H., & Bean, B. P. (2003). Subthreshold sodium currents and pacemaking of subthalamic neurons: Modulation by slow inactivation. *Neuron, 39*(1), 109–120.

Do, M. T. H., & Bean, B. P. (2004). Sodium currents in subthalamic nucleus neurons from nav1. 6-null mice. *Journal of Neurophysiology, 92*(2), 726–733.

Dodson, P. D., Larvin, J. T., Duffell, J. M., Garas, F. N., Doig, N. M., Kessaris, N., . . . Magill, P. J. (2015). Distinct developmental origins manifest in the specialized encoding of movement by adult neurons of the external globus pallidus. *Neuron, 86*(2), 501–513.

Dommett, E., Coizet, V., Blaha, C. D., Martindale, J., Lefebvre, V., Walton, N., . . . Redgrave, P. (2005). How visual stimuli activate dopaminergic neurons at short latency. *Science, 307*(5714), 1476–1479.

Dutar, P., & Nicoll, R. (1988). A physiological role for gabab receptors in the central nervous system. *Nature, 332*(6160), 156–158.

Exley, R., & Cragg, S. J. (2008). Presynaptic nicotinic receptors: A dynamic and diverse cholinergic filter of striatal dopamine neurotransmission. *British Journal of Pharmacology, 153* (Suppl 1), S283–297.

Fatt, P., & Katz, B. (1953). The effect of inhibitory nerve impulses on a crustacean muscle fibre. *Journal of Physiology, 121*(2), 374.

Ferré, S., Fuxe, K., Fredholm, B. B., Morelli, M., & Popoli, P. (1997). Adenosine–dopamine receptor–receptor interactions as an integrative mechanism in the basal ganglia. *Trends in Neurosciences, 20*(10), 482–487.

Ferre, S., Von Euler, G., Johansson, B., Fredholm, B. B., & Fuxe, K. (1991). Stimulation of high-affinity adenosine a2 receptors decreases the affinity of dopamine d2 receptors in rat striatal membranes. *Proceedings of the National Academy of Sciences, 88*(16), 7238–7241.

Fino, E., Glowinski, J., & Venance, L. (2005). Bidirectional activity-dependent plasticity at corticostriatal synapses. *Journal of Neuroscience, 25*(49), 11279–11287.

Fino, E., & Venance, L. (2010). Spike-timing dependent plasticity in the striatum. *Frontiers in Synaptic Neuroscience, 2*, 6.

Fredholm, B. B. (1995). Adenosine, adenosine receptors and the actions of caffeine. *Pharmacology & Toxicology, 76*(2), 93–101.

Fredholm, B. B., Chen, J.-F., Cunha, R. A., Svenningsson, P., & Vaugeois, J.-M. (2005). Adenosine and brain function. *International Review of Neurobiology, 63*(1), 191–270.

Frémaux, N., & Gerstner, W. (2016). Neuromodulated spike-timing-dependent plasticity, and theory of three-factor learning rules. *Frontiers in Neural Circuits, 9*, 85.

Freund, T., Powell, J., & Smith, A. (1984). Tyrosine hydroxylase-immunoreactive boutons in synaptic contact with identified striatonigral neurons, with particular reference to dendritic spines. *Neuroscience, 13*(4), 1189–1215.

Gauthier, J., Parent, M., Levesque, M., & Parent, A. (1999). The axonal arborization of single nigrostriatal neurons in rats. *Brain Research, 834*(1–2), 228–232.

Gerdeman, G. L., & Lovinger, D. M. (2001). Cb1 cannabinoid receptor inhibits synaptic release of glutamate in rat dorsolateral striatum. *Journal of Neurophysiology, 85*(1), 468–471.

Gerdeman, G. L., & Lovinger, D. M. (2003). Emerging roles for endocannabinoids in long-term synaptic plasticity. *British Journal of Pharmacology, 140*(5), 781–789.

Gerdeman, G. L., Partridge, J. G., Lupica, C. R., & Lovinger, D. M. (2003). It could be habit forming: Drugs of abuse and striatal synaptic plasticity. *Trends in Neurosciences, 26*(4), 184–192.

Gerdeman, G. L., Ronesi, J., & Lovinger, D. M. (2002). Postsynaptic endocannabinoid release is critical to long-term depression in the striatum. *Nature Neuroscience, 5*(5), 446–451.

Gerfen, C. R. (1988). Synaptic organization of the striatum. *Journal of Electron Microscopy Technique, 10*(3), 265–281.

Gerfen, C. R., & Surmeier, D. J. (2011). Modulation of striatal projection systems by dopamine. *Annual Review of Neuroscience, 34*, 441–466. doi: 10.1146/annurev-neuro-061010-113641.

Gittis, A. H., Nelson, A. B., Thwin, M. T., Palop, J. J., & Kreitzer, A. C. (2010). Distinct roles of gabaergic interneurons in the regulation of striatal output pathways. *Journal of Neuroscience, 30*(6), 2223–2234.

Goldberg, J. A., Ding, J. B., & Surmeier, D. J. (2012). Muscarinic modulation of striatal function and circuitry. In: Fryer, A. D., Christopoulos, A., Nathanson, N. M. (Eds.), *Muscarinic Receptors* (pp. 223–241). New York: Springer.

Grace, A. A., & Bunney, B. S. (1980). Nigral dopamine neurons: Intracellular recording and identification with l-dopa injection and histofluorescence. *Science, 210*(4470), 654–656.

Grace, A. A., & Bunney, B. S. (1983). Intracellular and extracellular electrophysiology of nigral dopaminergic neurons--1. Identification and characterization. *Neuroscience, 10*(2), 301–315.

Grace, A. A., & Bunney, B. S. (1984). The control of firing pattern in nigral dopamine neurons: Single spike firing. *Journal of Neuroscience, 4*(11), 2866–2876.

Grofova, I., Deniau, J., & Kitai, S. (1982). Morphology of the substantia nigra pars reticulata projection neurons intracellularly labeled with hrp. *Journal of Comparative Neurology, 208*(4), 352–368.

Guzman, J. N., Hernandez, A., Galarraga, E., Tapia, D., Laville, A., Vergara, R., . . . Bargas, J. (2003). Dopaminergic modulation of axon collaterals interconnecting spiny neurons of the rat striatum. *Journal of Neuroscience, 23*(26), 8931–8940.

Hebb, D. O. (1949). *The Organization of Behavior: A Neuropsychological Theory*: Hoboken, NJ: J. Wiley

Hegeman, D. J., Hong, E. S., Hernández, V. M., & Chan, C. S. (2016). The external globus pallidus: Progress and perspectives. *European Journal of Neuroscience, 43*(10), 1239–1265.

Herkenham, M., Lynn, A. B., de Costa, B. R., & Richfield, E. K. (1991). Neuronal localization of cannabinoid receptors in the basal ganglia of the rat. *Brain Research, 547*(2), 267–274.

Hernández, V. M., Hegeman, D. J., Cui, Q., Kelver, D. A., Fiske, M. P., Glajch, K. E., . . . Chan, C. S. (2015). Parvalbumin+ neurons and npas1+ neurons are distinct neuron classes in the mouse external globus pallidus. *Journal of Neuroscience, 35*(34), 11830–11847.

Hevers, W., & Lüddens, H. (1998). The diversity of gaba a receptors. *Molecular Neurobiology, 18*(1), 35–86.

Higley, M. J., Soler-Llavina, G. J., & Sabatini, B. L. (2009). Cholinergic modulation of multivesicular release regulates striatal synaptic potency and integration. *Nature Neuroscience, 12*(9), 1121–1128.

Hikosaka, O., Sakamoto, M., & Miyashita, N. (1993). Effects of caudate nucleus stimulation on substantia nigra cell activity in monkey. *Experimental Brain Research, 95*(3), 457–472.

Hohmann, A. G., & Herkenham, M. (2000). Localization of cannabinoid cb1 receptor mrna in neuronal subpopulations of rat striatum: A double-label in situ hybridization study. *Synapse, 37*(1), 71–80.

Holly, E. N., Davatolhagh, M. F., España, R. A., & Fuccillo, M. V. (2021). Striatal low-threshold spiking interneurons locally gate dopamine. *Current Biology, 31*(18), 4139–4147.

Horvitz, J. C. (2000). Mesolimbocortical and nigrostriatal dopamine responses to salient non-reward events. *Neuroscience, 96*(4), 651–656. doi: 10.1016/S0306-4522(00)00019-1.

Howlett, A., Barth, F., Bonner, T., Cabral, G., Casellas, P., Devane, W., . . . Martin, B. (2002). International union of pharmacology. Xxvii. Classification of cannabinoid receptors. *Pharmacological Reviews, 54*(2), 161–202.

Hunnicutt, B. J., Jongbloets, B. C., Birdsong, W. T., Gertz, K. J., Zhong, H., & Mao, T. (2016). A comprehensive excitatory input map of the striatum reveals novel functional organization. *Elife, 5*. doi: 10.7554/eLife.19103.

Jędrzejewska-Szmek, J., Damodaran, S., Dorman, D. B., & Blackwell, K. T. (2017). Calcium dynamics predict direction of synaptic plasticity in striatal spiny projection neurons. *European Journal of Neuroscience, 45*(8), 1044–1056.

Jiang, Z.-G., & North, R. (1991). Membrane properties and synaptic responses of rat striatal neurones in vitro. *The Journal of Physiology, 443*(1), 533–553.

Johansson, Y., & Silberberg, G. (2020). The functional organization of cortical and thalamic inputs onto five types of striatal neurons is determined by source and target cell identities. *Cell Reports, 30*(4), 1178–1194. e1173.

Jones, C. K., Byun, N., & Bubser, M. (2012). Muscarinic and nicotinic acetylcholine receptor agonists and allosteric modulators for the treatment of schizophrenia. *Neuropsychopharmacology, 37*(1), 16–42.

Jones, E., Coulter, J., Burton, H., & Porter, R. (1977). Cells of origin and terminal distrubution of corticostriatal fibers arising in the sensory-motor cortex of monkeys. *Journal of Comparative Neurology, 173*(1), 53–80.

Kaneda, K., Isa, K., Yanagawa, Y., & Isa, T. (2008). Nigral inhibition of gabaergic neurons in mouse superior colliculus. *The Journal of Neuroscience, 28*(43), 11071–11078.

Kawaguchi, Y. (1993). Physiological, morphological, and histochemical characterization of three classes of interneurons in rat neostriatum. *Journal of Neuroscience, 13*(11), 4908–4923.

Kawaguchi, Y., Wilson, C. J., Augood, S. J., & Emson, P. C. (1995). Striatal interneurones: Chemical, physiological and morphological characterization. *Trends in Neurosciences, 18*(12), 527–535.

Kemp, J. M., & Powell, T. P. (1971). The structure of the caudate nucleus of the cat: Light and electron microscopy. *Philosophical Transactions of the Royal Society of London. B, Biological Sciences, 262*(845), 383–401.

Kerr, J. N., & Wickens, J. R. (2001). Dopamine d-1/d-5 receptor activation is required for long-term potentiation in the rat neostriatum in vitro. *Journal of Neurophysiology, 85*(1), 117–124.

Khaliq, Z. M., & Bean, B. P. (2010). Pacemaking in dopaminergic ventral tegmental area neurons: Depolarizing drive from background and voltage-dependent sodium conductances. *Journal of Neuroscience, 30*(21), 7401–7413.

Khan, Z. U., Gutiérrez, A., Martín, R., Peñafiel, A., Rivera, A., & De La Calle, A. (1998). Differential regional and cellular distribution of dopamine d2-like receptors: An immunocytochemical study of subtype-specific antibodies in rat and human brain. *Journal of Comparative Neurology, 402*(3), 353–371.

Kita, H. (1993). Gabaergic circuits of the striatum. *Progress in Brain Research* (vol. 99, pp. 51–72). New York: Elsevier.

Kita, H., & Kita, S. (1994). The morphology of globus pallidus projection neurons in the rat: An intracellular staining study. *Brain Research, 636*(2), 308–319.

Kita, H., & Kitai, S. (1988). Glutamate decarboxylase immunoreactive neurons in rat neostriatum: Their morphological types and populations. *Brain Research, 447*(2), 346–352.

Kita, H., Kosaka, T., & Heizmann, C. (1990). Parvalbumin-immunoreactive neurons in the rat neostriatum: A light and electron microscopic study. *Brain Research, 536*(1–2), 1–15.

Kita, T., Kita, H., & Kitai, S. (1984). Passive electrical membrane properties of rat neostriatal neurons in an in vitro slice preparation. *Brain Research, 300*(1), 129–139.

Klug, J. R., Engelhardt, M. D., Cadman, C. N., Li, H., Smith, J. B., Ayala, S., . . . Jin, X. (2018). Differential inputs to striatal cholinergic and parvalbumin interneurons imply functional distinctions. *Elife, 7*, e35657.

Koós, T., & Tepper, J. M. (1999). Inhibitory control of neostriatal projection neurons by gabaergic interneurons. *Nature Neuroscience, 2*(5), 467.

Koós, T., & Tepper, J. M. (2002). Dual cholinergic control of fast-spiking interneurons in the neostriatum. *Journal of Neuroscience, 22*(2), 529–535.

Koos, T., Tepper, J. M., & Wilson, C. J. (2004). Comparison of ipscs evoked by spiny and fast-spiking neurons in the neostriatum. *Journal of Neuroscience, 24*(36), 7916–7922.

Kreitzer, A. C., & Malenka, R. C. (2005). Dopamine modulation of state-dependent endocannabinoid release and long-term depression in the striatum. *Journal of Neuroscience, 25*(45), 10537–10545.

Kreitzer, A. C., & Malenka, R. C. (2007). Endocannabinoid-mediated rescue of striatal ltd and motor deficits in parkinson's disease models. *Nature, 445*(7128), 643–647.

Kreitzer, A. C., & Regehr, W. G. (2001). Retrograde inhibition of presynaptic calcium influx by endogenous cannabinoids at excitatory synapses onto purkinje cells. *Neuron, 29*(3), 717–727.

Kultas-Ilinsky, K., & Ilinsky, I. (1990). Fine structure of the magnocellular subdivision of the ventral anterior thalamic nucleus (v amc) of macaca mulatta: Ii. Organization of nigrothalamic afferents as revealed with em autoradiography. *Journal of Comparative Neurology, 294*(3), 479–489.

Lahiri, A. K., & Bevan, M. D. (2020). Dopaminergic transmission rapidly and persistently enhances excitability of d1 receptor-expressing striatal projection neurons. *Neuron, 106*(2), 277–290.

Lapper, S., & Bolam, J. (1992). Input from the frontal cortex and the parafascicular nucleus to cholinergic interneurons in the dorsal striatum of the rat. *Neuroscience, 51*(3), 533–545.

Lee, C. R., Yonk, A. J., Wiskerke, J., Paradiso, K. G., Tepper, J. M., & Margolis, D. J. (2019). Opposing influence of sensory and motor cortical input on striatal circuitry and choice behavior. *Current Biology, 29*(8), 1313–1323. e1315.

Lee, F. J., Xue, S., Pei, L., Vukusic, B., Chéry, N., Wang, Y., . . . Liu, F. (2002). Dual regulation of nmda receptor functions by direct protein-protein interactions with the dopamine d1 receptor. *Cell, 111*(2), 219–230.

Lee, K., Bakhurin, K. I., Claar, L. D., Holley, S. M., Chong, N. C., Cepeda, C., . . . Masmanidis, S. C. (2019). Gain modulation by corticostriatal and thalamostriatal input signals during reward-conditioned behavior. *Cell Reports, 29*(8), 2438–2449. e2434.

Levey, A. I., Hersch, S. M., Rye, D. B., Sunahara, R. K., Niznik, H. B., Kitt, C. A., . . . Ciliax, B. J. (1993). Localization of d1 and d2 dopamine receptors in brain with subtype-specific antibodies. *Proceedings of the National Academy of Sciences, 90*(19), 8861–8865.

Li, H., Tornberg, J., Kaila, K., Airaksinen, M. S., & Rivera, C. (2002). Patterns of cation-chloride cotransporter expression during embryonic rodent cns development. *European Journal of Neuroscience, 16*(12), 2358–2370.

Lindvall, O., Björklund, A., & Skagerberg, G. (1984). Selective histochemical demonstration of dopamine terminal systems in rat di-and telecephalon: New evidence for dopaminergic innervation of hypothalamic neurosecretory nuclei. *Brain Research, 306*(1–2), 19–30.

Lovinger, D., & Mathur, B. (2016). *Handbook of Basal Ganglia Structure and Function*: New York: Academic Press.

Lovinger, D. M. (2010). Neurotransmitter roles in synaptic modulation, plasticity and learning in the dorsal striatum. *Neuropharmacology, 58*(7), 951–961. doi: 10.1016/j.neuropharm.2010.01.008.

Lovinger, D. M., & Choi, S. (1995). Activation of adenosine a1 receptors initiates short-term synaptic depression in rat striatum. *Neuroscience Letters, 199*(1), 9–12.

Lovinger, D. M., & Mathur, B. N. (2012). Endocannabinoids in striatal plasticity. *Parkinsonism & Related Disorders, 18* (Suppl 1), S132–134. doi: 10.1016/S1353-8020(11)70041-4.

Lovinger, D. M., Tyler, E. C., & Merritt, A. (1993). Short- and long-term synaptic depression in rat neostriatum. *Journal of Neurophysiology, 70*(5), 1937–1949.

Ma, T., Cheng, Y., Roltsch Hellard, E., Wang, X., Lu, J., Gao, X., . . . Wang, J. (2018). Bidirectional and long-lasting control of alcohol-seeking behavior by corticostriatal ltp and ltd. *Nature Neuroscience, 21*(3), 373–383.

MacLeod, N., James, T., Kilpatrick, I., & Starr, M. (1980). Evidence for a gabaergic nigrothalamic pathway in the rat. *Experimental Brain Research, 40*(1), 55–61.

Mahon, S., Vautrelle, N., Pezard, L., Slaght, S. J., Deniau, J.-M., Chouvet, G., & Charpier, S. (2006). Distinct patterns of striatal medium spiny neuron activity during the natural sleep–wake cycle. *Journal of Neuroscience, 26*(48), 12587–12595.

Malenka, R. C., & Bear, M. F. (2004). Ltp and ltd: An embarrassment of riches. *Neuron, 44*(1), 5–21.

Malinow, R., & Malenka, R. C. (2002). Ampa receptor trafficking and synaptic plasticity. *Annual Review of Neuroscience, 25*(1), 103–126.

Mallet, N., Micklem, B. R., Henny, P., Brown, M. T., Williams, C., Bolam, J. P., . . . Magill, P. J. (2012). Dichotomous organization of the external globus pallidus. *Neuron, 74*(6), 1075–1086.

Mallet, N., Schmidt, R., Leventhal, D., Chen, F., Amer, N., Boraud, T., & Berke, J. D. (2016). Arkypallidal cells send a stop signal to striatum. *Neuron, 89*(2), 308–316. doi: 10.1016/j.neuron.2015.12.017.

Martin, S. J., Grimwood, P. D., & Morris, R. G. (2000). Synaptic plasticity and memory: An evaluation of the hypothesis. *Annual Review of Neuroscience, 23*(1), 649–711.

Martinez-Mir, M., Probst, A., & Palacios, J. (1991). Adenosine a2 receptors: Selective localization in the human basal ganglia and alterations with disease. *Neuroscience, 42*(3), 697–706.

Mastro, K. J., Bouchard, R. S., Holt, H. A., & Gittis, A. H. (2014). Transgenic mouse lines subdivide external segment of the globus pallidus (gpe) neurons and reveal distinct gpe output pathways. *Journal of Neuroscience, 34*(6), 2087–2099.

McFarland, N. R., & Haber, S. N. (2000). Convergent inputs from thalamic motor nuclei and frontal cortical areas to the dorsal striatum in the primate. *Journal of Neuroscience, 20*(10), 3798–3813.

McGehee, D. S., Heath, M., Gelber, S., Devay, P., & Role, L. W. (1995). Nicotine enhancement of fast excitatory synaptic transmission in cns by presynaptic receptors. *Science, 269*(5231), 1692–1696.

Miller, R. (1981). *Meaning and Purpose in the Intact Brain*: New York: Oxford University Press.

Missale, C., Nash, S. R., Robinson, S. W., Jaber, M., & Caron, M. G. (1998). Dopamine receptors: From structure to function. *Physiological Reviews*, *78*(1), 189–225.

Nakanishi, H., Kita, H., & Kitai, S. (1991). Intracellular study of rat entopeduncular nucleus neurons in an in vitro slice preparation: Response to subthalamic stimulation. *Brain Research*, *549*(2), 285–291.

Nicoll, R. A. (2004). My close encounter with gabab receptors. *Biochemical Pharmacology*, *68*(8), 1667–1674.

Nosaka, D., & Wickens, J. R. (2022). Striatal cholinergic signaling in time and space. *Molecules*, *27*(4), 1202.

O'Hare, J. K., Li, H., Kim, N., Gaidis, E., Ade, K. K., Beck, J. M., . . . Calakos, N. (2017). Striatal fast-spiking interneurons drive habitual behavior. *Elife*, In press.

Omelchenko, N., & Sesack, S. R. (2009). Ultrastructural analysis of local collaterals of rat ventral tegmental area neurons: Gaba phenotype and synapses onto dopamine and gaba cells. *Synapse*, *63*(10), 895–906.

Oorschot, D. E. (2013). The percentage of interneurons in the dorsal striatum of the rat, cat, monkey and human: A critique of the evidence. *Basal Ganglia*, *3*(1), 19–24.

Paladini, C., & Roeper, J. (2014). Generating bursts (and pauses) in the dopamine midbrain neurons. *Neuroscience*, *282*, 109–121.

Parent, A., Charara, A., & Pinault, D. (1995). Single striatofugal axons arborizing in both pallidal segments and in the substantia nigra in primates. *Brain Research*, *698*(1–2), 280–284.

Parent, A., & Hazrati, L. N. (1995). Functional anatomy of the basal ganglia. Ii. The place of subthalamic nucleus and external pallidum in basal ganglia circuitry. *Brain Research*, *20*(1), 128–154. doi: 10.1016/0165-0173(94)00008-D.

Park, M. R., Lighthall, J. W., & Kitai, S. T. (1980). Recurrent inhibition in the rat neostriatum. *Brain Research*, *194*(2), 359–369.

Partridge, J. G., Tang, K. C., & Lovinger, D. M. (2000). Regional and postnatal heterogeneity of activity-dependent long-term changes in synaptic efficacy in the dorsal striatum. *Journal of Neurophysiology*, *84*(3), 1422–1429.

Pawlak, V., & Kerr, J. N. (2008). Dopamine receptor activation is required for corticostriatal spike-timing-dependent plasticity. *Journal of Neuroscience*, *28*(10), 2435–2446.

Pawlak, V., Wickens, J. R., Kirkwood, A., & Kerr, J. N. (2010). Timing is not everything: Neuromodulation opens the stdp gate. *Frontiers in Synaptic Neuroscience*, *2*, 146.

Percheron, G., Yelnik, J., & François, C. (1984). A Golgi analysis of the primate globus pallidus. III. Spatial organization of the striato-pallidal complex. *Journal of Comparative Neurology*, *227*, 214–227.

Piomelli, D. (2003). The molecular logic of endocannabinoid signalling. *Nature Reviews Neuroscience*, *4*(11), 873–884.

Pisani, A., Bernardi, G., Ding, J., & Surmeier, D. J. (2007). Re-emergence of striatal cholinergic interneurons in movement disorders. *Trends in Neurosciences*, *30*(10), 545–553.

Pitier, T., & Alger, B. (1994). Depolarization-induced suppression of gabaergic inhibition in rat hippocampal pyramidal cells: G protein involvement in a presynaptic mechanism. *Neuron*, *13*(6), 1447–1455.

Planert, H., Szydlowski, S. N., Hjorth, J. J., Grillner, S., & Silberberg, G. (2010). Dynamics of synaptic transmission between fast-spiking interneurons and striatal projection neurons of the direct and indirect pathways. *Journal of Neuroscience*, *30*(9), 3499–3507.

Plenz, D., & Kitai, S. T. (1998). Up and down states in striatal medium spiny neurons simultaneously recorded with spontaneous activity in fast-spiking interneurons studied in cortex-striatum-substantia nigra organotypic cultures. *Journal of Neurosciences*, *18*(1), 266–283.

Plotkin, J. L., Day, M., & Surmeier, D. J. (2011). Synaptically driven state transitions in distal dendrites of striatal spiny neurons. *Nature Neuroscience*, *14*(7), 881–888.

Prensa, L., & Parent, A. (2001). The nigrostriatal pathway in the rat: A single-axon study of the relationship between dorsal and ventral tier nigral neurons and the striosome/matrix striatal compartments. *Journal of Neurosciences*, *21*(18), 7247–7260.

Preston, R. J., Bishop, G. A., & Kitai, S. T. (1980). Medium spiny neuron projection from the rat striatum: An intracellular horseradish peroxidase study. *Brain Research*, *183*(2), 253–263.

Puopolo, M., Raviola, E., & Bean, B. P. (2007). Roles of subthreshold calcium current and sodium current in spontaneous firing of mouse midbrain dopamine neurons. *Journal of Neuroscience*, *27*(3), 645–656.

Radnikow, G., & Misgeld, U. (1998). Dopamine d1 receptors facilitate gabaasynaptic currents in the rat substantia nigra pars reticulata. *Journal of Neuroscience*, *18*(6), 2009–2016.

Rajakumar, N., Elisevich, K., & Flumerfelt, B. (1994). Parvalbumin-containing gabaergic neurons in the basal ganglia output system of the rat. *Journal of Comparative Neurology*, *350*(2), 324–336.

Ramanathan, S., Hanley, J. J., Deniau, J. M., & Bolam, J. P. (2002). Synaptic convergence of motor and somatosensory cortical afferents onto gabaergic interneurons in the rat striatum. *Journal of Neuroscience*, *22*(18), 8158–8169.

Raz, A., Feingold, A., Zelanskaya, V., Vaadia, E., & Bergman, H. (1996). Neuronal synchronization of tonically active neurons in the striatum of normal and parkinsonian primates. *Journal of Neurophysiology, 76*(3), 2083–2088.

Reynolds, J. N., & Wickens, J. R. (2002). Dopamine-dependent plasticity of corticostriatal synapses. *Neural Networks, 15*(4–6), 507–521.

Rice, M. E., Patel, J. C., & Cragg, S. J. (2011). Dopamine release in the basal ganglia. *Neuroscience, 198*, 112–137.

Robbe, D., Kopf, M., Remaury, A., Bockaert, J., & Manzoni, O. J. (2002). Endogenous cannabinoids mediate long-term synaptic depression in the nucleus accumbens. *Proceedings of the National Academy of Sciences, 99*(12), 8384–8388.

Role, L. W., & Berg, D. K. (1996). Nicotinic receptors in the development and modulation of cns synapses. *Neuron, 16*(6), 1077–1085.

Saunders, A., Huang, K. W., & Sabatini, B. L. (2016). Globus pallidus externus neurons expressing parvalbumin interconnect the subthalamic nucleus and striatal interneurons. *PLoS One, 11*(2), e0149798.

Schiffmann, S. N., Fisone, G., Moresco, R., Cunha, R. A., & Ferré, S. (2007). Adenosine a2a receptors and basal ganglia physiology. *Progress in Neurobiology, 83*(5), 277–292.

Schneider, J., Levine, M., Hull, C., & Buchwald, N. (1984). Effects of amphetamine on intracellular responses of caudate neurons in the cat. *Journal of Neuroscience, 4*(4), 930–938.

Schultz, W. (1998). Predictive reward signal of dopamine neurons. *Journal of Neurophysiology, 80*(1), 1–27.

Shabel, S. J., Proulx, C. D., Trias, A., Murphy, R. T., & Malinow, R. (2012). Input to the lateral habenula from the basal ganglia is excitatory, aversive, and suppressed by serotonin. *Neuron, 74*(3), 475–481.

Sharott, A., Doig, N. M., Mallet, N., & Magill, P. J. (2012). Relationships between the firing of identified striatal interneurons and spontaneous and driven cortical activities in vivo. *Journal of Neuroscience, 32*(38), 13221–13236.

Shen, W., Flajolet, M., Greengard, P., & Surmeier, D. J. (2008). Dichotomous dopaminergic control of striatal synaptic plasticity. *Science, 321*(5890), 848–851.

Shen, W., Tian, X., Day, M., Ulrich, S., Tkatch, T., Nathanson, N. M., & Surmeier, D. J. (2007). Cholinergic modulation of kir2 channels selectively elevates dendritic excitability in striatopallidal neurons. *Nature Neuroscience, 10*(11), 1458–1466. doi: 10.1038/nn1972.

Shink, E., Bevan, M., Bolam, J., & Smith, Y. (1996). The subthalamic nucleus and the external pallidum: two tightly interconnected structures that control the output of the basal ganglia in the monkey. *Neuroscience, 73*, 335–357.

Sidibe, M., & Smith, Y. (1999). Thalamic inputs to striatal interneurons in monkeys: Synaptic organization and co-localization of calcium binding proteins. *Neuroscience, 89*(4), 1189–1208.

Sivilotti, L., & Nistri, A. (1991). Gaba receptor mechanisms in the central nervous system. *Progress in Neurobiology, 36*(1), 35–92.

Smith, R., Musleh, W., Akopian, G., Buckwalter, G., & Walsh, J. (2001). Regional differences in the expression of corticostriatal synaptic plasticity. *Neuroscience, 106*(1), 95–101.

Smith, Y., Galvan, A., Ellender, T. J., Doig, N., Villalba, R. M., Ocampo, I. H., . . . Bolam, P. (2014). The thalamostriatal system in normal and diseased states. *Frontiers in Systems Neuroscience, 8*, 5.

Stephenson-Jones, M., Yu, K., Ahrens, S., Tucciarone, J. M., van Huijstee, A. N., Mejia, L. A., . . . Li, B. (2016). A basal ganglia circuit for evaluating action outcomes. *Nature, 539*(7628), 289.

Straub, C., Saulnier, J. L., Bègue, A., Feng, D. D., Huang, K. W., & Sabatini, B. L. (2016). Principles of synaptic organization of gabaergic interneurons in the striatum. *Neuron, 92*(1), 84–92.

Straub, C., Tritsch, N. X., Hagan, N. A., Gu, C., & Sabatini, B. L. (2014). Multiphasic modulation of cholinergic interneurons by nigrostriatal afferents. *Journal of Neuroscience, 34*(25), 8557–8569.

Surmeier, D. J., Carrillo-Reid, L., & Bargas, J. (2011). Dopaminergic modulation of striatal neurons, circuits, and assemblies. *Neuroscience, 198*, 3–18.

Surmeier, D. J., Plotkin, J., & Shen, W. (2009). Dopamine and synaptic plasticity in dorsal striatal circuits controlling action selection. *Current Opinion in Neurobiology, 19*(6), 621–628.

Svenningsson, P., Nishi, A., Fisone, G., Girault, J.-A., Nairn, A. C., & Greengard, P. (2004). Darpp-32: An integrator of neurotransmission. *Annual Review of Pharmacology and Toxicology, 44*, 269–296.

Szabo, B., Dörner, L., Pfreundtner, C., Nörenberg, W., & Starke, K. (1998). Inhibition of gabaergic inhibitory postsynaptic currents by cannabinoids in rat corpus striatum. *Neuroscience, 85*(2), 395–403.

Taverna, S., Ilijic, E., & Surmeier, D. J. (2008). Recurrent collateral connections of striatal medium spiny neurons are disrupted in models of parkinson's disease. *Journal of Neuroscience, 28*(21), 5504–5512.

Tepper, J. M., & Lee, C. R. (2007). Gabaergic control of substantia nigra dopaminergic neurons. *Progress in Brain Research, 160*, 189–208. doi: 10.1016/S0079-6123(06)60011-3.

Tepper, J. M., Martin, L., & Anderson, D. (1995). Gabaa receptor-mediated inhibition of rat substantia nigra dopaminergic neurons by pars reticulata projection neurons. *Journal of Neuroscience, 15*(4), 3092–3103.

Tepper, J. M., Tecuapetla, F., Koos, T., & Ibanez-Sandoval, O. (2010). Heterogeneity and diversity of striatal gabaergic interneurons. *Frontiers in Neuroanatomy, 4*, 150. doi: 10.3389/fnana.2010.00150.

Tozzi, A., de Iure, A., Di Filippo, M., Tantucci, M., Costa, C., Borsini, F., . . . Calabresi, P. (2011). The distinct role of medium spiny neurons and cholinergic interneurons in the d(2)/a(2)a receptor interaction in the striatum: Implications for parkinson's disease. *J Neurosci, 31*(5), 1850–1862. doi: 10.1523/JNEUROSCI.4082-10.2011.

Tunstall, M. J., Oorschot, D. E., Kean, A., & Wickens, J. R. (2002). Inhibitory interactions between spiny projection neurons in the rat striatum. *Journal of Neurophysiology, 88*(3), 1263–1269.

Uchimura, N., & North, R. A. (1990). Muscarine reduces inwardly rectifying potassium conductance in rat nucleus accumbens neurones. *Journal of Physiology, 422*, 369–380. doi: 10.1113/jphysiol.1990.sp017989.

Ueki, A., Uno, M., Anderson, M., & Yoshida, M. (1977). Monosynaptic inhibition of thalamic neurons produced by stimulation of the substantia nigra. *Cellular and Molecular Life Sciences, 33*(11), 1480–1482.

Urs, N. M., Peterson, S. M., & Caron, M. G. (2017). New concepts in dopamine d2 receptor biased signaling and implications for schizophrenia therapy. *Biological Psychiatry, 81*(1), 78–85.

von Krosigk, M., Smith, Y., Bolam, J. P., & Smith, A. D. (1992). Synaptic organization of gabaergic inputs from the striatum and the globus pallidus onto neurons in the substantia nigra and retrorubral field which project to the medullary reticular formation. *Neuroscience, 50*(3), 531–549.

Wallace, M. L., Saunders, A., Huang, K. W., Philson, A. C., Goldman, M., Macosko, E. Z., . . . Sabatini, B. L. (2017). Genetically distinct parallel pathways in the entopeduncular nucleus for limbic and sensorimotor output of the basal ganglia. *Neuron, 94*(1), 138–152. e135.

Wallmichrath, I., & Szabo, B. (2002). Cannabinoids inhibit striatonigral gabaergic neurotransmission in the mouse. *Neuroscience, 113*(3), 671–682.

Wang, Z., Kai, L., Day, M., Ronesi, J., Yin, H. H., Ding, J., . . . Surmeier, D. J. (2006). Dopaminergic control of corticostriatal long-term synaptic depression in medium spiny neurons is mediated by cholinergic interneurons. *Neuron, 50*(3), 443–452.

Wei, W., Ding, S., & Zhou, F.-M. (2017). Dopaminergic treatment weakens medium spiny neuron collateral inhibition in the parkinsonian striatum. *Journal of Neurophysiology, 117*(3), 987–999.

Wickens, J., & Kötter, R. (1995). Cellular models of reinforcement. In: Houk, J. C., Davis, J. L., & Beiser, D. G. (Eds.), *Models of Information Processing in the Basal Ganglia* (pp. 187–214). Cambridge, MA: The MIT Press.

Widrow, B., Kim, Y., & Park, D. (2015). The hebbian-lms learning algorithm. *Ieee ComputatioNal iNtelligeNCe magaziNe, 10*(4), 37–53.

Wilson, C. J. (2004). Basal ganglia. In Shephard, G. M. (Ed.), *The Synaptic Organization of the Brain* (5th ed., pp. 361–414). New York: Oxford University Press.

Wilson, C. J., & Goldberg, J. A. (2006). Origin of the slow afterhyperpolarization and slow rhythmic bursting in striatal cholinergic interneurons. *Journal of Neurophysiology, 95*(1), 196–204.

Wilson, C. J., & Groves, P. M. (1980). Fine structure and synaptic connections of the common spiny neuron of the rat neostriatum: A study employing intracellular injection of horseradish peroxidase. *Journal of Comparative Neurology, 194*(3), 599–615.

Wilson, C. J., & Groves, P. M. (1981). Spontaneous firing patterns of identified spiny neurons in the rat neostriatum. *Brain Research, 220*(1), 67–80.

Wilson, C. J., & Kawaguchi, Y. (1996). The origins of two-state spontaneous membrane potential fluctuations of neostriatal spiny neurons. *Journal of Neuroscience, 16*(7), 2397–2410.

Wilson, R. I., Kunos, G., & Nicoll, R. A. (2001). Presynaptic specificity of endocannabinoid signaling in the hippocampus. *Neuron, 31*(3), 453–462.

Witten, I. B., Lin, S.-C., Brodsky, M., Prakash, R., Diester, I., Anikeeva, P., . . . Deisseroth, K. (2010). Cholinergic interneurons control local circuit activity and cocaine conditioning. *Science, 330*(6011), 1677–1681.

Wu, Y., Richard, S., & Parent, A. (2000). The organization of the striatal output system: A single-cell juxtacellular labeling study in the rat. *Neuroscience Research, 38*(1), 49–62.

Wu, Y.-W., Kim, J.-I., Tawfik, V. L., Lalchandani, R. R., Scherrer, G., & Ding, J. B. (2015). Input-and cell-type-specific endocannabinoid-dependent ltd in the striatum. *Cell Reports, 10*(1), 75–87.

Yan, Z., Flores-Hernandez, J., & Surmeier, D. (2001). Coordinated expression of muscarinic receptor messenger rnas in striatal medium spiny neurons. *Neuroscience, 103*(4), 1017–1024.

Yan, Z., & Surmeier, D. J. (1996). Muscarinic (m2/m4) receptors reduce n-and p-type ca2+ currents in rat neostriatal cholinergic interneurons through a fast, membrane-delimited, g-protein pathway. *Journal of Neuroscience, 16*(8), 2592–2604.

Yin, H. H., Mulcare, S. P., Hilario, M. R., Clouse, E., Holloway, T., Davis, M. I., . . . Costa, R. M. (2009). Dynamic reorganization of striatal circuits during the acquisition and consolidation of a skill. *Nature Neuroscience, 12*(3), 333–341. doi: 10.1038/nn.2261.

Yin, H. H., Park, B. S., Adermark, L., & Lovinger, D. M. (2007). Ethanol reverses the direction of long-term synaptic plasticity in the dorsomedial striatum. *European Journal of Neuroscience, 25*(11), 3226–3232.

Yung, K., Bolam, J., Smith, A., Hersch, S., Ciliax, B., & Levey, A. (1995). Immunocytochemical localization of d1 and d2 dopamine receptors in the basal ganglia of the rat: Light and electron microscopy. *Neuroscience, 65*(3), 709–730.

Zadeh, A. A., Turner, B. D., Calakos, N., & Brunel, N. (2021). Non-monotonic effects of gabaergic synaptic inputs on neuronal firing. *PLoS Computational Biology 18*(6), e1010226.

Zhang, W., & Linden, D. J. (2003). The other side of the engram: Experience-driven changes in neuronal intrinsic excitability. *Nature Reviews Neuroscience, 4*(11), 885–900.

Zhang, Y.-F., Reynolds, J. N., & Cragg, S. J. (2018). Pauses in cholinergic interneuron activity are driven by excitatory input and delayed rectification, with dopamine modulation. *Neuron, 98*(5), 918–925. e913.

Zheng, T., & Wilson, C. J. (2002). Corticostriatal combinatorics: The implications of corticostriatal axonal arborizations. *Journal of Neurophysiology, 87*(2), 1007–1017.

Zhou, F. W. (2016). The substantia nigra pars reticulata. In: Steiner, H. (Ed.), *Handbook of Behavioral Neuroscience* (vol. 24, pp. 293–316). New York: Elsevier.

Zhou, F. W., Jin, Y., Matta, S. G., Xu, M., & Zhou, F. M. (2009). An ultra-short dopamine pathway regulates basal ganglia output. *Journal of Neuroscience, 29*(33), 10424–10435. doi: 10.1523/JNEUROSCI.4402-08.2009.

Zhou, F. M., Liang, Y., & Dani, J. A. (2001). Endogenous nicotinic cholinergic activity regulates dopamine release in the striatum. *Nature Neuroscience, 4*(12), 1224–1229. doi: 10.1038/nn769 nn769.

Zhou, F. M., Wilson, C. J., & Dani, J. A. (2002). Cholinergic interneuron characteristics and nicotinic properties in the striatum. *Journal of Neurobiology, 53*(4), 590–605. doi: 10.1002/neu.10150.

Zucker, R. S., & Regehr, W. G. (2002). Short-term synaptic plasticity. *Annual Review of Physiology, 64*(1), 355–405.

4 Current Ideas on BG Function

In the 1980s, a number of influential ideas on basal ganglia (BG) function were introduced, based on a synthesis of modern anatomical and physiological knowledge (Albin, Young, & Penney, 1989; Alexander, DeLong, & Strick, 1986). They are not only widely taught but have also had a major impact on clinical practice, especially in the treatment of movement disorders. In this chapter, we shall review these ideas on how the BG mediate action selection and reinforcement learning and how they malfunction in various disorders.

4.1 PARALLEL LOOPS

A key feature of BG organization is the existence of parallel cortico-BG-thalamocortical networks. The output of the BG reaches multiple thalamic nuclei, which in turn project to the cortical regions where corticostriatal projections originate, thus forming a closed loop (Alexander et al., 1986). This pattern of connectivity is sometimes called reentry, as the BG outputs target the thalamic nuclei that project back, or "reenter" the cortex. Such reentrant projections were also found in the ventral striatopallidal system. For example, the olfactory tubercle projects to the ventral pallidum, which projects to the mediodorsal thalamus, which projects to the piriform cortex, forming a loop involved in the processing of olfactory signals (De Olmos & Heimer, 1999).

The concept of reentrant BG loops as basic functional units in behavior was first proposed by Bucy (1942). Bucy observed that athetosis, a neurological condition defined by involuntary and repetitive writhing movements of the extremities, could be produced by damage to any component in a larger circuit that includes striatum, globus pallidus, ventrolateral thalamus, or premotor cortex. According to him, athetosis might involve a closed circuit in which cortex projects to the striatum, which in turn projects to the globus pallidus, which then projects to the thalamus, which finally projects back to the cortex (Bucy, 1942). But conclusive anatomical evidence did not become available until decades later, when Strick and colleagues demonstrated that the thalamocortical portion of the "motor" circuit mainly terminates within a restricted region in the supplementary motor cortex that also projects to the striatum (Schell & Strick, 1984). Additional BG output channels were later found, and the loop structure was then applied to other cortical areas in subsequent work (Alexander, DeLong, & Strick, 1986; Middleton & Strick, 2000). A similar organization is also found in mice (Foster et al., 2021).

Recurrent or reverberating projections have long been known in neuroscience, starting with De No's observation that a cortical neuron excites another cortical neuron, which in turn sends a reciprocal excitatory projection (Lorente de Nó, 1934). This organization is thought to be the basis of persistent excitation in a reverberating neural circuit (Hebb, 1949). The cortico-BG loop, however, is a larger circuit with at least four synapses (corticostriatal, striatonigral, nigrothalamic, and thalamocortical), two excitatory and two inhibitory.

It has been suggested that, at the level of behavioral function, the relevant unit is a closed cortico-BG-loop and that there are multiple such loops, each with cortical, striatal, pallidal, and thalamic components (Alexander et al., 1986). This "parallel loops" model draws an important distinction between operation and content. Different loops perform similar operations but differ in content, i.e. the type of signals processed and what they represent (Alexander et al., 1986). Within each loop, different components have unique contributions, but together they form a circuit that performs a specific operation.

Based on primate research, five loops were originally proposed (Figure 4.1). For example, in the oculomotor loop, the frontal eye field and visual cortical areas send projections to a specific striatal

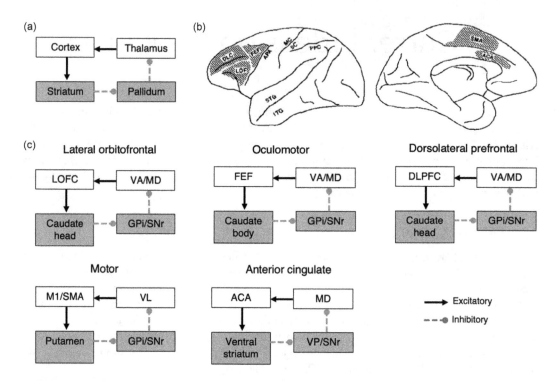

FIGURE 4.1 Parallel loops and reentry. (a) The most basic reentrant circuit has four components: two excitatory synapses and two inhibitory synapses. (b) Five major cortical areas that receive BG output via the thalamus. DLC, dorsolateral prefrontal cortex; FEF, frontal eye field; LOF, lateral orbitofrontal cortex; SMA, supplementary motor area; ACA, anterior cingulate area. (c) Five loops were initially identified based on monkey research. Different loops target different subregions of GPi, SNr, and thalamus, which are not indicated here. MD, mediodorsal nucleus of thalamus; VA, ventral anterior nucleus of thalamus; VL, ventral lateral nucleus of thalamus. Based on Alexander, DeLong, and Strick (1986), with permission.

region (in the caudate body of primates), which projects to a corresponding substantia nigra pars reticulata (SNr) region, which projects to the superior colliculus as well as the ventral thalamus (Hikosaka, Takikawa, & Kawagoe, 2000). A similar organization is found in the skeletomotor loop, where the somatotopic organization in the corticostriatal projections is thought to reflect distinct channels for the control of different body parts.

The parallel loops model predicts that BG damage within a given loop will produce similar symptoms. This prediction is supported by results from lesion studies in monkeys (Middleton & Strick, 2000). For example, lesions in the inferior temporal cortex produce deficits in visual recognition learning similar to those produced by lesions in the associative striatal region that receives inferior temporal cortical projections (Divac, Rosvold, & Szwarcbart, 1967). Lesions of the prefrontal cortex and its target in the caudate nucleus both impair performance on working memory tasks, which require the ability to maintain some representation online during a delay (Goldman & Nauta, 1977; Goldman-Rakic, 2011; Künzle, 1978; Levy, Friedman, Davachi, & Goldman-Rakic, 1997).

According to the parallel loops model, different loops process different types of signals, but there has been a debate on the extent to which these cortico-BG loops are truly segregated. Some argue that, given the striking pattern of convergence in the BG circuits, the information reaching the BG is being funneled. Note that the debate between parallel processing and information funneling is not over whether there is convergence of inputs to the BG but whether convergence occurs within or between loops (Alexander et al., 1986; Parent & Hazrati, 1995a; Percheron & Filion, 1991). According to the original parallel loop model, convergence occurs within each of the functionally segregated loops. By contrast, according to the information funneling view, convergence can occur both within and across domains, so there is no strictly parallel and segregated processing of signals from the cortex.

While the existence of reentrant networks is now well-established, there is evidence supporting interaction between loops. For example, in the globus pallidus internus (GPi), the dendritic arbors are very large and oriented at right angles to striatal axons, suggesting that converging inputs come from multiple striatal domains (Percheron, Yelnik, & François, 1984). In the SNr, there are distinct domains based on patterns of striatonigral connectivity, but dendrites of SNr neurons near the domain border often extend into adjacent domains, so each domain is also in a position to receive inputs targeting a neighboring domain (Foster et al., 2021). Moreover, projections from the sensorimotor and associative striatal regions can converge in the same SNr region. At the next stage, in the nigrothalamic projections, there is also convergence of limbic, associative, and sensorimotor inputs from both medial and lateral SNr onto thalamic cells that project to the motor cortex (Aoki et al., 2019). Figure 4.2 summarizes the patterns of convergence and divergence.

Percheron and colleagues argue that anatomical convergence reflects the maximal aperture of a focusing system. According to them, the degree of segregation between loops (focusing) could be adjusted dynamically by neuromodulatory inputs (Percheron & Filion, 1991). They suggest that the conventional descriptions of strict somatotopic organization and functional segregation can be attributed to limitations in standard experimental designs, in which the monkeys are highly restrained when the neurons are mapped: "in monkeys left free to use their full motor repertoire the same pallidal or nigral neuron might respond to signals from different parts of the body, and might

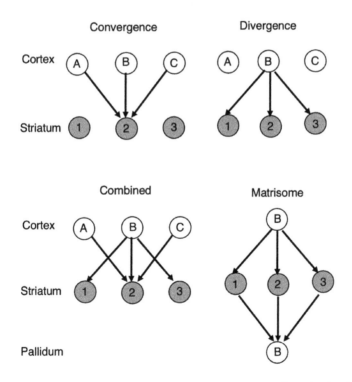

FIGURE 4.2 Patterns of corticostriatal inputs. In the convergence example, striatal cell 2 receives inputs from cortical units A, B, and C. If all three are required to generate spiking, then cell 2 output can stand for the cortical pattern ABC. In the divergence example, a single cortical cell B projects to striatal cells 1, 2, and 3. Depending on the synaptic weights, activation of B could lead to activation of some or all three striatal neurons simultaneously or in a sequence. The "combined" example is more representative of the corticostriatal projections. In the matrisome organization, a given cortical region sends divergent projections to dispersed striatal areas, which then reconverge in the pallidum.

integrate "motor," "oculomotor," "limbic" and two types of "prefrontal" signals, according to context." However, this hypothesis remains to be tested, as no study has directly compared BG activity in the freely moving condition with that found in the restrained condition.

4.2 CONVERGENCE AND DIVERGENCE

In principle, there is no contradiction between the existence of parallel networks and their complex patterns of convergence and divergence. There can be closed portions of the loops as well as overlapping and interactive portions. Both can characterize the BG circuits, so long as we do not assume that elements of each network are fixed and strictly localized.

Yeterian and Van Hoesen observed that cortical areas that are connected with each other project to a common striatal domain (Yeterian & Van Hoesen, 1978). According to them, each striatal region may receive inputs not just from one cortical region but rather from an extensive cortical network of regions reciprocally connected with each other. Recent work shows that this rule applies more to the IT pathway than the PT pathway (Hooks et al., 2018).

Cortical areas that are functionally related also tend to be interconnected. Often these related areas target a common striatal area. For example, projections from two interconnected frontal cortical areas, the frontal and supplementary eye fields, converge in the matrix compartments in the sensorimotor striatum (Flaherty & Graybiel, 1991; Parthasarathy, Schall, & Graybiel, 1992). In the associative striatum, the oculomotor region receives inputs from three interconnected areas implicated in the control of eye movements: frontal eye fields (area 8), posterior parietal cortex, and dorsolateral prefrontal cortex (areas 9 and 10). All three areas converge onto the caudate, though their detailed organization at the cellular level remains unclear. Moreover, corticostriatal projections from the hand-related area in the primary somatosensory cortex and from the hand-related area in the primary motor cortex also overlap with one another (Flaherty & Graybiel, 1993).

In corticostriatal projections from the somatosensory cortex, most terminals target the dorsolateral striatum, but there are radially oriented projections that also extend into the medial striatum (Brown, Smith, & Goldbloom, 1998). It has been suggested that body parts that naturally move together (e.g. jaw, lips, and tongue) could project to a single lamina in the striatum (Brown et al., 1998). Brown and colleagues argued that this lattice-like arrangement could permit the monitoring of body part position during behavior, as well as the recombination of tactile inputs. For example, there may be one set of locations for the hindlimb and forelimb combination, another for whisker and hindlimb, still another for hindlimb and trunk, and so on. In rats, the forelimb, hindlimb, and whisker areas interdigitate in the anterior dorsolateral striatum, and lesions to this region can impair the performance of grooming sequences, in which the corresponding body parts may be coordinated to produce behavior. This observation suggests the presence of a striatal circuit that coordinates different body parts required in a behavioral sequence like grooming (Cromwell & Berridge, 1996). Like combining different letters from the alphabet into words, different body parts are therefore combined into specific modules for action.

Thalamic and cortical areas that are functionally related and interconnected can also send convergent projections to the striatum, even onto individual spiny projection neurons (SPNs) (Dubé, Smith, & Bolam, 1988; Kemp & Powell, 1970; Kocsis, Sugimori, & Kitai, 1977; McFarland & Haber, 2000, 2001). Thalamic nuclei that target a given striatal region usually projects to the cortical region that also project to that striatal region (Hunnicutt et al., 2016). Thalamostriatal projections are often axon collaterals of the thalamocortical projections (Royce, 1983). Projections from interconnected ventral thalamic and motor cortical areas also converge in the sensorimotor striatum, and the projection field of corticothalamic inputs often overlaps with the thalamocortical or thalamostriatal projecting nuclei. (Hintiryan et al., 2016; McFarland & Haber, 2000).

Since thalamic neurons and cortical neurons projecting to a given striatal area are often reciprocally connected, thalamostriatal inputs and corticostriatal inputs to a given SPN could be active at the same time. This organization suggests that these inputs may play complementary roles. For example, via more proximal synaptic contacts, the thalamostriatal inputs may promote up-state transitions, while cortical inputs can more effectively generate firing in SPNs. Thalamic inputs are thus in a position to independently modulate the striatal output in response to cortical inputs.

Finally, divergence is also well-established anatomically (Alloway, Mutic, & Hoover, 1998; Parent & Hazrati, 1995b; Swanson, 2000). A given cortical area may project to multiple striatal regions. For example, a single cortical column in the rat somatosensory cortex may project to multiple regions in the dorsolateral striatum (Alloway et al., 1998). The prelimbic cortex can project to most associative and limbic striatal regions. Likewise, a small population of striatal neurons may project to several distinct and dispersed groups of pallidal or nigral neurons (Swanson, 2000).

Whereas convergence allows many cortical areas to activate a limited set of striatal neurons and perhaps an even smaller set of pallidal neurons, divergence allows a cortical representation to be sent to multiple striatal areas, so that it is capable of recruiting multiple striatal modules. The strength of the corticostriatal connection can determine the order or latency with which a given cortical neuron may recruit the postsynaptic striatal neurons, and lateral inhibition may also allow some units, once activated, to suppress others.

4.3 INTERACTION BETWEEN LOOPS

Joel and Weiner proposed a split-circuit organization that emphasizes the interconnectivity between networks and a hierarchical arrangement (Joel & Weiner, 1994, 2000). For example, a given cortical area projects to one striatal region, which targets a pallidal region that eventually projects back to a closed or reentrant pathway as well as an open pathway to a neighboring network. The general direction of signal propagation seems to be from limbic to associative and finally to sensorimotor, providing a hierarchical interpretation of the tripartite organization.

Haber and colleagues have reported reciprocal and nonreciprocal connections between the striatum and pallidum. As shown in Figure 4.3, a given striatal region can project to a pallidal region that in turn influences dopaminergic projection to a neighboring striatal region (Haber, 2003; Haber, Fudge, & McFarland, 2000). In addition, a closed loop that reenters the limbic prefrontal cortex can also give rise to an open limbic pathway that terminates in the associative prefrontal cortex. The nucleus accumbens shell can influence the neighboring core region through striato-pallido-thalamic pathways (Zahm & Brog, 1992). Finally, recent work has demonstrated that interactions also occur at the output stage of the BG: projections from the medial SNr can target limbic thalamic regions that reenter the medial prefrontal cortex, but at the same time also reach motor thalamic regions that reenter the motor cortex as well (Aoki et al., 2019).

As a description of anatomical organization, the parallel loops model is useful, even if interactions between loops are now well documented. But as an explanation of BG function it is inadequate. In some ways, it is an extension of the popular belief that the brain generates actions by activating the corticospinal pathway, merely adding a closed loop with subcortical components. The underlying assumption is that, ultimately, the circuit is able to influence effectors through projections to the motor cortex. Yet this model does not explain the computation performed at each step in the loop and how each loop contributes to behavior. Its narrow focus on reentrant projections back to the cortical area of origin has also led to the neglect of descending projections to the brainstem. Many, if not most, of the pallidal projections to the thalamus are collaterals of projections to the tectum and brainstem, which could ultimately influence the reticulospinal pathway (McElvain et al., 2021; Takakusaki, Saitoh, Harada, & Kashiwayanagi, 2004). The exclusive focus on the thalamocortical targets of BG output reflects the standard behavioral tasks used in primate studies, which focused on arm and hand movements commonly associated with corticospinal projections. But the role of head and torso movements in these tasks and the relationship between BG output and the reticulospinal pathway have largely been neglected.

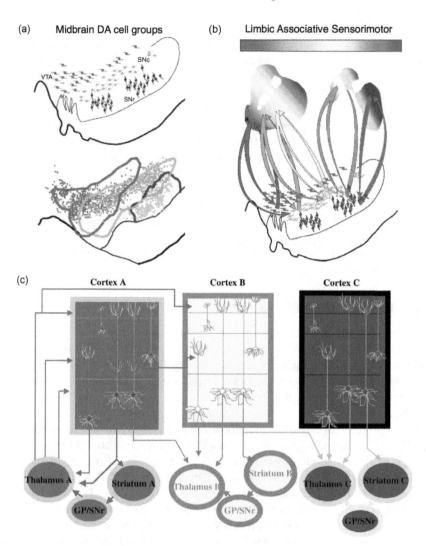

FIGURE 4.3 Anatomical evidence for open connections that allow interactions between different basal ganglia networks. (a) Schematic showing the organization of midbrain dopamine (DA) cell groups. The dorsal tier (red) includes the ventral tegmental area and retrorubral DA cells; the ventral tier includes the densocellular neurons (orange) and the cell columns (blue). (b) Organization of striato-nigro-striatal projections. This organization allows ventral striatal regions to influence more dorsal striatal regions via spiraling striato-nigro-striatal projections. (c) Layer 5 cortical pyramidal neurons may project to a thalamic region that is part of another circuit, thus allowing limbic areas to influence associative and sensorimotor areas. From Haber (2003) with permission.

Another limitation of the parallel loops model is that its functional classification also reflects traditional localizationist assumptions (Chapter 1). Even if we accept the functional designation of an oculomotor loop for eye movements and a skeletomotor loop for limb movements, it is unclear how these loops are engaged in normal behavior (e.g. reaching) that involves the use of both eyes and arms. The parallel loops model does not address the dynamic coordination of distinct functional modules in real-time during behavior. Likewise, the analysis of the patterns of convergence and divergence in BG connectivity has not yielded much insight into function. The interpretations are based mostly on the investigation of static anatomical relationships rather than neural dynamics during behavior. Other than emphasizing the complexity of the organization, the analysis in itself does not make detailed predictions at the level of behavior. To appreciate the functional contributions, we must turn to physiological studies that have elucidated the nature of computation performed at each stage in the BG circuit.

4.4 DISINHIBITION

Seminal studies by Deniau and colleagues established disinhibition as a fundamental physiological operation of the BG circuits (Chevalier, Vacher, Deniau, & Desban, 1985; Deniau, Chevalier, & Feger, 1978; Deniau, Feger, & Le Guyader, 1976). They found that striatal stimulation inhibited identified nigrothalamic neurons and that nigral stimulation inhibited thalamic and tectal (collicular) neurons. In the direct pathway, there are two successive GABAergic synapses: striatonigral and nigrocollicular. Striatonigral output inhibits the SNr activity; when dSPNs are activated by cortical input, the direct pathway will transiently reduce tonic inhibition of the superior colliculus (Chevalier, Deniau, Thierry, & Feger, 1981; Yoshida & Precht, 1971). At the level of the superior colliculus or thalamus, the net effect is excitatory. If we only consider the direct pathway, then excitatory input to the striatum can be converted to net excitation of a target region of BG outputs (Ueki & Yoshida, 1976).

Disinhibition is often used to explain BG contribution to behavior. BG output neurons are often tonically active, capable of firing in the absence of synaptic inputs (Nakanishi, Kita, & Kitai, 1987). Such tonic inhibition is assumed to suppress behavior. Disinhibition would allow the behavior in question to be released by a pause in the BG output, leading to generation of the appropriate behavior (Albin, Young, & Penney, 1989; Chevalier & Deniau, 1990). This is summarized in Figure 4.4.

At first glance, disinhibition appears to be superfluous. Direct excitatory projections from the cortex to the superior colliculus, or any other target of BG projections, would seem to be sufficient. If the whole circuit simply allows the cortex to activate the superior colliculus, why are the two additional inhibitory synapses in the direct pathway needed? Two answers to this question have been proposed.

First, BG circuits do not simply relay signals from the cortex or thalamus but allow considerable transformation of their inputs. Different combinations of cortical inputs from multiple cortical areas can converge on striatal neurons, and the BG circuit can transform and recombine

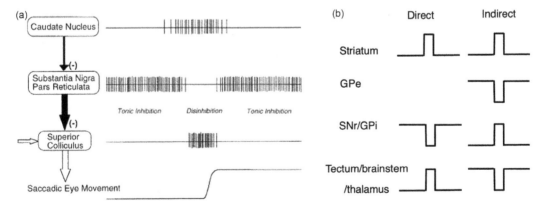

FIGURE 4.4 Disinhibition. (a) Illustration of activity in different components of the direct pathway disinhibition circuit, starting from the oculomotor striatum (part of the caudate in monkeys). From Hikosaka, Takikawa, and Kawagoe (2000) with permission. (b) Schematic summary of firing rates in different components of the basal ganglia (BG) and their targets. Direct and indirect pathways have opposite effects on BG output neurons and their targets. Only the simple indirect pathway is shown here (see Figure 2.8).

the cortical inputs in different ways. Thus, the net excitatory effect from disinhibition is different from direct excitation, which would not have allowed the considerable transformation of the cortical signals.

Secondly, disinhibition and direct excitation have different effects. Disinhibition is thought to support a "gating" function for the BG. For example, consider a typical target neuron in the deep/intermediate layer of the superior colliculus. This neuron receives BG output projections. In direct excitation, the target neuron is usually assumed to be quiet, and when an excitatory input arrives, the output will be some function of the input. The assumption is that the cortex is inactive in the absence of movements. Disinhibition would allow for an active cortex, which by itself does not necessarily produce movement. BG output neurons tonically inhibit their target neurons. According to the common interpretation, such active inhibition is needed to prevent the target neuron from responding to all the excitatory inputs. A transient suppression of the tonically active BG output neuron by another inhibitory input would then open the gate, allowing the selective expression of only one channel of output, which presumably would result in the generation of some movement downstream. Consequently, both independent input from another source as well as an appropriately timed pause in BG output are required in the generation of the action.

Coincidentally, a similar principle of disinhibition was proposed by students of animal behavior. As Lorentz noted in his Nobel lecture (Lorenz, 1974): "I regard as the most important break-through of all our attempts to understand animal and human behavior the recognition of the following fact: the elemental neural organization underlying behavior does not consist of a receptor, an afferent neuron stimulating a motor cell and of an effector activated by the latter. Holst's hypothesis, which we confidently can make our own, says that the basic central nervous organization consists of a cell permanently producing endogenous stimulation, but prevented from activating its effector by another cell which, also producing endogenous stimulation, exerts an inhibiting effect. It is this inhibiting cell which is influenced by the receptor and ceases its inhibitory activity at the biologically 'right' moment." Based on the studies of animal behavior, Lorenz and von Holst were thus able to deduce the principle of disinhibition. Neurons that receive tonic BG output, for example the output neurons in the superior colliculus, are prevented from activating effectors. They are now transiently enabled to activate the effectors when BG output pauses at the right moment. When this is coupled with some excitatory input, a specific action can be selected.

Perhaps the strongest empirical support for the disinhibition hypothesis comes from work by Hikosaka and colleagues, who studied the role of the striatonigral and nigrocollicular pathways in monkey eye movements (Hikosaka et al., 2000). In primates, part of the caudate receives cortical inputs related to visual cortical areas and projects to the dorsolateral SNr, which in turn projects to the deep/intermediate layers of the superior colliculus. Many SNr neurons have response fields, usually centered in the contralateral hemifield, and inhibit saccade-related collicular neurons (Hikosaka & Wurtz, 1983a, b, c). Some SNr neurons show a pause in firing in response to a spot of light to which a monkey makes a saccade, as a pre-saccadic response, or during a delay period before the saccade. Once SNr neurons stop firing, disinhibition generates a strong drive for collicular neurons to fire and generate the saccade.

These experiments suggest that, without the BG output, the superior colliculus may generate saccades to any visual input, but the nigrocollicular projection prevents such random saccade generation. With direct excitation, multiple excitatory inputs could excite the same areas and produce conflicting commands. For action selection, some additional mechanism must open the gate for the appropriate action. The BG circuit thus prevents convergent excitatory signals from triggering collicular output, and the gate can be opened by transiently pausing the tonic inhibition from the SNr (Hikosaka et al., 2000). On this account, unwanted behaviors are somehow suppressed by maintaining or increasing tonic inhibition, while desired behaviors are released by decreasing or removing such inhibition. In support of this idea, when the nigrocollicular inhibition was blocked with a GABA-A receptor agonist (muscimol), uncontrollable saccades were observed (Hikosaka &

Wurtz, 1985a, b). Thus, disinhibition is often interpreted as a gating mechanism, making it possible to generate movement only with the appropriate sensory inputs.

4.5 RATE MODEL

Disinhibition explains the action of the direct pathway. But another pathway, the indirect pathway, has the opposite effect as the direct pathway (Albin et al., 1989; Gerfen et al., 1990). Whereas the direct pathway produces a transient pause in BG output, the indirect pathway increases the inhibitory BG output (Gerfen et al., 1990; Hikosaka et al., 2000). Consequently, the direct pathway is often thought to promote behavior via disinhibition, whereas the indirect pathway is thought to suppress behavior. In textbooks, these two pathways are sometimes called "Go" and "No-Go" pathways.

One attempt to explain the function of this organization is the "rate model," which was developed to explain the symptoms of movement disorders. Delong proposed that the average firing rate of pallidal neurons (especially GPi and SNr) determines how much behavior is produced. In particular, there is an inverse relationship between the *rate* of BG output nuclei and behavior: as firing rate is increased, less behavior is expressed. According to the rate model, the direct and indirect pathways can bidirectionally modulate the mean firing rate of BG output neurons (e.g. in the GPi and SNr), the direct pathway reducing it, while the indirect pathway increasing it (DeLong, 1990). Together, they can scale the amplitude and speed of movements.

The rate model, shown in Figure 4.5, has served as the basic conceptual framework for clinical interventions in the treatment of movement disorders. According to this model, the imbalance between direct and indirect pathways leads to hypokinetic or hyperkinetic symptoms. To explain the symptoms of Parkinson's disease, this model populates that direct pathway activity decreases and indirect pathway activity increases as a result of DA depletion. This is because DA

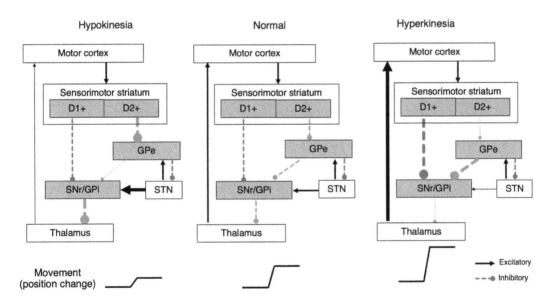

FIGURE 4.5 Rate model. The motor cortico-BG network is shown. In hypokinesia, there is excessive output from indirect pathway spiny projection neurons (D2+) in the striatum, as well as excessive output from the subthalamic nucleus (STN). The net result is reduced activation of the thalamus and downstream motor cortex. In hyperkinesia, there is greater output from direct pathway spiny projection neuron (D1+) as well as globus pallidus externus (GPe) neurons, while STN output is reduced. Consequently, thalamic output is disinhibited, resulting in increased motor cortical output and excessive movement.

has opposite effects on direct and indirect pathways. In the direct pathway, by acting on D1R, DA enhances SPN excitability, potentiates glutamatergic signaling, and enhances GABA release from dSPNs (Cepeda, Colwell, Itri, Chandler, & Levine, 1998; Floran, Aceves, Sierra, & Martinez-Fong, 1990; Lahiri & Bevan, 2020). In the indirect pathway, DA is expected to have the opposite effect by activating Gi-coupled D2 receptors. The net effect of DA is to increase direct pathway activation and reduce indirect pathway activation. After DA depletion, D2 suppression of indirect pathway spiny projection neurons (iSPNs) is expected to be reduced. This is thought to result in an imbalance in direct and indirect pathway activity, producing increased activity in GPi/SNr neurons (Mallet, Ballion, Le Moine, & Gonon, 2006). Consequently, there is increased inhibition of thalamic and cortical neurons, and reduced movement inhibition (bradykinesia or akinesia). In contrast, in hyperkinetic disorders, such as dystonia and hemiballismus, the opposite pattern is found due to inadequate BG inhibition of the motor thalamocortical network (Gernert, Bennay, Fedrowitz, Rehders, & Richter, 2002).

Since the rate model predicts that increased indirect output will increase BG output and akinesia, one implication is that manipulations that reduce indirect pathway activity (and reduce net BG output) can restore movement. In support of this claim, lesioning the STN or GPi can ameliorate parkinsonian symptoms (Bergman, Wichmann, & DeLong, 1990; Laitinen, 1995).

The rate model offers a conceptual framework for understanding the basic symptoms of movement disorders, but its limitations are increasingly recognized (Nambu, 2008). First, the idea that the BG output can be captured by a single measure, the average rate of firing, is simplistic. It assumes that BG output is uniform and monolithic. Influenced by this model, many studies only quantified the average firing rate while neglecting how the activity of each neuron changes over time and its relationship with behavior. As we shall see in later chapters, the average firing rate of BG output neurons is not a meaningful measure.

Another limitation is that the explanation of clinical symptoms is based on simplistic descriptions of behavior. For example, bradykinesia is considered hypokinetic, whereas uncontrollable actions, as in dyskinesia or Tourette's syndrome, are considered hyperkinetic. Categories like "hypo" or "hyper" imply a standard: too little or too much of what? Is the implied measure frequency, amplitude, force, or a combination of these, as in the concept of action vigor?

More importantly, the rate model also makes several predictions that are not supported by experiments (Nambu, 2008). For example, it predicts hyperactive iSPNs and hypoactive globus pallidus externus (GPe) neurons after DA depletion, but there is no clear support for this claim. It predicts that excessive striatopallidal inhibition results in disinhibition of the STN, but overactivity of the STN in PD is not due to reduced GPe activity (Vila et al., 1997). Finally, according to the rate model, hypokinetic symptoms are due to abnormally strong inhibition of the thalamus and cortex by excessive BG outputs from the GPi. It therefore predicts increased GPi activity after DA depletion, but this was not found (Wichmann et al., 1999).

The rate model also assumes that the BG generate behavior via projections to the thalamus: for example, the GPi output influences movement by targeting the ventral thalamus, which in turn projects to the motor cortical areas, allowing BG output to influence the effectors through the corticospinal pathway (Alexander et al., 1986; Hoover & Strick, 1993). Thus, reduced BG output is expected to increase ventral thalamic drive to the motor cortex, whereas increased BG output is expected to suppress thalamocortical activity. This account predicts that GPi and thalamic activity should be anticorrelated, but this prediction is not supported by direct comparisons of activity in GPi and its target in the ventrolateral-anterior nucleus of the thalamus (Schwab et al., 2020). In monkeys performing a reaching task, Turner and colleagues found that increases in firing at the time of reaching were very common in both GPi and thalamus neurons. Surprisingly, thalamic activity often changes before GPi activity. Motor cortical projections to both the ventrolateral-anterior nucleus and GPi may be responsible for the thalamic excitation observed (Prasad, Carroll, & Sherman, 2020). It is also possible that the direct projections from the GPe to the thalamus may disinhibit thalamic targets (iSPN inhibits GPe, which inhibits thalamus) (Foster et al., 2021). Consequently, indirect pathway activation could have a net excitatory effect on the thalamus via disinhibition.

Another prediction of the rate model is that lesions of the thalamic nuclei receiving BG projections (thalamotomy) should reduce movements since the thalamus can no longer excite the motor cortex to initiate action. But this prediction is not supported by clinical observations; on the contrary, thalamotomy is often effective for improving motor symptoms in PD (Benabid, Pollak, Louveau, Henry, & De Rougemont, 1987). In addition, the rate model predicts that pallidotomy (lesion of the GPi) should increase movement, since the GPi can no longer inhibit the thalamocortical network. Yet, pallidotomy can actually be an effective treatment for hyperkinetic symptoms such as hemiballismus and dyskinesia. It can reduce levodopa-induced dyskinesia contralateral to the lesion, suggesting that dyskinetic movements are due to GPi outputs to the thalamocortical network. In normal monkeys, GPi inactivation causes hypometria (reduced movement amplitude) and bradykinesia (slowness in movement) without impairing movement initiation or accuracy (Desmurget & Turner, 2008; Turner & Desmurget, 2010).

4.6 ACTION SELECTION MODELS

Another class of models focuses on action selection. These models share some of the basic assumptions of the rate model, such as disinhibition and opponent effect of direct and indirect pathways, but provide a different interpretation of their behavioral significance.

4.6.1 Focused Selection

Based on clinical observations, Denny-Brown proposed that the BG can facilitate some behaviors while suppressing others (Denny-Brown & Yanigisawa, 1976). In the more modern formulation, the focused selection model, the direct pathway is used for selecting desired actions while the indirect pathway is used for suppressing competing behaviors (Mink, 1996). This model assumes that posture and action are mediated by distinct mechanisms. The indirect pathway is also hypothesized to promote postural fixation, whereas the direct pathway turns off the fixation to initiate action. After DA depletion, focused selection is impaired, and simultaneous activation of these mechanisms produces conflict, as manifested in postural instability or akinesia. This account differs from the rate model in the relative roles assigned to the two pathways. According to the rate model, they act on the same BG output neurons to adjust their average firing rate, which regulates the amplitude and speed of actions. According to the focused selection model, the two pathways do not act on the same BG output neurons but rather on different sets of neurons to select desired actions and suppress postural control or competing actions. For example, during a reaching movement, it is assumed that one must turn off postural control just for the reaching arm and turn on the motor pattern generators. When the reach is completed, the pattern generators must be turned off and the postural control turned on again.

Mink argued that the GPi receives two opponent inputs: a focused dSPN inhibition from the direct pathway and a more diffused excitation from the subthalamic nucleus (STN) from the indirect pathway (Mink & Thach, 1993). This organization is shown in Figure 4.6. Striatal and STN projections both converge on single GPi neurons, but STN–GPi fibers arborize more widely than those from dSPNs. A single striatal efferent would inhibit a small region of the GPi or SNr (Bolam, Smith, Ingham, von Krosigk, & Smith, 1993), but a single STN neuron is in a position to excite many GPi neurons (Hazrati & Parent, 1993). The indirect pathway is therefore expected to activate many GPi/SNr neurons and thus inhibit large areas of the thalamus that generate competing motor programs. On the other hand, the direct pathway is thought to disinhibit thalamic neurons only in the center area corresponding to the channel for the desired action. Hikosaka argues that this

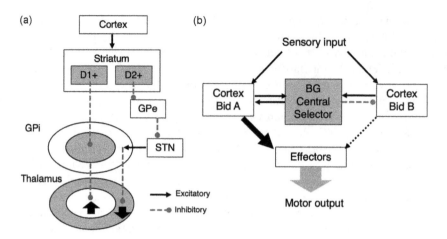

FIGURE 4.6 Action selection models. (a) In the focused selection model, STN inputs excite the GPi region surrounding a functional channel. The center region corresponding to desired action is thus activated, while the surrounding region corresponding to competing action programs is suppressed, as indicated by the black arrows in the thalamus. (b) Central selection model. A central selector allows sensory input to gain access to effectors. The selector detects the most salient input and provides return facilitation by reentrant projection, thus further reinforcing this bid to amplify its access to effectors.

center-surround organization enables two modes of action. In the sequential mode, direct and indirect pathways could be involved in starting and stopping behavior: at the time of transition from one element of a movement sequence to the next, it is necessary to suppress the previous one and initiate the next element. On the other hand, in the simultaneous mode, these pathways have a focusing effect: selecting desired actions and suppressing competing actions (Hikosaka et al., 2000).

4.6.2 Central Selection

Another variant of the action selection model was proposed by Redgrave and Gurney. According to this "central selection" model, the BG solve a selection problem, which arises when multiple motivational systems seek simultaneous access to motor output (Redgrave, Prescott, & Gurney, 1999). The choice is not between action and posture control, as in the focused selection model, but between different goals. For example, to escape harm or to obtain rewards a similar set of muscles may be used; such conflicts arise when both sets of goals are activated. The BG is hypothesized to resolve conflicts over access to limited motor resources. This idea can also be extended to the cognitive domain. Depending on their targets, the outputs of the BG can generate movements or influence internal cognitive processes.

According to this model, BG implement a centralized circuit for selection, with an architecture that reduces connection costs. To appreciate the advantage of central selection, let us consider the simplest winner-take-all network, in which each unit competes with all other units and the unit with the highest output is selected as the winner. The connection cost is high because, for selection to take place, all competitors must be reciprocally connected with each other. For example, using lateral inhibition, each unit has an inhibitory link to every other, so that the activation of one unit could suppress activity in all other units. Increased activity in one competitor suppresses the others, thereby also reducing their inhibitory effects on the winner. This requires an extremely well-connected network: selection among n competing units requires $n(n-1)$ connections. Although lateral inhibition in the striatum is often considered to implement a winner-take-all mechanism for selection (Tunstall, Oorschot, Kean, & Wickens, 2002), the connectivity of the SPNs does not support a fully connected system. A given SPN is not connected with all other

SPNs: its dendritic arbor is too small to share inputs with many other SPNs, and its collaterals can only inhibit a small subset of SPNs within a small area. Because neighboring SPNs have few inputs in common, they cannot compete with each other (Zheng & Wilson, 2002). Alternatively, there could at first be a local competition in a subset of SPNs, and the winner from the local group would then compete with winners from other groups. This organization, analogous to a representative democracy, would require any one of the neurons within a group to represent the output of the whole group, and this representative neuron must in turn be fully connected with other representatives.

On the other hand, in the central selection organization, different units do not directly compete with each other (Figure 4.6b). This design requires only two connections for each competitor (to and from the central selector). Only 2n connections are needed, much less than the $n(n-1)$ connections required by a fully distributed architecture. The most salient bid is selected and the winning bid is returned to the unit sending the bid, using the reentrant BG outputs to the thalamocortical networks. Consequently, the BG can arbitrate between competitors bidding for the use of an output channel for the effector. To add a potential action to the pool for selection, it is only necessary to provide input to the BG, and also to receive the result of arbitration from the BG output. Different cortical regions may send bids, and only one bid from the successful channel would be reinforced via the reentrant projections from the thalamocortical network. Redgrave and colleagues speculated that the original role of the BG was to arbitrate between the demands of multiple midbrain systems, as there are similar loop organizations connecting midbrain areas and the BG. For example, the superior colliculus projects to the thalamus, which projects to the BG, which then projects back to the superior colliculus (McHaffie, Stanford, Stein, Coizet, & Redgrave, 2005). With the expansion of the cerebral cortex, however, it became the dominant source for bids.

4.6.3 PROBLEMS WITH ACTION SELECTION

Action selection models beg the question of how an action is defined and what constitutes the pool of competing behaviors. The action is assumed to be categorical, either selected or suppressed, and the BG merely open the gate for actions, but the parameters of action are determined elsewhere. If the alternative to action is posture control, as assumed in the focused selection model (Mink, 1996), then it is not explained how broad inhibition of target regions can generate posture control or why the activation of the indirect pathway can generate ipsiversive turning movements (Kravitz et al., 2010; Tai, Lee, Benavidez, Bonci, & Wilbrecht, 2012; Zhang et al., 2022).

4.7 REINFORCEMENT LEARNING

In the models discussed so far, there is the implicit assumption that the BG generate specific actions given a particular state. The state-action association can be implemented by corticostriatal projections. The organization of the BG has also been compared to classic neural networks that classify different input patterns (Houk, 1995). According to Houk, the massive convergence in the corticostriatal projections, abrupt transition between up and down states in SPNs, and modifiable weights by neuromodulation suggest that the striatum may function as a pattern detector, similar to the original perceptron neural network (Rosenblatt, 1958).

A noteworthy feature of BG organization is the relatively independent sorting of inputs. The different compartments (striosome and matrix) and pathways (direct/indirect) represent different ways of classifying corticostriatal inputs (Gerfen, 1992). The corticostriatal connectivity would allow specific SPN populations to be activated by a combination of independent criteria, like sorting college students based on their year, hometown, and area of study. The same neuron can belong to distinct groups based on these criteria. For example, the striosome compartment appears to preferentially process interoceptive inputs from inside the body, whereas matrix processes proprioceptive inputs from muscles and joints. On the other hand, the distinction between direct and indirect pathways is less concerned with the types of inputs received, as they often receive similar inputs. Rather, they appear to perform distinct operations on the content, as they have opposing influences on BG output nuclei.

The sorting of corticostriatal inputs can be dynamically adjusted. Neuromodulation or thalamic inputs could provide conditional activation of SPNs by specific patterns of inputs. Moreover, long-lasting modifications in the synaptic weights can shape and redefine the classification performed by the SPNs. Using feedback from the action, it is possible to alter the weights, so that the probability of selecting that action given the state is altered. This process is known as reinforcement learning (RL), which is usually considered "unsupervised," because there are no explicit instructions on the correct behavior.

A common experimental example is a rat learning to press a lever for a food reward. The rat is not told what to do, but it is able to quickly discover the action that earns reward. The lever press, which has never been exhibited, is repeated frequently as a result of learning. The reinforcement from the food reward is believed to modify the connection between state and action, and thus change the probability of behavior given a particular state. Unfortunately, the concept of a state is deliberately vague, even more so than the conventional concept of a stimulus, so that it could include any combination of internal and environmental inputs.

According to a common account, the BG circuit and dopaminergic pathways together implement the actor-critic model, with DA serving as the main reinforcement signal (Barto, 1995; Miller, 1981). Given the role of DA in neural plasticity, state-action connections can be modified (Calabresi, Pisani, Centonze, & Bernardi, 1997; Kerr & Wickens, 2001; Miller, 1981; Pawlak & Kerr, 2008). According to a popular hypothesis, phasic DA activity reflects the difference between actual reward and predicted reward, a reward prediction error (RPE) signal used to update synaptic weights (Schultz, 1998). Through iterative comparisons between predicted and actual rewards, the critic uses the RPE to update reward predictions and allows the actor to change its weights accordingly and adjust behavioral output to maximize future rewards (Sutton & Barto, 2018).

If the assumed performance model is simply some S–R (or state-action) association, then the content of learning is simply the strength of association between the S element and the R element, with the R element representing some behavioral policy. The BG have been implicated in procedural learning and habit formation, which are assumed to rely on S–R RL (Mishkin, Malamut, and Bachevalier, 1984; Saint-Cyr, Taylor, & Lang, 1988). However, other work has also shown that the content of learning cannot be described simply by S–R associations (Balleine & O'Doherty, 2010; Yin & Knowlton, 2006).

As shown in Figure 4.7, the actor in the popular actor-critic model is essentially a more abstract description of the action selection system. The actor selects an action based on the action value (Q value). Action value reflects the weight that converts a given state to the production of some action. The higher the value for a given action, the more likely it is to be selected (Collins & Frank, 2014). If these inputs represent some state, then changing the weight of the corticostriatal synapse can alter the possibility that a given action is selected in a particular state. The RPE is expected to alter the weight of the state-action connection. Just like the action selection model of the BG, the behavioral policy in RL is also traditionally assumed to be categorical. How the action is generated is outsourced to downstream "motor" structures, which will somehow transform the action value into behavior. RL only changes the probability of selecting a discrete action by adjusting the strength of the corticostriatal synapse, which changes the activation function of the SPN.

A key assumption of RL is that what is learned is a prediction of future value and that this prediction is directly translated into performance. But this assumption is problematic, as learning cannot be equated with changes in performance (Kimble, 1961; Rescorla, 1980). Performance may change due to fatigue, or a change in motivational state, or development and maturation. It is possible to produce changes in performance without any long-lasting changes in neural connections. For example, a rat has already learned to press a lever for food but may not press it when it is not hungry. As it becomes hungry, it may increase lever pressing. On the other hand, a naïve but hungry rat will also increase lever pressing as a result of learning. The increase in lever pressing per se is not evidence of

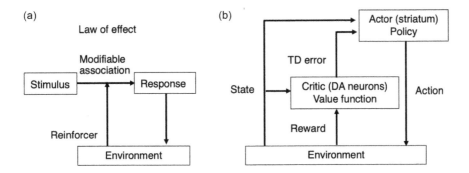

FIGURE 4.7 Reinforcement learning (RL). (a) The original S–R reinforcement model proposed by Thorndike and Hull. (b) Illustration of the actor-critic model, a popular RL model that is thought to be implemented by the basal ganglia (BG). In particular, the actor is thought to be implemented by the striatum, or the BG as a whole, whereas the critic is thought to be implemented by dopamine neurons. The agent can visit a finite number of states, and by visiting a state, a numerical reward is collected, where negative numbers may represent punishment. From every state, there are subsequent states that can be reached by means of actions. The value of a given state is defined by the averaged future reward when selecting actions from this particular state. The goal of an RL algorithm is to select actions that maximize the expected cumulative reward to the agent.

learning. Indeed, there has long been a debate on whether the phasic DA signal is to be interpreted as a prediction error signal or as a signal involved in modulating performance (Salamone & Correa, 2012). We shall return to a discussion of this problem in Chapter 12.

4.8 THE CHALLENGE OF BEHAVIORAL ANALYSIS

The foregoing review of current models of BG function shows that they all share certain assumptions about behavior. The description of behavior in these models is remarkably abstract. Actions are either selected or not selected. They only vary along one dimension: the probability of selection. The question of how much, when, and how fast is not considered.

According to the dominant paradigm, behavior is a "response" to inputs. The brain is assumed to receive inputs and transform these inputs into behavior outputs (Yin, 2020). The dominant approach to studying the brain–behavior relationship is the input/output analysis, in which sensory input is manipulated while motor output is measured. In a typical experiment, the subject is restrained, often with head fixation, and instructed to perform specific actions when prompted by discrete cues. Neural activity is studied in relation to some arbitrarily defined behavioral events, such as cue, response, and reward, all recorded as a series of discrete time stamps. Conventional experimental designs are therefore "event-based." What happens between events, as defined by the observer, is assumed to be irrelevant.

An example of behavior that is not addressed by conventional models is turning following unilateral manipulations of the BG. It has long been known that unilateral stimulation of the striatum often results in body-turning to the opposite side (contraversive) as well as movements on the contralateral side of the body, usually flexion of the contralateral foreleg and hind leg (Ferrier, 1876; Forman & Ward, 1957; Laursen, 1962). While unilateral direct pathway activation using optogenetics produces contraversive turning behavior, indirect pathway activation produces behavioral suppression or ipsiversive turning (Kravitz et al., 2010; Yu et al., 2022; Zhang et al., 2022). Likewise, unilateral SNr inhibition results in contraversive movement. In agreement with BG projections to the tectum and brainstem, turning can also be produced by stimulating target regions of BG projections (Cools, 1985). These observations cannot be explained by the rate model or the action selection model, as neither has anything to say about the direction of turning, or the direction of any behavior.

There are other puzzling clinical features that cannot be explained by conventional models. For example, BG damage often produces deviant or asymmetric postures. In torticollis, a tilted neck is found, often associated with rigidity or stiffness. The traditional perspective highlights the absence of behavior (i.e. akinesia) but neglects the continuous final common path outputs generated while the patient is frozen, which results in physical exhaustion despite the lack of movement. Nor does it explain why rigidity is reduced when the body is supported (Martin & Hurwitz, 1962).

Recent studies showed that the assumption of a discrete action varying only in its probability of selection is simplistic. In freely moving animals, there is a striking relationship between BG activity and continuous behavioral variables. Both striatal and pallidal neurons represent continuous kinematic variables like position and velocity (Yin, 2016; Yin, 2017). In later chapters, we shall review experimental evidence showing that the BG output quantitatively and continuously determines specific behavioral parameters.

4.9 SUMMARY

A major organizational principle of the BG circuits is the existence of multiple parallel cortico-BG circuits, each consisting of a cortico-BG-cortical loop characterized by reentrant projections. Parallel loops are defined by distinct functional territories in the cortex and their corresponding BG targets. A basic loop has a cortical component, a striatal component, a pallidal component, and a thalamic component that projects back to the cortical component. There is also considerable overlap and interaction between different loops. This complex pattern of convergence and divergence deviates from strict parallelism and could allow for the recombination of distinct cortical elements in generating learned behavioral patterns and sequences.

In the simplest BG circuit, as found in the direct pathway, there are two consecutive inhibitory synapses, which produce disinhibition (net excitation) of target neurons. According to the rate model, because the BG output is inhibitory and tonically active, at rest it serves to prevent action initiation. If the tonic inhibitory output is excessive, then there is too much behavioral suppression, or hypokinesia. If the output is not enough, then there is uncontrollable behavior, or hyperkinesia. The direct and indirect pathways serve to regulate the average BG output. According to the focused selection model, the direct pathway achieves action selection using disinhibition, whereas the indirect pathway serves to suppress competing behaviors such as postural fixation. According to the central selection model, the BG serve to select different bids that share common effectors. Different goals or purposes must recruit specific final common pathways for satisfaction. These models have key limitations. Above all, they ignore the continuity of behavior, focusing exclusively on discrete measures at the event level. Consequently, they lack a satisfactory definition of action, and fail to account for the relationship between continuous behavioral variables and BG activity or the key symptoms following damage to the BG.

REFERENCES

Albin, R. L., Young, A. B., & Penney, J. B. (1989). The functional anatomy of basal ganglia disorders. *Trends in Neurosciences*, *12*(10), 366–375.

Alexander, G. E., DeLong, M. R., & Strick, P. L. (1986). Parallel organization of functionally segregated circuits linking basal ganglia and cortex. *Annual Review of Neuroscience*, *9*, 357–381.

Alloway, K. D., Mutic, J. J., & Hoover, J. E. (1998). Divergent corticostriatal projections from a single cortical column in the somatosensory cortex of rats. *Brain Research*, *785*(2), 341–346.

Aoki, S., Smith, J. B., Li, H., Yan, X., Igarashi, M., Coulon, P., . . . Jin, X. (2019). An open cortico-basal ganglia loop allows limbic control over motor output via the nigrothalamic pathway. *Elife*, *8*, e49995.

Balleine, B. W., & O'Doherty, J. P. (2010). Human and rodent homologies in action control: Corticostriatal determinants of goal-directed and habitual action. *Neuropsychopharmacology*, *35*(1), 48–69. doi: 10.1038/npp.2009.131.

Barto, A. G. (1995). Adaptive critics and the basal ganglia. In Houk, J. C., Davis, J., & Beiser, D. (Eds.), *Models of Information Processing in the Basal Ganglia* (pp. 215–232). Cambridge, MA: MIT Press.

Benabid, A.-L., Pollak, P., Louveau, A., Henry, S., & De Rougemont, J. (1987). Combined (thalamotomy and stimulation) stereotactic surgery of the vim thalamic nucleus for bilateral parkinson disease. *Stereotactic and Functional Neurosurgery, 50*(1–6), 344–346.

Bergman, H., Wichmann, T., & DeLong, M. R. (1990). Reversal of experimental parkinsonism by lesions of the subthalamic nucleus. *Science, 249*(4975), 1436–1438.

Bolam, J. P., Smith, Y., Ingham, C. A., von Krosigk, M., & Smith, A. D. (1993). Convergence of synaptic terminals from the striatum and the globus pallidus onto single neurones in the substantia nigra and the entopeduncular nucleus. *Progress in Brain Research, 99*, 73–88.

Brown, L. L., Smith, D. M., & Goldbloom, L. M. (1998). Organizing principles of cortical integration in the rat neostriatum: Corticostriate map of the body surface is an ordered lattice of curved laminae and radial points. *Journal of Comparative Neurology, 392*(4), 468–488.

Bucy, P. C. (1942). The neural mechanisms of athetosis and tremor. *Journal of Neuropathology & Experimental Neurology, 1*(2), 224–239.

Calabresi, P., Pisani, A., Centonze, D., & Bernardi, G. (1997). Synaptic plasticity and physiological interactions between dopamine and glutamate in the striatum. *Neuroscience & Biobehavioral Reviews, 21*(4), 519–523.

Cepeda, C., Colwell, C. S., Itri, J. N., Chandler, S. H., & Levine, M. S. (1998). Dopaminergic modulation of nmda-induced whole cell currents in neostriatal neurons in slices: Contribution of calcium conductances. *Journal of Neurophysiology, 79*(1), 82–94.

Chevalier, G., & Deniau, J. M. (1990). Disinhibition as a basic process in the expression of striatal functions. *Trends in Neurosciences, 13*(7), 277–280.

Chevalier, G., Deniau, J. M., Thierry, A. M., & Feger, J. (1981). The nigro-tectal pathway. An electrophysiological reinvestigation in the rat. *Brain Research, 213*(2), 253–263.

Chevalier, G., Vacher, S., Deniau, J. M., & Desban, M. (1985). Disinhibition as a basic process in the expression of striatal functions. I. The striato-nigral influence on tecto-spinal/tecto- diencephalic neurons. *Brain Research, 334*(2), 215–226.

Collins, A. G., & Frank, M. J. (2014). Opponent actor learning (opal): Modeling interactive effects of striatal dopamine on reinforcement learning and choice incentive. *Psychological Review, 121*(3), 337.

Cools, A. R. (1985). Brain and behavior: Hierarchy of feedback systems and control of input. In: Bateson, P., Klopfer, P. (Eds.), *Perspectives in Ethology* (pp. 109–168). New York: Springer.

Cromwell, H. C., & Berridge, K. C. (1996). Implementation of action sequences by a neostriatal site: A lesion mapping study of grooming syntax. *Journal of Neuroscience, 16*(10), 3444–3458.

De Olmos, J. S., & Heimer, L. (1999). The concepts of the ventral striatopallidal system and extended amygdala. *Annals of the New York Academy of Sciences, 877*(1), 1–32.

DeLong, M. R. (1990). Primate models of movement disorders of basal ganglia origin. *Trends in Neurosciences, 13*(7), 281–285.

Deniau, J. M., Chevalier, G., & Feger, J. (1978). Electrophysiological study of the nigro-tectal pathway in the rat. *Neuroscience Letters, 10*(3), 215–220. doi: 10.1016/0304-3940(78)90228-8.

Deniau, J. M., Feger, J., & Le Guyader, C. (1976). Striatal evoked inhibition of identified nigro-thalamic neurons. *Brain Research, 104*(1), 152–156. doi: 10.1016/0006-8993(76)90656-9.

Denny-Brown, D., & Yanigisawa, N. (1976). The role of the basal ganglia in the initiation of movement. In Yahr, M. D. (Ed.), *The Basal Ganglia* (pp. 115–149). New York: Raven Press.

Desmurget, M., & Turner, R. S. (2008). Testing basal ganglia motor functions through reversible inactivations in the posterior internal globus pallidus. *Journal of Neurophysiology, 99*(3), 1057–1076.

Divac, I., Rosvold, H. E., & Szwarcbart, M. K. (1967). Behavioral effects of selective ablation of the caudate nucleus. *Journal of Comparative and Physiological Psychology, 63*(2), 184–190.

Dubé, L., Smith, A. D., & Bolam, J. P. (1988). Identification of synaptic terminals of thalamic or cortical origin in contact with distinct medium-size spiny neurons in the rat neostriatum. *Journal of Comparative Neurology, 267*(4), 455–471.

Ferrier, D. (1876). *The Functions of the Brain*: New York: GP Putnam's Sons.

Flaherty, A. W., & Graybiel, A. M. (1991). Corticostriatal transformations in the primate somatosensory system. Projections from physiologically mapped body-part representations. *Journal of Neurophysiology, 66*(4), 1249–1263.

Flaherty, A. W., & Graybiel, A. M. (1993). Two input systems for body representations in the primate striatal matrix: Experimental evidence in the squirrel monkey. *Journal of Neuroscience, 13*(3), 1120–1137.

Floran, B., Aceves, J., Sierra, A., & Martinez-Fong, D. (1990). Activation of d1 dopamine receptors stimulates the release of gaba in the basal ganglia of the rat. *Neuroscience Letters, 116*(1–2), 136–140.

Forman, D., & Ward, J. W. (1957). Responses to electrical stimulation of caudate nucleus in cats in chronic experiments. *Journal of Neurophysiology 20*, 230–244.

Foster, N. N., Barry, J., Korobkova, L., Garcia, L., Gao, L., Becerra, M., . . . Dong, H. W. (2021). The mouse cortico-basal ganglia-thalamic network. *Nature, 598*(7879), 188–194. doi: 10.1038/s41586-021-03993-3.

Gerfen, C. R. (1992). The neostriatal mosaic: Multiple levels of compartmental organization in the basal ganglia. *Annual Review of Neuroscience, 15*, 285–320.

Gerfen, C. R., Engber, T. M., Mahan, L. C., Susel, Z., Chase, T. N., Monsma, F. J., Jr., & Sibley, D. R. (1990). D1 and d2 dopamine receptor-regulated gene expression of striatonigral and striatopallidal neurons. *Science, 250*(4986), 1429–1432.

Gernert, M., Bennay, M., Fedrowitz, M., Rehders, J. H., & Richter, A. (2002). Altered discharge pattern of basal ganglia output neurons in an animal model of idiopathic dystonia. *Journal of Neuroscience, 22*(16), 7244–7253.

Goldman, P. S., & Nauta, W. J. (1977). An intricately patterned prefronto-caudate projection in the rhesus monkey. *Journal of Comparative Neurology, 72*(3), 369–386.

Goldman-Rakic, P. S. (2011). Circuitry of primate prefrontal cortex and regulation of behavior by representational memory. In: Mountcastcle, V. (Ed.), *Handbook of Physiology, the Nervous System* (pp. 373–417). Bethesda, MD: American Physiological Society.

Haber, S. N. (2003). The primate basal ganglia: Parallel and integrative networks. *Journal of Chemical Neuroanatomy, 26*(4), 317–330.

Haber, S. N., Fudge, J. L., & McFarland, N. R. (2000). Striatonigrostriatal pathways in primates form an ascending spiral from the shell to the dorsolateral striatum. *Journal of Neuroscience, 20*(6), 2369–2382.

Hazrati, L. N., & Parent, A. (1993). Striatal and subthalamic afferents to the primate pallidum: Interactions between two opposite chemospecific neuronal systems. *Progress in Brain Research, 99*, 89–104.

Hebb, D. O. (1949). *The Organization of Behavior: A Neuropsychological Theory*: Hoboken, New Jersey: J. Wiley.

Hikosaka, O., Takikawa, Y., & Kawagoe, R. (2000). Role of the basal ganglia in the control of purposive saccadic eye movements. *Physiological Reviews, 80*(3), 953–978.

Hikosaka, O., & Wurtz, R. H. (1983a). Visual and oculomotor functions of monkey substantia nigra pars reticulata. I. Relation of visual and auditory responses to saccades. *Journal of Neurophysiology, 49*(5), 1230–1253.

Hikosaka, O., & Wurtz, R. H. (1983b). Visual and oculomotor functions of monkey substantia nigra pars reticulata. Ii. Visual responses related to fixation of gaze. *Journal of Neurophysiology, 49*(5), 1254–1267.

Hikosaka, O., & Wurtz, R. H. (1983c). Visual and oculomotor functions of monkey substantia nigra pars reticulata. Iv. Relation of substantia nigra to superior colliculus. *Journal of Neurophysiology, 49*(5), 1285–1301.

Hikosaka, O., & Wurtz, R. H. (1985a). Modification of saccadic eye movements by gaba-related substances. I. Effect of muscimol and bicuculline in monkey superior colliculus. *Journal of Neurophysiology, 53*(1), 266–291.

Hikosaka, O., & Wurtz, R. H. (1985b). Modification of saccadic eye movements by gaba-related substances. Ii. Effects of muscimol in monkey substantia nigra pars reticulata. *Journal of Neurophysiology, 53*(1), 292–308.

Hintiryan, H., Foster, N. N., Bowman, I., Bay, M., Song, M. Y., Gou, L., . . . Dong, H. W. (2016). The mouse cortico-striatal projectome. *Nature Neuroscience*. doi: 10.1038/nn.4332.

Hooks, B. M., Papale, A. E., Paletzki, R. F., Feroze, M. W., Eastwood, B. S., Couey, J. J., . . . Gerfen, C. R. (2018). Topographic precision in sensory and motor corticostriatal projections varies across cell type and cortical area. *Nature Communication, 9*(1), 3549. doi: 10.1038/s41467-018-05780-7.

Hoover, J. E., & Strick, P. L. (1993). Multiple output channels in the basal ganglia. *Science, 259*(5096), 819–821.

Houk, J. (1995). Information processing in modular circuits linking basal ganglia. In: Houk, J., Davis, J. L, Beiser, D. G. (Eds.), *Models of Information Processing in the Basal Ganglia* (pp. 3–9). Cambridge, MA: MIT Press.

Hunnicutt, B. J., Jongbloets, B. C., Birdsong, W. T., Gertz, K. J., Zhong, H., & Mao, T. (2016). A comprehensive excitatory input map of the striatum reveals novel functional organization. *Elife, 5*. doi: 10.7554/eLife.19103.

Joel, D., & Weiner, I. (1994). The organization of the basal ganglia-thalamocortical circuits: Open interconnected rather than closed segregated. *Neuroscience, 63*(2), 363–379.

Joel, D., & Weiner, I. (2000). The connections of the dopaminergic system with the striatum in rats and primates: An analysis with respect to the functional and compartmental organization of the striatum. *Neuroscience, 96*(3), 451–474.

Kemp, J. M., & Powell, T. P. (1970). The cortico-striate projection in the monkey. *Brain, 93*(3), 525–546.

Kerr, J. N., & Wickens, J. R. (2001). Dopamine d-1/d-5 receptor activation is required for long-term potentiation in the rat neostriatum in vitro. *Journal of Neurophysiology, 85*(1), 117–124.

Kimble, G. A. (1961). *Hilgard and Marquis' Conditioning and Learning* (2nd edition): New York: Appleton-Century-Crofts.

Kocsis, J., Sugimori, M., & Kitai, S. (1977). Convergence of excitatory synaptic inputs to caudate spiny neurons. *Brain Research*, *124*(3), 403–413.

Kravitz, A. V., Freeze, B. S., Parker, P. R., Kay, K., Thwin, M. T., Deisseroth, K., & Kreitzer, A. C. (2010). Regulation of parkinsonian motor behaviours by optogenetic control of basal ganglia circuitry. *Nature*, *466*(7306), 622–626. doi: 10.1038/nature09159.

Künzle, H. (1978). An autoradiographic analysis of the efferent connections from premotor and adjacent prefrontal regions (areas 6 and 9) in macaca fascicularis; pp. 210–234. *Brain, Behavior and Evolution*, *15*(3), 210–234.

Lahiri, A. K., & Bevan, M. D. (2020). Dopaminergic transmission rapidly and persistently enhances excitability of d1 receptor-expressing striatal projection neurons. *Neuron*, *106*(2), 277–290.

Laitinen, L. V. (1995). Pallidotomy for parkinson's disease. *Neurosurgery Clinics of North America*, *6*(1), 105–112.

Laursen, A. M. (1962). Movements Evoked from the Region of the Caudate Nucleus in Cats. *Acta Physiologica Scandinavica*, *54*, 175–184.

Levy, R., Friedman, H. R., Davachi, L., & Goldman-Rakic, P. S. (1997). Differential activation of the caudate nucleus in primates performing spatial and nonspatial working memory tasks. *Journal of Neuroscience*, *17*(10), 3870–3882.

Lorente de Nó, R. (1934). Studies on the structure of the cerebral cortex. Ii. Continuation of the study of the ammonic system. *Journal für Psychologie und Neurologie*.

Lorenz, K. (1973). Autobiography. In: Odelberg, W. (Ed.), *Les Prix Nobel*. Stockholm: Nobel Foundation.

Mallet, N., Ballion, B., Le Moine, C., & Gonon, F. (2006). Cortical inputs and gaba interneurons imbalance projection neurons in the striatum of parkinsonian rats. *Journal of Neuroscience*, *26*(14), 3875–3884.

Martin, J. P., & Hurwitz, L. (1962). Locomotion and the basal ganglia. *Brain*, *85*(2), 261–276.

McElvain, L. E., Chen, Y., Moore, J. D., Brigidi, G. S., Bloodgood, B. L., Lim, B. K., . . . Kleinfeld, D. (2021). Specific populations of basal ganglia output neurons target distinct brain stem areas while collateralizing throughout the diencephalon. *Neuron*, *109*, 1–18.

McFarland, N. R., & Haber, S. N. (2000). Convergent inputs from thalamic motor nuclei and frontal cortical areas to the dorsal striatum in the primate. *Journal of Neuroscience*, *20*(10), 3798–3813.

McFarland, N. R., & Haber, S. N. (2001). Organization of thalamostriatal terminals from the ventral motor nuclei in the macaque. *Journal of Comparative Neurology*, *429*(2), 321–336.

McHaffie, J. G., Stanford, T. R., Stein, B. E., Coizet, V., & Redgrave, P. (2005). Subcortical loops through the basal ganglia. *Trends in Neurosciences*, *28*(8), 401–407. doi: 10.1016/j.tins.2005.06.006.

Middleton, F. A., & Strick, P. L. (2000). Basal ganglia and cerebellar loops: Motor and cognitive circuits. *Brain Research*, *31*(2–3), 236–250.

Miller, R. (1981). *Meaning and Purpose in the Intact Brain*: New York: Oxford University Press.

Mink, J. W. (1996). The basal ganglia: Focused selection and inhibition of competing motor programs. *Progress in Neurobiology*, *50*(4), 381–425.

Mink, J. W., & Thach, W. T. (1993). Basal ganglia intrinsic circuits and their role in behavior. *Current Opinion in Neurobiology*, *3*(6), 950–957.

Mishkin, M., Malamut, B., & Bachevalier, J. (1984). Memories and habits: Two neural systems. In Lynch, G., McGaugh, J. L., & Weinberger, N. (Eds.), *Neurobiology of Learning and Memory* (pp. 65–77). New York: Guilford Press.

Nakanishi, H., Kita, H., & Kitai, S. (1987). Intracellular study of rat substantia nigra pars reticulata neurons in an in vitro slice preparation: Electrical membrane properties and response characteristics to subthalamic stimulation. *Brain Research*, *437*(1), 45–55.

Nambu, A. (2008). Seven problems on the basal ganglia. *Current Opinion in Neurobiology*, *18*(6), 595–604.

Parent, A., & Hazrati, L. N. (1995a). Functional anatomy of the basal ganglia. I. The cortico-basal ganglia-thalamo-cortical loop. *Brain Research Reviews*, *20*(1), 91–127. doi: 10.1016/0165-0173(94)00007-C.

Parent, A., & Hazrati, L. N. (1995b). Functional anatomy of the basal ganglia. Ii. The place of subthalamic nucleus and external pallidum in basal ganglia circuitry. *Brain Research Reviews*, *20*(1), 128–154. doi: 10.1016/0165-0173(94)00008-D.

Parthasarathy, H., Schall, J., & Graybiel, A. M. (1992). Distributed but convergent ordering of corticostriatal projections: Analysis of the frontal eye field and the supplementary eye field in the macaque monkey. *Journal of Neuroscience*, *12*(11), 4468–4488.

Pawlak, V., & Kerr, J. N. (2008). Dopamine receptor activation is required for corticostriatal spike-timing-dependent plasticity. *Journal of Neuroscience*, *28*(10), 2435–2446.

Percheron, G., & Filion, M. (1991). Parallel processing in the basal ganglia: Up to a point. *Trends in Neurosciences*, *14*(2), 55–56.

Percheron, G., Yelnik, J., & François, C. (1984). A golgi analysis of the primate globus pallidus. Iii. Spatial organization of the striato-pallidal complex. *Journal of Comparative Neurology, 227*(2), 214–227.

Prasad, J. A., Carroll, B. J., & Sherman, S. M. (2020). Layer 5 corticofugal projections from diverse cortical areas: Variations on a pattern of thalamic and extrathalamic targets. *Journal of Neuroscience, 40*(30), 5785–5796.

Redgrave, P., Prescott, T. J., & Gurney, K. (1999). The basal ganglia: A vertebrate solution to the selection problem? *Neuroscience, 89*(4), 1009–1023.

Rescorla, R. A. (1980). *Pavlovian Second-Order Conditioning: Studies in Associative Learning*. New York : Psychology Press.

Rosenblatt, F. (1958). The perceptron: A probabilistic model for information storage and organization in the brain. *Psychological Review, 65*(6), 386.

Royce, G. J. (1983). Single thalamic neurons which project to both the rostral cortex and caudate nucleus studied with the fluorescent double labeling method. *Experimental Neurology, 79*(3), 773–784.

Saint-Cyr, J., Taylor, A. E., & Lang, A. (1988). Procedural learning and neostriatal dysfunction in man. *Brain, 111*(4), 941–960.

Salamone, J. D., & Correa, M. (2012). The mysterious motivational functions of mesolimbic dopamine. *Neuron, 76*(3), 470–485.

Schell, G. R., & Strick, P. L. (1984). The origin of thalamic inputs to the arcuate premotor and supplementary motor areas. *Journal of Neuroscience, 4*(2), 539–560.

Schultz, W. (1998). Predictive reward signal of dopamine neurons. *Journal of Neurophysiology, 80*(1), 1–27.

Schwab, B. C., Kase, D., Zimnik, A., Rosenbaum, R., Codianni, M. G., Rubin, J. E., & Turner, R. S. (2020). Neural activity during a simple reaching task in macaques is counter to gating and rebound in basal ganglia–thalamic communication. *PLoS Biology, 18*(10), e3000829.

Sutton, R. S., & Barto, A. G. (2018). *Reinforcement Learning: An Introduction*: Cambridge, MA: MIT Press.

Swanson, L. W. (2000). Cerebral hemisphere regulation of motivated behavior. *Brain Research, 886*(1–2), 113–164.

Tai, L.-H., Lee, A. M., Benavidez, N., Bonci, A., & Wilbrecht, L. (2012). Transient stimulation of distinct subpopulations of striatal neurons mimics changes in action value. *Nature Neuroscience, 15*(9), 1281.

Takakusaki, K., Saitoh, K., Harada, H., & Kashiwayanagi, M. (2004). Role of basal ganglia–brainstem pathways in the control of motor behaviors. *Neuroscience Research, 50*(2), 137–151.

Tunstall, M. J., Oorschot, D. E., Kean, A., & Wickens, J. R. (2002). Inhibitory interactions between spiny projection neurons in the rat striatum. *Journal of Neurophysiology, 88*(3), 1263–1269.

Turner, R. S., & Desmurget, M. (2010). Basal ganglia contributions to motor control: A vigorous tutor. *Current Opinion in Neurobiology, 20*(6), 704–716.

Ueki, A., & Yoshida, M. (1976). Some physiological aspects of the basal ganglia. *Stereotactic and Functional Neurosurgery, 39*(3–4), 296–301.

Vila, M., Levy, R., Herrero, M.-T., Ruberg, M., Faucheux, B., Obeso, J. A., . . . Hirsch, E. C. (1997). Consequences of nigrostriatal denervation on the functioning of the basal ganglia in human and nonhuman primates: An in situ hybridization study of cytochrome oxidase subunit i mrna. *Journal of Neuroscience, 17*(2), 765–773.

Wichmann, T., Bergman, H., Starr, P. A., Subramanian, T., Watts, R. L., & DeLong, M. R. (1999). Comparison of mptp-induced changes in spontaneous neuronal discharge in the internal pallidal segment and in the substantia nigra pars reticulata in primates. *Experimental Brain Research, 125*(4), 397–409.

Yeterian, E. H., & Van Hoesen, G. W. (1978). Cortico-striate projections in the rhesus monkey: The organization of certain cortico-caudate connections. *Brain Research, 139*(1), 43–63. doi: 10.1016/0006-8993(78)90059-8.

Yin, H. H. (2016). The role of opponent basal ganglia outputs in behavior. *Future Neurology, 11*(2), 149–169.

Yin, H. H. (2017). The basal ganglia in action. *Neuroscientist, 23*(3), 299–313. doi: 10.1177/1073858416654115.

Yin, H. H. (2020). The crisis in neuroscience. In: Mansell, W. (Ed.), *The Interdisciplinary Handbook of Perceptual Control Theory* (pp. 23–48). New York: Elsevier.

Yin, H. H., & Knowlton, B. J. (2006). The role of the basal ganglia in habit formation. *Nature Review of Neuroscience, 7*(6), 464–476.

Yoshida, M., & Precht, W. (1971). Monosynaptic inhibition of neurons of the substantia nigra by caudato-nigral fibers. *Brain Research, 32*(1), 225–228. doi: 10.1016/0006-8993(71)90170-3.

Yu, C., Jiang, T. T., Shoemaker, C. T., Fan, D., Rossi, M. A., & Yin, H. H. (2022). Striatal mechanisms of turning behaviour following unilateral dopamine depletion in mice. *European Journal of Neuroscience*. doi: 10.1111/ejn.15764.

Zahm, D., & Brog, J. (1992). On the significance of subterritories in the "accumbens" part of the rat ventral striatum. *Neuroscience, 50*(4), 751–767.

Zhang, J., Hughes, R. N., Kim, N., Fallon, I. P., Bakhurin, K., Kim, J., . . . Yin, H. H. (2022). A one-photon endoscope for simultaneous patterned optogenetic stimulation and calcium imaging in freely behaving mice. *Nature Biomedical Engineering*, 7, 499–510.

Zheng, T., & Wilson, C. J. (2002). Corticostriatal combinatorics: The implications of corticostriatal axonal arborizations. *Journal of Neurophysiology*, *87*(2), 1007–1017.

5 Behavior and Control

A key limitation in the previous attempts to explain basal ganglia (BG) function is their inadequate description of behavior. They implicitly assume that the brain receives sensory inputs and converts them into motor outputs. Here, we shall examine the limitations of this linear causation paradigm, and introduce an alternative paradigm based on the principles of hierarchical control. Although this chapter does not discuss the BG directly, it addresses the basic problem of how behavior is generated and provides a conceptual framework within which BG function can be understood. Later chapters will develop a model of BG function, the transition control model, based on this framework.

5.1 INSUFFICIENCY PRINCIPLE

To appreciate the basic challenge of understanding behavior, it is useful to start with Wittgenstein's question: "What is left over if I subtract the fact that my arm goes up from the fact that I raise my arm? §621" (Wittgenstein, 1953). A conventional answer would be as follows: when the arm is raised passively, some external force moves the arm, but when a man raises his own arm volitionally, his own brain and muscles are responsible for moving the arm. But this answer is misleading, as there is always some external force acting on the arm, whether or not it is raised volitionally. The force generated by the muscles is not the only force involved. Other forces, less apparent but always present, also act on the arm at the same time. For example, if one raises the hand and holds it there, gravity is pulling the arm downwards. In order to counter the effect of gravity, muscle output must be produced to generate just the right amount of torque so that the arm stays still, yet this has to be done without sensing gravity directly. Muscle output is the only way in which neural activity can be 'actuated' to affect the external environment. It is roughly proportional to the output from alpha motor neurons, i.e. the final output of the nervous system known as the 'final common path' (Sherrington, 1906).

According to the traditional view, behavior is simply the output of the nervous system. This view was challenged by Bernstein, who wrote (p. 19): "The muscles were considered completely passive and driven by central impulses. However, the same impulse may produce completely different effects because of the interplay of external forces and because of variations in the initial conditions. A determinate effect is possible for a movement only in a case where the central impulse E is very different under different conditions, being a function of the positions and the velocities of the limbs and operating very differently in the differential equation with various initial conditions" (Bernstein, 1935). Muscle output alone is insufficient to specify what type of behavior is being generated. If output commands to the muscles are recorded and played back, the consequent behavior would not be the same for each playback. Given the variable and unpredictable environmental forces that act on the body, to produce consistent movements, the final common path output must *vary* every time. That repeating the same neural output at the final common path is not sufficient to repeat the same behavior may be called the insufficiency principle (Yin, 2014; Yin, 2017).

To hold the arm still, the relevant motor neurons must increase their activity to generate the torque needed to balance environmental forces like gravity pulling the arm downward. Even when the arm is not moving, neural output from the final common path can vary significantly. After a while, as the muscles start to show fatigue, more neural output from the final common path is needed to maintain the same position. Pre-programmed variations in muscle output will fail to keep the arm still because the forces that act on the arm, as well as the properties of the arm muscles, can change unpredictably. This problem becomes more acute when one moves the arm: the arm accelerates toward the final position and then starts decelerating before it gets there, exhibiting a distinct velocity profile. For the arm to reach the target consistently, the neural output from the final common path must also vary every time. While muscle contraction is necessary for movement, repeating the same behavior actually requires *variable* neural outputs from the final common path.

5.2 SOLUTIONS TO THE CALCULATION PROBLEM

Bernstein's insight poses a challenge to conventional theories of the brain. If the final product of the neural signaling to the muscles does not guarantee a particular behavior, how does the brain determine how much output to produce without knowing how much disturbance there is? This is the "calculation problem" (Yin, 2013).

In motor control and robotics, the most common proposed solutions to the calculation problem are known as the inverse and forward models. Such models basically attempt to pre-program behavior in an open-loop fashion. The inverse model attempts to determine "the input to some system we wish to control which will achieve some desired output" (Wolpert & Kawato, 1998). The aim is to compute the appropriate motor command given the goal state and the environmental context. In the example of a reaching movement, the arm is the plant, the input is the control signal reaching the arm muscles, and the desired output is a specific arm movement. The environmental context would include the geometry and inertia of the object, the center of mass of the arm, the orientation of the body relative to gravity, and so on. Using all such parameters, the inverse approach aims to adjust the input–output relationship of the system to issue the right command, or so-called "control signals," to the final common path. On the other hand, the forward model predicts the consequences of actions. It is based on the assumption that actual sensory feedback is inadequate due to delays in neural signal conduction (transport lags) and that it would be necessary to use internal predictions of feedback instead of actual feedback.

Transport lags are inevitable in any controller. The effects of delays resemble the effects of lowering the sampling frequency. In both cases, the controller is forced to make use of old data in determining the output. Tuning the system parameters as well as some form of transient memory may be necessary for handling feedback delays, but the presence of such delays does not necessarily require predicting sensory input or building a detailed model of the environment.

The brain is often assumed to perform both inverse and forward calculations and then use the results to send signals to muscles. For example, given the desired position of the arm and the starting position, the brain may compute the trajectory needed and how much torque must be generated. To do so, one needs a set of equations to describe the causal relationship (Wolpert & Kawato, 1998). In the forward approach, to predict the effect of behavioral outputs, it would be necessary to take into account all properties of the nervous system and muscles, the exact functions that convert neural output into muscle tension, the time-varying relationship between muscle tension and load, and the dynamics of the interactions between the body and the environment in real time for any movement.

The modern engineering approach views the calculation problem as a problem in mechanics to be solved by some ideal observer. Using powerful computers, engineers with knowledge of mechanics can compute forward and inverse solutions, as long as the system is simple and the environment is free of unpredictable disturbances. But in any realistic environment, the forward/inverse modeling approach is overwhelmed by computational complexity.

This approach is based on a false assumption and a false inference. The assumption is that the system is using the same strategy as the external observer. The inference is that the products of this computation will then be the signals needed to generate behavior. The key problem is posed from the perspective of the observer, not from the perspective of the actor. Just because an engineer, equipped with powerful computers and knowledge of physics, can perform matrix calculations to obtain the inverse models, does not mean that the brain must perform similar computations in real time. The mechanics problem posed does not exist for the actor, who is merely following physical laws, not computing them on the fly. Unfortunately, this false assumption is also common in contemporary motor neuroscience.

5.3 THE DEFINITION OF CONTROL

The above discussion raises an obvious question: if consistent behavior cannot be achieved by pre-programmed outputs based on inverse and forward models, how does the organism solve the calculation problem? The brain of the organism does not solve the calculation problem in the same way as the observer's brain witnessing the same action. It solves this problem using a far simpler approach, namely negative feedback control. Because this solution is frequently misunderstood and often dismissed, we now turn to a closer examination of control theory.

Control is defined as the process by which a variable is reached or maintained at a certain value or a range of values (Powers, 1973a). To illustrate the phenomenon of control, let us consider the behavior of holding one's arm up in front of the body. This is an example of position control. The body produces systematic resistance to environmental disturbances, mostly gravity, and the force exerted exactly cancels the downward pull. If we push down on the arm, we encounter proportional resistance, which reflects the control effort of a position controller. The control system generates an action that specifically opposes the effect of the disturbance on the controlled variable. What is counterintuitive is that the disturbance is not perceived directly, and the input variable, being controlled, does not change much, precisely because the controller output minimizes the effect of the disturbance.

Contrary to the conventional view, what is controlled is not behavioral output but perceptual inputs. The control system can measure the states of the environment with its sensors, compare them with reference states, and generate outputs to control the value of some input variable. The reference signal is an internal representation of the desired state. It requests a specific input by generating variable outputs. At the highest levels, reference signals correspond to what are commonly called goals or purposes.

For the control system to function, it must contain internal representations of desired states, the appropriate sensors for detecting physical changes in the environment, and the circuitry for comparing the two and for converting the difference into some output, which through the environment affects the input (Figure 5.1).

There are three major components in any negative feedback control system: an input function, a comparison function, and an output function. The input function converts perceptual inputs into some signal that reaches the comparator. The internal reference signals dictate how much signal the sensor should acquire. The feedback is negative, which does not mean that it is aversive or inhibitory, but only that when the loop is closed, the error continuously reduces itself. The comparator computes an 'error' signal by taking the difference between the input and the reference signal. This error signal is then converted to an output by the output function. For example, with a multiplicative gain, the error is multiplied by a constant to generate a larger signal. However, because error is self-reducing, this does not mean that the output is simply larger. Rather, it means that the error is more quickly reduced when the gain is high.

The reference signal is not a cause in the traditional sense. The reason for action, in this sense, is not the cause for action, because cause–effect explanation fails to explain the closed-loop mechanism. By itself, the reference cannot determine the course of action. The actual action will always be jointly determined by ongoing feedback and references. Nor can perceptual input determine action. Output can be produced in one of two ways: either a sensory input alters the present perception and moves it beyond the range specified by the reference state, thus generating a disturbance, or by top–down adjustment of the reference signals, which also generates error from the comparator. The former scenario is found in reflexes, which reflect the actions of lower control systems, when their perceptual input deviates from the reference setting. The traditional account of reflexes as stimulus–response ignores the key variables being controlled and the control organization that can explain the behavior quantitatively.

Negative feedback control solves the calculation problem by opposing the disturbance with an output that 'mirrors' the disturbance. Its output is not a function of the input alone but of the

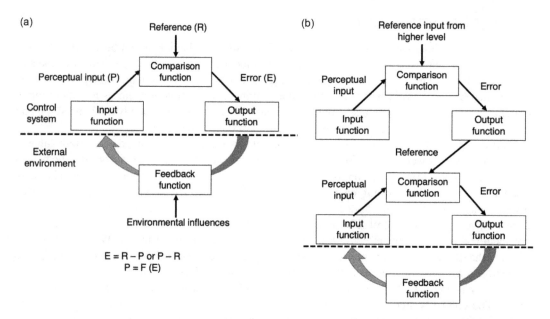

FIGURE 5.1 Negative feedback control. (a) A simple control system. The comparator produces a signal representing the discrepancy between the perceptual input and the reference signal. Environmental influences are possible sources of disturbance for the controlled variable. Note that the perceptual input signal is the controlled variable. The dotted line represents the boundary between the control system and the environment. Anything outside the system is considered the environment. (b) Hierarchical organization of control systems. Only two levels are shown here. The output of the higher level system converts the error signal to a reference signal for the lower-level system.

difference between the reference signal and the input. The disturbance is not merely a property of the environment but an emergent property of the organism interacting with the environment, generated by a dynamic comparison process between internal goals and sensory inputs.

5.4 COMPUTING IN A CONTROL SYSTEM

The comparator performs subtraction and generates the difference between perceptual input and reference input. To compute the difference, these quantities must be opposite in sign. In the simplest case, this operation can be implemented by an excitatory input and an inhibitory input to a neuron. The output will reflect the difference as the inhibitory signal is subtracted from the excitatory one.

The sign of the reference signal has implications for system behavior. If the perceptual input (p) is positive, then the reference input (r) must be negative. The error (e) is then $p - r$. Output is generated when perceptual input is greater than reference, effectively setting up a threshold. System output will reduce the perception to the reference level, exhibiting "avoidance." On the other hand, if p is negative, then r must be positive ($e = r - p$). In such a system, there is output (positive error) when perceptual input is less than reference. Such a system will not stop producing output unless the level of perceptual input matches the reference signal. Output is generated to increase perceptual input until it reaches the desired level, exhibiting "seeking" behavior.

In neural control systems, signals (firing rates) cannot be negative. This presents a challenge for bidirectional system control, in which the value of the controlled variable can be increased or decreased. A bidirectional control requires two one-way control systems, or one system with two distinct output functions with opposite effects on the value of the perceptual input variable. For example, in controlling temperature, the "too hot" error goes to the cooling element, whereas the "too cold" error is sent to the heating element. The activation of either cooling or heating element is not a function of current temperature, as the absolute temperature value does not determine the output. What determines output is the deviation from reference. As we shall see in later chapters, the use of opponent output functions for bidirectional control is common at all levels of the control hierarchy.

A drawback of feedback controllers is that they are prone to oscillations under certain conditions. They must be tuned to avoid oscillations. If the feedback is delayed such that the control effort will further increase the feedback, then positive feedback is created, generating oscillations. For example, turning the knob in the shower does not cause the hot water to come out instantaneously. If the delay is significant, the water feels cold during the interval before the arrival of hot water. If this "too cold" error keeps generating the turning action for hot water, the water will be scalding when it finally arrives. This generates the "too hot" error which activates the control system in the opposite direction to cool, and so on. The culprit is the delay between turning the knob and the temperature change. The internal control organization should be sufficiently slowed so that the action does not occur too quickly relative to the delay or the sampling frequency of the input function. To tune the system to avoid oscillations, the gain can be set so that the change in controller output during the delay does not generate a comparable or larger error in the opposite direction.

5.5 MISUNDERSTANDING CONTROL

Despite its simplicity, negative feedback control is widely misunderstood. Perhaps the main error in applying control theory to the study of living organisms is the misleading identification of system components in modern engineering. Control engineers typically adopt the third-person perspective of an external user, treating the control system as a slave ("servo"). Since a servo must obey the commands of the user, it is not supposed to have intrinsic purposes of its own.

By convention, what is called the input is the reference signal introduced by the user. The user introduces this input, which then brings some output to the desired value. For example, set the thermostat to 25°, and after some time it will reach 25°. To the user, the thermostat appears to be some input/output device that takes the reference input and produces the right temperature output. This view, while useful to the engineer, is misleading when applied to the study of living organisms. The so-called output is in fact the perceptual input (sensed temperature) to the system. Thus, in modern engineering convention, the controlled variable is often misleadingly labeled output, and the intrinsic reference is labeled input. The reference signals are assumed to be supplied by the user rather than the organism itself. These assumptions have led to an "inside-out" view of control with unfortunate consequences. By adopting the perspective of the user, modern engineering theory has deliberately abolished the autonomy of the controller (Figure 5.2). The objective is to inject the right reference to achieve the desired result. But what is actually controlled can only be some sensor reading of the relevant process variable that is of interest to the engineer who designed the system. It is only viewed as a control of output from the perspective of an external observer. As we shall see, this "objectivist fallacy" has led to major conceptual confusions in fields like optimal control and reinforcement learning.

The fallacy is to attribute to the system being observed properties that only reflect the capability of the observer. The stones reportedly dropped by Galileo from the Tower of Pisa do not compute the gravitational acceleration constant as they fall, even though their trajectory can be predicted by an observer using the equation. It is responsible for the common assumption, discussed earlier, that forward and inverse computations of kinematics and dynamics are needed to generate movements.

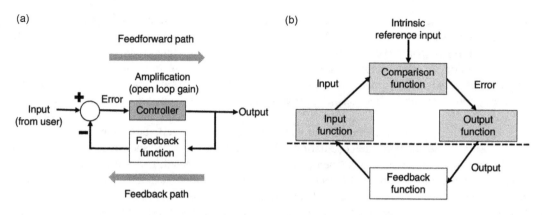

FIGURE 5.2 Misunderstanding control. (a) The negative feedback control system as commonly illustrated in engineering textbooks. (b) The control system as illustrated according to perceptual control theory. Compare (a) with (b). Shaded boxes are parts of the control system. Note that the intrinsic reference is labeled as 'input' in the conventional illustration, but the reference is internal to the organism not a signal injected by the user or engineer. What is injected is a representation of the desired state, and what is labeled output is actually the measured process variable, corresponding to the perceptual input to the system.

The failure of this approach is illustrated by traditional robotics. The more joints or degrees of freedom a robot has, the more difficult it is to control using the forward approach. An arm with 14 degrees of freedom can be modeled by a 14 by 14 Jacobean matrix. To compute the commands or 'control signals' needed to perform a simple reaching action would require performing complex matrix calculations in real time. Yet even with a very powerful computer and knowledge of mechanics, it is impossible to find the inverse of the environmental equation for a realistic environment, where the physical dynamics can only be described by matrices of nonlinear differential equations that cannot be solved analytically.

In contrast, control systems do not calculate the desired effect of an output and then produce the signals that will create that desired effect. Because forward calculations or predictions of future states are not needed within the system, negative feedback control does not suffer from the so-called curse of dimensionality. Additional degrees of freedom can be handled by using multiple systems that control orthogonal variables (Powers, 2008). Two independent dimensions, for example, can be controlled with a composite of two one-dimensional systems. Generalizing to many dimensions, an N-dimensional control system can be constructed by using N one-dimensional control systems.

A related misunderstanding is to equate control with physical equilibrium. In physical equilibrium, a dynamical system can reach a stable state, such as a pendulum restoring its original position or a raindrop falling down a windowpane. But the equilibrium state should not be confused with the state of a controlled variable, despite their superficial resemblance. A falling raindrop has no perception, reference, protection from disturbance, or negative feedback. It is not on the ground because its internal reference specifies it; it is there due to the constant action of gravity. Equilibrium draws on the energy supplied by the disturbance to generate the restoring force, whereas the energy used by a control system to resist disturbance is supplied by a separate source. Biological control systems acquire and store energy to replenish the independent "battery" that supplies this power. Normally, the energy responsible for the system output far exceeds the energy supplied by the input.

Misunderstanding of control systems is responsible for the myth that negative feedback systems are slow and reactive (Rack, 1981; Sterling, 2012). Indeed transport lags are common in the nervous system due to synaptic transmission, conduction of action potentials, etc. Since all physical systems have lags, the relevant question is not whether the system can tolerate lags but how much lag there is and what the consequences are for the quality of control. If the presence of significant lags does not allow high gains without instability, then the question is exactly how fast the system can act and what its natural behavior looks like with the permissible gain. Previous criticism of negative feedback is largely based on qualitative claims that biological control systems are too sluggish without attempting to model the behavior in question. In fact, when comparable components are used, a system with negative feedback is faster than one without negative feedback. In a stimulus–response system, to obtain a response proportional to a stimulus, it would be necessary to adjust the transfer function so that the output reaches some desired steady state value. In a closed-loop system, the gain amplifies the error signal, but such amplification does not generate a huge output because the output is limited by negative feedback. Without feedback, amplification would result in an enormous overshoot of the desired output value, but with negative feedback, as the output increases, so does the negative feedback that is canceling it. Instead of generating a huge output, the high gain allows the controlled variable to reach the reference value more quickly.

5.6 HIERARCHICAL CONTROL

An organism contains more than one control system. At any given moment, many controllers could be acting to control their inputs, though not all sensory inputs are being controlled. Many biological control systems are fully functional at birth. Homeostatic processes typically refer to control systems of this type in which the controlled variable, like body temperature, must stay in a relatively constant range. In addition, there are intermittent control systems that operate more transiently and sporadically, controlling variables that are learned or refined by experience (Craik, 1947).

A given controller occupies a particular level in a hierarchy (Powers, 1973b). Higher levels command lower levels by adjusting the reference signals of lower levels. The lower level compares its own perceptual input with top–down references. Descending reference signals do not directly alter lower-level output. Changing the output directly without adjusting the reference would not produce the desired effects. Doing so could affect lower-level perceptual inputs, which would deviate from the reference value. Consequently, the changes would be treated as a disturbance and resisted by the controller. The descending reference signal should not dictate how much output the lower levels should produce but how much input they should generate. The actual outputs produced would be jointly determined by the reference and input at each level, not preprogrammed at the higher levels.

For any behavior, there could be multiple descriptions. For example, flipping the light switch, alerting the prowler, and contracting the biceps while relaxing the triceps of the right arm are all descriptions of the same event (Davidson, 1963). They may even share common circuits in the lower levels of the control hierarchy, which explains their superficial similarity, but they are not generated in the same way if we ultimately trace it to the level where error originates first. Each description implies a unique reference state. This lead level is where the termination of the error completes the action.

The same action, as defined by a high-level purpose, can be executed in different ways. With respect to the higher levels, many actions are therefore equivalent. Such motor equivalence is a consequence of the organization of the control hierarchy. Different levels of the control hierarchy are associated with different definitions of the controlled variable and the possible means of achieving them. The higher the level, the more options are possible, though the higher levels can only access the options though learning. To reduce temperature control error from heat, walking into an air-conditioned room and taking off a sweater are equivalent from the perspective of the temperature controller. Eliminating one of the possible actions does not eliminate goal achievement because alternatives are available. But for an infant that has yet to learn many of the instrumental actions for achieving temperature control, the options are more limited.

5.7 BEYOND SENSORIMOTOR TRANSFORMATIONS

According to the traditional paradigm, the primary function of the nervous system is sensorimotor transformation. As Hull explained (p. 18): "Neural impulses set in motion by the action of these receptors pass along separate nerve fibers to the central ganglia of the nervous system, notably the brain. The brain, which acts as a kind of automatic switchboard, together with the remainder of the central nervous system, routes and distributes the impulses to individual muscles and glands in rather precisely graded amounts and sequences. When the neural impulse reaches an effector organ (muscle or gland) the organ ordinarily becomes active, the amount of activity usually varying with the magnitude of the impulse" (Hull, 1943).

There has long been a debate on whether the BG have sensory or motor functions (Brown, Schneider, & Lidsky, 1997). The striatum receives inputs from cortical areas that are traditionally labeled sensory, e.g. primary somatosensory cortex. Striatal neurons are responsive to sensory input in multiple modalities, e.g. sudden taps, bright flashes of light, loud clicks or handclaps, as well as cutaneous stimulation and passive manipulation (Carelli & West, 1991; Krauthamer, 1979). From the present perspective, however, the debate on whether any brain region is sensory or motor is futile because there are no sensorimotor transformations in control systems. Rather, inputs are compared with reference to produce outputs. Of course, anatomically, there are sensors and effectors, afferent pathways and efferent pathways, but signaling in these pathways is no longer sequential once the loop is closed. The traditional assumption ignores the feedback function that incorporates the effects of the behavioral output on the afferent stimulation. This feedback function closes the loop and renders linear causation untenable since the efference acts to adjust the afference at the same time as afference is affecting output. Sensory and motor are superficial descriptions of a larger and partly invisible process of control, which is characterized by circular, rather than linear, causation.

Analyzing control systems as input/output devices can actually lead to the wrong conclusion. Because the feedback path inverts the conventional understanding of a system, negative feedback can create a powerful illusion (Powers, 1973b). In the presence of control, the output is not related to the input but rather mirrors the pattern of environmental disturbances. The apparent function relating input to output is merely the inverse of the feedback function, a property of the environment rather than the organism.

5.8 REINFORCEMENT AND TELEOLOGY

Under the influence of the linear causation paradigm, purpose is either denied or considered a helpful metaphor at best. In the control system, reference signals are literally necessary for the system to work. There is no need for any homunculus, no need to violate the time arrow of causality.

The control system analysis demonstrates that behavior is teleological. This key point is often misunderstood, even by those who believe that actions are purposeful. For example, Fuster argues that teleology would invert the "temporal direction of causality. Of course, this inversion is not real in physical terms. It is only real in cognitive, thus neural terms, inasmuch as the presentation of the goals of future actions antecede and cause those actions to occur..." (Fuster, 2015). But he fails to grasp that, even when the goal precedes the action, it is not causal in the traditional sense. The reference command is not the cause because it alone does not determine the output. The output is jointly determined by reference and input.

The concept of reinforcement is an attempt to explain control phenomena while avoiding teleology. Hull explains his reasoning clearly: "teleology is the name of the belief that the terminal stage of certain environmental-organismic interaction cycles somehow is at the same time one of the antecedent determining conditions which bring the behavior cycle about. This approach... involves a kind of logical circularity: to deduce the outcome of any behavioral situation in the sense of the deductive predictions here under consideration, it is necessary to know all the relevant antecedent conditions, but these cannot be determined until the behavioral outcome has been deduced" (Hull, 1943, p. 26).

Behavior and Control

The standard reinforcement learning model is a description of the behavior but sheds no light on how that behavior is generated. Traditionally, reinforced behavior is often viewed as response substitution, as more responses are added to the same stimulus or state via synaptic plasticity. This idea has had a major influence on thinking about the BG. For example, according to Mishkin, stimulus input is processed by the striatum, the pallidum generates outputs, and the S–R association is strengthened by reinforcement (Mishkin, 1984). But what is reinforced or repeated is actually a consistent consequence, not a consistent output. The action will vary when a disturbance is imposed. If the normal path to the lever is blocked, a different trajectory can be produced. What is "reinforced" can only be high-level reference signals. As we shall see in later chapters, only a working control hierarchy can explain the phenomenon of reinforcement learning.

5.9 NEURAL SIGNALING AND CONTROL SYSTEMS

All control systems, biological or artificial, require signaling from one component to another. Specific circuit arrangements make it possible to change these signals, both in size and in timing (Grodins, 1963). How can neural transmission implement the signaling required in a control system?

There is much conceptual confusion on the topic of neural signaling. This confusion stems from original findings on the all-or-none nature of action potentials. Taking advantage of vacuum tube amplifiers for recording tiny electrical signals from neurons, Adrian and colleagues performed the first recordings of action potentials and showed that their waveforms are highly stereotyped (Adrian & Bronk, 1929; Adrian & Zotterman, 1926). They also found that the muscle spindle output from the frog toe is proportional to the amount of stretch. This finding suggests that neural signals, in the form of firing rate, carry an analog quantity, i.e. the stretch of a muscle fiber. This is an example of analog signaling.

Analog is often equated with continuous, whereas digital is considered discrete, but this distinction is superficial. What matters is the representational format of a signal (Maley, 2011). In analog signaling, the quantity is all that matters. For example, the amount of money can have discrete units like cents and dollars. If we wish to represent any arbitrary variable, say generosity, with some amount of money, more generosity must be represented with more money, even if each unit is discrete, say one cent or one dollar. On the other hand, digital signaling uses a different representational format altogether, requiring arbitrary symbolic representations to signal quantity.

An important assumption in this book is that the representational format in the nervous system is analog in nature, even though it uses discrete units in the form of action potentials. The magnitude of a neural signal is the firing rate, or the number of action potentials generated within some time window. This rate may have a monotonic relationship with some behavioral variable, though it is by no means straightforward to discover this relationship. As we shall see, the failure to discover such relationships using traditional experimental methods is perhaps the key reason for doubting the analog nature of neural representations.

Historically, the all-or-none feature of action potentials influenced the first designers of computers, who decided that a digital code was being used by the brain (Von Neumann, 1958). McCulloch and Pitts proposed that neurons perform logical operations (McCulloch & Pitts, 1943). In classical logic, any proposition can have two values: true or false. If we view each neuron as firing or not firing, neural activity can be viewed as a proposition, e.g. A or not A. With a collection of such 'truth values' regarding certain propositions with a digital code of 1 and 0. then a collection of three neurons will have eight possible configurations: 000, 001, 010, 011, 100, 101, 110, and 111. However, for a collection of truth values to encode a proposition, they must be precisely synchronized, and extracted by the observer, who has access to all these states. This scenario is biologically implausible.

The human observer, accustomed to symbolic manipulation learned at schools, uses categories to interpret the perceptual observations, but it does not follow that logical categories are related to the all-or-none property of individual neurons. A proposition about temperature might be: it is cold and the truth value could be either true (1) or false (0). For continuous control, however, our temperature sensors must convey more than hot and cold. Neural representation of temperature must use a roughly linear scale with firing rate as the analog of temperature.

The digital computer, being an input/output device rather than a control system, only changes its output when instructed to do so. Symbolic operations in digital computing involve a coding and decoding process. Quantities must first be converted into symbols (1 and 0), which are then manipulated according to rules that convert input to output. To add two signals, one must first convert each into a pattern of 1s and 0s, and then uses a circuit to follow a specific set of instructions for input/output conversion, e.g. $0+0=0$, $0+1=1+0=1$, and $1+1=1$ with a carry to the next place. For such a system to function well, great precision in the components is needed, but in biology such precision is neither necessary nor achievable.

In contrast, analog computations are not performed by manipulating symbols according to rules; an analog computer 'embodies' the computation. To find out the sum of 200 mL and 350 mL, we can pour 200 ml of water from one cup into another with 350 ml and measure the total. Likewise, a neuron receiving two signals may produce an output that is roughly the sum of the two inputs. Symbols or rules for symbol manipulation are not required in this operation. Signals can be transformed via physical effects implemented by the pulsed output of neurons and chemical transmission across synapses. There is no other message hidden in the patterns of firing, no encoding or decoding.

Firing rate is an analog signal that can vary along a single dimension. For example, the analog of a muscle length is a number, and so is the roundness of a ball, or resemblance to one's grandmother. That the firing rate signal is one dimensional is not a limitation. Multiple orthogonal dimensions are needed, with distinct neuronal populations each corresponding to a single dimension. Complex representations can be formed using vectors or matrices of values.

Analog circuits can generate output signals that are specific functions of the input signals and solve linear or nonlinear differential equations or matrices of equations without symbol manipulation using 0s and 1s (Figure 5.3). To solve differential equations, it is necessary to add and subtract variables, change algebraic signs, multiply by constants, differentiate, and integrate. For example, one of the major technological challenges during the Second World War was battleship ballistics. To shoot things with a gun from a battleship with any accuracy, it is necessary to determine how high to aim 20-m gun barrels to reach a moving target far away, shooting from a platform that moves with the ocean waves, using shells weighing as much as a car, while compensating for the wind and the Coriolis effect. All this is done without a digital computer. Instead of using discrete symbols, the computation is performed with gears, cams, racks, pins, and other mechanical parts, taking

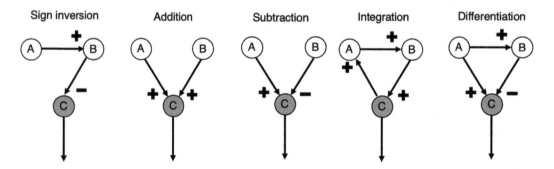

FIGURE 5.3 Analog computing. Illustrations of how common analog operations can be performed by neural circuits. Signal inversion can be achieved by the addition of an inhibitory interneuron. Addition and subtraction can be implemented by excitatory and inhibitory neurotransmitters. Integration can be implemented by processes like recurrent excitation. Finally, differentiation can be achieved when a neuron receives an excitatory input and an inverted version of the same input with a slight delay.

advantage of the "embodied" geometric and trigonometric relationships. For example, a potentiometer can multiply a specific variable by a constant, and a gear can perform addition, subtraction, multiplication, and division.

In electric circuits, voltages can be quite similar, and what determines their functional significance is their position within a circuit and their relationship with other components. By itself, a 5 V change somewhere in a circuit tells us little about what the circuit is doing or what type of device it is found. What endows the signal with significance or meaning are its source and destination, as well as the physical effects it can have within a network. To understand that, we must understand the circuit as a whole.

5.10 INFORMATION AND CODING

The relationship between neural activity and behavior has been interpreted in the framework of information coding. The correlation between firing rate and the amount of stretch became known as 'rate coding,' which is considered one of several neural coding strategies. The coding analogy has produced conceptual confusion about the relationship between neural activity and behavior.

"Coding" is sometimes used merely to mean that the neural activity is correlated with some variable. In this sense, it is harmless though hardly necessary. If statistical significance is the only criterion used, then most brain areas can show activity that is significantly correlated with most variables. It is commonplace to say that correlation does not imply causation, but what is usually neglected is the strength of correlation and the nature of the variables being correlated. In a control system with negative feedback, it is possible for two variables to be linked by causal mechanisms without being correlated at all (Kennaway, 2020). For example, there is no correlation between the input and output of a working thermostat. However, since the controlled input approximates the reference value of a particular level, analog representations of the controlled variables must be common in the nervous system. In other words, we would expect high correlations between the neural signals and either the controlled input variable or the reference signal. We would expect a monotonic relationship between the neural activity and the controlled variable. As we shall see in later chapters, this prediction is supported by recent findings on the BG.

Although there have been many attempts to find correlations between neural activity and behavioral variables, most have failed, especially in regions beyond the peripheral sensory receptors. In the brain, such attempts often produced ambiguous results, as neural activity was found to be weakly correlated with nearly every imaginable variable. Instead of concluding that there may be something wrong with how conventional experiments were conducted, many concluded instead that more sophisticated mathematics are needed to read the code hidden in neural activity. It is assumed that information contained in the neural activity is sent to another region, which then decodes the code. The target neuron is literally acting as a homunculus (Rieke, 1999). Using a decoding strategy, weak correlations can now be analyzed using some mathematical transformation of the signal. Information that can be extracted from the firing pattern is thought to be encoded by neurons. If some algorithm can convert this neural code into some objective variable, then it is assumed that the transformation needed for decoding is performed by downstream neurons in the same way as the observer (Brette, 2019; Granit, 1955; Perkel & Bullock, 1968). This approach resulted in a proliferation of possible neural codes, e.g. precise timing of individual spikes (Rieke, 1999), the oscillatory pattern of brain activity (Buzsaki, 2009), or Bayesian inference on probability distributions (Ma, Beck, Latham, & Pouget, 2006).

According to this paradigm, neurons transmit information to other neurons in much the same way that information is transmitted from the sender to the receiver via a telecommunications channel. When Shannon attempted to quantify the transmission of a signal from source to destination as information, he assumed that both sender and receiver knew the code and shared the same lookup table, a set of context-free symbols. The amount of information contained in the signal sets the upper bound on what a receiver could receive. One might quantify such information as a decrease in uncertainty or entropy (Gallistel & King, 2011). Information flow to the receiver reduces the entropy.

Information is not an intrinsic property of the system being observed. It is observer-dependent. Falling leaves could be viewed as carriers of information about the gravitational constant and air drag, but it would be absurd to say that they are encoding these features. According to the information coding analogy, the mind of the observer is simply collapsed into the neuron (Nizami, 2017, 2019). It is assumed that neurons are observers of other neurons and that information comes to each neuron in some pre-labeled form. But the observer's capacity to read any code is an emergent property of the brain, not the property of a single neuron. The information coding and decoding approach requires a homunculus, a little man inside with the same knowledge as the computational neuroscientist (Rieke, 1999).

The objectivist fallacy can give rise to misinterpretations of neural signals. Many aspects of neural activity are irrelevant for signal transmission, yet they are often treated as neural codes. For example, although subthreshold membrane potential is also an analog quantity correlated with the relevant behavioral variables does not mean it is the signal being transmitted, unless nonsynaptic transmission of signals can be demonstrated to play a key functional role. To the neurons downstream that receive projections from neuron A, excitatory postsynaptic potentials (EPSPs) in neuron A do not exist, though they can be observed by the electrophysiologist performing intracellular recording. Likewise, the main reason that frequency domain measures like spectral power are believed to be neural signals coding for various variables is that the observer can discern some relationship between them and behavior (Buzsaki, 2009). To demonstrate that they are relevant neural signals used in the brain rather than epiphenomena of other processes, it would be necessary to show how these signals are transformed by neural circuits and ultimately generate behavior.

Suppose we record the reference signal in a control system. It would resemble the controlled variable. At the initiation of the control process, it could occur before the actual input reaches the reference value and appear to predict the future state of some variable. But the significance of the reference signal is only revealed once we understand the whole control system in which it is found.

To understand how any neural circuit works, we must consider its content and operation. What the signal represents is its content, which is determined by its relationship with the rest of the circuit. On the other hand, operation refers to the type of transformation performed or implemented by the neural circuit, e.g. $y=2x$. A TV can display any type of content, but inside it, similar operations are performed regardless of which programs are being shown.

The functional heterogeneity typically associated with the BG can be attributed to diversity in content. The signals operated on by the BG circuits come from cortical and thalamic inputs. The basic operations performed by these circuits are similar. They include addition, subtraction, integration, differentiation, multiplication and division, and sign inversion. Each class of variables appears to undergo similar types of analysis and transformation within the BG circuits, and each component of the circuit could perform the same operation regardless of the content of the signal. For example, the recurrent excitation circuit may implement a function similar to time integration.

5.11 SUMMARY

It is often assumed that behavior is simply the output of the nervous system, but this assumption is false. Repeating final common path outputs and specific patterns of muscle contractions does not repeat behavior. Rather, neural command signals to the muscles must vary precisely the right amount to offset the effects of environmental disturbances, in order to achieve the highly consistent behavior that we observe and take for granted.

Achieving consistent results by generating variable outputs is made possible by closed-loop negative feedback control. Although the process of control appears to be simple, it has in fact counterintuitive properties that are often misunderstood due to misleading engineering conventions that ignore the autonomy of the control system. The control system model itself does require inverse or forward computations, as is assumed in modern control engineering. It does not calculate the desired effect of an output and then produce the signals that will create that desired effect. Control systems, by generating output as a function of the difference between sensory input and internal reference, produce variable actions on the environment in order to match inputs to internally specified desired states.

Understanding the control hierarchy has many implications for neuroscience. Above all, there is no need for complex inverse/forward computations to be performed inside the brain; rather, these computations are embodied in the entire loop when inputs are controlled. In the modern engineering approach, on the other hand, the observer implicitly uses physical laws and powerful computers to perform the necessary computations but attributes these functions performed by the observer to the system being observed. The objectivist fallacy replaces the perspective of the behaving organism with the perspective of the external observer. Consequently, the "control of output" assumption reverses the organism/environment relationship, resulting in the inside-out view of control, according to which the comparison function is placed in the environment, or in the mind of the observer. Modern control engineering is based on the premise that the user must inject reference signals to generate the needed control signals for the plant. The engineer is unwittingly part of the control loop as the ideal observer. But biological organisms are autonomous because they possess intrinsic reference signals. Their behaviors are outward manifestations of the actions of a large collection of control systems organized hierarchically. In a control hierarchy, commands sent to lower levels are requests for input, not output; they specify what and how much to sense, not how much output to produce.

The phenomenon of control violates the traditional paradigm of linear causation. The function of the brain is not sensorimotor transformation or generation of behavior but rather the control of specific input variables. Many common assumptions about how neural signaling enables behavior are influenced by information theory and digital computing, which offer misleading analogies. To the observer, the neural activity seems to be a code that stands for something. The fact that a human observer can decode neural activity does not mean that the neural tissue that receives the observed signal will also interpret or decode it in the same way.

In analog computing, there are no separate encoding and decoding processes. The computational processes needed for control systems to work are embodied by the physical interactions between components of the system. The neural signal, usually measured as the rate of firing, is an analog signal. Logical and symbolic functions are not intrinsic features of neural signaling, but rather emergent properties of large neural networks. To understand neural signaling, we must consider how it is used in the circuit where it is found, the physical effects it has on other signals, and the emergent function from these interactions. By recording identified neurons during their normal functioning and analyzing the relationship between their activity and controlled variables that are continuously measured, it is possible to identify these neurons as specific parts of the control system.

REFERENCES

Adrian, E. D., & Bronk, D. W. (1929). The discharge of impulses in motor nerve fibres part ii. The frequency of discharge in reflex and voluntary contractions. *The Journal of Physiology*, *67*(2), i3.

Adrian, E. D., & Zotterman, Y. (1926). The impulses produced by sensory nerve-endings: Part ii. The response of a single end-organ. *The Journal of Physiology*, *61*(2), 151–171.

Bernstein, N. A. (1935). The problem of the interrelation of coordination and localization. *Archives of Biological Sciences*, *38*, 15–59.

Brette, R. (2019). Is coding a relevant metaphor for the brain? *Behavioral and Brain Sciences*, *42*, 1–44.

Brown, L. L., Schneider, J. S., & Lidsky, T. I. (1997). Sensory and cognitive functions of the basal ganglia. *Current Opinion in Neurobiology*, *7*(2), 157–163.

Buzsaki, G. (2009). *Rhythms of the Brain*: Oxford: Oxford University Press.

Carelli, R. M., & West, M. O. (1991). Representation of the body by single neurons in the dorsolateral striatum of the awake, unrestrained rat. *Journal of Comparative Neurology*, *309*(2), 231–249.

Craik, K. J. (1947). Theory of the human operator in control systems1. *British Journal of Psychology. General Section*, *38*(2), 56–61.
Davidson, D. (1963). Actions, reasons, and causes. *Journal of Philosophy*, *60*, 685–700.
Fuster, J. (2015). *The Prefrontal Cortex*: Cambridge, MA: Academic Press.
Gallistel, C. R., & King, A. P. (2011). *Memory and the Computational Brain: Why Cognitive Science will Transform Neuroscience* (vol. 6): Hoboken, NJ: John Wiley & Sons.
Granit, R. (1955). *Receptors and Sensory Perception*: New Haven, CT: Yale University Press.
Grodins, F. S. (1963). *Control Theory and Biological Systems*: New York: Columbia University Press.
Hull, C. (1943). *Principles of Behavior*: New York: Appleton-Century-Crofts.
Kennaway, R. (2020). When causation does not imply correlation: Robust violations of the faithfulness axiom. In: Mansell, W. (Ed.), *The Interdisciplinary Handbook of Perceptual Control Theory* (pp. 49–72). New York: Elsevier.
Krauthamer, G. M. (1979). Sensory functions of the neostriatum. In: Divac, I. (Ed.), *The Neostriatum* (pp. 263–289). New York: Elsevier.
Ma, W. J., Beck, J. M., Latham, P. E., & Pouget, A. (2006). Bayesian inference with probabilistic population codes. *Nature Neuroscience*, *9*(11), 1432–1438.
Maley, C. J. (2011). Analog and digital, continuous and discrete. *Philosophical Studies*, *155*(1), 117–131.
McCulloch, W. S., & Pitts, W. (1943). A logical calculus of the ideas immanent in nervous activity. *The Bulletin of Mathematical Biophysics*, *5*(4), 115–133.
Mishkin, M., Malamut, B., & Bachevalier, J. (1984). Memories and habits: Two neural systems. In Lynch, G., McGaugh, J. L., & Weinberger, N. (Eds.), *Neurobiology of Learning and Memory* (pp. 65–77). New York: Guilford Press.
Nizami, L. (2017). I, neuron: The neuron as the collective. *Kybernetes*, *46*(9), 1508–1526.
Nizami, L. (2019). Information theory is abused in neuroscience. *Cybernetics & Human Knowing*, *26*(4), 47–97.
Perkel, D. H., & Bullock, T. H. (1968). Neural coding. *Neurosciences Research Program Bulletin*, *6*(3), 221–348.
Powers, W. T. (1973a). *Behavior: Control of Perception*: New Canaan, CT: Benchmark Publications.
Powers, W. T. (1973b). Feedback: Beyond behaviorism. *Science*, *179*(71), 351–356.
Powers, W. T. (2008). *Living Control Systems iii: The Fact of Control*. Available at https://cepa.info/263.
Rack, P. M. (1981). Limitations of somatosensory feedback in control of posture and movement. In: Partridge, L. D., Benton, L. A., & Brooks, V. B. (Eds.), *Handbook of Physiology: Nervous System* (vol. 2, pp. 229–256). Bethesda, MD: American Physiological Society.
Rieke, F. (1999). *Spikes: Exploring the Neural Code*: Cambridge, MA: The MIT Press.
Sherrington, C. S. (1906). *The Integrative Action of the Nervous System*: New Haven, CT: Yale University Press.
Sterling, P. (2012). Allostasis: A model of predictive regulation. *Physiology & Behavior*, *106*(1), 5–15.
Von Neumann, J. (1958). *The Computer and the Brain*: New Haven: Yale University Press.
Wittgenstein, L. (1953). *Philosophical Investigations*: London: Blackwell.
Wolpert, D. M., & Kawato, M. (1998). Multiple paired forward and inverse models for motor control. *Neural Networks*, *11*(7), 1317–1329.
Yin, H. H. (2013). Restoring purpose in behavior. In: Baldassarre, G., & Mirolli, M. (Eds.), *Computational and Robotic Models of the Hierarchical Organization of Behavior* (pp. 319–347). Berlin: Springer.
Yin, H. H. (2014). How basal ganglia outputs generate behavior. *Advances in Neuroscience*, *2014*, 768313.
Yin, H. H. (2017). The basal ganglia in action. *Neuroscientist*, *23*(3), 299–313. doi: 10.1177/1073858416654115.

6 The Place of the BG in the Hierarchy

The basal ganglia (BG) do not usually send direct projections to motor neurons; rather, they project to the brainstem and ventral thalamus, which ultimately influence the motor neurons via reticulospinal and corticospinal pathways. To understand how BG outputs contribute to behavior, we must first appreciate the contributions of structures that receive descending BG projections.

In this chapter, we shall first review the lower levels in the action hierarchy, the controllers for muscle tension and length, which are located in the spinal cord. We shall then examine how the midbrain and brainstem nuclei implement control over body configuration and orientation. Finally, we shall consider the relationship between posture and movement, and how BG outputs may send descending signals to command the brainstem and midbrain position control systems.

6.1 POSTURE AND MOVEMENT

Postural control is the foundation for locomotion and other movements. Developmentally, postural control systems mature before those for locomotion. In the human newborn, for example, more than 6 months of development is required before sitting is possible, followed by crawling, standing, and walking much later.

As Magnus explains (p. 341): "If we try to put the body of a dead animal upon its feet, the carcass immediately falls down to the ground, because the relaxed muscles cannot carry the weight of the body against the action of gravity. The same happens with a living animal after total extirpation of the brain .... The centers of the spinal cord can indeed cause and regulate every complicated combinations of movements, but they are unable to give to the muscles that steady and enduring tone which is necessary for simple standing" (Magnus, 1925). What Magnus fails to appreciate is that it is not possible to control a posture simply by sending a tonic descending signal to muscles. When standing, the body has no intrinsic stability. To maintain balance, it is necessary to generate just the right outputs from hundreds of muscles in real time. This poses a tremendous computational challenge.

According to the conventional model of postural reflexes, some input reaches the postural control center, and following a sensorimotor transformation, some compensatory output is generated to maintain posture (Magnus, 1924; Roberts, 1967). This account had a major influence on ideas about BG function. Denny-Brown, for example, argued that descending BG outputs can modify the sensorimotor transformation that occurs in postural reflexes (Denny-Brown, 1962). The focused selection model (Chapter 4) also assumes that the BG suppress the postural control mechanism to enable action selection (Mink, 1996). But as we shall see, such suppression is not necessary when the relationship between postural control and volitional action is properly understood.

Because the input does not determine output in a control system, the same input could be associated with very different postural adjustments (von Holst & Mittelstaedt, 1950). There is no consistent relationship between sensory input and effector output in postural control since the relevant inputs are controlled by varying the outputs. It is not possible to produce consistent effects (i.e. maintain a particular posture) by choosing some amount of tension in each muscle and repeating that tension because consistent tension cannot produce the same posture given the influence of environmental disturbances. This is another illustration of the calculation problem discussed in Chapter 5.

If a rat is turned over on its side, righting reflexes are observed, yet it is also able to assume any arbitrary posture volitionally without producing righting reflexes. Since the volitional action can also generate similar vestibular and proprioceptive inputs, why is it not canceled by the righting reflex? Why are there no outputs generated in response to similar sensory inputs when the action is self-generated? According to the conventional account, volitional behavior can only be achieved by suppressing the normal reflex so that the sensory input can no longer evoke the motor output, but postural reflexes are not suppressed during volitional behavior (von Holst & Mittelstaedt, 1950). What the postural reflex account neglects is the comparison process between input and internal reference signals, which determines controller output. Consequently, the output is proportional to the disturbance.

A body configuration can be described as a position vector, and postural control requires control of this multidimensional position vector. The same controllers used to control posture are used for volitional movements. Instead of suppressing postural control, it is possible to change posture during volitional movements by changing the reference signals of postural controllers via top–down projections. Both posture and movement can be explained by the control hierarchy.

6.1.1 Parkinsonian Rigidity

Position control explains many common neurological symptoms. For example, when the neurologist pushes the patient to assess resistance to the push or asks the patient to maintain some arbitrary posture, what is being tested is position control. The neurological exam unwittingly tests for the controlled variable, using disturbance to find the mirroring output from the nervous system.

A common symptom related to postural control is rigidity, often associated with Parkinson's disease. In rigidity, the body of the patient is stiff and unyielding to a push, and the posture is "frozen" (Rushworth, 1960). Standard models of the BG fail to explain these observations adequately, but they can be explained by position control. Rigidity actually reflects the output of position control systems with high gain. Position control still works normally, but their reference signals fail to change. Consequently, the posture is fixed, while output varies according to the amount of disturbance. In support of this account, rigidity is not unconditional; it is reduced when the body is supported, for example when the patient is supported or suspended in water (Denny-Brown, 1962). Physical support of the body reduces disturbance to position controllers; likewise, when suspended in water, the buoyancy of water reduces the major source of disturbance from gravity. As a result, reduced muscle output (and rigidity) is generated by the position controllers.

6.2 CONTROL OF MUSCLE TENSION AND LENGTH

Position controllers must send outputs to the final common path to continuously adjust muscle tension. The final common path refers to the projection from the alpha motor neurons to muscles. It represents the output function of a muscle tension controller, converting a tension error signal from the alpha motor neurons into muscle torque (Henneman, Clamann, Gillies, & Skinner, 1974). The Golgi tendon organs sense tension, acting like a strain gauge between the extrafusal muscle fibers and their attachments. Muscle contraction stretches the attached tendon and stimulates the Golgi tendon organs. In the Golgi tendon reflex or autogenic reflex, muscle contraction activates Ib inhibitory interneurons, which inhibits alpha motor neurons to stop the contraction. In other words, whenever actual tension exceeds the reference tension, the error can generate output that reduces sensed tension by reducing muscle contraction. Generation of tension is thus limited by negative feedback from sensed tension, and the amount of inhibition is roughly proportional to the amount of contraction (Houk & Henneman, 1967). Normally, the tension controller maintains muscle tone as specified by descending tension reference signals (Figure 6.1).

The tension controller is the lowest level of the action hierarchy (Powers, 1973). It is used for all behaviors, but it is rarely the lead level for any behavior, with the exception of reflexes like the

The Place of the BG in the Hierarchy

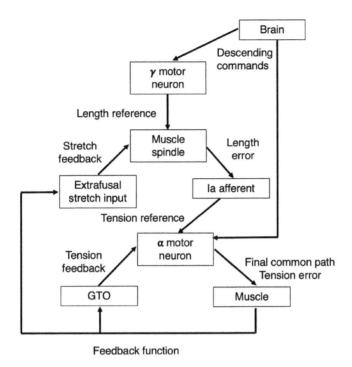

FIGURE 6.1 Control of muscle tension and length. A highly simplified illustration of the relationship between a higher muscle length controller and a lower tension controller. Tension feedback is provided by Golgi tendon organ (GTO), with a sign inversion provided by Ib interneurons (not shown here) to produce negative feedback. The comparator in the tension controller is implemented by the alpha motor neuron, which receives tension feedback from the Golgi tendon organ and a reference tension signal from higher levels. The difference between tension reference and some representation of sensed tension is the error signal that produces output from alpha motor neurons. When tension exceeds the reference tension, the muscle will stop contracting. Length feedback is provided by stretch input at the muscle spindle. The muscle spindle acts as a mechanical comparator. When the gamma-induced stretch of the intrafusal fibers simulates the effect of stretching the main muscle fibers, the Ia signal represents the difference between current length (Ia activation) and gamma activation (length reference). The length error signal is then converted into a force reference that enters the alpha motor neuron.

autogenic reflex, which generates compensatory output when there is a disturbance to muscle tone. All higher levels, however, must use the tension controller to generate torque and act on the external environment. They can do so via multiple descending pathways that dictate the tension reference signals.

6.2.1 Muscle Length Control and the Fusimotor System

The next level in the hierarchy, just above tension control, is a muscle length controller (Powers, 1973; Yin, 2014). To change muscle length, one must contract or relax muscles. The length controller senses muscle length and the rate of change in length using muscle spindles, which are proprioceptive receptors found in most muscles (McMahon, 1984). Unlike Golgi tendon organs, which are activated by muscle contraction, muscle spindles are activated by muscle stretch. They are located deep inside the muscle, in parallel with the extrafusal muscle fibers that are responsible for muscle contraction. A typical muscle spindle contains three types of intrafusal muscle fibers: the nuclear chain fibers and the static nuclear bag fibers signal the static length of the muscle, whereas the dynamic nuclear bag fibers carry signals representing rate of change in muscle length. These intrafusal muscle fibers, being much shorter than the main extrafusal muscle fibers, do not directly contribute to muscle contraction.

The main sensory innervation of the muscle comes from primary (group Ia) afferent fibers, which innervate both nuclear bag and nuclear chain fibers, and secondary (group II) afferent fibers, which innervate nuclear chain fibers. The primary afferents represent the rate of change in muscle length. The smaller secondary afferent fibers represent muscle length. When the extrafusal fibers are stretched, the parallel spindle fibers are also stretched, activating the primary and secondary afferents. These afferents in turn excite alpha motor neurons to produce contraction of the main muscle fibers.

A key feature of the muscle spindles is that the intrafusal fibers can be contracted by descending signals from various motor pathways. They are innervated by gamma motor neurons, which innervate contractile elements at their ends. This arrangement is known as the fusimotor system. Activation of gamma motor neurons does not directly cause contraction of extrafusal muscle fibers; rather, it results in contraction of intrafusal fibers at the two ends of the spindle. When the end portions of the intrafusal muscle fibers contract, they elongate the noncontractile central portion (Granit, 1955). During such contractions, the length of the muscle spindle does not change significantly because it is anchored at both ends (Leksell, 1945). However, because the central (equatorial) region of the spindle contains all the stretch-sensitive channels, elongation results in the activation of these channels and the primary afferent signal. Although intrafusal fibers do not contribute directly to the contraction of the main extrafusal muscle fibers, they can do so indirectly by activating the primary afferent, the main source of excitatory drive to the alpha motor neurons. Since the primary afferent activates the final common path, it carries a reference signal for muscle tension. It can be activated either by stretching of the main muscle or top–down fusimotor stretching of the intrafusal fibers. Either route can lead to the activation of alpha motor neurons, as what matters is the elongation of the central noncontractile portion of the spindle where the stretch receptors are located. The descending fusimotor commands mimic the effects of muscle stretch on spindle output (Figure 6.1).

Descending gamma activation leads to muscle contraction using top–down adjustment of tension reference, while muscle contraction shortens the spindles. As the amount of shortening due to contraction is subtracted from gamma-induced stretching, the spindle acts as a mechanical comparator. It compares the top–down length reference with the sensed length signal and produces an output that reflects the difference between the two. The length error signal, then, is proportional to the shortening of the intrafusal muscles minus the shortening of the main muscle, and it is this error signal that activates the alpha motor neuron to produce muscle contraction. The Ia afferent signal that drives alpha motor neurons is proportional to the length error. The fusimotor system is often thought to keep the spindle responsive to stretching even when the muscle is relaxed. What this account ignores is the comparison process at the spindle. The muscle spindle output is not only a sensory signal representing muscle length. Rather, it can represent a length error signal independent of actual muscle length.

Likewise, the spindle can also compare the rate of change signals. The dynamic (derivative) and static (proportional) components of muscle length can be signaled by primary and secondary afferents. The descending fusimotor signals also contain dynamic and static components. This organization makes it possible for the descending signal to control the rate of change in muscle length as well as the static muscle length. The proportional and derivative components are also common in engineered control systems. It is possible that, depending on the demands of the movement, either one or the other or some combination of both components are used. This is similar to the tuning of PD (proportional-derivative) control systems.

The isolated action of the length controller is the well-known stretch reflex, which increases alpha motor neuron output to cause muscle contraction. For example, in the knee jerk reflex, a tap to the knee (patellar ligament) stretches the muscle spindles for the quadriceps femoris muscle, an extensor whose contraction produces the kicking action and simultaneous relaxation of the antagonistic flexor, the hamstring muscle. The above review suggests that the same mechanism is used for the simple stretch reflex and for voluntary movements. The tension and length controllers are the two lowest levels of the action hierarchy. They can be used by multiple descending projections from the brain and brainstem.

6.3 BIDIRECTIONAL CONTROL

Since muscles can only contract, relaxing is passive. In the skeletomotor system, the arrangement of muscles around joints allows them to pull in opposite directions by contracting, thereby increasing or decreasing the joint angle. Each motor neuron pool represents the output function of a one-way tension controller. Since firing rate cannot be negative, only positive signals, namely Ia afferent activity, is used to produce output for a given tension controller. But it is possible to achieve bidirectional control using a composite system with antagonistic outputs. The neural implementation of this mechanism is the well-known reciprocal inhibition organization (McDougall, 1903). While the primary afferent excites alpha motor neurons for a given muscle, it also projects to Ia interneurons, which inverts the sign of the signal and inhibits motor neurons for the antagonistic muscle. Consequently, as one muscle is activated, the other is relaxed, and the difference between the two is proportional to the net torque applied to the joint. The degree of reciprocal inhibition is subject to regulation by descending projections to Ia interneurons.

For the sake of simplicity, consider a pair of muscles that regulate a particular joint angle. Both are maintained at some baseline tension by a "common-mode" signal, which consists of two reference signals, one sent to each of the antagonist pairs, producing two equal and opposing forces for a net torque of zero. The common-mode signal represents zero net torque but not the zero-reference state of each tension controller. The reference tension can vary to maintain zero net torque as long as the outputs of the two antagonistic tension controllers oppose each other. Although the firing rate is always positive, a given signal can be negative *relative to* the effective zero. As we shall see in later chapters, analogous arrangements are also found in the BG.

6.3.1 STIFFNESS CONTROL

Houk correctly argues that, since muscle length and force co-vary in normal movements, it is not possible to keep both variables constant (Houk, 1979). But he fails to realize that the relevant controllers do not have constant reference signals for muscle length or tension. In a control hierarchy, the lower controlled variable must vary to keep the higher variable at specific levels. Tension can be varied to reach a specific length, and length can be varied in the service of a higher-level reference.

Failing to understand the hierarchical and independent control of muscle length and tension, Houk proposed that the key variable being controlled is stiffness, defined as force change divided by length change (Nichols & Houk, 1976). Indeed, stiffness can be controlled, but it is distinct from tension control or length control. Reciprocal inhibition prevents conflicts between antagonistic effectors by generating two antiphase signals to these effectors. On the other hand, in stiffness control, descending signals specify muscle tone by adjusting agonist and antagonist tensions up and down together, which is tantamount to controlling the sum or average of agonist and antagonist tensions.

Stiffness control requires excitatory descending influences that affect alpha motor neurons of both agonist and antagonist muscles. For example, when one holds a "plank" position, both dorsal and ventral muscles stiffen to resist disturbances to position from gravity. Since muscles behave like nonlinear springs with spring constants that increase with tension, simultaneous activation of these muscles increases stiffness and the gain of the position control system. As muscle tone increases, the body can better resist disturbance and maintain its position. With fatigue, however, output function gain is reduced as muscles fail to generate sufficient torque, and the position control system starts to oscillate at a fixed frequency. The body starts to shake, and such shaking can be reduced simply by providing support, i.e. less disturbance from gravity.

6.3.2 BANDWIDTH LIMITATIONS

Biological position control systems have limited bandwidth, mainly due to delays in neural signaling and relatively low gain compared to engineered systems. Consequently, their capacity to control is much reduced with sudden changes in the direction of disturbance, even if they can resist a strong push in one direction. This is because their effective gain depends on the frequency of the disturbance. For example, if we push against someone who is trying to maintain his posture, there is usually proportional resistance to the push and successful maintenance of the posture. But if instead of pushing we shake the body rapidly, he will not be able to resist the high-frequency disturbance.

This property of position control can be tested also in patients with Parkinsonian rigidity. If rigidity reflects position control, one would expect a similar limit in bandwidth, since frequency dependence of the gain is a key property of the underlying neural controller. Without changing the magnitude of the disturbance, we would predict a significant reduction in rigidity when the disturbance applied exceeds a certain frequency.

6.4 RETICULOSPINAL PATHWAY

Bidirectional control systems for muscle tension, length, and joint angle are directly commanded by higher levels of postural control. During voluntary locomotion or movement, top–down commands can alter the reference position vector. When the reference signals remain the same, the posture or body configuration is maintained.

The brain can influence the final common path via the reticulospinal pathway, which targets many regions in the spinal cord and all types of motor neurons, producing coordinated contraction or relaxation of muscles in many body parts (Peterson, 1979; Peterson, Pitts, & Fukushima, 1979). In lampreys, it coordinates the alternation of muscles on two sides of the body, for example, left–right alternation or dorsal–ventral alternation, which allows them to swim (Sten Grillner & Wallen, 2002; Kozlov et al., 2001). Being evolutionarily conserved, the reticulospinal pathway is fundamental for movements in all vertebrates (Grillner, Wallen, Saitoh, Kozlov, & Robertson, 2008). It is especially important for postural control and axial movements, supplemented by the vestibulospinal pathway (Lawrence & Kuypers, 1968a, b). It is responsible for movements that counteract the effects of postural disturbances in any direction (Deliagina, Zelenin, & Orlovsky, 2002).

The reticulospinal pathway is modulated by BG outputs, either directly or indirectly via projections to areas like the mesencephalic locomotor region (MLR), including the pedunculopontine nucleus (PPN). Substantia nigra pars reticulata (SNr) projections to the PPN can regulate muscle tone via descending reticulospinal pathways. These projections can regulate stiffness and the gain of the body configuration controller (Takakusaki, 2008; Kaoru Takakusaki, 2017; Takakusaki, Habaguchi, Ohtinata-Sugimoto, Saitoh, & Sakamoto, 2003).

The MLR includes a number of nuclei, including the PPN and cuneiform nucleus (Martinez-Gonzalez, Bolam, & Mena-Segovia, 2011; Sherman et al., 2015; Shik, Severin, & Orlovsky, 1966; Shik & Orlovsky, 1976; Takakusaki, 2017). A related region is the diencephalic locomotor region, analogous to the subthalamus, including the zona incerta and neighboring regions (El Manira, Pombal, & Grillner, 1997; Menard & Grillner, 2008). These regions also receive strong projections from the BG and, in turn, target the reticulospinal neurons.

Takakusaki has shown two distinct reticulospinal pathways with opposite effects on muscle tone (Takakusaki, 2017). Neurons in the dorsomedial ponto-medullary reticular formation (PMRF) reduce muscle tone via medial reticulospinal projections to the ventral spinal cord. In contrast, neurons in the ventromedial PMRF can increase muscle tone through the lateral reticulospinal pathway.

6.5 POSITION CONTROLLERS FOR ORIENTATION

In goal-directed locomotion, continuous steering is needed to adjust one's relationship with respect to the goal. Orienting and steering require movement of the rostral part of the body, especially the

The Place of the BG in the Hierarchy

head, where various sensory receptors are located. For example, visual input can be used to align the gaze with the distal targets in the environment, like pointing a camera at some object. On the other hand, even without vision, it is possible to achieve orientation control using vestibular inputs. For example, even in the dark, one may orient the body toward environmental targets. Whereas body configuration control mainly relies on proprioceptive and vestibular inputs, orientation control relies on the use of vestibular inputs or distal senses like vision.

It takes at least four neural control systems to point to anything in space: up, down, left, and right. Masino and colleagues found distinct regions in the mesencephalic tegmentum (medial reticular formation) that are specialized in horizontal and vertical components of head movements (Masino, 1992; Masino & Grobstein, 1989a, b; Masino & Knudsen, 1990). After lesions to the nucleus of the medial longitudinal fasciculus (nMLF), frogs could still respond to the prey, but their movements lacked the horizontal directional component (Masino & Grobstein, 1989a, b). If the right nMLF tract is lesioned, when prey is on the right side, the animal only moves forward without the rightward component. Consequently, although the frog could still see the prey and generate a prey capture movement, it would miss the target. As shown in Figure 6.2, studies by Masino and colleagues on visually elicited head saccades in barn owls further demonstrated separate representation of horizontal, vertical, and rotational components in the brainstem (Masino & Knudsen, 1990).

Crawford and colleagues showed that the interstitial nucleus of Cajal and surrounding structures such as the rostral interstitial medial longitudinal fasciculus and nucleus of Darkschewitsch generate outputs that command specific head positions (Klier, Wang, Constantin, & Crawford, 2002). Outputs from these brainstem nuclei represent variable outputs from position controllers

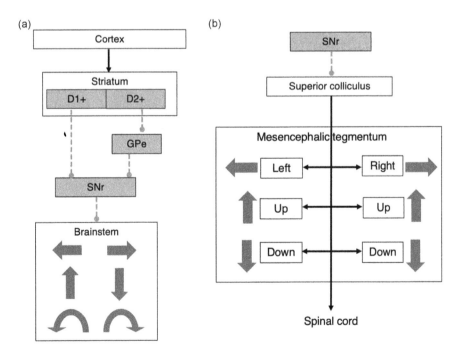

FIGURE 6.2 Hierarchical control of orientation. (a) Neural circuit involved in top–down control of orientation. The basal ganglia (BG) project directly targets brainstem modules for different vector component of the movements. For rotational movements, only roll movements (clockwise and counterclockwise) are shown, as they have been demonstrated by brainstem stimulation experiments, but it is assumed that there are pitch and yaw components as well. (b) Nigrocollicular control of gaze. Arrows indicate the magnitude and direction of eye movements caused by stimulation at that location. The superior colliculus does not appear to be critical for rotational movements. Based on Masino (1992). SNr, substantia nigra pars reticulata.

and directly influence motor neurons (Fukushima, Pitts, & Peterson, 1979). For example, unilateral stimulation and inactivation of the interstitial nucleus of Cajal produce head tilts in opposite directions. At the neutral position, there is zero head torsion, reflecting balanced activation of clockwise and counterclockwise neuronal populations. The outputs from these opponent populations represent a pair of tonic-descending reference signals. Stimulation of the right side produces counterclockwise turning, whereas stimulation on the left side produces clockwise turning (Farshadmanesh, Byrne, Wang, Corneil, & Crawford, 2012).

In the primate oculomotor systems, there are also separate circuits at the brainstem level controlling ipsilateral horizontal components of eye movements. In the brainstem nuclei, there are distinct areas responsible for the generation of ipsiversive, contraversive, upward, and downward saccades. Outputs from the superior colliculus reach the interstitial nucleus of Cajal, nucleus Darkschewitsch, red nucleus, and midbrain reticular formation, which are responsible for generating different components of head and eye movements (King, Fuchs, & Magnin, 1981; Luschei & Fuchs, 1972). In primate eye movements, the effectors are the extraocular muscles, which have groups specializing in mainly vertical and horizontal movements (e.g. medial and lateral rectus muscles). For head movements, the effectors involved are far more complex; for example, in the barn owl, the neck alone has at least 31 muscle pairs, none of which are arranged to pull in purely horizontal or vertical directions. Consequently, each position vector component specified by the descending commands from the brainstem must recruit a variable set of effectors with the net effect of achieving the corresponding reference. These results suggest the presence of distinct position controllers for vertical and horizontal components of orienting and steering movements. Because normal movements are not strictly vertical or horizontal, multiple controllers are used. For example, a diagonal movement would involve the rightward and upward components.

6.6 MIDBRAIN CONTRIBUTIONS TO ORIENTING

As described in Chapter 2, the tectum (superior and inferior colliculi) is a key structure for orienting and steering, and receives massive BG output projections, especially from the SNr and SNl (Huerta & Harting, 1984). When the tectum is lesioned, animals fail to respond to visual stimuli presented to the receptive field location of the lesioned area (Kostyk & Grobstein, 1987). In the superior colliculus, the superficial layer receives inputs from the retina, which project to the intermediate and deep layers, which also receive inputs from multiple modalities, including somatosensory and auditory sources, and project to the midbrain tegmentum and spinal cord (Huerta & Harting, 1984; Isa & Hall, 2009). Through these descending projections, the superior colliculus is able to move exteroceptive sensors in multiple modalities (e.g., eyes, ears, and whiskers) and maintain a certain orientation toward some location in the environment (Isa & Naito, 1995). It also sends ipsilateral descending projections to the reticulospinal pathway, which contributes to turning and torso movements (Dean, Redgrave, & Mitchell, 1988; Redgrave, Dean, Mitchell, Odekunle, & Clark, 1988; Redgrave, Mitchell, & Dean, 1987).

In foveating animals, the location of the neutral "center" or default origin is located in the rostral portion, where so-called fixation neurons are found. The neutral position is the center of the retina. These neurons fire tonically during fixation and decrease firing during saccades (Munoz & Wurtz, 1992). In monkeys, most units in the intermediate layer show low firing rates, with the exception of those that receive inputs from the foveal region, which have high tonic firing rates (Goffart, Hafed, & Krauzlis, 2012; Munoz & Wurtz, 1992). Inactivation of the rostral superior colliculus does not prevent fixation but shifts the fixation location (Hafed, Goffart, & Krauzlis, 2008). This is what we expect from a lasting change in the descending reference signal.

An eccentric stimulus not in the center of the receptive field can generate a gaze shift as the system is recentered. The target stimulus is once again captured by the center after the gaze shift, which requires output from the intermediate and deep layers (Munoz, Pelisson, & Guitton, 1991).

Deviation from the center (fovea region for the primate oculomotor system) can generate an error signal that is ultimately sent to multiple output functions activating multiple effector systems, e.g. both head and eye movements (Meyer, O'Keefe, & Poort, 2020).

6.7 NIGROCOLLICULAR PATHWAY AND EYE MOVEMENTS

In addition to generating postural changes critical for movement, BG output also adjusts the rostral part of the body, especially the head and eyes, for steering and orienting. Nigrocollicular projections are hypothesized to be the key source of descending signals that adjust the reference signals of the orientation controllers (Beckstead, Domesick, & Nauta, 1979; Di Chiara, Porceddu, Morelli, Mulas, & Gessa, 1979; McElvain et al., 2021; Redgrave, Marrow, & Dean, 1992). The nigrocollicular pathway projects to the deep/intermediate layers of the superior colliculus. It is critical for target orientation control using visual and somatosensory inputs. The SNl also projects strongly to the inferior colliculus (Chapter 2), so presumably a similar process occurs for auditory inputs, i.e. orienting to a sound.

In primates, the nigrocollicular projection is critical for self-initiated or memory-guided saccades. When the monkey must move its eyes toward some arbitrary target, some SNr neurons reduce their tonic firing rates transiently (Hikosaka & Wurtz, 1983; Joseph & Boussaoud, 1985). These saccade-related nigral neurons project to the intermediate layers of the superior colliculus, which show burst firing at the time of saccades (Hikosaka & Wurtz, 1983). According to this account, a pause in SNr output would allow a burst in superior colliculus activity and a subsequent saccade (Hikosaka, Takikawa, & Kawagoe, 2000). However, SNr neurons do not simply show decreases in firing rate. Both increases and decreases in SNr single unit activity at the time of saccade or smooth pursuit have been reported (Basso, Pokorny, & Liu, 2005; Sato & Hikosaka, 2002). We shall return to the functional significance of such outputs below.

The deep/intermediate collicular layers receive two major inputs: glutamatergic inputs from the superficial layer and GABAergic inputs from the SNr, which represent descending reference signals from the BG (Isa & Hall, 2009). Nigrocollicular projections target not only the intermediate layer projection neurons but also the GABAergic interneurons (Kaneda, Isa, Yanagawa, & Isa, 2008). This anatomical organization suggests that sensory inputs and descending nigral signals may converge to implement a comparison function. The superficial layer input represents the input function of the orientation controller, whereas the nigrocollicular projection can adjust the reference signal. This arrangement makes it possible to compare reference and perceptual input, generating an error signal that is then sent downstream to adjust the oculomotor and skeletomotor configurations (Yin, 2014).

On the collicular map, the high tonic activity in the foveal region represents the default reference setting, which allows the eyes to be centered at rest. The inhibitory SNr projection can adjust the threshold for orienting movements, as more excitatory input is needed to offset the inhibitory top–down reference signal (Chapter 5). A large input is needed to offset the reference and thus eliminate errors. Activation of the peripheral units can produce an error signal proportional to the eccentricity of the stimulus. This error signal can be converted to multiple types of outputs, including movements of the head and neck and eyes, to "capture" the salient input. In principle, by reducing the inhibitory output to a specific location in the deep/intermediate layer of the superior colliculus, the nigrocollicular projections can increase the firing rate of neurons at that location. The nigrocollicular projection thus achieves top–down selection of arbitrary locations in the visual field by mimicking the salience effect using disinhibition. Consequently, one can volitionally orient toward any stimulus.

Many SNr neurons send uncrossed projections to the superior colliculus. Like their target collicular neurons, the SNr neurons have small visual receptive fields centered in contralateral space (Jiang, Stein, & McHaffie, 2003). The retinal input to the superficial collicular layers is excitatory and contralateral. On the other hand, the crossed nigrocollicular neurons are activated by visual stimuli, whereas the descending uncrossed nigrocollicular projections, which show high spontaneous activity, are inhibited by visual stimuli. The uncrossed neurons have their receptive fields in the ipsilateral visual field, unlike their target collicular neurons. They can inhibit the activation of collicular neurons that move the eyes to any ipsilateral location. As the uncrossed SNr neurons pause, the target collicular neurons increase firing. Normally high nigrocollicular inhibition must be reduced selectively for specific visual input locations in the collicular map in order to acquire visual inputs.

6.8 BG REGULATION OF STEERING AND ORIENTING

The excessive emphasis on BG projections to the thalamocortical network in primate studies has led to the common misconception that the BG contribute to behavior by acting on the corticospinal pathway. But BG outputs reach the midbrain and brainstem nuclei involved in position control, as just described. For example, the SNr, implicated in movements of the head and orofacial musculature, projects directly to the interstitial nucleus of Cajal and the nucleus Darkschewitsch (McElvain et al., 2021). The lateral SNr contains orofacial regions, whereas the medial SNr is often associated with eye and head movement (Redgrave et al., 1992). GABAergic projection neurons from the ventral tegmental area (VTA), SNr, and globus pallidus internus all project to the PPN. These outputs are in a position to provide top–down regulation of steering and propulsion during locomotion (Grillner, Wallen, Saitoh, Kozlov, & Robertson, 2008).

In the clinical condition known as cervical dystonia or spasmodic torticollis, there is a fixed and rigid bending of the neck muscles. In monkeys, this condition can be experimentally created by asymmetric inactivation of the SNr with muscimol. SNr inactivation mimics the GABAergic striatonigral inhibition, resulting in disinhibition of the superior colliculus and other major target regions of the SNr (Burbaud, Bonnet, Guehl, Lagueny, & Bioulac, 1998; Dybdal et al., 2013). It can produce not only contralaterally directed torticollis but also dyskinesias of the contralateral limbs.

Turning is one of the most robust effects observed after experimental manipulations of the BG. Figure 6.3 provides a summary of turning effects after unilateral stimulation. Unilateral DA depletion in the striatum produces ipsiversive turning (toward the depleted side). However, as the depleted side becomes more sensitive to DA, when DA or DA agonists are given, animals show contraversive turning. DA-induced turning is mediated by striatonigral output and disinhibition of the superior colliculus and other target regions (Di Chiara, Morelli, Porceddu, & Gessa, 1978). Likewise, unilateral activation of the striatonigral pathway produces contraversive turning (Kravitz et al., 2010; Rossi et al., 2015). Unilateral striatal lesions can produce a posture in which the neck and body are curved toward the side of the lesion. The laterally curved posture results from an imbalance in striatal outputs from the two hemispheres; the animal turns away from the side of the greater striatal activity (Ferrier, 1876; Sten Grillner & Robertson, 2016; Jung & Hassler, 1960). The turning effect requires intact target regions like the tectum and the reticulospinal pathway.

Unilateral SNr lesions produce sustained contralateral turning (Di Chiara, Oianas, Del Fiacco, Spano, & Tagliamonte, 1977). Unilateral inhibition of the SNr can also produce contraversive quadrupedal rotation and neck flexion (Holmes et al., 2012; Kilpatrick, Collingridge, & Starr, 1982). Reduced SNr output could increase the output of downstream neurons in the superior colliculus. Indeed, unilateral collicular stimulation leads to contraversive head turning away from the stimulated side (Anderson, Yoshida, & Wilson, 1971). Reduced nigrocollicular transmission can be mimicked by blocking GABA receptors in the superior colliculus bilaterally, which results in a condition called "explosive running," characterized by uncontrollable running interspersed with jumping (Cools, Coolen, Smit, & Ellenbroek, 1984). A rat in this condition would often run until hitting a wall, only to turn and run again. Running does not stop until the rat is completely exhausted, lying flat on its belly and often trying to crawl forward with clear signs of hypoxia (Bart Ellenbroek, personal communications). The superior colliculus also projects to the cuneiform nucleus, and fast running has been observed when glutamate is injected into the cuneiform nucleus, suggesting that

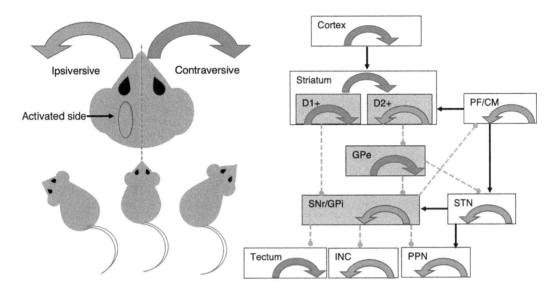

FIGURE 6.3 Summary of turning behavior after unilateral stimulation of different cortical, basal ganglia, brainstem, and thalamic regions. In the example given here, excitatory stimulation on the left side is shown. Inhibition of the same areas usually results in turning in the opposite direction. Ipsiversive is leftward and counterclockwise and contraversive is rightward and clockwise. The arrows indicate the net effect of stimulating a particular area. Other effects of stimulation are not summarized here. PF/CM, parafascicular/centromedial thalamus; PPN, pedunculopontine nucleus; INC, interstitial nucleus of Cajal.

this projection is at least partly responsible for explosive running (Mitchell, Dean, & Redgrave, 1988; Redgrave et al., 1988).

Cools described the turning behavior of cats after unilateral muscimol injection into the SNr (Cools, 1985). The injected cat, with presumably reduced SNr output, repeatedly attempted to bridge the gap between the current position and some egocentrically defined position on the contralateral side of the body. The turning radius depended on the origin of the egocentric reference frame. Each body part was associated with a distinct origin located along the midline. The turning started from the vertical axis of the egocentric coordinate system, and gradually moved from the rostral to more caudal positions (oculus, auriculum, cranium, scapula, and pelvis). The injected cats initially moved their ears, then eyes, head, neck, shoulders, and forelimbs. The recruitment of these body parts did not require visual input, as it was observed even when the cat wore a bandage covering its eyes. For example, the cat anteroflexed its head, kept it there for a while, and then returned its head to the original starting point. When it could not bend its head any further, it started to bend its torso in the same direction.

These observations suggest that SNr inhibition can generate an egocentric distance error between the current position and the target position. The different points of departure along the body reflect different comparison functions in distinct neural controllers. Symmetric posture is the neutral position. Drug-induced movements stop only when the body has reached a position marked by fixed deviations from the axes of the egocentric coordinate system. Each point reflects a neutral position in a comparison function.

Based on these results, Cools suggested a *propriotopic* organization in the nigrotectal output, based on a set of egocentric coordinates. Attempts to bridge the gap between given points suggest the presence of continuous position error signals. The progressive recruitment of the rostral to caudal segments of the body may reflect different thresholds of activation for the position controllers associated with these different body parts.

6.9 SNr AND POSTURAL CONTROL

Proprioceptive and vestibular feedback information can reach the BG nuclei as well as their targets (Stiles & Smith, 2015). For example, the vestibular nucleus projects to the parafascicular nucleus of the thalamus, which in turn projects to the sensorimotor (dorsolateral) striatum as well as the tectum and PPN.

Barter et al. examined the relationship between BG outputs from the SNr and continuous postural disturbances in mice (Barter, Castro, Sukharnikova, Rossi, & Yin, 2014). Each mouse stood on a platform, while continuous disturbances were introduced in the roll plane, i.e. tilting to the left and right. Behavioral output is required to cancel the effect of the tilt disturbance on the controlled variable, the configuration of perceptual signals that collectively report the state of being upright. Signals representing the controlled variable (from vestibular, visual, and proprioceptive inputs) are compared with the reference, and an error signal is generated to generate behavioral output that corrects the deviation.

The SNr output was highly correlated with postural disturbance. When the platform was level, most neurons were tonically active, and increases and decreases from this baseline firing rate represented specific directions of tilt. As shown in Figure 6.4, two major types of neurons can be identified. The first type (L+R−) increased firing with a tilt to the left and decreased firing with a tilt to the right, whereas the second type showed the opposite pattern (L−R+).

These results resemble results from reticulospinal pathway neurons. Previous work has shown that the reticulospinal output can generate movements that resist the effects of postural disturbances (Zelenin, Orlovsky, & Deliagina, 2007). If a reticulospinal neuron is excited by the nose-up pitch tilts, it can produce compensatory downward bending of the body by activating muscles on the ventral side. Opponent reticulospinal neurons were found to control posture in a particular axis of rotation; they were activated by rotation in opposite directions and produced movements generating torques counteracting the postural disturbances (Deliagina, Orlovsky, Zelenin, & Beloozerova, 2006; Deliagina et al., 2002). In the lamprey, reticulospinal neurons receive inputs

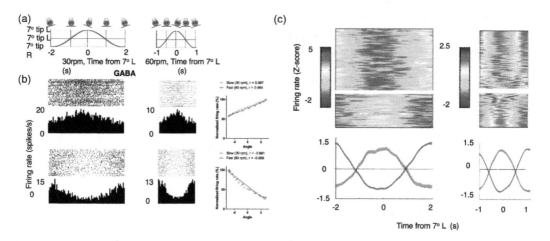

FIGURE 6.4 Substantia nigra pars reticulata (SNr) activity is modulated by postural disturbances. (a) The mouse stands on an elevated platform, and continuous postural disturbances are introduced in the roll plane. Left, slow disturbance (30rpm, 7° to left and right); right, fast disturbance (60rpm). (b) Neural activity in the SNr during postural disturbances. Time zero indicates 7 degrees of tilt to the left of the mouse. Raster plots of two representative SNr GABAergic projection neurons. One is positively correlated with the tilt angle, whereas the other is negatively correlated. (c) Heat maps display spike density functions of all SNr neurons that are modulated by tilt disturbance. Bottom, population summary of the two opponent populations (blue and red traces). Based on Barter, Castro, Sukharnikova, Rossi, and Yin (2014) and Yin (2016).

from the vestibular nuclei and send projections to motor neurons (Pavlova, Popova, Orlovsky, & Deliagina, 2004). Given the similarities in reticulospinal activity and SNr activity during postural disturbances and the known anatomical connectivity, it is possible that BG output may adjust the reference settings of reticulospinal neurons, either directly or indirectly through projections to the PPN.

These results show that the BG output is not all or none, coding action or no action, as assumed in conventional models. During the continuous tilt disturbances, the BG output is never on or off but is modulated directly by the postural disturbance as the same posture was maintained throughout. The BG output does not simply pause to open the gate for action but rather continuously determines action parameters.

6.10 BG AND LOCOMOTION

Central pattern generators are neural circuits capable of generating intrinsic rhythms like flexor–extensor alternation (Brown, 1911; Grillner & Wallen, 1985). Spinal CPGs, especially those responsible for stepping movements, have been extensively studied (Kiehn, 2016; McCrea & Rybak, 2008; Rossignol, Dubuc, & Gossard, 2006).

Normal locomotion, however, is far more than just pattern generation. In addition to stepping, locomotion requires continuous postural control. Animals with only an intact spinal cord but no descending signals from the brain are capable of stepping when supported, but they cannot maintain balance or move around in any natural movement (Grillner & Wallen, 1985). Stepping can be demonstrated even before infants start walking, but to walk successfully, appropriate postural adjustments are needed to accompany stepping. For example, leaning or tilting the body forward creates a shift in the center of gravity. To maintain balance, one must catch oneself by stepping forward. The shifts in the center of gravity generally precede stepping. During such shifts, self-generated disturbances are automatically resisted by the postural control systems. They trigger stepping, which then catches the body. Balancing requires orientation of the body relative to the world and position of the body relative to the support polygon and ground surface.

Although the BG do not generate the stepping rhythm, they could regulate the center of gravity through outputs to the PPN and related regions (Austin & Kalivas, 1991; Garcia-Rill, 1986; Roseberry et al., 2016). In decerebrate cats, such descending signals for postural adjustments are presumably absent. If a decerebrate cat is placed on the treadmill, it will start to run, showing the intact spinal stepping mechanism, but it is unable to walk around normally without support (Edgerton, Grillner, Sjöström, & Zangger, 1976).

The MLR also projects to reticulospinal neurons and commands the spinal pattern generators for locomotion (Grillner et al., 2008). Although the MLR is traditionally associated with locomotion based on stimulation results, it may not be directly involved in stepping but rather regulates body tilts responsible for top–down generation of disturbances needed for stepping initiation. Selective activation of glutamatergic projection neurons in the PPN, a major component of the MLR receiving BG projections, can initiate locomotion in mice (Roseberry et al., 2016). Activation of the direct pathway can also initiate locomotion, whereas activating the indirect pathway stops locomotion. However, there is a significant lag between stimulation onset and onset of locomotion, suggesting that BG output to the PPN does not directly trigger stepping, but causes postural adjustments that can then lead to stepping.

6.11 SNr AND POSITION COORDINATES

Given direct projections from the BG to a variety of position controllers reviewed above, one hypothesis is that the BG output can alter position reference. Their descending signals must be multidimensional and continuously varying according to position coordinates to specify a continuous position vector.

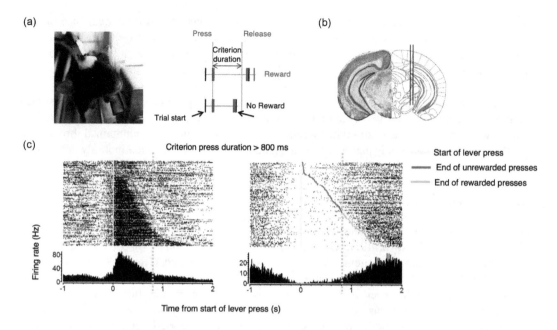

FIGURE 6.5 Opponent populations of substantia nigra pars reticulata (SNr) projection neurons. (a) Mice were required to press a lever with a minimum duration of 800 ms. Releasing the lever prematurely resulted in no reward, but presses exceeding 800 ms were rewarded. (b) Placement of electrodes in the SNr. (c) Two representative examples of putative gamma-Aminobutyric accid GABA) neurons from the SNr. The neuron on the left increased firing while the lever was held down, whereas the neuron on the right paused firing during the same period. From Fan, Rossi, and Yin (2012) with permission.

Fan et al. examined the activity of SNr output neurons in mice trained to press a lever for reward (Fan, Rossi, & Yin, 2012). To earn rewards, mice had to hold down the lever for a minimum duration (e.g. 800 ms) before releasing. There was no cue indicating reward availability, and releasing the lever early would result in no reward. This design uses the hold period to tag neural activity while the animal maintains a steady posture, since there is little movement when the animal is holding the lever down. The position reference hypothesis predicts a tonic and stable reference signal during the holding period as a position is held. In support of the position reference hypothesis, the firing rates of most SNr neurons were stable during the hold period. In other words, a fixed posture was associated with a fixed firing rate in the BG output. However, opponent types of GABAergic projection neurons were found. As shown in Figure 6.5, the first type increased firing at the onset of pressing, maintaining the high firing rate for the entire duration of the holding period. In contrast, the second type reduced firing rates at the onset of the press, maintaining the reduced rate until lever release. Their firing rates exhibited a step-like change at the onset of lever holding. This pattern was independent of whether a given trial was rewarded: even on trials where the hold duration was too short to earn a reward, the same pattern was observed.

6.11.1 SNR AND REPRESENTATION OF POSITION VECTORS

Results from the study by Fan et al. suggest a relationship between tonic SNr activity and posture maintenance. Although they are in accord with the position reference hypothesis, they do not provide direct evidence for the representation of position vectors in BG output because the actual position coordinates were not measured. In another study, Barter et al. combined wireless *in vivo* electrophysiology and continuous video tracking to study the relationship between behavior and

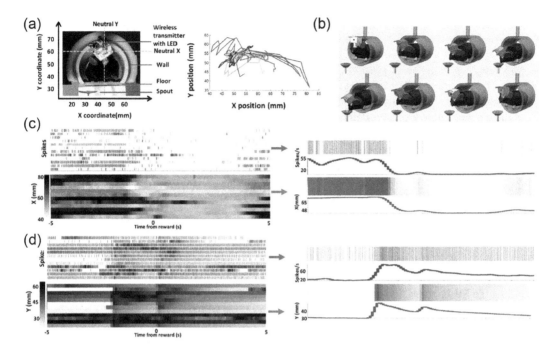

FIGURE 6.6 SNr neurons represent instantaneous position coordinates using Cartesian coordinates. (a) A photo of a mouse in the wireless electrophysiology setup. Head position is monitored with a light-emitting diode (LED) on the headstage. Typical head movement trajectories are shown. Each color shows movement from a single trial. (b) Illustration of the behavioral task. Mice are trained to approach a sucrose spout after the presentation of an auditory cue. (c) Raster plots showing data from individual trials (rows). This neuron increases firing when moving to the left and decreases firing when moving to the right. The firing rate at any time reflects instantaneous position coordinates (x coordinates). The overall change reflects movement amplitude as well as direction. (d) Another SNr neuron that increases firing when the head is elevated, with firing rate correlated with y-coordinates. From Barter et al. (2015) with permission.

SNr activity (Barter et al., 2015). Instantaneous head position coordinates for x- and y-coordinates were measured at the same time as neural activity. They discovered different classes of SNr neurons that represented distinct components of the position vector (Figure 6.6).

SNr neurons can be further divided into opponent populations based on movement direction. For example, for horizontal movement along the x-axis, two types of neurons changed their firing rates according to the direction of movement: one type increased firing during leftward movement and decreased firing during rightward movement, and a second type increased firing during rightward movement and decreased firing during leftward movement. The same was true of the vertical component of the movement along the y-axis. These results suggest that the SNr projection neurons provide an analog representation of a position vector for egocentric movements, with distinct populations for x- and y-coordinates. Each type of neuron could represent a specific position vector component, and at least four types of SNr neurons change their firing rates during any normal orienting behavior.

Since SNr activity generally leads the instantaneous position variable, it does not seem to provide perceptual representations of head position. Rather, the results reviewed above support the hypothesis that SNr output provides reference signals for lower-level position controllers. There appear to be independent controllers, each responsible for a particular direction of movement: leftward, rightward, upward, and downward. These position vectors have default reference states corresponding to neutral postures. For example, a neutral position for the orienting system is the "straight ahead" posture, reflecting balanced reference signals with stable values when the animal is at rest. During volitional behavior, the descending commands do not dictate output but specify position-related perceptual inputs for the lower level by altering reference signals.

Movement is a change in position. Such a change could be achieved in two ways. First, in the absence of descending BG output, lower-level changes could be generated due to resistance to lower-level disturbances. When any environmental change forces the position-related inputs to deviate from the current reference values, control effort is generated. When a position is maintained using negative feedback control, the perceptual inputs to the position controller do not vary much, while the outputs mirror the disturbance. Secondly, in volitional actions, the BG send descending signals that alter the position reference signals, thereby commanding the target position controllers to reach a new position vector. To maintain the reference position, the SNr output neurons simply maintain their current firing rates. When their firing rates change, a new position vector is requested.

6.12 VTA OUTPUT AND HEAD POSITION

The VTA also contains GABAergic projection neurons, which are similar to those in the SNr. VTA GABA neurons are a major source of the limbic BG outflow and synapse in a variety of regions, including the MLR, the reticular formation, and the hypothalamus (Mogenson, Jones, & Yim, 1980; Morales & Margolis, 2017; Shank, Seitz, Bubar, Stutz, & Cunningham, 2007).

Recent work revealed a role of the VTA output in orienting behavior. Hughes et al. recorded VTA GABA neurons in mice performing a continuous tracking task (Hughes et al., 2019b). On this task, mice must follow a reward spout that is constantly moving either horizontally or vertically. A sucrose solution was delivered from the moving spout. Mice were able to follow the reward spout, turning their heads continuously. When the target was moving from left to right, the head would rotate more about the roll and yaw axes. When the target was moving up and down, it would rotate more around the pitch axis.

Using three-dimensional motion capture, Hughes et al. studied the relationship between self-initiated rotational movements and VTA output. They measured the head angle of the animal about three independent principal axes of motion: pitch, yaw, and roll. They found distinct populations of neurons representing head angle about a principal axis of rotation. For each axis, opponent cell groups were found. For example, for the roll axis, some neurons increased firing when the head moved in a clockwise direction, and decreased firing during counterclockwise movement (Figure 6.7). Thus, the VTA output can represent instantaneous head angles using three vector components.

These neurons also showed angle representations that slightly leading the actual position changes, suggesting that the VTA output is responsible for causing the rotational motion. When VTA GABA neurons were excited using optogenetics, the mouse lowered its head and showed ipsiversive turning: it moved along the yaw and roll axes toward the stimulated hemisphere. In contrast, when they were inhibited, the opposite was observed: the mouse showed contraversive turning, raising its head, and moving away from the stimulated side. There was a linear relationship between frequency of stimulation and head angle. Because stimulation was not restricted to any functional class of neurons, neurons corresponding to all three axes were probably affected. These results suggest that distinct populations of VTA output independently control three orthogonal axes of rotation, allowing independent control of rotation along the pitch, yaw, and roll axes. Using VTA output activity, it is possible to place the head of the mouse in any orientation.

As no study has been performed to determine the relationship between SNr activity and rotational kinematics, it is difficult to compare VTA output and SNr output directly. We cannot rule out the possibility that the SNr activity is also related to rotational kinematics. Moreover, the different classes of neurons related to kinematics only represent a subset of the BG output, so further studies are needed to fully characterize the BG output from these regions. Nevertheless, clearly there is a striking relationship between BG output and head angle. Together, these findings

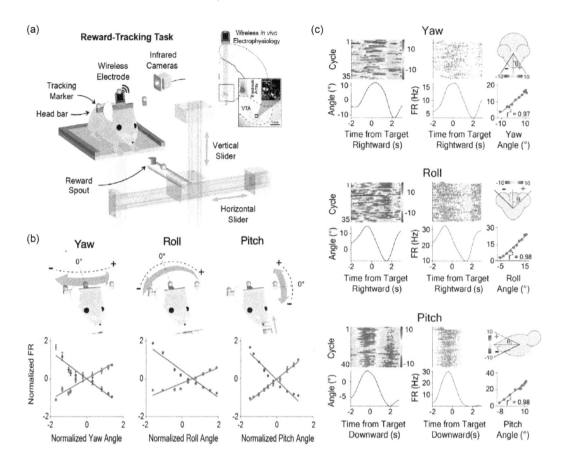

FIGURE 6.7 Ventral tegmental area (VTA) glutamate decarboxylase (GABA) output and rotational kinematics of the head. (a) Schematic of behavioral tasks and wireless in vivo electrophysiology. (b) Opponent VTA GABA neuronal populations for yaw, roll, and pitch neurons. They represent instantaneous head angles along three independent axes of rotation. (c) Representative VTA GABA neurons that are positively correlated with head angle (indicated by red in (b)). From Hughes et al. (2019b) with permission.

provide evidence for analog and continuous neural representation of position-related variables in the BG and demonstrate that such signals are responsible for generating specific movements. They support the hypothesis that the BG output sends position reference signals to to lower-level position controllers.

6.13 FUNCTIONAL SIGNIFICANCE OF BG OUTPUTS

In principle, there are 6 degrees of freedom for head movement: three translational (x, y, z) and three rotational (pitch, yaw, roll). A point in space can be described with three values, each representing the distance from the origin (x, y, z) in Cartesian coordinates. Likewise, head angle is a vector with three values: pitch, yaw, and roll. Position and orientation can be described by the position of the origin of the reference frame and the orientation of its axes relative to the reference frame.

Each vector is an ordered set of magnitudes. If each neuron carries a one-dimensional signal with its firing rate, then different classes of neurons are required to create a multidimensional vector. If the BG output can represent position vectors, they cannot possibly be uniform, as traditionally assumed. A summary of the proposed classes of BG output neurons that command lower position controllers is shown in Figure 6.8.

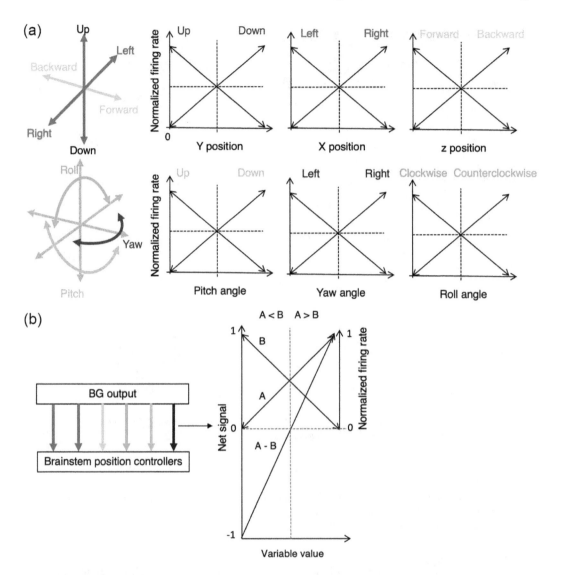

FIGURE 6.8 How BG output alters the reference position vector. (a) There are six major degrees of freedom in motion, three translational (forward/backward, up/down, left/right), and three rotational (pitch, yaw, roll). Each axis of motion is represented by two opponent classes of BG output neurons. (b) As neural signals carried by neurons (firing rate) cannot be negative, they cannot signal negative errors when the input is more than error. To signal both positive and negative errors, it is necessary to create a pair of systems, one for positive and the other for negative, but both using positive neural outputs. The sign is not indicated by the signal they carry but by the net effect of the circuit, as the pair of signals activate effectors with opposite effects.

To command the position controllers, one must alter the reference position vector. For the sake of simplicity, let us consider a hypothetical change in one of the vector components: if the head moves upward and the position changes from position 5 to position 8, the sequence will be 5, 6, 7, 8, and when the head moves back to the original position, the following sequence is observed: 8, 7, 6, 5. Each number indicates the instantaneous position coordinate along the y-axis. The actual position changes are generated by BG output neurons, especially from the SNr. How quickly the position reference changes are equivalent to movement velocity. An important implication, which we will discuss in more detail in Chapter 7, is that the rate of change in BG output reflects movement velocity.

6.13.1 Two-Way Comparison Functions

To achieve bidirectional control, we need a pair of opponent signals, one for positive errors and the other for negative errors. Firing rate, as an analog signal, cannot be negative in the absolute sense, but it can be negative relative to some arbitrary zero point. One way to build a composite two-way controller is to set the zero point to some default value, so that it generates a positive signal on one channel for A>B and another positive signal on the other channel for B>A (Figure 6.8). When the position is at the default value, no output is generated. However, when the input deviates from this value, a signal will be generated, and its sign depends on the direction, e.g. left of center or right of center. To generate bidirectional signals with positive neural signals, the effective zero point can be set close to the middle of the dynamic range for each neuron. For example, if a neuron can fire from 0 to 40 Hz, then 20 Hz would be the zero point. In Cartesian position coordinates, this corresponds to the x- or y-component of the origin of some egocentric reference frame. Likewise, for rotational kinematics, the corresponding measures are head angle about a particular axis of rotation. For every axis of rotation, the middle of the range is the effective zero, and the negative and positive signs of the signals reflect different directions. In the VTA GABA output neurons, they correspond to increases or decreases from the tonic firing rates (Hughes et al., 2019a).

The meaning of a zero-reference signal depends on the perceptual input function. For example, in the default position of facing straight ahead, corresponding to the effective zero, two opponent systems produce equal output. During a left turn, one channel increases as the other decreases, and vice versa when one turns right. The range of the neural signal reflects the range of egocentric position coordinates, i.e. maximum movement amplitude. The BG output neuron shows the lowest firing rate at one end of the range and the highest rate at the other end.

At the middle of the range, the head is not tilted or turned in any direction. The average output value is the bias in the system, corresponding to the neutral position. If the tonic output from the BG is set to zero, this two-way position control mechanism will cease to function. This is only possible when no movement at all is needed, perhaps in a state of deep sleep. In an awake behaving animal, the tonic signal increases to bring the pair into the middle of their operating range. The tonic signal (bias) enables bidirectional control.

According to traditional accounts, an action is either selected or suppressed, and the output of the BG is either suppressing or disinhibiting target regions. The tonic GABAergic output is therefore responsible for tonic suppression of behavior, through uniform high inhibition of multiple potential behavioral systems. Only one channel is selected via disinhibition. The assumption is that tonic inhibition does not carry a signal. The results discussed above suggest that it is misleading to view tonic inhibition as merely a mechanism for suppressing behavior. BG output sends a set of signals representing vectorial quantities to specify actions in real time by requesting specific position vectors through the use of negative feedback position controllers. The significance of inhibition can only be understood in the overall analog operation performed by the circuit. However, the results discussed above suggest that the tonic GABAergic output introduces a bias that allows bidirectional control of the position variable in a push–pull circuit.

6.14 SUMMARY

At the lowest level of the action hierarchy, the final common path generates muscle contraction. It is the output function of the muscle tension control system. The BG outputs do not directly affect the final common pathway but rather exert control over effectors mainly through the reticulospinal pathway. Below the BG, there are lower-level controllers for muscle tension, muscle length, rate of change in muscle length, and body configuration, implemented by the spinal cord and reticulospinal pathway. The outputs of position control systems adjust reference signals to the lowest levels, regulating muscle length and joint angle. Position errors are ultimately converted to changes in muscle tension.

In locomotion, the center of gravity can be controlled by adjusting the leaning of the body, which can generate disturbance to lower postural controllers to induce correctional stepping. Descending outputs from the BG are critical for initiating and stopping locomotion and for postural adjustments during locomotion.

Turning and steering reflect the coordinated outputs of multiple overlapping control systems for different parts of the body axis. BG outputs from SNr and VTA are critical for orientation, turning, and head position control. The output neurons from these regions can represent instantaneous position coordinates. According to the working hypothesis, the BG outputs send descending reference signals to position controllers. Studies have shown that the output neurons from the SNr and VTA provide reference position vectors, which are the reference signals requesting the corresponding input vectors from lower-level position controllers in the brainstem. When descending reference signals are stable, as indicated by tonic firing rates in BG output neurons, there is no overt movement, even though lower-level position controllers can continuously vary muscle outputs to control the position. Descending commands from the BG can generate volitional movements by specifying how much of the position input should be acquired by lower-level controllers.

The BG output can have many distinct channels for top–down adjustment of reference states the desired sensory states of many different types of variables. In the absence of descending commands from the BG, the tectum and lower levels can still achieve position control of perceptual inputs from multiple modalities. During the generation of voluntary actions, continuous and time-varying BG output can specify action parameters, in particular the instantaneous position vector.

REFERENCES

Anderson, M., Yoshida, M., & Wilson, V. (1971). Influence of superior colliculus on cat neck motoneurons. *Journal of Neurophysiology, 34*(5), 898–907.

Austin, M. C., & Kalivas, P. W. (1991). Dopaminergic involvement in locomotion elicited from the ventral pallidum/substantia innominata. *Brain Research, 542*(1), 123–131.

Barter, J. W., Castro, S., Sukharnikova, T., Rossi, M. A., & Yin, H. H. (2014). The role of the substantia nigra in posture control. *European Journal of Neuroscience, 39 (9)*, 1465–1473.

Barter, J. W., Li, S., Sukharnikova, T., Rossi, M. A., Bartholomew, R. A., & Yin, H. H. (2015). Basal ganglia outputs map instantaneous position coordinates during behavior. *Journal of Neuroscience, 35*(6), 2703–2716.

Basso, M. A., Pokorny, J. J., & Liu, P. (2005). Activity of substantia nigra pars reticulata neurons during smooth pursuit eye movements in monkeys. *European Journal of Neuroscience, 22*(2), 448–464.

Beckstead, R. M., Domesick, V. B., & Nauta, W. J. (1979). Efferent connections of the substantia nigra and ventral tegmental area in the rat. *Brain Research, 175*(2), 191–217.

Burbaud, P., Bonnet, B., Guehl, D., Lagueny, A., & Bioulac, B. (1998). Movement disorders induced by gamma-aminobutyric agonist and antagonist injections into the internal globus pallidus and substantia nigra pars reticulata of the monkey. *Brain Research, 780*(1), 102–107.

Cools, A. R. (1985). Brain and behavior: Hierarchy of feedback systems and control of input. In: Bateson, P., & Klopfer, P. (Eds.), *Perspectives in Ethology* (pp. 109–168). New York: Springer.

Cools, A. R., Coolen, J. M., Smit, J. C., & Ellenbroek, B. A. (1984). The striato-nigro-collicular pathway and explosive running behaviour: Functional interaction between neostriatal dopamine and collicular gaba. *European Journal of Pharmacology, 100*(1), 71–77.

Dean, P., Redgrave, P., & Mitchell, I. J. (1988). Organisation of efferent projections from superior colliculus to brainstem in rat: Evidence for functional output channels. *Progress in Brain Research, 75*, 27–36.

Deliagina, T. G., Orlovsky, G. N., Zelenin, P. V., & Beloozerova, I. N. (2006). Neural bases of postural control. *Physiology (Bethesda), 21*, 216–225. doi: 10.1152/physiol.00001.2006.

Deliagina, T. G., Zelenin, P. V., & Orlovsky, G. N. (2002). Encoding and decoding of reticulospinal commands. *Brain Research Reviews, 40*(1–3), 166–177.

Denny-Brown, D. (1962). *The Basal Ganglia and their Relation to Disorders of Movement*: Oxford: Oxford University Press.

Di Chiara, G., Morelli, M., Porceddu, M., & Gessa, G. (1978). Evidence that nigral gaba mediates behavioural responses elicited by striatal dopamine receptor stimulation. *Life Sciences, 23*(20), 2045–2051.

Di Chiara, G., Oianas, M., Del Fiacco, M., Spano, P., & Tagliamonte, A. (1977). Intranigral kainic acid is evidence that nigral non-dopaminergic neurones control posture. *Nature, 268,* 743–745.

Di Chiara, G., Porceddu, M., Morelli, M., Mulas, M., & Gessa, G. (1979). Evidence for a gabaergic projection from the substantia nigra to the ventromedial thalamus and to the superior colliculus of the rat. *Brain Research, 176*(2), 273–284.

Dybdal, D., Forcelli, P. A., Dubach, M., Oppedisano, M., Holmes, A., Malkova, L., & Gale, K. (2013). Topography of dyskinesias and torticollis evoked by inhibition of substantia nigra pars reticulata. *Movement Disorders, 28*(4), 460–468.

Edgerton, V., Grillner, S., Sjöström, A., & Zangger, P. (1976). Central generation of locomotion in vertebrates. In: Herman, R. M., Grillner, S., Stein, P., Stuart, D. G. (Eds.), *Neural Control of Locomotion* (pp. 439–464). New York: Springer.

El Manira, A., Pombal, M., & Grillner, S. (1997). Diencephalic projection to reticulospinal neurons involved in the initiation of locomotion in adult lampreys lampetra fluviatilis. *Journal of Comparative Neurology, 389*(4), 603–616.

Fan, D., Rossi, M. A., & Yin, H. H. (2012). Mechanisms of action selection and timing in substantia nigra neurons. *Journal of Neuroscience, 32*(16), 5534–5548. doi: 10.1523/JNEUROSCI.5924-11.2012.

Farshadmanesh, F., Byrne, P., Wang, H., Corneil, B. D., & Crawford, J. D. (2012). Relationships between neck muscle electromyography and three-dimensional head kinematics during centrally induced torsional head perturbations. *Journal of Neurophysiology, 108*(11), 2867–2883.

Ferrier, D. (1876). *The Functions of the Brain*: New York: GP Putnam's Sons.

Fukushima, K., Pitts, N. G., & Peterson, B. W. (1979). Interstitiospinal action on forelimb, hindlimb, and back motoneurons. *Experimental Brain Research, 37*(3), 605–608.

Garcia-Rill, E. (1986). The basal ganglia and the locomotor regions. *Brain Research, 396*(1), 47–63.

Goffart, L., Hafed, Z. M., & Krauzlis, R. J. (2012). Visual fixation as equilibrium: Evidence from superior colliculus inactivation. *Journal of Neuroscience, 32*(31), 10627–10636.

Graham Brown, T. (1911). The intrinsic factors in the act of progression in the mammal. *Proceedings of the Royal Society of London, 84*(572), 308–319.

Granit, R. (1955). *Receptors and Sensory Perception*: New Haven: Yale University Press.

Grillner, S., & Robertson, B. (2016). The basal ganglia over 500 million years. *Current Biology, 26*(20), R1088–R1100.

Grillner, S., & Wallen, P. (1985). Central pattern generators for locomotion, with special reference to vertebrates. *Annual Review of Neuroscience, 8*(1), 233–261.

Grillner, S., & Wallen, P. (2002). Cellular bases of a vertebrate locomotor system–steering, intersegmental and segmental co-ordination and sensory control. *Brain Research Reviews, 40*(1), 92–106.

Grillner, S., Wallen, P., Saitoh, K., Kozlov, A., & Robertson, B. (2008). Neural bases of goal-directed locomotion in vertebrates--an overview. *Brain Research Reviews, 57*(1), 2–12. doi: 10.1016/j.brainresrev.2007.06.027.

Hafed, Z. M., Goffart, L., & Krauzlis, R. J. (2008). Superior colliculus inactivation causes stable offsets in eye position during tracking. *Journal of Neuroscience, 28*(32), 8124–8137.

Henneman, E., Clamann, H. P., Gillies, J. D., & Skinner, R. D. (1974). Rank order of motoneurons within a pool: Law of combination. *Journal of Neurophysiology, 37*(6), 1338–1349.

Hikosaka, O., Takikawa, Y., & Kawagoe, R. (2000). Role of the basal ganglia in the control of purposive saccadic eye movements. *Physiological Reviews, 80*(3), 953–978.

Hikosaka, O., & Wurtz, R. H. (1983). Visual and oculomotor functions of monkey substantia nigra pars reticulata. Iv. Relation of substantia nigra to superior colliculus. *Journal of Neurophysiology, 49*(5), 1285–1301.

Holmes, A. L., Forcelli, P. A., DesJardin, J. T., Decker, A. L., Teferra, M., West, E. A., . . . Gale, K. (2012). Superior colliculus mediates cervical dystonia evoked by inhibition of the substantia nigra pars reticulata. *Journal of Neuroscience, 32*(38), 13326–13332.

Houk, J., & Henneman, E. (1967). Responses of golgi tendon organs to active contractions of the soleus muscle of the cat. *Journal of Neurophysiology, 30*(3), 466–481.

Houk, J. C. (1979). Regulation of stiffness by skeletomotor reflexes. *Annual Review of Physiology, 41*(1), 99–114.

Huerta, M. F., & Harting, J. K. (1984). Connectional organization of the superior colliculus. *Trends in Neurosciences, 7*(8), 286–289.

Hughes, R. N., Watson, G. D., Petter, E. A., Kim, N., Bakhurin, K. I., & Yin, H. H. (2019a). Precise coordination of three-dimensional rotational kinematics by ventral tegmental area gabaergic neurons. *Current Biology, 29*(19), 3244–3255. e3244.

Hughes, R. N., Watson, G. D., Petter, E. A., Kim, N., Bakhurin, K. I., & Yin, H. H. (2019b). Precise coordination of three-dimensional rotational kinematics by ventral tegmental area gabaergic neurons. *Current Biology, 29*(19), 3244–3255.

Isa, T., & Hall, W. C. (2009). Exploring the superior colliculus in vitro. *Journal of Neurophysiology, 102*(5), 2581.

Isa, T., & Naito, K. (1995). Activity of neurons in the medial pontomedullary reticular formation during orienting movements in alert head-free cats. *Journal of Neurophysiology, 74*(1), 73–95.

Jiang, H., Stein, B. E., & McHaffie, J. G. (2003). Opposing basal ganglia processes shape midbrain visuomotor activity bilaterally. *Nature, 423*(6943), 982–986.

Joseph, J., & Boussaoud, D. (1985). Role of the cat substantia nigra pars reticulata in eye and head movements i. Neural activity. *Experimental Brain Research, 57*(2), 286–296.

Jung, R., & Hassler, R. (1960). The extrapyramidal motor system. *Handbook of Physiology, 2*, 863–927.

Kaneda, K., Isa, K., Yanagawa, Y., & Isa, T. (2008). Nigral inhibition of gabaergic neurons in mouse superior colliculus. *Journal of Neuroscience, 28*(43), 11071–11078.

Kiehn, O. (2016). Decoding the organization of spinal circuits that control locomotion. *Nature Reviews Neuroscience, 17*(4), 224.

Kilpatrick, I., Collingridge, G., & Starr, M. (1982). Evidence for the participation of nigrotectal γ-aminobutyrate-containing neurones in striatal and nigral-derived circling in the rat. *Neuroscience, 7*(1), 207–222.

King, W., Fuchs, A., & Magnin, M. (1981). Vertical eye movement-related responses of neurons in midbrain near intestinal nucleus of cajal. *Journal of Neurophysiology, 46*(3), 549–562.

Klier, E. M., Wang, H., Constantin, A. G., & Crawford, J. D. (2002). Midbrain control of three-dimensional head orientation. *Science, 295*(5558), 1314–1316.

Kostyk, S. K., & Grobstein, P. (1987). Neuronal organization underlying visually elicited prey orienting in the frog—i. Effects of various unilateral lesions. *Neuroscience, 21*(1), 41–55.

Kozlov, A. K., Aurell, E., Orlovsky, G. N., Deliagina, T. G., Zelenin, P. V., Hellgren-Kotaleski, J., & Grillner, S. (2001). Modeling postural control in the lamprey. *Biological Cybernetics, 84*(5), 323–330.

Kravitz, A. V., Freeze, B. S., Parker, P. R., Kay, K., Thwin, M. T., Deisseroth, K., & Kreitzer, A. C. (2010). Regulation of parkinsonian motor behaviours by optogenetic control of basal ganglia circuitry. *Nature, 466*(7306), 622–626. doi: 10.1038/nature09159.

Lawrence, D. G., & Kuypers, H. G. (1968a). The functional organization of the motor system in the monkey i. The effects of bilateral pyramidal lesions. *Brain, 91*(1), 1–14.

Lawrence, D. G., & Kuypers, H. G. (1968b). The functional organization of the motor system in the monkey ii. The effects of lesions of the descending brain-stem pathways. *Brain, 91*(1), 15–36.

Leksell, L. (1945). The action potential and excitatory effects of the small ventral root fibres to skeletal muscle. *Acta Physiologica Scandinavica, 10*, Suppl. 31.

Luschei, E. S., & Fuchs, A. F. (1972). Activity of brain stem neurons during eye movements of alert monkeys. *Journal of Neurophysiology, 35*(4), 445–461.

Magnus, R. (1924). *Körperstellung*: Berlin: Springer.

Magnus, R. (1925). Croonian lecture.—animal posture. *Proceedings of the Royal Society of London. Series B, Containing Papers of a Biological Character, 98*(690), 339–353.

Martinez-Gonzalez, C., Bolam, J. P., & Mena-Segovia, J. (2011). Topographical organization of the pedunculopontine nucleus. *Frontiers in Neuroanatomy, 5*, 22.

Masino, T. (1992). Brainstem control of orienting movements: Intrinsic coordinate systems and underlying circuitry. *Brain, Behavior and Evolution, 40*(2–3), 98–111.

Masino, T., & Grobstein, P. (1989a). The organization of descending tectofugal pathways underlying orienting in the frog, rana pipiens. *Experimental Brain Research, 75*(2), 245–264.

Masino, T., & Grobstein, P. (1989b). The organization of descending tectofugal pathways underlying orienting in the frog, rana pipiens. Ii. Evidence for the involvement of a tecto-tegmento-spinal pathway. *Experimental Brain Research, 75*(2), 245–264.

Masino, T., & Knudsen, E. I. (1990). Horizontal and vertical components of head movement are controlled by distinct neural circuits in the barn owl. *Nature, 345*, 434–437.

McCrea, D. A., & Rybak, I. A. (2008). Organization of mammalian locomotor rhythm and pattern generation. *Brain Research Reviews, 57*(1), 134–146.

McDougall, W. (1903). The nature of inhibitory processes within the nervous system. *Brain, 26*(2), 153–191.

McElvain, L. E., Chen, Y., Moore, J. D., Brigidi, G. S., Bloodgood, B. L., Lim, B. K., . . . Kleinfeld, D. (2021). Specific populations of basal ganglia output neurons target distinct brain stem areas while collateralizing throughout the diencephalon. *Neuron, 109*, 1–18.

McMahon, T. A. (1984). *Muscles, Reflexes, and Locomotion*: Princeton, NJ: Princeton University Press.

Menard, A., & Grillner, S. (2008). Diencephalic locomotor region in the lamprey--afferents and efferent control. *Journal of Neurophysiology, 100*(3), 1343–1353. doi: 10.1152/jn.01128.2007.

Meyer, A. F., O'Keefe, J., & Poort, J. (2020). Two distinct types of eye-head coupling in freely moving mice. *Current Biology, 30*(11), 2116–2130.

Mink, J. W. (1996). The basal ganglia: Focused selection and inhibition of competing motor programs. *Progress in Neurobiology, 50*(4), 381–425.

Mitchell, I., Dean, P., & Redgrave, P. (1988). The projection from superior colliculus to cuneiform area in the rat ii. Defence-like responses to stimulation with glutamate in cuneiform nucleus and surrounding structures. *Experimental Brain Research, 72*(3), 626–639.

Mogenson, G. J., Jones, D. L., & Yim, C. Y. (1980). From motivation to action: Functional interface between the limbic system and the motor system. *Progress in Neurobiology, 14*(2–3), 69–97.

Morales, M., & Margolis, E. B. (2017). Ventral tegmental area: Cellular heterogeneity, connectivity and behaviour. *Nature Review Neuroscience, 18*(2), 73–85. doi: 10.1038/nrn.2016.165.

Munoz, D. P., Pelisson, D., & Guitton, D. (1991). Movement of neural activity on the superior colliculus motor map during gaze shifts. *Science, 251*(4999), 1358–1360.

Munoz, D. P., & Wurtz, R. H. (1992). Role of the rostral superior colliculus in active visual fixation and execution of express saccades. *Journal of Neurophysiology, 67*(4), 1000–1002.

Nichols, T., & Houk, J. (1976). Improvement in linearity and regulation of stiffness that results from actions of stretch reflex. *Journal of Neurophysiology, 39*(1), 119–142.

Pavlova, E. L., Popova, L. B., Orlovsky, G. N., & Deliagina, T. G. (2004). Vestibular compensation in lampreys: Restoration of symmetry in reticulospinal commands. *Journal of Experimental Biology, 207*(Pt 26), 4595–4603. doi: 10.1242/jeb.6247.

Peterson, B. W. (1979). Reticulospinal projections to spinal motor nuclei. *Annual Review of Physiology, 41*, 127–140. doi: 10.1146/annurev.ph.41.030179.001015.

Peterson, B. W., Pitts, N. G., & Fukushima, K. (1979). Reticulospinal connections with limb and axial motoneurons. *Experimental brain research, 36*(1), 1–20.

Powers, W. T. (1973). *Behavior: Control of Perception*: New Canaan, CT: Benchmark Publications.

Redgrave, P., Dean, P., Mitchell, I., Odekunle, A., & Clark, A. (1988). The projection from superior colliculus to cuneiform area in the rat. I. Anatomical studies. *Experimental Brain Research, 72*(3), 611–625.

Redgrave, P., Marrow, L., & Dean, P. (1992). Topographical organization of the nigrotectal projection in rat: Evidence for segregated channels. *Neuroscience, 50*(3), 571–595.

Redgrave, P., Mitchell, I., & Dean, P. (1987). Descending projections from the superior colliculus in rat: A study using orthograde transport of wheatgerm-agglutinin conjugated horseradish peroxidase. *Experimental Brain Research, 68*(1), 147–167.

Roberts, T. D. M. (1967). *Neurophysiology of Postural Mechanisms*: London: Butterworths.

Roseberry, T. K., Lee, A. M., Lalive, A. L., Wilbrecht, L., Bonci, A., & Kreitzer, A. C. (2016). Cell-type-specific control of brainstem locomotor circuits by basal ganglia. *Cell, 164*(3), 526–537. doi: 10.1016/j.cell.2015.12.037.

Rossi, M. A., Go, V., Murphy, T., Fu, Q., Morizio, J., & Yin, H. H. (2015). A wirelessly controlled implantable led system for deep brain optogenetic stimulation. *Frontiers in Integrative Neuroscience, 9*, 8.

Rossignol, S., Dubuc, R., & Gossard, J.-P. (2006). Dynamic sensorimotor interactions in locomotion. *Physiological Reviews, 86*(1), 89–154.

Rushworth, G. (1960). Spasticity and rigidity: An experimental study and review. *Journal of Neurology, Neurosurgery, and Psychiatry, 23*(2), 99.

Sato, M., & Hikosaka, O. (2002). Role of primate substantia nigra pars reticulata in reward-oriented saccadic eye movement. *Journal of Neuroscience, 22*(6), 2363–2373.

Shank, E. J., Seitz, P. K., Bubar, M. J., Stutz, S. J., & Cunningham, K. A. (2007). Selective ablation of gaba neurons in the ventral tegmental area increases spontaneous locomotor activity. *Behavioral Neuroscience, 121*(6), 1224.

Sherman, D., Fuller, P. M., Marcus, J., Yu, J., Zhang, P., Chamberlin, N. L., . . . Lu, J. (2015). Anatomical location of the mesencephalic locomotor region and its possible role in locomotion, posture, cataplexy, and parkinsonism. *Frontiers in Neurology, 6*, 140. doi: 10.3389/fneur.2015.00140.

Shik, M., Severin, F., & Orlovsky, G. N. (1966). Control of walking and running by means of electrical stimulation of mid-brain. *BIOPHYSICS-USSR, 11*(4), 756–&.

Shik, M. L., & Orlovsky, G. N. (1976). Neurophysiology of locomotor automatism. *Physiological Reviews, 56*(3), 465–501.

Stiles, L., & Smith, P. F. (2015). The vestibular–basal ganglia connection: Balancing motor control. *Brain Research, 1597*, 180–188.

Takakusaki, K. (2008). Forebrain control of locomotor behaviors. *Brain Research Review, 57*(1), 192–198. doi: 10.1016/j.brainresrev.2007.06.024.

Takakusaki, K. (2017). Functional neuroanatomy for posture and gait control. *Journal of Movement Disorders, 10*(1), 1.

Takakusaki, K., Habaguchi, T., Ohtinata-Sugimoto, J., Saitoh, K., & Sakamoto, T. (2003). Basal ganglia efferents to the brainstem centers controlling postural muscle tone and locomotion: A new concept for understanding motor disorders in basal ganglia dysfunction. *Neuroscience, 119*(1), 293–308.

von Holst, E., & Mittelstaedt, H. (1950). The reafference principle. In: Martin, R. (Ed.), *The Collected Papers of Erich Von Holst* (pp. 139–173). Coral Gables: University of Miami Press.

Yin, H. H. (2014). How basal ganglia outputs generate behavior. *Advances in Neuroscience, 2014*, 768313.

Yin, H. H. (2016). The role of opponent basal ganglia outputs in behavior. *Future Neurology, 11*(2), 149–169.

Zelenin, P. V., Orlovsky, G. N., & Deliagina, T. G. (2007). Sensory-motor transformation by individual command neurons. *Journal of Neuroscience, 27*(5), 1024–1032. doi: 10.1523/JNEUROSCI.4925-06.2007.

7 Transition Control

In the previous chapter, we reviewed evidence for the descending basal ganglia (BG) command of position controllers for orientation and body configuration. In particular, we advanced the hypothesis that projection neurons from the SNr and ventral tegmental area (VTA) can send position reference signals to comparators of various position controllers, which are mostly found in the brainstem. As the firing rates of many SNr neurons represent the instantaneous position vector, it follows that the rate of change in such BG output reflects the movement velocity. In this chapter, we shall review evidence for the presence of a velocity controller in the BG, which is situated just above the position controllers and used to adjust the position reference signals at a specified rate. It is hypothesized that the striatonigral circuit implements a neural integrator that converts the striatal output signal into the rate of change in the nigral output.

In volitional behavior, movement velocity can be varied arbitrarily. Velocity control is only one example of a broader category of transition control, defined as the control of the rate of change in perceptual representations. It is hypothesized that the BG circuits implement a variety of transition control systems by generating outputs that alter the reference signals of lower-level controllers.

7.1 VELOCITY CONTROL

The most common symptoms of Parkinson's disease include rigidity, tremor, and akinesia, which result from the death of dopamine (DA) neurons that innervate the BG. In the most extreme cases, patients are frozen (akinesia), unable to initiate volitional actions (Sacks, 1991). Striatal degeneration or DA depletion is usually associated with slowed movements, and the type of movement affected depends on the striatal region damaged (Leigh et al., 1983; Hotson et al., 1986; Rascol et al., 1989; Kori et al., 1995; Lasker & Zee, 1997; Vermersch et al., 1999). An old idea, which remains popular today, is that the BG provide motor energy or vigor (Schwab et al., 1959; Turner & Desmurget, 2010; Dudman & Krakauer, 2016).

The most common measure of motor output in primate studies is electromyography (EMG), which measures muscle activity, but this measure is only useful for studying the lowest level of tension control. Kinematic variables are largely independent of muscle contraction as measured by EMG. A given posture can be associated with a wide range of muscle outputs. Velocity or position can be controlled by varying muscle tension.

In one early study, Crutcher and DeLong found that many striatal neurons (from the putamen) show direction-specific activity that is relatively independent of EMG (Crutcher & DeLong, 1984a). They concluded that the BG may specify movement parameters independent of the activity of specific muscles. Studies also showed that globus pallidus internus (GPi) inactivation reduced the velocity and amplitude of reaching movements (Mink & Thach, 1991; Inase et al., 1996). Using positron emission tomography (PET) imaging in humans, Turner and colleagues showed that activity in the putamen (sensorimotor striatum) and GPi is strongly correlated with the average speed and amplitude of arm movements during a continuous tracking task (Turner et al., 2003). In another study, recording striatal activity from mice on a rotarod task, Costa et al. (2004) also found correlations between the speed of rotation on the rotarod and single-unit activity in the sensorimotor striatum. As mice usually move more quickly with a higher speed of rotation, this finding suggests a correlation between running speed and striatal activity. By contrast, during the baseline rest period, when mice were not moving, striatal neurons showed a much lower firing rate (Costa et al., 2004; Yin et al., 2009). Together, these results suggest that sensorimotor striatal activity is independent of lower-level muscle output but is instead related to kinematics (position and its derivatives). However, due to technical limitations in these early studies, there was no direct measure of kinematic variables in freely moving animals. Consequently, the precise relationship between striatal activity and kinematics remains obscure.

Using simultaneous movement tracking and *in vivo* electrophysiology (as described in Chapter 6), Kim et al. examined the relationship between striatal activity and movement velocity (Kim et al., 2014). They used a Pavlovian stimulus-reward task in which a sucrose reward was delivered following an auditory cue. The mouse had to stand on an elevated platform facing the reward spout. Once trained, the mouse moved toward the spout immediately following the auditory cue to collect the sucrose reward. As shown in Figure 7.1, many spiny projection neurons (SPNs) were correlated with vector components of velocity. Firing rates reflect either horizontal velocity or vertical velocity (rate of change in x or y position coordinates). They can be divided into different classes corresponding to velocity in different directions (leftward, rightward, upward, and downward). For example, the firing rate is positively correlated with leftward movement velocity but negatively correlated with velocity in the opposite direction.

Since many studies reported that striatal activity could be modulated by reward expectancy (Kawagoe et al., 1998; Stalnaker et al., 2012; Wang et al., 2013), Kim et al. tested whether the velocity-related SPNs also showed this property. They found that SPNs showed velocity tuning regardless of whether a reward was delivered. On some trials, they delivered an aversive air puff

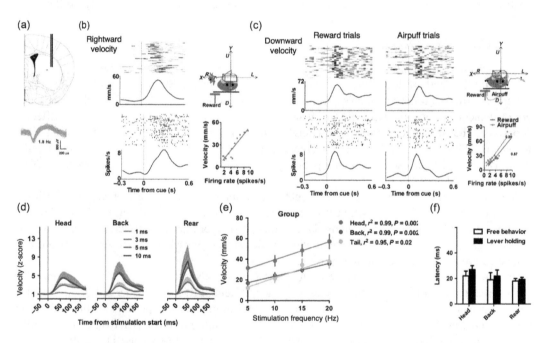

FIGURE 7.1 Sensorimotor spiny projection neurons (SPNs) represent movement velocity. (a) Top, placement of electrode array in the dorsolateral or sensorimotor striatum is shown in a coronal section of the mouse brain. Bottom, representative spike waveform of an SPN. (b) Top, movement velocity in the rightward direction. Bottom, firing rate of a SPN recorded in the dorsolateral striatum. There is a high correlation between neural activity and velocity. (c) A representative SPN showing a correlation with downward velocity. The relationship between firing rate and velocity is not affected by the outcome of the trial, whether it is a sucrose reward or an aversive air puff. (a–c) Reproduced from Kim, Barter, Sukharnikova, and Yin. (2014) with permission. (d) Normalized movement velocity (z-scored) from three body markers using 3D motion capture following a single pulse of optogenetic stimulation of direct pathway spiny projection neurons (dSPNs). (e) Movement velocity as a function of stimulation frequency. (f) Movement latency following dSPN stimulation. Free behavior, mouse is moving freely when stimulation begins; lever holding, mouse is trained to hold a lever down, and stimulation begins during the holding period. (d–f) From Bartholomew et al. (2016) with permission.

Transition Control

from the same location as the sucrose reward, but this manipulation in outcome valence did not change the relationship between SPN firing rate and velocity.

Stimulation experiments provide further support for the role of the striatum in velocity control (Figure 7.1d–f). Selective stimulation of direct pathway spiny projection neurons (dSPN) neurons using optogenetics generates movements with a velocity that depends on stimulation frequency (Rossi et al., 2015; Bartholomew et al., 2016; Kim et al., 2019). Even a single pulse of light generated movement with a short latency (~20 ms).

Taken together, these findings show that SPNs in the sensorimotor striatum can represent the movement velocity vector. This conclusion is also supported by studies using very different behavioral measures like locomotion and joystick manipulation (Rueda-Orozco and Robbe, 2015; Yttri & Dudman, 2016). Since dSPNs project to the SNr, which contains neurons that represent the position vector (see Chapter 6), these results suggest the presence of a neural integrator in the striatonigral circuit that converts velocity to position. The firing rate of dSPNs can represent the velocity error signal that enters the integrator, where it is converted into a sequence of position reference signals. The rate of change in the SNr output is proportional to the dSPN output.

7.2 DA AND KINEMATICS

The nigrostriatal projection is the main source of DA to the dorsal striatum. As we have seen in Chapter 3, DA can strongly modulate the output of SPNs (Lahiri & Bevan, 2020). Increased DA signaling generally increases locomotion and stereotyped behaviors like gnawing, sniffing, and licking (Costall et al., 1977; Giros et al., 1996).

An obvious question is how the activity of DA neurons is related to kinematic variables. Using a similar experimental design as described above, Barter et al. recorded DA neurons in the SNc (Barter et al., 2015a). As shown in Figure 7.2, although most DA neurons showed phasic activity at the time of cue presentation or reward delivery, as previously reported (Schultz, 1998), their activity was

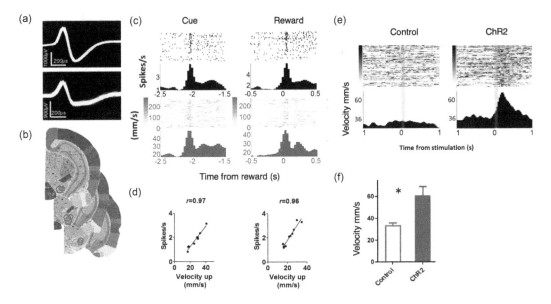

FIGURE 7.2 Activity of SNc dopamine (DA) neurons represents movement kinematics. (a) Representative spike waveforms from SNc DA neurons. (b) Locations of electrodes shown in coronal brain sections. (c) Top, raster plots of representative DA neurons showing correlation with upward velocity. Burst activity at both cue (left) and reward (right) is shown. Bottom, corresponding velocity plots. (d) Correlation between firing rate and upward velocity. (e) Optogenetic stimulation of DA neurons using channelrhodopsin 2 (ChR2) expressed in TH+ neurons. Stimulation increased velocity. (f) Summary of optogenetic results. From Barter et al. (2015a) with permission.

actually correlated with movement velocity and acceleration, independent of learning. There was a contraversive preference, e.g., left SNc DA neurons were more commonly correlated with the velocity of rightward movement. Likewise, optogenetic stimulation of DA neurons produced contraversive movements. Just like the SPNs, the DA neurons also showed the same correlation with movement kinematics regardless of outcome valence—whether aversive (air puff trials) or rewarding (sucrose trials). The activity of many SNc DA neurons is thus similar to their target neurons in the striatum.

According to a prominent view of DA function, phasic DA activity signals reward prediction errors (RPE), the difference between actual and expected rewards (Schultz et al., 1997; Schultz, 1998). But the results on kinematic representation in DA neurons do not support this conclusion. We shall postpone a detailed discussion of the role of DA in learning until Chapter 12. For now, it suffices to point out some key methodological differences. Most studies that purportedly showed DA encoding in RPE were performed in head-fixed animals (Waelti et al., 2001; Fiorillo et al., 2003; Cohen et al., 2012; Eshel et al., 2015). When the animals are prevented from moving, neither subtle movements nor the neural signals necessary for movement generation are abolished. Results showing the relationship between phasic DA activity and kinematics suggest that the previously observed shifts in phasic DA activity over time could reflect changes in movement. In previous experiments that tested the RPE hypothesis, the actual behavior generated to acquire or consume rewards was never carefully measured and represented a major confound. In Chapter 12, we shall critically examine previous results in support of the RPE hypothesis and present findings that are incompatible with this interpretation.

7.3 POSITION CONTROL VERSUS VELOCITY CONTROL

To explain the results reviewed above, the hypothesis advanced here is that the sensorimotor cortico-BG circuit contains a movement velocity controller. Velocity, in the sense defined here, is the rate of change in position-related signals from proprioception. For example, the velocity of a hand being waved is roughly the first derivative of a joint angle. Velocity in this sense is not the same as changes in spatial location—being transported in a car would not activate the velocity controller, yet running on a treadmill would. It is also distinct from the overall rate of behavior in some time window, e.g. the rate of lever pressing, which is a higher-order transition variable that we will consider in Chapter 8.

According to the present hypothesis, the velocity controller is hierarchically above the position controller. The velocity controller is implemented by the sensorimotor BG circuit. Its output from the SNr and presumably other BG output nuclei sends reference signals to the position controllers described in Chapter 6. To appreciate the functional significance of this organization, we must first consider the difference between velocity control and position control. If we are only concerned with start and end positions, then a position controller would be sufficient. Such a controller can move its position sensors to acquire the values specified by the reference signal, but it cannot control velocity. The actual velocity is determined largely by loop gain. If we assume, for the sake of simplicity, that both the environment and the effector are stable, a certain amount of position errors will result in a certain rate of change in position as it is corrected. Since velocity is not directly controlled, however, it is not possible to reach some arbitrary velocity. If we wish to vary velocity, it would be necessary to sense the rate of change in position separately, as an analog variable, and control it using negative feedback. The output from the velocity controller is a sequence of position reference signals that change at the desired rate. Note that this does not mean that a precise velocity value must be reached for each movement. Rather, a velocity reference signal can serve as a general signal for action initiation, like stepping on the gas pedal of a car, even if movement velocity is not explicitly controlled for its own sake. The reference signal to the velocity controller could come from any higher-level control system, so that the instantaneous velocity reference could reflect the error in controlling another variable. Consequently, a velocity reference at any given time can be considered a general-purpose command signal for volitional behavior.

Transition Control

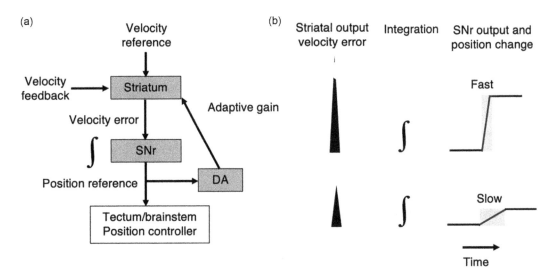

FIGURE 7.3 Velocity control. (a) Illustration of the velocity controller found in the sensorimotor cortico-BG network. The velocity loop is above the position loop, in a cascade organization. Dopamine (DA) acts as an adaptive gain in this circuit. (b) The striatal velocity error at each time step enters an integrator. The rate of change in the integrator's output, as illustrated by the slope, is proportional to the magnitude of its input. Thus, a large velocity error signal would result in a fast movement, whereas a small velocity error would result in a slow movement.

Suppose we have a position controller with three possible positions: 1, 2, and 3. A reference signal of 3 is a command to sense position 3. If the reference signal is fixed at 3, this position controller will maintain position 3, producing variable outputs as needed to resist the environmental disturbance. If we plot position (a string of 3s) over time, we will see a flat line. To control how *quickly* the value changes from 3 to 1, we must represent the velocity variable explicitly. The velocity error must be converted to the rate of change in the position reference, regardless of the start or end position. The relationship between velocity control and position control is illustrated in Figure 7.3.

In short, by placing the velocity controller just above the position controller and using an integrator in the output function of the velocity controller, it is possible to generate the appropriate reference signals for position controllers in volitional movement. In velocity control, the rate of change in sensed body position is controlled by generating variable descending outputs to command position controllers. This is known as cascade control in engineering.

7.4 VECTOR INTEGRATION TO ENDPOINT (VITE) MODEL

An early model of velocity control, inspired by work on the BG (Anderson & Horak, 1985), is the vector integration to endpoint (VITE) model proposed by Bullock and Grossberg (Bullock & Grossberg, 1988, 1989). According to this model, when generating a reaching movement toward a target, the difference between the current effector position and the target position (called the difference vector) is multiplied by a separate GO signal, which specifies the speed of movement. In other words, a distance error signal is used to scale the velocity command, which is integrated to yield the position command, which is then sent to spinal motor neurons. Downstream, Bullock and Grossberg also proposed a separate model, called FLETE (Factorization of LEngth and Tension), to explain the function of the downstream spinal circuits. Like the description of the lower levels from Chapter 6, the FLETE model also assumes that muscle and tension can be independently controlled, and that these controllers are commanded by the output of the VITE model (Bullock & Grossberg, 1989).

The VITE model is a pioneering effort in explaining velocity control. It has three noteworthy features. First, it assumes correctly that there is a hierarchical relationship between kinematic variables and muscle length and tension, and that trajectories can be generated independently of force, as variable force is generated even when the same movement is repeated. Second, it uses a separate multiplicative gain (the GO signal) to scale the velocity command. Even if a new command is selected, there will be no output unless the GO signal is sufficient. In this way, a new target can be primed, approaching the threshold for activation, before the GO signal becomes sufficient to generate a suprathreshold command. Third, it uses a comparison function to compare the current position and the desired position, as well as an integrator to generate commands that are sent to lower levels of muscle length and muscle force control. These features are similar to the model of velocity control presented here.

However, the VITE model also has a number of limitations and differs from the present model in important ways. It uses a comparison between target and current position from visual perception to generate a command that is then scaled to activate the muscle length and tension controllers, but this is not a necessary feature of velocity control per se. It is only needed to explain behaviors that require the control of a spatial relationship, which we shall discuss separately in Chapter 10. Moreover, it fails to describe the action hierarchy that is needed to achieve velocity control, or the appropriate closed loop negative feedback loop for velocity control and position control. It also neglects the role of position control (as opposed to spatial relationship control) and the hierarchical relationship between velocity control and position control. It also assumes that integration is achieved by projections to the cortex, whereas the present model is based on evidence for local neural integrators in the BG.

7.5 A NEURAL INTEGRATOR IN THE BG

As already mentioned, in both striatum and SNr, distinct classes of neurons were found that are selective for opposite directions of movement along a particular axis (Kim et al., 2014; Barter et al., 2015b). The difference is that SNr output signals the position vector, whereas SPNs signal the rate of change in the position vector. These results suggest that the striatonigral circuit implements a neural integrator in the output function of the velocity controller. This integrator transforms velocity errors into a sequence of position reference signals. Ignoring the indirect pathway for now for the sake of simplicity, the position reference signal is sent from the dSPN to the target of SNr projections (e.g. superior colliculus or brainstem nuclei) via the basic disinhibition circuit (Chapter 4). Consequently, dSPN activation can have a net excitatory effect on the target neurons. The integrator converts the instantaneous velocity error, which is reflected in a firing rate of velocity-related SPN within a short window, into the rate of change of SNr output and thus also the rate of change in the net excitatory signal reaching targets in the brainstem.

Integrators are often used in control systems to achieve stability. A control system can generate output even when the current error signal is zero because the output at any time reflects the total error accumulated in some time window in the past. This feature allows the recently achieved position to be maintained by sending a steady reference command to downstream systems. The output of the integrator represents the sum of its inputs accumulated in a recent time window, and the rate of change in the output is proportional to the input magnitude—which is represented by striatal output and modulated by DA (Yin, 2014).

7.5.1 Integrator Dynamics

A perfect integrator exists only in mathematics. In nature, integrators are more or less leaky. A common example of a leaky integrator is a bucket of water with a hole in the bottom (Figure 7.4). The rate at which the water flows out is proportional to the amount of water in the bucket. Using the bucket analogy, the water level represents the BG output firing rate and the current egocentric position coordinate. The inflow to the bucket is the direct pathway activity at any time. It is possible to

Transition Control

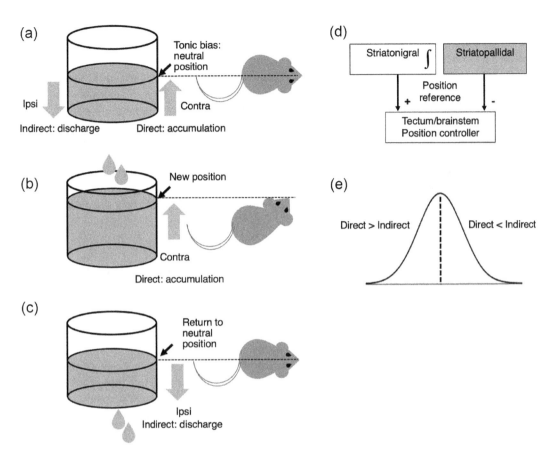

FIGURE 7.4 Neural integration and behavior. (a) A water bucket is used to illustrate the integrator. The water level at any time indicates the instantaneous position coordinate of a reference position vector, as represented by the BG output. At rest, there is water in the bucket, reflecting the neutral position. (b) Direct pathway activity represents inflow to the bucket, generating a contraversive movement. (c) Indirect pathway activity discharges the integrator, resulting in a return to the neutral position. (d) Illustration of the impact of direct and indirect pathway activity on structures that receive BG output projections. The position reference reflects the input received by regions targeted by BG output projections, e.g. nigrotectal projections. While the BG output is inhibitory, the net effect of disinhibition is excitatory. For example, striatonigral activation results in tectal activation. Signs indicate net effect of the circuit. (e) Illustration of a movement velocity profile. When direct activity exceeds indirect activity for a given direction of movement, there is net acceleration. When indirect activity exceeds direct activity there is deceleration.

use the leak to slow down the system's output, achieving an effect similar to damping (Yin, 2017). The size of the leak can also be adjusted in real time.

The finding that opponent types of neurons are common in the BG suggest the presence of a phase splitter circuit (Fan et al., 2012; Barter et al., 2014, 2015b; Hughes et al., 2019). In a phase splitter, a particular input signal is converted to a pair of output signals that are opposite in polarity. These antiphase signals can be used to generate opponent classes of output. A classic example is the reciprocal inhibition circuit in the spinal cord, where the same primary afferent signal is used to create a pair of opponent signals for agonist and antagonist muscles, e.g. the biceps contract while the triceps are relaxed. This circuit requires an inverter, implemented by an inhibitory interneuron, to change the sign of a signal. At present, the circuit mechanisms for the phase splitter in the BG remain unclear. As shown in Chapter 3, the striatum also contains inhibitory interneurons as well as recurrent inhibition circuits that could implement reciprocal inhibition circuits. Another possible implementation involves the bridging collaterals, which are axon branches from the striatonigral pathway that target the GPe (Cazorla et al., 2014). Through these collaterals to the GPe, dSPNs can also disinhibit another set of SNr neurons. Whereas direct striatonigral projections suppress SNr neurons, it is possible for bridging collaterals to exert a net excitatory effect on SNr neurons.

In movement velocity control, distinct vector components of velocity are controlled independently. For example, moving to the left and moving to the right involve two different populations of striatal neurons and presumably two distinct neural integrators. These integrators have opponent functions. At the neutral starting position of facing straight ahead, the two integrators are assumed to be half full, which is reflected in the tonic firing rates of SNr neurons near half maximum. When turning to the left, the left integrator is filled while the right integrator is emptied or discharged. To stop this ongoing movement or to move in the opposite direction, the integrator must stop accumulation. When the left integrator is discharged, the right controller is filled. The net change in position (displacement or movement amplitude) is therefore equal to the inflow minus the outflow or leak, as given by the following equation:

$$dQ_o = G_o * e - K_d * Q_o$$

Here dQ_o denotes the change in the system output (Q_o). This is the amount that is added to the integrator. G_o is the multiplicative gain, and e is the velocity error. The leak is a product of the damping constant (K_d) and the current output (Q_o). Inflow minus damping equals the rate of accumulation in the integrator or displacement in a single time step. For each time step, the water level can increase or decrease depending on the inflow and the leak. If inflow equals outflow, then there is no movement. If inflow is greater than outflow, the change will be positive, causing movement in one direction; if outflow exceeds inflow, the change will be negative, corresponding to movement in the opposite direction. Such a mechanism can be used for damping, stopping a movement, and returning to a neutral position.

There are multiple versions of the indirect pathway that can implement the damping mechanism (Figure 2.8). The STN component allows a reverberation circuit that amplifies the indirect pathway activity by exciting the BG output neurons in the SNr and GPi. Because the STN receives inputs from many brain regions, this organization allows it to adjust the discharging process independently. For example, cortical input can activate or enhance the discharging process as needed to terminate a movement or change direction. In addition, STN activation can also generate ipsiversive movement via direct projections to the brainstem, bypassing the BG circuit (Watson et al., 2021).

The present model explains why SPNs are usually very quiet and fire sparsely, whereas output neurons in the GPi and SNr fire at high rates, often tonically. Velocity-related neurons typically have low baseline firing rates with short bursts of activity, whereas position-related neurons are tonically active with higher firing rates. Since the rate of change in the output is proportional to the magnitude of its input, a large velocity signal will produce a fast movement, and a small signal will produce a slow movement. The resultant displacement, which reflects movement amplitude, is determined by the total amount of signal accumulated in the integrator. Consequently, the commanded position controller generates position changes in a specific direction and at a specific rate.

The level of accumulation in the integrator is to be interpreted as the value of a component in the instantaneous egocentric position vector. Tonic BG output represents the neutral position of different body parts, from which both increases and decreases represent signals sent to target neurons. When the animal is resting with no apparent voluntary behavior, there is still position control due to tonic reference signals sent to the downstream structures. As reviewed in Chapter 6, in tonically active SNr neurons, the effective zero point is the midpoint of the dynamic range, reflecting the neutral or symmetric egocentric position (straight ahead for the orienting system). The BG output does not change in the absence of voluntary behaviors.

7.6 DIRECT AND INDIRECT PATHWAYS

The leak or damping mechanism has the opposite effect as accumulation in the integrator. A similar mechanism can also implement the "discharging" function to remove the accumulated signal. Since the direct pathway is hypothesized to be the main integrator in the BG, the most obvious candidate for damping and discharging is the indirect pathway. The key difference between damping and discharging is the timing and perhaps the magnitude of the leak. If striatonigral transmission fills the integrator, striatopallidal activation during this process could act as damping. It can slow down the accumulation, resulting in a slower movement in that direction. The velocity profile of the movement often has a bell-shaped function, so the damping process may not be uniform throughout the movement. During the initial acceleration, there could be very little or no damping, while during deceleration, it could increase until the target is reached. To return to the starting position, the discharging of the integrator may be needed. For example, one turns to the left and returns to the initial position, but this discharging process is delayed. The leak mechanism could be adjusted so that it is very low during the accumulation process but becomes active once the movement in one direction is complete. In this case, the integrator is filled first and then discharged.

The direct pathway is traditionally called the "go" pathway, whereas the indirect pathway is the "no-go" pathway (Albin et al., 1989; Gerfen et al., 1990; Collins & Frank, 2014). This account predicts that dSPNs and indirect pathway spiny projection neurons (iSPNs) cannot be activated at the same time, as their outputs would conflict with each other. This prediction was not supported by initial attempts to measure activity in these pathways simultaneously (Cui et al., 2013; Isomura et al., 2013; Tecuapetla et al., 2014; Meng et al., 2018). These studies found that neurons in both pathways are simultaneously activated during behavior. One interpretation is that the indirect pathway may play a complementary role by suppressing irrelevant behaviors, as proposed by the focused selection model (Mink, 1996). However, the temporal and spatial resolution in both neural and behavioral measures used in the initial studies were not sufficient to determine the relationship between striatal activity and behavior. For example, fiber photometry, which collects signals from a population of neurons rather than from individual neurons, was used to measure neural signals, and accelerometers, which cannot accurately sense the direction of movement, were used to measure behavior (Cui et al., 2013). In freely moving animals, for example, what appears to be concurrent activation may simply be due to asynchronous or alternating activity that is averaged across neurons or sampled at a low rate. Simultaneous dual-color calcium imaging in the striatum showed that the activity of iSPNs often lags that of dSPNs. iSPNs are often active during deceleration or at the start of ipsiversive movements.

In the velocity control model, opponent effects on the integrator determine the direction of motion. The velocity profile of the movement reflects the actions of the opponent pathways, with the direct pathway contributing to acceleration and the indirect pathway to deceleration. Thus, the direct pathway could initiate the action, and the indirect pathway could act as a brake through a delayed discharge of the integrator. Note that this interpretation of integrator dynamics only applies to velocity control. As we shall see, for higher-order transitions, what is accumulated and discharged in the integrator could reflect quantities other than changes in position reference.

Increasing striatal output on one side produces contraversive turning (Koshikawa, 1994; Yu et al., 2022). This is largely due to direct pathway activation, because unilateral direct pathway stimulation produces the same effect. By contrast, indirect pathway stimulation can produce the opposite effect of ipsiversive turning (Kravitz et al., 2010). Stimulation of many SPNs, including those that are unrelated to turning behavior, and stimulation of only a few SPNs specifically related to turning produced similar effects: in both conditions, there is contraversive turning (Zhang et al., 2022). This suggests that a subset of SPNs is sufficient for the generation of the movement in question and that SPNs located next to the turning-related SPNs do not usually produce conflicting outputs.

7.7 NEUROBIOLOGICAL IMPLEMENTATION OF INTEGRATION

One possible mechanism for integration in the striatonigral circuit is synaptic facilitation. Striatonigral synapses are strongly facilitating—the postsynaptic response grows with repeated inputs within a time window (Connelly et al., 2010). Synaptic facilitation is usually found at synapses with low release probability (Zucker & Regehr, 2002). The increasing GABAergic output with each additional pulse leads to more inhibition of SNr, but the net effect on target neurons (e.g. those in the superior colliculus) is the opposite due to the inhibitory effect of SNr outputs (Figure 7.4d). To see how the integration is achieved in this circuit, we must focus on the firing rate of neurons receiving SNr projections. The output from dSPNs is the input to the integrator. Through disinhibition, the integrator output will be reflected in the collicular firing rate, as the nigrocollicular synapse is neither facilitating nor depressing (Kaneda et al., 2008).

In a position controller, the capacity of the integrator has a specific physical interpretation: the dynamic range reflects the range of position coordinates spanning the movement of a body part, how quickly the integrator gets filled reflects velocity, and how much is accumulated reflects displacement. For a leftward head turn, if a given dSPN represents a velocity command, which is ultimately converted into collicular or brainstem output, the total displacement is determined by the number of striatonigral pulses and the velocity by the firing rate of dSPNs that reflect leftward velocity. If the integrator capacity is 5 pulses, then that would also correspond to the movement amplitude from the neutral position to the extreme left position of the head when it is fully turned. Using disinhibition, the superior colliculus would receive a signal reflecting the integral of the dSPN signal; its rate of change is proportional to the firing rate.

Increasing striatonigral transmission simply increases collicular output via disinhibition. To appreciate the accumulation process, it is important to focus on the net disinhibitory effect of striatonigral output on targets of BG projections. As the SNr neurons are inhibitory, the integration of inhibitory inputs results in more disinhibition or net excitation of target structures (Figure 7.4d). Thus, the plot of synaptic facilitation at the striatonigral synapse in Figure 7.5b is similar to the actual response of the target neurons in the brainstem or superior colliculus.

Recovery from facilitation shows how long the signal can be held in the integrator. Connelly et al. showed that, in postnatal day 20 rats, the response returns to the baseline level after 3 seconds; 10, 50, and 100 Hz all produced comparable facilitation. The response quickly asymptotes with additional pulses. If we use the bucket analogy, the integrator's output is proportional to the water level. The striatonigral integrator (bucket) appears to have limited capacity and reaches the maximum quickly, in most cases after 5 pulses, roughly as many pulses as expected from a short burst of SPN activity (Figure 7.5b). This explains why sustained firing in SPNs is rarely observed, as it is rarely needed to fill the integrator. Indeed, optogenetic experiments showed that just a few brief pulses of light delivered to dSPNs could generate significant movement (Bartholomew et al., 2016). The effect of striatonigral activation is similar to briefly stepping on the gas pedal; a car is an example of a velocity controller.

Synaptic facilitation is traditionally considered a high-pass filter (Zucker & Regehr, 2002). The underlying assumption is that high-frequency inputs increase release probability at the axon terminal and elicit postsynaptic spiking more reliably, so that they are more likely to be transmitted at facilitating synapses. But this view makes the unwarranted assumption that low firing rates at the presynaptic terminal would result in little postsynaptic response. In contrast, according to the present view, facilitation is a mechanism for neural integration (more similar to a low pass filter). Each subsequent pulse generates a larger output than the one before, reflecting the sum or accumulation of recent inputs. What makes the striatonigral integrator counterintuitive is its use of disinhibition with two GABAergic synapses instead of direct excitation, so we have to focus on the downstream nuclei that receive BG projections to appreciate the integration of signals.

On the other hand, synaptic depression can be a mechanism for differentiation (like a high-pass filter). Interestingly, both the STN–SNr synapse and the GPe–SNr synapse are depressing. At the GPe–SNr synapse, what is being differentiated is the disinhibitory signal from the indirect pathway. At the STN–SNr synapse, it could be input from the motor cortex to the STN (hyperdirect pathway) or indirect pathway input. Differentiation, being the inverse of integration, may allow the indirect or hyperdirect pathway output to offset the direct pathway output.

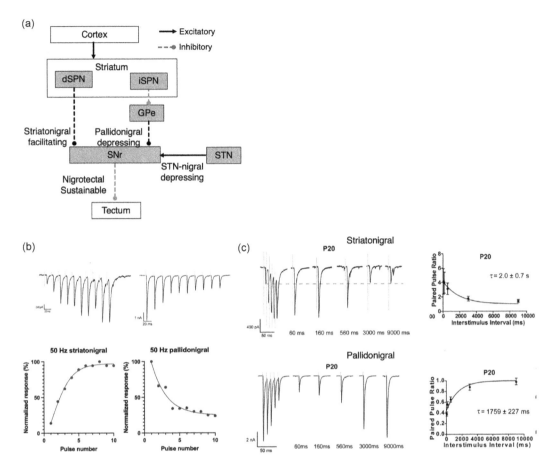

FIGURE 7.5 Synaptic facilitation and integration. Potential synaptic mechanisms for integration and discharging. (a) The striatonigral synapse is facilitating, whereas the pallidonigral synapse is depressing. The subthalamic nucleus (STN)-nigral synapse is excitatory and also depressing. The nigrotectal synapse is neither facilitating nor depressing. Shown are all inhibitory postsynaptic currents (IPSCs) with different voltage clamp configurations. (b) The facilitating striatonigral synapse allows the direct pathway to serve as a leaky integrator, whereas the depressing pallidonigral synapses allow the indirect pathway to exert the exact opposite effect of discharging the integrator. Each pallidonigral pulse would roughly discharge one pulse from the striatonigral integrator. Note that, given the disinhibitory circuit, the normalized response measure here is also proportional to the net output of downstream neurons receiving nigral projections, e.g. in the tectum. Stimulation artifacts shown in gray. (c) Recovery from facilitation at the striatonigral synapse or depression at the pallidonigral synapse can be fitted with an exponential function. The time constant (τ) is the time it takes to reach ~0.37 (1/e) value of the initial value. Based on Connelly, Schulz, Lees, and Reynolds (2010) with permission.

7.8 DOPAMINE AND GAIN CONTROL

As a neuromodulator, DA does not produce current in the postsynaptic neuron directly but changes the postsynaptic response to excitatory and inhibitory inputs (Hjelmstad, 2004; Gerfen & Surmeier, 2011). For example, it can potentiate dSPN output in response to glutamatergic input by activating D1 receptors (Lahiri & Bevan, 2020) and enhance GABA release from striatonigral terminals (Ding et al., 2015). These effects on dSPNs increase the rate of accumulation in the integrator. In contrast, DA activation of D2 receptors suppresses the excitability of striatopallidal neurons, which is assumed to reduce damping. DA can increase accumulation in the integrator while simultaneously reducing the leak, thus increasing the gain of velocity control.

7.8.1 Bradykinesia and Akinesia

The velocity control model has important implications for our understanding of classic movement disorders. As already mentioned, the most obvious symptoms related to velocity control are bradykinesia, slowness in movement, and akinesia, lack of volitional movement. Both are considered core symptoms of Parkinson's disease. According to the influential rate model (Chapter 4), bradykinesia or akinesia is due to excessive BG output (Albin et al., 1989; Alexander and Crutcher, 1990; DeLong, 1990). DA depletion results in reduced direct output and increased indirect pathway output, and thus increased BG output. In turn, this is responsible for excessive inhibition of the thalamus and reduced activity in motor cortical regions that receive the thalamic projections. But as we have seen, higher BG output does not necessarily lead to more inhibition of behavior, since BG output is not monolithic. There are multiple classes of neurons corresponding to different dimensions of movement, and for each dimension there are opponent classes of neurons reflecting movement direction.

According to the present model, DA depletion reduces the responsiveness of striatal neurons to the velocity reference, resulting in reduced striatal output, especially from dSPNs. As a result, the SNr projection neurons cannot change their firing rates quickly or at all, and either bradykinesia or akinesia is observed (Panigrahi et al., 2015).

DA depletion has opposite effects on direct and indirect neurons. It reduces the direct pathway velocity error entering the integrator, but increases damping because the striatopallidal output is higher without D2 receptor activation. Mice with genetic deletion of D2Rs showed bradykinesia (slowness in movement) despite relatively normal DA release (Lemos et al., 2016). The lack of D2 signaling, which presumably increases overall iSPN output, could also reduce movement velocity.

Instead of producing a general increase in BG output neurons, as predicted in the rate model, DA depletion reduces the rate of change in multiple classes of BG output neurons corresponding to different directions of motion in different body parts. In the striatum, the reduction in peak firing rate reflects a reduction in peak velocity. In akinesia, no striatal signal enters the integrator, and the position reference does not change. Akinesia is usually accompanied by rigidity, so that the current position or body configuration is continuously defended against disturbances given the fixed position reference. Although the lower position controllers can still produce output to resist disturbances, what is lacking are descending commands, represented by changes in BG outputs, to change the position reference signals.

7.8.2 Adaptive Gain

As reviewed earlier, there are different classes of DA neurons, similar to the different classes of SPNs representing different velocity vector components. Since GABAergic projection neurons in the SNr send collaterals to DA neurons, DA neurons can be disinhibited by the striatonigral pathway, just like any target of SNr outputs (Tepper & Lee, 2007). In other words, when dSPNs fire, they can also exert a net excitatory effect on the DA neurons. When a set of dSPNs for leftward velocity are activated, they can also disinhibit a set of DA neurons that project back to the striatum. Because SNr GABA neurons represent position, whereas nearby SNc DA neurons represent velocity, the SNr–SNc pathway appears to perform differentiation (Barter et al., 2015a).

The output of the BG also project to DA neurons, which then project back to the striatum, where DA signaling can adjust the gain of velocity control (Figure 7.3). This circuit can implement an online parameter estimator, a direct method for achieving adaptive gain. In adaptive gain, the gain of the controller is not fixed but changes according to current demands. For example, DA neurons that fire during leftward movements are hypothesized to project to striatal neurons that also fire during leftward movements. This organization allows DA to further bias the activation of those dSPNs that activate the DA neurons via disinhibition.

Of course, the striatonigral pathway is not the only input that drives DA activity. There are multiple excitatory inputs to DA neurons, for example from the prefrontal cortex. These inputs could potentially increase striatal DA even before action generation. In other words, salient cues, which are known to elicit synchronized bursting of DA neurons, can prime striatal activity before action initiation (Comoli et al., 2003; Redgrave & Gurney, 2006; Rossi et al., 2013a). This priming effect could be more global, perhaps increasing excitability across many striatal regions. On the other hand, the adaptive gain mechanism is expected to be more specific to the striatal modules initiating the movement, providing a mechanism for sustained activation of such units.

The present model predicts that DA depletion should result in a degraded representation of velocity in striatal neurons since DA signaling is needed to reliably generate striatal output in response to excitatory inputs. This prediction is supported by a study on the relationship between movement and striatal neural activity in a mouse model of Parkinson's disease (MitoPark mice). Dudman and colleagues found that MitoPark mice showed clear bradykinesia and a selective loss of DA neurons. Moreover, velocity representation in striatal neurons was degraded in these mice (Panigrahi et al., 2015). DA replacement treatment with levodopa (L-DOPA) not only rescued bradykinesia but also restored the striatal representation of velocity. These results suggest that the velocity representation found in SPNs is enabled by the modulatory influence of DA. Without DA, the subthreshold membrane potential may still be related to velocity, but this activity does not reliably reach the threshold for action potentials. Because DA serves as a multiplicative gain, DA depletion and dSPN inhibition could have similar effects on peak movement velocity.

7.9 ADAPTIVE GAIN AND REINFORCEMENT OF ACTION PARAMETERS

Adaptive gain uses recent efference copy signals to adjust the gain. As we shall see in Chapter 8, this mechanism also promotes the repetition of the last action variant by biasing the corticostriatal input that is responsible for the most recent command. Experiments by Yttri and Dudman provide support for this hypothesis (Yttri & Dudman, 2016, 2018). They trained mice to move a joystick to earn rewards, and delivered stimulation that was contingent on movement velocity thresholds. As striatal stimulation can elicit movements, they carefully adjusted stimulation parameters to avoid generating overt movements. Stimulation of dMSNs in the dorsomedial striatum during the fastest third of movements resulted in a gradual increase in movement velocity on subsequent trials. By contrast, stimulation of iMSNs during the fastest third of movement resulted in slower movements. Changes in movement velocity are transient; once stimulation stopped, they gradually returned to baseline levels within 50 trials.

These results are incompatible with standard models of BG function. Action selection models predict that iSPN stimulation will reduce the probability of selecting actions that lead to iSPN activation, but stimulation changed the velocity of future movements without changing the probability of action selection. If iSPN activation simply stops or prevents actions, then we would expect a reduction in the number of movements, but that is not what was observed. These findings show that action selection and action parameter specification are dissociable. Stimulation of the direct and indirect pathways can change action parameters without changing the probability of action selection.

When stimulation is made contingent upon the slowest third of movements, dSPN stimulation results in slower movements on subsequent trials, whereas iSPN stimulation results in faster movements on subsequent trials. Once again, the movement velocity returned to baseline levels once stimulation stopped.

According to the adaptive gain hypothesis, when fast movements are accompanied by dSPN stimulation, the DA neurons receive an efference copy of the velocity reference and update the gain of those velocity controllers receiving the same velocity reference signal, which are responsible for the high velocity. DA selectively increases the excitability of dSPNs by activating D1 receptors. This is sufficient to make the fast-velocity command signals more easily generated. DA can selectively and transiently potentiate and depotentiate cortical inputs to the SPNs. If a fast movement is followed by increased adaptive gain, then the last reference signal is promoted briefly after the movement. On the other hand, iSPN stimulation following a fast movement is expected to have the opposite effect on DA neurons. It is expected to inhibit DA neurons (iSPNs disinhibit SNr GABA neurons, which inhibit DA neurons). Thus, DA neurons that are activated by the fast-velocity command are inhibited, reducing the adaptive gain for the striatal cells responsible for generating the command. In contrast, stimulating dSPNs during slower movements would increase adaptive gain for striatal units responsible for slow movements. Stimulating iSPNs during slow movements is expected to reduce adaptive gain for such movements, thus increasing velocity on future trials.

If all movements, regardless of their velocity, are accompanied by stimulation, the neurons for both fast and slow movements receive a comparable adaptive gain signal. Because there is no detailed quantification of the behavior other than velocity, it is unclear whether fast and slow movements can be distinguished in other ways, and whether they are generated by distinct controllers. It is also worth noting that Yttri and Dudman performed stimulation in the dorsomedial striatum, part of the associative BG network, rather than the sensorimotor network. As we shall see in later chapters, the dorsomedial striatum is critical for higher-order transition control, including repetition control. As we shall see in Chapter 9, a similar mechanism is responsible for many observations on self-stimulation (Wise, 2004; Kravitz et al., 2012; Rossi et al., 2013b).

7.10 CORTICOSTRIATAL CIRCUIT AND TRANSITION CONTROL

In most actions, the goal is not to reach a particular velocity. Rather, velocity is varied according to other needs and environmental contingencies, just as a car may be driven for many purposes. This raises the question of where the velocity reference comes from.

According to the present model, velocity control is merely one example of transition control. More generally, the rate of change in any higher-level perceptual representation can also be controlled by varying behavior. This type of control may be called transition control. According to the present model, transition control is a central function of the BG. The striatum receives reference inputs representing desired rates of change and perceptual inputs representing ongoing transitions in higher-order variables and generates the requisite transition command signals. It could implement a comparison function for transition control.

The major source for the transition reference inputs and perceptual inputs seems to be the cerebral cortex. Cortical circuits can generate abstract and invariant representations, which are critical for control (Rolls, 2016). For example, a monkey wishing to grab an apple must be able to recognize it as such despite changes in lighting or viewing angle. This type of object invariance can be generated using multiple stages of signal processing, with neurons at a given stage responding to just a limited set of neurons from an earlier stage and generating new representations at each stage. A single neuron in the last stage can respond to a combination of inputs from earlier processing stages (Rolls, 2016). The final product of the perceptual hierarchical processing, the invariant representation, can reach the BG via corticostriatal projections and enter a comparison function.

The BG do not directly contribute to the transformation of perceptual signals at different stages. BG neurons lack the type of detailed sensory tuning often found in cortical and thalamic neurons. The generation of invariant representations is mainly achieved by the posterior sensory cortices in forward projections from primary cortical regions to higher-order regions. From the perspective of the transition control model, the cortex can be divided into two major parts: an anterior (rostral) division and a posterior (caudal) area, which are also connected by long-range intracortical projections. In primates, this division is clearly separated by the central sulcus. The anterior division (frontal cortex) contains the executive network, and the posterior cortex contains the perceptual network. Not only do the sensory cortical regions influence frontal cortical regions via direct projections, but the key top-down influence of frontal regions on posterior sensory regions is also well established (Teuber, 1964; Buschman & Miller, 2007; Passingham & Wise, 2012).

Inputs from functionally related regions in the anterior and posterior cortex often converge upon the same area in the striatum (see Chapter 2). For example, the primary motor cortex (M1) in

Transition Control

the frontal division is highly connected with the primary somatosensory cortex (S1), and M1 and S1 projections both converge in the sensorimotor striatum. The frontal and posterior association regions both project to the association striatum. Each division has both primary and association areas, which are also connected with corresponding regions in the other division. Cortical areas that are strongly connected also tend to share striatal targets (Yeterian & Van Hoesen, 1978).

What is the functional significance of this organization? According to the present model, the frontal "executive" division is prescriptive, specifying the desired state, whereas the posterior sensory division is descriptive, representing the current state (Figure 7.6). These divisions are also reciprocally connected with the relevant thalamic nuclei. For example, the ventral tier "motor" thalamic nuclei are connected with the motor cortex; the posterior medial thalamic nuclei are connected with the posterior somatosensory cortex. Both send convergent projections to the same striatal region.

An open window is both a perception and potentially a goal state. Whether it "should be" open or closed also depends on still higher reference signals as well as other factors such as the weather and time of day. The command "open the window" specifies a desired reference state, retrieved from memory. If the window is currently closed, then matching the reference perception of an open window would require a particular action. Note the action of opening the window cannot be defined by muscle tension, joint angle, velocity, or even sequence of movements, but only by a well-defined goal state. It can be achieved in many different ways. Exactly how it is reached can be variable, as the signals from the final common pathway will vary each time a window is opened. The action is initiated by a top-down command, proportional to the higher-level error, to the transition controller from the corticostriatal projections. By directing outputs from related anterior executive and posterior sensory representations to the same striatal region, the organization of the corticostriatal projections can enable the comparison between reference input and perceptual input. A comparison of desired state and current state is achieved in the striatum, which issues the reference signals needed to recruit the needed effectors in achieving the desired state.

For any given striatal region implicated in the generation of specific actions, we expect to find corresponding anterior and posterior cortical inputs for reference and perceptual inputs. The frontal cortical division is expected to contain a collection of goals or reference states, which are memories of specific perceptual states. The posterior division houses the basic sensory hierarchical input functions discussed above. It is responsible for the transformations necessary for producing invariant representations and contains all the basic perceptual input functions that are being controlled at the level of transition control.

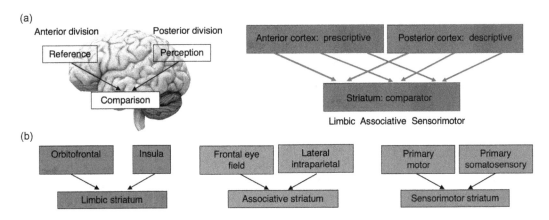

FIGURE 7.6 Convergence of anterior and posterior cortical projections in the striatum. (a) The anterior or frontal division of the cortex contains libraries for reference states, whereas the posterior or sensory division contains the higher-level sensory pathways. Corticostriatal projections from a given frontal area and a related posterior area can converge in the same striatal area. (b) Three examples of such convergent corticostriatal projection patterns.

Two excitatory corticostriatal inputs converging onto the same SPN cannot form a working comparison function since both reference and input signals have the same sign. A microcircuit is required to invert the sign of the perceptual signal, possibly using inhibitory interneurons. The output of the comparator can provide a descending signal for action generation. Alternatively, it can also reinforce or select a cortical command to reactivate itself or to generate the next component (address) in the sequence. This latter possibility will be discussed in more detail in Chapter 8. Either way, BG output is continuously updated to alter perceptual representations.

In "volitional" actions, descending commands change the reference signal. The animal can select any arbitrary part of its perceptual field and orient toward it, regardless of the input value from that location in the perceptual map. In the visual system, this means selecting any arbitrary location in the visual field and orienting toward that location, even if there is no salient input from it. In the bottom–up or "reflexive" mode, a salient stimulus may trigger an orienting movement because it exceeds the current reference signal of a lower-level controller.

At present, the detailed functional organization of different corticostriatal projections remains poorly defined, but the present account explains the convergence of corresponding anterior and posterior cortical divisions in the striatum, a prominent feature of corticostriatal projections. The present account does not preclude the possibility that some comparison functions may be performed within the cortex.

7.11 COMPARED WITH OTHER MODELS OF BG FUNCTION

Previous models failed to explain the differences between different types of BG output neurons. The rate model, for example, only considers average firing rates: increased BG output is responsible for hypokinesia or akinesia, whereas decreasing BG output increases behavior, resulting in hyperkinesia (DeLong, 1990). But our results suggest that increases and decreases in individual BG output neurons cannot be interpreted as increases and decreases in behavior or action selection. To understand how the BG work, it is necessary to elucidate the signals being sent from the output neurons.

According to the standard rate model of the BG, a transient pause in BG output initiates movement (DeLong, 1990). But this model cannot explain why the same neuron can have opposite responses depending on the direction of motion (i.e. increase during leftward motion but decrease during rightward motion) or why there are neurons with opposite preferences (i.e. decrease during leftward and increase during rightward motion). A change in body configuration is not achieved by a pause in SNr nuclei or by either increasing or decreasing the BG output. To generate a specific action, multiple functional classes of neurons must coordinate their activity at the same time, some increasing and others decreasing their firing rates depending on the direction of motion. A movement is associated with coordinated changes in at least four different types of BG output neurons: each increases in firing rate during movement in a particular direction (up, down, left, right) and decreases firing during movement in the opposite direction. BG output does not simply pause when a behavior is produced. Just as often, neurons can increase firing at the onset of movement. As reviewed in Chapter 6, the amount of position change in one axis is reflected by the total change in firing rate in the relevant SNr neurons.

In the SNr and VTA GABAergic neurons, the effective zero appears to be close to the middle of their dynamic range, so increases and decreases above and below this level are the actual signals sent to lower levels. The sign determines the direction of movement along a single axis, and the magnitude determines displacement along that direction. Changes in firing rate reflect the direction of motion, not the degree of behavioral suppression.

According to the focused selection model, the tonic inhibitory output from the BG is responsible for muscle co-contraction during postural control. It is assumed that such inhibition interferes with the commands for voluntary movement, so to enable action, one must transiently remove the inhibition (Mink & Thach, 1991). For example, at rest, one maintains arm position via postural control, but to perform a reaching movement, it is necessary to turn off postural control and activate neurons responsible for reaching (Mink, 1996). The antagonism is thought to be between some system that keeps the animal still and a different system that causes movement, yet our results suggest that there is no antagonism between posture and movement. Rather, the antagonism is between units that move the body in opposite directions.

Yttri and Dudman proposed a history-dependent gain hypothesis on BG function that takes into account results on the striatal representation of movement velocity (Yttri & Dudman, 2018). They argued that the BG circuit regulates the gain of cortical output. This hypothesis is based on a key anatomical organization. The same cortical regions not only project to the BG but also to midbrain and brainstem premotor regions. For example, a single cortical pyramidal neuron may project to the superior colliculus and also send axon collaterals to the striatum, which ultimately influence collicular output via the striatonigral pathway and SNr outputs. Since the descending cortical commands can activate the superior colliculus and other structures directly, the question is why the BG circuit is needed. According to the history-dependent gain hypothesis, using the BG circuit, the cortex can modify the gain of its own output. Cortical commands for movement are sent to the premotor structures, but the gain of this command is adjusted by the BG output. This signal will then amplify the cortical output to premotor structures. As we shall see in later chapters, the BG can indeed further potentiate or depotentiate specific cortical commands via reentrant projections, though their role is not limited to altering the gain of cortical output.

Cerebral cortex and BG projections to the brainstem are opposite in sign, as cortical projections are glutamatergic and pallidal projections are GABAergic. Yttri and Dudman argue that this arrangement makes it possible for the target structure to compare the cortical and BG inputs; in the target area (e.g. superior colliculus), the BG output, being inhibitory, is subtracted from the excitatory cortical output. Normally, BG output and descending motor commands are calibrated to match, but when the BG output is larger than the cortical command, a negative movement speed could be possible, similar to akinesia. However, this model neglects the basic disinhibitory effect of BG activation. Via the direct pathway, any cortical input to the BG can be turned into an excitatory input to the superior colliculus, which also receives direct excitatory projections from the same set of cortical neurons. If a cortical signal activates the superior colliculus directly via glutamatergic projections, it is also able to activate it indirectly via disinhibition. Consequently, the fact that the BG output is inhibitory does not necessarily mean that its output is being subtracted from the cortical command.

Another key limitation of the history-dependent gain hypothesis is its assumption that the BG produce a scalar quantity, a gain signal that simply adjusts the magnitude of cortical commands. By focusing exclusively on behavioral vigor, rather than vector quantities like velocity, this hypothesis fails to explain observations on kinematic variables reviewed in this chapter. It also ignores the role of position control and the relationship between posture and movement (Barter et al., 2015b; Hughes et al., 2019).

7.12 LIMITATIONS IN PREVIOUS EXPERIMENTAL DESIGNS

It is important to consider why results on the BG representation of kinematic variables could not have been collected using conventional experimental designs. The classic paradigm for studying single-neuron activity in behaving animals relies on restrained monkeys. These studies focused on hand and arm movements to study the contribution of the motor cortex to movement in monkeys (Evarts, 1968). They usually recorded activity from head-fixed monkeys sitting in a chair, while the head and body were assumed to remain still because the animal was restrained. Using this preparation, researchers largely neglected the basic movement repertoire that has been evolutionarily conserved and often associated with the reticulospinal pathway (Grillner et al., 2008).

Using the classic experimental designs, primate studies found that movement-related neural activity in the BG appears to follow that of the cortical motor areas, often after muscle activity (Aldridge et al., 1980; Crutcher & DeLong, 1984a, b). For example, in the monkey striatum, most movement-related activity was reported to occur ~50 ms after the onset of relevant muscle activity (Crutcher & DeLong, 1984a). These results suggest that BG activity occurs too late to initiate action. But the experimental limitations make it difficult to interpret these results. For example, if the neuron in question is responsible for sniffing, head movements, or torso movements, the EMG measures from the forearm would not have revealed its role in behavior. If the reticulospinal system is indeed more critical for head or torso movements than for finger and arm movements, exclusive focus on the latter category of movements is unlikely to reveal the role of the BG. Studies have also found striatal activity before movement initiation, though such activity can also be difficult to interpret (Neafsey et al., 1978; Schultz & Romo, 1988; Lauwereyns et al., 2002). It is difficult to determine which striatal areas are activated, and when they are activated, in relation to motor cortical output.

There are limitations in the behavioral measures used in the studies reviewed in this chapter, such as the limited body parts measured and often the lack of quantification for rotational movements. But even with these limitations, the studies were able to uncover a striking correspondence between neural activity and behavior, which allows the development of a working model for the circuit mechanism underlying how the BG generate behavior. These results suggest that it is impossible to determine the relationship between neural activity and behavior if the behavior is not adequately measured. The traditional focus on restrained animals and the lack of continuous behavioral measures have prevented the discovery of the relationship between BG activity and behavior.

7.13 SUMMARY

The activity of many sensorimotor striatal neurons reflects velocity, while the SNr neurons they target show position-related activity. There is a hierarchical relationship between velocity and position in the action control hierarchy. The output of the velocity controller is a sequence of position reference signals sent to the comparison functions of the lower-level position controllers. According to the transition control model, these reference signals are generated by an integrator circuit in which a velocity vector is converted to a position vector. The amount of velocity error enters the integrator, and the integrator output reflects the accumulation (integration) of the recent error. The amount of input to the integrator within a time window is therefore proportional to the rate of change in integrator output. The "instantaneous" velocity error is converted into a rate of change in position reference to generate movement. Once a new position is achieved, it is held for some time in the absence of new velocity errors.

The sensorimotor BG network is hypothesized to implement a movement velocity controller that sends reference signals to the brainstem position controllers to command voluntary movements. Opponent BG outputs command distinct position controllers that move parts of the body in different directions along a particular axis of motion. In the sensorimotor striatum, neural activity can represent velocity, the rate of change in position. The striatonigral pathway implements a neural integrator that converts the velocity signals to position vectors, which act as position reference signals for position controllers. The striatopallidal (indirect) pathway implements damping or discharging of the integrator. Stimulation of the sensorimotor striatonigral pathway generates contraversive movement, whereas activation of iSPNs suppresses can generate ipsiversive movement.

Velocity control, the control of the rate of change in proprioceptive inputs, is a particular type of transition control. Other types of transitions, defined as the rate of change in other perceptual representations, can also be controlled using negative feedback. More broadly, the BG are critical for transition control. The striatum is in a position to receive convergent inputs from the anterior and posterior divisions of the cerebral cortex. The frontal cortex sends reference signals, and the posterior sensory cortex sends perceptual input signals to a specific striatal region, where these signals may be compared. The difference represents a transition error signal, which is integrated in a leaky integrator and ultimately converted to descending reference signals to generate volitional actions. In later chapters, we shall consider other types of transition control and how they can also be implemented by the BG.

REFERENCES

Albin, R. L., Young, A. B., & Penney, J. B. (1989). The functional anatomy of basal ganglia disorders. *Trends in Neurosciences*, *12*, 366–375.

Aldridge, J. W., Anderson, R. J., & Murphy, J. T. (1980). The role of the basal ganglia in controlling a movement initiated by a visually presented cue. *Brain Research*, *192*, 3–16.

Alexander, G. E., & Crutcher, M. D. (1990). Functional architecture of basal ganglia circuits: Neural substrates of parallel processing. *Trends in Neurosciences*, *13*, 266–271.

Anderson, M., & Horak, F. (1985). Influence of the globus pallidus on arm movements in monkeys. III. Timing of movement-related information. *Journal of Neurophysiology*, *54*, 433–448.

Barter, J. W., Castro, S., Sukharnikova, T., Rossi, M. A., & Yin, H. H. (2014). The role of the substantia nigra in posture control. *European Journal of Neuroscience*, *39*(9), 1465–1473.

Barter, J. W., Li, S., Lu, D., Rossi, M., Bartholomew, R., Shoemaker, C. T., Salas-Meza, D., Gaidis, E., & Yin, H. H. (2015a). Beyond reward prediction errors: The role of dopamine in movement kinematics. *Frontiers in Integrative Neuroscience*, *9*, 39.

Barter, J. W., Li, S., Sukharnikova, T., Rossi, M. A., Bartholomew, R. A., & Yin, H. H. (2015b). Basal ganglia outputs map instantaneous position coordinates during behavior. *Journal of Neuroscience*, *35*, 2703–2716.

Bartholomew, R. A., Li, H., Gaidis, E. J., Stackmann, M., Shoemaker, C. T., Rossi, M. A., & Yin, H. H. (2016). Striatonigral control of movement velocity in mice. *European Journal of Neuroscience*, *43*, 1097–1110.

Bullock, D., & Grossberg, S. (1988). Neural dynamics of planned arm movements: Emergent invariants and speed-accuracy properties during trajectory formation. *Psychological Review*, *95*, 49.

Bullock, D., & Grossberg, S. (1989). VITE and FLETE: Neural modules for trajectory formation and postural control. In: Hershberger, W. A. (Ed.), *Advances in Psychology* (pp. 253–297). New York: Elsevier.

Buschman, T. J., & Miller, E. K. (2007). Top-down versus bottom-up control of attention in the prefrontal and posterior parietal cortices. *Science*, *315*, 1860–1862.

Cazorla, M., de Carvalho, F. D., Chohan, M. O., Shegda, M., Chuhma, N., Rayport, S., Ahmari, S. E., Moore, H., & Kellendonk, C. (2014). Dopamine d2 receptors regulate the anatomical and functional balance of Basal Ganglia circuitry. *Neuron*, *81*, 153–164.

Cohen, J. Y., Haesler, S., Vong, L., Lowell, B. B., & Uchida, N. (2012). Neuron-type-specific signals for reward and punishment in the ventral tegmental area. *Nature*, *482*, 85–88.

Collins, A. G., & Frank, M. J. (2014). Opponent actor learning (OpAL): Modeling interactive effects of striatal dopamine on reinforcement learning and choice incentive. *Psychological Review*, *121*, 337.

Comoli, E., Coizet, V., Boyes, J., Bolam, J. P., Canteras, N. S., Quirk, R. H., Overton, P. G., & Redgrave, P. (2003). A direct projection from superior colliculus to substantia nigra for detecting salient visual events. *Nature Neuroscience*, *6*, 974–980.

Connelly, W. M., Schulz, J. M., Lees, G., & Reynolds, J. N. (2010). Differential short-term plasticity at convergent inhibitory synapses to the substantia nigra pars reticulata. *Journal of Neuroscience*, *30*, 14854–14861.

Costa, R. M., Cohen, D., & Nicolelis, M. A. (2004). Differential corticostriatal plasticity during fast and slow motor skill learning in mice. *Current Biology*, *14*, 1124–1134.

Costall, B., David Marsden, C., Naylor, R. J., & Pycock, C. J. (1977). Stereotyped behaviour patterns and hyperactivity induced by amphetamine and apomorphine after discrete 6-hydroxydopamine lesions of extrapyramidal and mesolimbic nuclei. *Brain Research*, *123*, 89–111.

Crutcher, M. D., & DeLong, M. R. (1984a). Single cell studies of the primate putamen. II. Relations to direction of movement and pattern of muscular activity. *Experimental Brain Research*, *53*, 244–258.

Crutcher, M. D., & DeLong, M. R. (1984b). Single cell studies of the primate putamen. I. Functional organization. *Experimental Brain Research*, *53*, 233–243.

Cui, G., Jun, S. B., Jin, X., Pham, M. D., Vogel, S. S., Lovinger, D. M., & Costa, R. M. (2013). Concurrent activation of striatal direct and indirect pathways during action initiation. *Nature*, *494*, 238–242.

DeLong, M. R. (1990). Primate models of movement disorders of basal ganglia origin. *Trends in Neurosciences*, *13*, 281–285.

Ding, S., Li, L., & Zhou, F. M. (2015). Nigral dopamine loss induces a global upregulation of presynaptic dopamine D1 receptor facilitation of the striatonigral GABAergic output. *Journal of Neurophysiology*, *113*, 1697–1711.

Dudman, J. T., & Krakauer, J. W. (2016). The basal ganglia: From motor commands to the control of vigor. *Current opinion in Neurobiology*, *37*, 158–166.

Eshel, N., Bukwich, M., Rao, V., Hemmelder, V., Tian, J., & Uchida, N. (2015). Arithmetic and local circuitry underlying dopamine prediction errors. *Nature*, *525*, 243–246.

Evarts, E. V. (1968). Relation of pyramidal tract activity to force exerted during voluntary movement. *Journal of Neurophysiology, 31*, 14–27.

Fan, D., Rossi, M. A., & Yin, H. H. (2012). Mechanisms of action selection and timing in substantia nigra neurons. *Journal of Neuroscience: The Official Journal of the Society for Neuroscience, 32*, 5534–5548.

Fiorillo, C. D., Tobler, P. N., & Schultz, W. (2003). Discrete coding of reward probability and uncertainty by dopamine neurons. *Science, 299*, 1898–1902.

Gerfen, C. R., Engber, T. M., Mahan, L. C., Susel, Z., Chase, T. N., Monsma, F. J., Jr., & Sibley, D. R. (1990). D1 and D2 dopamine receptor-regulated gene expression of striatonigral and striatopallidal neurons. *Science, 250*, 1429–1432.

Gerfen, C. R., & Surmeier, D. J. (2011). Modulation of striatal projection systems by dopamine. *Annual review of Neuroscience, 34*, 441–466.

Giros, B., Jaber, M., Jones, S. R., Wightman, R. M., & Caron, M. G. (1996). Hyperlocomotion and indifference to cocaine and amphetamine in mice lacking the dopamine transporter. *Nature, 379*, 606–612.

Grillner, S., Wallen, P., Saitoh, K., Kozlov, A., & Robertson, B. (2008). Neural bases of goal-directed locomotion in vertebrates--an overview. *Brain Research Review, 57*, 2–12.

Hjelmstad, G. O. (2004). Dopamine excites nucleus accumbens neurons through the differential modulation of glutamate and GABA release. *Journal of Neuroscience, 24*, 8621–8628.

Hotson, J., Langston, E., & Langston, J. (1986). Saccade responses to dopamine in human MPTP-induced parkinsonism. *Annals of Neurology: Official Journal of the American Neurological Association and the Child Neurology Society, 20*, 456–463.

Hughes, R. N., Watson, G. D., Petter, E. A., Kim, N., Bakhurin, K. I., & Yin, H. H. (2019). Precise coordination of three-dimensional rotational kinematics by ventral tegmental area GABAergic neurons. *Current Biology, 29*, 3244–3255. e3244.

Inase, M., Buford, J. A., & Anderson, M. E. (1996). Changes in the control of arm position, movement, and thalamic discharge during local inactivation in the globus pallidus of the monkey. *Journal of Neurophysiology, 75*, 1087–1104.

Isomura, Y., Takekawa, T., Harukuni, R., Handa, T., Aizawa, H., Takada, M., & Fukai, T. (2013). Reward-modulated motor information in identified striatum neurons. *Journal of Neuroscience, 33*, 10209–10220.

Kaneda, K., Isa, K., Yanagawa, Y., & Isa, T. (2008). Nigral inhibition of GABAergic neurons in mouse superior colliculus. *Journal of Neuroscience, 28*, 11071–11078.

Kawagoe, R., Takikawa, Y., & Hikosaka, O. (1998). Expectation of reward modulates cognitive signals in the basal ganglia. *Nature Neuroscience, 1*, 411–416.

Kim, N., Barter, J. W., Sukharnikova, T., & Yin, H. H. (2014). Striatal firing rate reflects head movement velocity. *European Journal of Neuroscience, 40*, 3481–3490.

Kim, N., Li, H. E., Hughes, R. N., Watson, G. D. R., Gallegos, D., West, A. E., Kim, I. H., & Yin, H. H. (2019). A striatal interneuron circuit for continuous target pursuit. *Nature Communications, 10*, 2715.

Kori, A., Miyashita, N., Kato, M., Hikosaka, O., Usui, S., & Matsumura, M. (1995). Eye movements in monkeys with local dopamine depletion in the caudate nucleus. II. Deficits in voluntary saccades. *Journal of Neuroscience, 15*, 928–941.

Koshikawa, N. (1994). Role of the nucleus accumbens and the striatum in the production of turning behaviour in intact rats. *Reviews in the Neurosciences, 5*, 331–346.

Kravitz, A. V., Freeze, B. S., Parker, P. R., Kay, K., Thwin, M. T., Deisseroth, K., & Kreitzer, A. C. (2010). Regulation of parkinsonian motor behaviours by optogenetic control of basal ganglia circuitry. *Nature, 466*, 622–626.

Kravitz, A. V., Tye, L. D., & Kreitzer, A. C. (2012). Distinct roles for direct and indirect pathway striatal neurons in reinforcement. *Nature Neuroscience, 15*, 816–818.

Lahiri, A. K., & Bevan, M. D. (2020). Dopaminergic transmission rapidly and persistently enhances excitability of D1 receptor-expressing striatal projection neurons. *Neuron, 106*(2), 277–290.

Lasker, A. G., & Zee, D. S. (1997). Ocular motor abnormalities in Huntington's disease. *Vision Research, 37*, 3639–3645.

Lauwereyns, J., Watanabe, K., Coe, B., & Hikosaka, O. (2002). A neural correlate of response bias in monkey caudate nucleus. *Nature, 418*, 413–417.

Leigh, R. J., Newman, S. A., Folstein, S. E., Lasker, A. G., & Jensen, B. A. (1983). Abnormal ocular motor control in Huntington's disease. *Neurology, 33*, 1268–1268.

Lemos, J. C., Friend, D. M., Kaplan, A. R., Shin, J. H., Rubinstein, M., Kravitz, A. V., & Alvarez, V. A. (2016). Enhanced GABA transmission drives bradykinesia following loss of dopamine D2 receptor signaling. *Neuron, 90*, 824–838.

Meng, C., Zhou, J., Papaneri, A., Peddada, T., Xu, K., & Cui, G. (2018). Spectrally resolved fiber photometry for multi-component analysis of brain circuits. *Neuron, 98*, 707–717. e704.

Mink, J. W. (1996). The basal ganglia: Focused selection and inhibition of competing motor programs. *Progress in Neurobiology, 50,* 381–425.

Mink, J. W., & Thach, W. T. (1991). Basal ganglia motor control. III. Pallidal ablation: Normal reaction time, muscle cocontraction, and slow movement. *Journal of Neurophysiology, 65,* 330–351.

Neafsey, E., Hull, C., & Buchwald, N. (1978). Preparation for movement in the cat. II. Unit activity in the basal ganglia and thalamus. *Electroencephalography and Clinical Neurophysiology, 44,* 714–723.

Panigrahi, B., Martin, K. A., Li, Y., Graves, A. R., Vollmer, A., Olson, L., Mensh, B. D., Karpova, A. Y., & Dudman, J. T. (2015). Dopamine Is Required for the Neural Representation and Control of Movement Vigor. *Cell, 162,* 1418–1430.

Park, J., Coddington, L. T., & Dudman, J. T. (2020). Basal ganglia circuits for action specification. *Annual Review of Neuroscience, 43,* 485–507.

Passingham, R. E., & Wise, S. P. (2012). *The Neurobiology of the Prefrontal Cortex: Anatomy, Evolution, and the Origin of Insight*: Oxford: Oxford University Press.

Rascol, O., Clanet, M., Montastruc, J.-L., Simonetta, M., Soulier-Esteve, M., Doyon, B., & Rascol, A. (1989). Abnormal ocular movements in Parkinson's disease: Evidence for involvement of dopaminergic systems. *Brain, 112,* 1193–1214.

Redgrave, P., & Gurney, K. (2006). The short-latency dopamine signal: A role in discovering novel actions? *Nature Reviews Neuroscience, 7,* 967–975.

Rolls, E. T. (2016). *Cerebral Cortex: Principles of Operation*: Oxford University Press.

Rossi, M. A., Fan, D., Barter, J. W., & Yin, H. H. (2013a). Bidirectional modulation of substantia nigra activity by motivational state. *PLoS One, 8,* e71598.

Rossi, M. A., Go, V., Murphy, T., Fu, Q., Morizio, J., & Yin, H. H. (2015). A wirelessly controlled implantable LED system for deep brain optogenetic stimulation. *Frontiers in Integrative Neuroscience, 9,* 8.

Rossi, M. A., Sukharnikova, T., Hayrapetyan, V. Y., Yang, L., & Yin, H. H. (2013b). Operant self-stimulation of dopamine neurons in the substantia nigra. *PLoS One, 8,* e65799.

Rueda-Orozco, P. E., & Robbe, D. (2015). The striatum multiplexes contextual and kinematic information to constrain motor habits execution. *Nature Neuroscience, 18*(3), 453–460.

Sacks, O. (1991). *Awakenings*:London: Duckworth & Co.

Schultz, W. (1998). Predictive reward signal of dopamine neurons. *Journal of Neurophysiology, 80,* 1–27.

Schultz, W., Dayan, P., & Montague, P. R. (1997). A neural substrate of prediction and reward. *Science, 275,* 1593–1599.

Schultz, W., & Romo, R. (1988). Neuronal activity in the monkey striatum during the initiation of movements. *Experimental Brain Research, 71,* 431–436.

Schwab, R. S., England, A. C., & Peterson, E. (1959). Akinesia in Parkinson's disease. *Neurology, 9,* 65–65.

Stalnaker, T. A., Calhoon, G. G., Ogawa, M., Roesch, M. R., & Schoenbaum, G. (2012). Reward prediction error signaling in posterior dorsomedial striatum is action specific. *Journal of Neuroscience, 32,* 10296–10305.

Tecuapetla, F., Matias, S., Dugue, G. P., Mainen, Z. F., & Costa, R. M. (2014). Balanced activity in basal ganglia projection pathways is critical for contraversive movements. *Nature Communications, 5*(1), 4315.

Tepper, J. M., & Lee, C. R. (2007). GABAergic control of substantia nigra dopaminergic neurons. *Progress in Brain Research, 160,* 189–208.

Teuber, H.-L. (1964). The riddle of frontal lobe function in man. In: Warren, J. M., & Akert, K. (Eds.), *The Frontal Granular Cortex And Behavior* (pp. 410–444). New York: McGraw Hill.

Turner, R. S., & Desmurget, M. (2010). Basal ganglia contributions to motor control: A vigorous tutor. *Current Opinion in Neurobiology, 20,* 704–716.

Turner, R. S., Desmurget, M., Grethe, J., Crutcher, M. D., & Grafton, S. T. (2003). Motor subcircuits mediating the control of movement extent and speed. *Journal of Neurophysiology, 90,* 3958–3966.

Vermersch, A., Gaymard, B., Rivaud-Pechoux, S., Ploner, C., Agid, Y., & Pierrot-Deseilligny, C. (1999). Memory guided saccade deficit after caudate nucleus lesion. *Journal of Neurology, Neurosurgery & Psychiatry, 66,* 524–527.

Waelti, P., Dickinson, A., & Schultz, W. (2001). Dopamine responses comply with basic assumptions of formal learning theory. *Nature, 412,* 43–48.

Wang, A. Y., Miura, K., & Uchida, N. (2013). The dorsomedial striatum encodes net expected return, critical for energizing performance vigor. *Nature Neuroscience, 16,* 639–647.

Watson, G. D., Hughes, R. N., Petter, E. A., Fallon, I. P., Kim, N., Severino, F. P. U., & Yin, H. H. (2021). Thalamic projections to the subthalamic nucleus contribute to movement initiation and rescue of parkinsonian symptoms. *Science Advances, 7,* eabe9192.

Wise, R. A. (2004). Dopamine, learning and motivation. *Nature Reviews Neuroscience, 5,* 483–494.

Yeterian, E. H., & Van Hoesen, G. W. (1978). Cortico-striate projections in the rhesus monkey: The organization of certain cortico-caudate connections. *Brain Research, 139,* 43–63.

Yin, H. H. (2014). Action, time and the basal ganglia. *Philosophical Transactions of the Royal Society B: Biological Sciences, 369*(1637), 20120473.

Yin, H.H. (2017). The basal ganglia in action. *Neuroscientist, 23*, 299–313.

Yin, H. H., Mulcare, S. P., Hilario, M. R., Clouse, E., Holloway, T., Davis, M. I., Hansson, A. C., Lovinger, D. M., & Costa, R. M. (2009). Dynamic reorganization of striatal circuits during the acquisition and consolidation of a skill. *Nature Neuroscience, 12*, 333–341.

Yttri, E. A., & Dudman, J. T. (2016). Opponent and bidirectional control of movement velocity in the basal ganglia. *Nature, 533*, 402–406.

Yttri, E. A., & Dudman, J. T. (2018). A proposed circuit computation in basal ganglia: History-dependent gain. *Movement Disorders, 33*, 704–716.

Yu, C., Jiang, T. T., Shoemaker, C. T., Fan, D., Rossi, M. A., & Yin, H. H. (2022). Striatal mechanisms of turning behaviour following unilateral dopamine depletion in mice. *European Journal of Neuroscience, 56*(5), 4529-4545.

Zhang, J., Hughes, R. N., Kim, N., Fallon, I. P., Bakhurin, K., Kim, J., Severino, F. P. U., & Yin, H. H. (2022). A one-photon endoscope for simultaneous patterned optogenetic stimulation and calcium imaging in freely behaving mice. *Nature Biomedical Engineering, 7*, 499–510.

Zucker, R. S., & Regehr, W. G. (2002). Short-term synaptic plasticity. *Annual Review of Physiology, 64*, 355–405.

8 Higher-Order Transitions and Cognition

In the last chapter, we considered how velocity control may be implemented by basal ganglia (BG) circuits. Movement velocity, the rate of change in proprioceptive inputs, is only one example of a transition variable. More generally, transitions can be defined as the rate of change in any perceptual variable. Regardless of *what* is being changed, the *rate* at which it is being changed can be represented independently and controlled using negative feedback.

In this chapter, we shall consider how simple transitions may be concatenated to form higher-order transitions and how higher-order transitions may be controlled using the BG circuit. This requires the introduction of several concepts—event, repetition, time, and serial order—that are indispensable in explaining BG function. We shall also consider how the control of higher-order transitions contributes to what are ordinarily called "cognitive" functions, such as mental imagery and working memory.

8.1 EVENT REPETITION AND CONTROL OF TEMPO

When we knock on a door, each knock involves a simple proprioceptive transition, and the velocity of the hand can be varied volitionally, as discussed in Chapter 7. This is done by sending a sequence of hand position reference signals for each knock. Controlling the rate of knocking, however, is not the same as controlling movement velocity. To vary the rate, we do not just vary how quickly the hand moves from one position to another. To knock more quickly, we can reduce the pause between knocks. That is, we can control the tempo of knocking by varying the inter-knock interval without significantly changing hand velocity.

The change in hand position during a knock is a simple transition with a cyclical structure. In each cycle, the hand travels to the door and returns to the starting position. Such a cycle is a repeatable unit of behavior. This level of description can be considered the "event" level, which is a higher-order transition. It is often taken for granted and used as the basic unit in the verbal description of behavior, e.g. knock, kick, or wave. It is also the source of the perennial bias that our experience consists of discrete categories for which we have words. Action selection models, for example, assume that the individual action being selected is an event, as if the event level were the only relevant level. They fail to recognize the multiple levels of control supporting a single knock, because they do not take into account the detailed action parameters, such as the change in the angle of the elbow, bending of the fingers, flick of the wrist, and so on. The subsidiary components, whether force or joint angle, are protected from disturbances by the respective control systems.

In velocity control, the smallest unit is the minimum detectable position change. However, the entire transition can also be treated as a single unit, like a knock. This higher-order transition unit can be repeated. A significant chunk of behavior with a cyclical structure is now treated as a basic unit. The reference signal represents a certain number of repetitions. It is hypothesized that such repetition control is also implemented by the BG circuit.

If the BG circuit described in Chapters 6 and 7 is important for controlling movement velocity, other BG circuits may be used for other transition variables. For example, what is repeated may be a reference signal that enters the striatum via corticostriatal projections, and this signal can be replayed using the reentrant projections back to the frontal cortex. The BG output can reactivate the cortical reference signal that is responsible for initiating the simple transition. Figure 8.1 illustrates the relationship between movement velocity and event repetition.

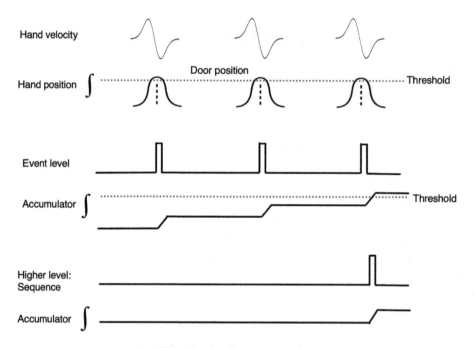

FIGURE 8.1 Event and repetition. Event and repetition control are illustrated by the example of knocking on a door. Schematic showing how simple transitions like hand movement can be used to generate higher-order transitions using a coupling mechanism. Each knock can be discretized as an individual event (pulse), which can enter another integrator/accumulator, and some arbitrary threshold can be set for the integrator so that another pulse can be generated downstream when the threshold is reached.

Clinical observations suggest that the BG are also critical for the control of higher-order transitions. Parkinsonian bradykinesia is not only the slowing of the simple movement transition but also the increased pause between transitions (Benecke, Rothwell, Dick, Day, & Marsden, 1987; Cools, van den Bercken, Horstink, van Spaendonck, & Berger, 1984). Not only are the individual movements slower, but the latencies and intervals between movements are also longer (Cools, 1981).

8.2 REGULATION OF RHYTHMIC BEHAVIOR

In mammals, behaviors like locomotion, breathing, licking, and chewing depend on innately organized central pattern generators (CPG) that are located in the brainstem and spinal cord (Kiehn, 2016; McElvain et al., 2018; Moore, Kleinfeld, & Wang, 2014; Nakamura & Katakura, 1995). For these movements, the BG can provide top-down regulation of the pattern generators without directly dictating the rhythm or tempo of the pattern. On the other hand, in other types of more arbitrary movements, like knocking or pressing a lever, there are no innate pattern generators that determine the rhythm. Their rhythm or order are learned through experience. To repeat these behaviors the descending projections from the BG output to the brainstem are not sufficient. Rather, it may engage reentrant projections to the thalamocortical network to reactivate specific reference signals for the next component. Below, we shall first start with a discussion of BG regulation of pattern generators before considering learned behavioral sequences.

8.2.1 LICKING

In rodents and many other mammals, isolated licks are rare. More commonly, licking behavior is expressed as variable bouts of licking at 6–10 Hz with a relatively constant inter-lick interval (Travers, Dinardo, & Karimnamazi, 1997). Licking involves orofacial movements generated by

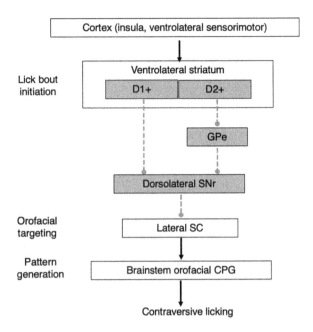

FIGURE 8.2 Descending basal ganglia (BG) regulation of licking behavior. Illustration of the BG circuit for top-down regulation of orofacial movements. GPE, globus pallidus externus; SNr, substantia nigra pars reticulata; CPG, central pattern generators.

brainstem motor neurons in the trigeminal, facial, ambiguous, and hypoglossal nuclei (Lowe, 1984; Travers & Norgren, 1983). The lateral medullary reticular formation, which projects to the motor neurons for orofacial musculature, apparently contains the relevant pattern generators (Chandler & Goldberg, 1988; Chen, Travers, & Travers, 2001; Nakamura & Katakura, 1995).

Although BG projections to lower brainstem pattern generators are not required for reflexive orofacial and tongue movements, they are needed for volitional licking (Bignall & Schramm, 1974; Grill & Norgren, 1978; Rossi et al., 2016; Toda et al., 2017). The BG circuit for top-down regulation of licking is shown in Figure 8.2. BG can receive sensory feedback from licking via multiple channels, and in turn BG outputs can regulate licking via descending projections. For example, SNr neurons send bilateral projections, with ipsilateral predominance, to the lateral pontine and medullary reticular formations, which contain orofacial pattern generators (Krosigk & Smith, 1991; von Krosigk, Smith, Bolam, & Smith, 1992; Yasui et al., 1992; Yasui, Tsumori, Ono, & Kishi, 1997). In anesthetized rats, electrical stimulation of SNr can also activate orofacial muscles (Inchul et al., 2005).

Another pathway via which BG output can influence orofacial movements is the nigro-tecto-reticular projections (Yasui et al., 1992). As discussed in Chapter 6, the superior colliculus (tectum) contains multiple target acquisition systems for distinct sensory modalities, and its output ultimately adjusts and moves the various sensors. Selective activation of nigrocollicular axon terminals can suppress licking (Rossi et al., 2016). Superior colliculus inactivation with muscimol (a GABA-A agonist) also suppresses licking. Both SNr and SC exhibit oscillatory activity, reflecting the licking pattern.

Further upstream, the ventrolateral striatum contains the orofacial region that receives orofacial cortex and projects to the dorsolateral SNr (Delfs & Kelley, 1990; Hintiryan et al., 2016; Redgrave, Marrow, & Dean, 1992; Yasui, Tsumori, Ando, & Domoto, 1995). For example, in the "striatal jaw region," which receives inputs from the lateral sensorimotor cortex and insula, rhythmic jaw movements can be induced by electrical stimulation (Sasamoto, Zhang, & Iwasaki, 1990; Satoda et al., 2002). In addition, some neurons in the ventrolateral striatum modulate their firing in relation to licking (Chen et al., 2021; Kelley, Lang, & Gauthier, 1988).

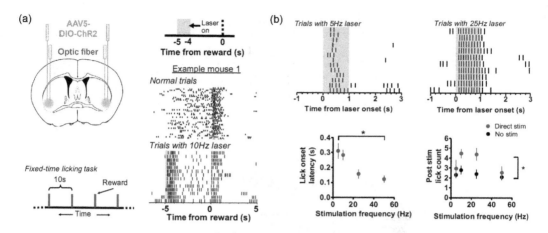

FIGURE 8.3 Direct pathway stimulation in the ventrolateral striatum elicits licking. (a) Mice were trained on a fixed-time licking task. There is normally no licking at 5 seconds before the expected time of reward delivery, but stimulation of direct pathway spiny projection neurons reliably generates licking. Rasters show individual licks. (b) Increasing stimulation frequency reduces latency to lick bout onset (left) and increases lick count after the termination of stimulation. From Bakhurin et al. (2020) with permission.

Activation of direct and indirect pathways can produce opposite effects on licking (Figure 8.3). Selective stimulation of direct pathway spiny projection neuron (dSPN) increased licking, especially in the contraversive direction. Activation of indirect pathway spiny projection neurons (iSPNs) suppressed contraversive licking but increased ipsiversive licking. This effect cannot be explained by the standard action selection model, according to which indirect activation should suppress licking. Optogenetic activation of iSPNs in the ventrolateral striatum suppresses output from the ipsilateral lateral superior colliculus but has a net excitatory effect on the contralateral colliculus (Lee, Wang, & Sabatini, 2020).

As the relevant dSPNs increase their outputs, a longer pause in the SNr is predicted, resulting in prolonged activation of the pattern generator without significantly increasing tongue velocity or the intrinsic rhythm of the licking pattern. By modulating the duration of the bout, the BG output can regulate the overall rate of licking, which is limited by the intrinsic rhythm of the CPG for licking. Normally, the frequency of licking is about 5–6 Hz for anticipatory licking before delivery of fluid rewards and 6–8 Hz for consummatory licking. Lick frequency does not linearly scale with direct pathway stimulation frequency (Bakhurin et al., 2020). Stimulation-evoked licking can reach a maximum of ~10 Hz, which appears to be the limit of the pattern generator (Bakhurin et al., 2020).

This type of top-down regulation is different from the continuous velocity control discussed in Chapter 7. Continuous control appears to characterize behaviors like turning and orienting, whereas for the regulation of licking, there are more discretized settings. For velocity control, a larger striatal output results in faster movement. To modulate the licking CPG, larger striatal output results in a longer bout of licking. The command turns on the CPG for some time, and the magnitude of the signal is proportional to the duration of suprathreshold "on" time (similar to duty cycle). This mechanism is similar to the so-called bang–bang control system, in which a suprathreshold error signal generates the same output, which will persist until the controlled variable comes back into range and the error drops below the threshold. If we zoom out and count the number of licks within a larger time window, we can still see a monotonic relationship between direct pathway output and the overall rate of repetition. The overall lick rate is still proportional to striatal output.

Ventrolateral striatal dopamine (DA) has also been implicated in ingestive behaviors (Roitman, Stuber, Phillips, Wightman, & Carelli, 2004; Salamone et al., 1996). Local infusion of amphetamine, a DA agonist, can produce orofacial movements. On the other hand, DA antagonists can reduce force, duration, and number of licks (Fowler & Mortell, 1992). As adaptive gain, DA modulates the amount of striatal signal being integrated by the striatonigral circuit. It is therefore predicted to regulate the duty cycle of a relatively fixed pattern. This hypothesis is supported by data from mice with a genetic deletion of the DA transporter (DAT). Since DAT is critical for DA reuptake and normally terminates DA signaling once it is released, the DAT knockout mice showed dramatically increased DA signaling. Compared to control mice, these mice showed a higher number of licks within a bout, reduced inter-lick intervals, and increased contact duration of individual licks (Rossi & Yin, 2015). Increased DA signaling therefore increases the "duty cycle" when regulating a stereotyped behavior like licking.

8.3 RELATED RATES AND GEAR COUPLING

BG output can also regulate the performance of repetitive movements that are not generated by CPGs. The hypothesis advanced here is that the reentrant projections back to the cortical region reactivate the reference signal for action components in a learned behavioral sequence. The mechanism for repetition control is similar to that for descending control of licking by adjusting the "on time," except that the neural circuit involved is hypothesized to involve BG outputs to the thalamocortical network. A large and sustained signal results in many repetitions. Likewise, DA signaling in the repetition control region may determine whether the action will be repeated. As what is repeated is a high-level transition command sent to position controllers, the movement that is actually achieved will be variable as the detailed action parameters may be determined by lower-level comparisons with sensory input. It is hypothesized that the sensorimotor and perhaps the associative cortico-BG networks also contain controllers for higher-order transitions at the event level. These controllers regulate repetition of specific actions, especially those that are learned. In later chapters, we shall consider how such learning is achieved.

As discussed in Chapter 7, the neural integrator implemented by the striatonigral circuit appears to have limited capacity. As it can only hold a few pulses for a few seconds, it is suitable for simple transitions like turning the head or reaching. To combine simple transitions into more complex sequences, multiple integrators may be coupled. Now each "pulse" to be accumulated represents a simple transition like a knock, and a larger integrator can accumulate the pulses (e.g. knock, knock, knock). Little is known about how such a larger integrator is implemented neurobiologically. In principle, any circuit with recurrent excitation can also serve as an integrator, so a plausible candidate is the reentrant loops via the thalamocortical network.

For example, the different hands of a mechanical clock move at different speeds. Sixty ticks of the second hand will move the minute hand once, and 60 ticks of the minute hand will move the hour hand just once. This is achieved using gear coupling, in which the gear ratio (60:1) determines the relative rate of change (Figure 8.1). Biologically, it can be implemented by the equivalent neural gear coupling: accumulation of some number of pulses at the simple transition level is converted to a single pulse at the next order of transitions. The completion of a full cycle in one transition controller activates the input function of another controller.

At present, the neural implementation of such a gear coupling mechanism remains unclear. BG projections to the thalamus may contribute to the coupling of different transition controllers. Because these projections often consist of collaterals of axons that target the superior colliculus and brainstem, they can provide an efference copy of descending reference signals (Faull & Mehler, 1978; McElvain et al., 2021; Redgrave et al., 1992).

One major thalamic target of BG projections is the ventromedial thalamus. In response to a current step, ventromedial thalamus neurons show burst firing that is rapidly desensitized (Kase, Uta, Ishihara, & Imoto, 2015). The nigrothalamic projection can therefore produce brief bursts of activity whenever there is a change in SNr firing, reflecting a change in position reference that normally results in a simple movement (e.g. a single knock). Such a mechanism can also "discretize" the signal for a separate controller. Neurons in the ventromedial thalamus are known to show rebound excitation following a pause in BG output, especially when there is little excitatory input from the corticothalamic projections (Goldberg, Farries, & Fee, 2013). How these neurons respond to pallidal input depends on concurrent cortical drive. When the cortical input is high, instead of rebound excitation, there is entrainment—both pallidal and thalamic neurons fire together at high rates. Without concurrent cortical drive, a reduction in SNr output can produce a net excitation of thalamic neurons. This may signal the completion of one simple transition, and such a signal is then sent to another higher-order transition controller (Figure 8.4). While this account is admittedly speculative, it illustrates the type of circuit mechanism needed to convert a simple transition commanded by BG output to a pulse in a higher-order transition controller.

The use of efference copy does not rule out a role for sensory feedback in the successful execution of the movement. Rather, it allows a very rapid transition to the next element. If the execution of the current movement is prevented by disturbances after the issuing of the descending command, there could be additional inputs that interrupt the execution of the sequence. But such corrective mechanisms, being based on slightly delayed sensory feedback, might be slower. For example, the command for pressing a key while typing may prompt the command for the next key press before feedback about the successful completion of the action (e.g. seeing the letter on the screen). Currently, there is no working model for controlling higher-order transitions that combine many simple transitions.

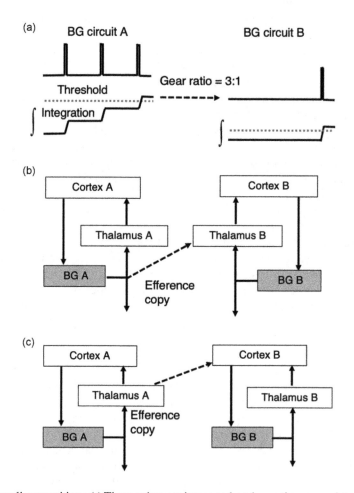

FIGURE 8.4 Coupling transitions. (a) Three pulses are integrated, and once the accumulated quantity exceeds a threshold, a single pulse is generated by the "detector," which can be accumulated in another integrator. Different basal ganglia (BG) circuits can be coupled to control higher-order transitions. (b) One possible circuit implementation of the mechanism in A. The output of one BG circuit not only reenters the corresponding thalamocortical circuit to complete the loop but also enters a thalamic region that projects to a different cortical area. (c) Another possible circuit that can implement the gear coupling mechanism. Here, the thalamic region receiving the BG output projects to a different cortical region. The dotted arrow indicates efference copy.

8.4 INTERVAL TIMING

In our description of repetition control, we used a clock analogy. In any time-keeping device, elapsed time is measured by state changes, such as the position of a pendulum. In fact, any regularly repeating pattern can be used as a timer because we judge the lapse of time relative to some other set of transitions and repetitions. An examination of transition control therefore also sheds light on the neural mechanisms of timing since the sense of time is an emergent phenomenon at the level of transition control.

The relationship between transitions and time is described by William James with characteristic clarity: "we can no more intuit a duration than we can intuit an extension, devoid of all sensible content…. Our heart-beats, our breathing, the pulses of our attention, fragments of words or sentences that pass through our imagination, are what people this dim habitat. Now, all these processes are rhythmical, and are apprehended by us, as they occur, in their totality; the breathing and pulses of attention, as coherent successions, each with its rise and fall; the heartbeats similarly, only relatively far more brief; the words not separately, but in connected groups. In short, empty our minds as we may, some form of *changing process* remains for us to feel, and cannot be expelled. And along with the sense of the process and its rhythm goes the sense of the length of time it lasts. Awareness of change is thus the condition on which our perception of time's flow depends …" (James, 1890).

Although there is no dedicated perceptual organ for time, animals are capable of timing their behaviors (Buhusi & Meck, 2005; Treisman, 1963; Yin, 2014). There is a long tradition of studying timing in the seconds-to-minutes range, known as interval timing (Gibbon & Church, 1990). In a typical timing experiment, a mouse is trained on a fixed-time schedule in which food is delivered after a certain interval, say 10 seconds. On probe trials, the reward is omitted, and the time of the peak rate of lever pressing, or peak time, is used as a measure for the rat's internal estimation of time. If the duration to be timed is 10 seconds, then after training, the animal would show a peak time with a mean of roughly 10 seconds with some spread. If the animal is trained on a longer interval, then the mouse will show a different peak time roughly corresponding to the interval but with a larger spread (Figure 8.5). If the peak time data is normalized, measured as

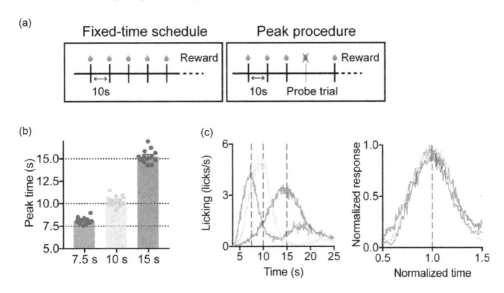

FIGURE 8.5 Interval timing. (a) Illustration of fixed-time training schedule and the peak procedure in which occasional probe trials without rewards were given. The peak time is the time at which response is highest on peak trials. It is used as a measure of the animal's internal representation of the time interval. (b) Average peak times shown by mice after three different criterion fixed-time training schedules (7.5, 10, and 15 seconds, indicated by dotted lines). When plotted with normalized time on the x axis, the three distributions are highly similar, showing the scalar property. (c) Left, data from a representative mouse showing licking on peak trials for the three different intervals. Right, the same data with the time axis normalized, showing the scalar property in interval timing. From Toda et al. (2017) with permission.

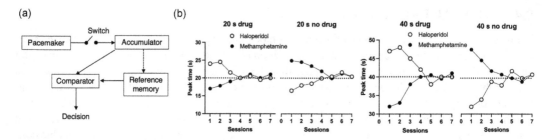

FIGURE 8.6 DA and clock speed. (a) Schematic of the pacemaker–accumulator model. See the text for a detailed explanation. (b) Mean peak times measured on peak probe trials from four groups of rats. Rats were initially trained on either a 20- or 40-second timing procedure. They were then tested under the influence of dopaminergic drugs or with no drug (data shown here). Methamphetamine (1.5 mg/kg i.p.) is a nonselective dopamine agonist. Haloperidol (0.12 mg/kg i.p.) is a D2 antagonist. Based on Meck (1996).

standard deviation (deviation from mean divided by mean), then the peak response function will look similar for the two different intervals (Treisman, 1963). According to Weber's law, the perceived change in stimulus magnitude (just noticeable difference) is proportional to the stimulus magnitude. The ratio of the change in quantity to the quantity itself is known as the Weber fraction. For interval timing, the Weber fraction is usually 5%–15%. That the Weber fraction remains relatively constant despite changes in stimulus magnitude is known as the scalar property.

A popular model of interval timing is the scalar expectancy model, a type of "pacemaker–accumulator" model (Gibbon, 1977, Gibbon & Church, 1990). This model contains a Poisson pacemaker that provides repetitive pulses that enter an integrator/accumulator (Figure 8.6). Upon feedback (e.g. reward presentation), the accumulated number of pulses is stored in reference memory. The number of pulses currently accumulated in the accumulator indicates time passed, and the rate of accumulation is called clock speed (Gibbon & Church, 1990). A similar mechanism can also explain counting or representation of numerosity (Meck & Church, 1983).

Pacemaker–accumulator models share some features with the transition control model described in Chapter 7. Most notably, they both use a linear scale as well as an integrator. But these models are open-loop, focusing mainly on cognitive aspects of time perception. In transition control, however, the same accumulator is not just used for timing but for action generation using negative feedback control, and interval timing is an emergent property of action generation in repetition control. Instead of a single accumulator, multiple coupled accumulators are needed, which could explain timing in different ranges. The efference copy of the action combined with feedback generates the pulses entering the accumulator.

8.4.1 BG and Timing

A variety of evidence suggests that the BG play a key role in timing (Merchant, Harrington, & Meck, 2013). PD patients often report that time stands still (Sacks, 1991). Striatal lesions impair or eliminate interval timing behavior. *In vivo* recordings from the striatum showed that the response times of striatal neurons scaled with the interval being timed (Mello, Soares, & Paton, 2015). Human neuroimaging studies also found activation of a sensorimotor cortico-BG circuit during timing, including the pre-supplementary motor area, dorsal premotor cortex, and putamen (Coull, Vidal, Nazarian, & Macar, 2004). Interestingly, this circuit is also activated in generating behavior with a regular beat, an example of repetition control (Hugo Merchant, Grahn, Trainor, Rohrmeier, & Fitch, 2015). These results support the hypothesis that the repetition control system is crucial for the control of tempo as well as interval timing.

8.4.2 DA Modulation of Timing

In studies of interval timing, DA has been implicated in regulating "clock speed." Injections of DA agonists such as methamphetamine can increase clock speed, whereas DA antagonists such as haloperidol decrease clock speed (Maricq, Roberts, & Church, 1981; Meck, 1983). In one study, Meck trained some rats on an operant peak-interval timing task under the influence of DA agonists and antagonists and then tested them without these drugs (drug/no-drug group) (Meck, 1996). He also trained some rats without any drug first and then tested them under the influence of the drug (no-drug/drug group). The time of peak response was used as a measure of clock speed.

The scalar expectancy model makes an interesting prediction. Suppose a pulse is emitted by the hypothetical pacemaker every 200 ms on average, and 100 pulses are required to time a duration of 20 seconds. For the no-drug/drug group, the pacemaker–accumulator model would provide the following account. After training with no drug, the reference memory comparator will be set at 100 hypothetical pulses, each with a duration of 200 ms. On probe trials, the peak time would normally average 100 pulses, or 20 seconds. However, if amphetamine makes the clock run faster, then when tested under the influence of amphetamine, 100 pulses will accumulate more quickly, as each pulse is now less than 200 ms. Consequently, the peak time will be less than 20 seconds, and there will be a leftward shift in the peak time. In contrast, if haloperidol makes the clock run slower, 100 pulses will accumulate later than during baseline training, and the peak time will occur later than 20 seconds (rightward shift in peak time).

The results of Meck's experiments are shown in Figure 8.6. Indeed, the predicted pattern was found. A peak time of ~17s under methamphetamine suggests that the internal clock is about 15% faster, and a peak time of ~24 seconds under haloperidol suggests that the clock is about 20% slower. With continued training under drug influence, the model predicts that peak time will return to 20 seconds because the reference memory has been updated. Indeed, the injected rats eventually learned to time accurately. The model also makes another interesting prediction: since the reference memory is updated with the shorter pulses under methamphetamine, more pulses are needed to represent 20 seconds. In contrast, with haloperidol, fewer pulses will be needed in order to represent the interval. But what happens when the same animals are tested later without any drug, when the clock speed is normal? According to the model, this would produce the opposite pattern. Those that learned to time under amphetamine would now show a rightward shift, and those that learned to time under haloperidol would show a leftward shift. Again, this was also observed (Figure 8.6, no-drug condition).

These results support the idea that DA modulates clock speed. They cannot be explained by common models of DA function such as incentive salience or reward prediction error (Berridge & Robinson, 1998; Schultz, Dayan, & Montague, 1997). However, they can be explained by the transition control model. By determining the gain of transition controllers, DA is in a position to modulate the rate of change in transition variables. A single pulse in the pacemaker–accumulator model is the equivalent of a simple transition in the transition control model, roughly the completion of a simple movement such as turning one's head to the left. It is treated as a single unit at the level of events. The basic unit of timing is this simple transition being accumulated, and DA is expected to determine the rate of accumulation in the integrator.

In the transition control model, DA can increase the size of the signal entering the integrator by activating D1 receptors as well as reduce the leak by activating D2 receptors. Blocking D2 receptors or increasing the activity of the indirect pathway is expected to reduce the rate of accumulation. In support of this prediction, it has been found that blocking D2 receptors reduces clock speed; indeed, the ability of a DA antagonist to reduce clock speed appears to be proportional to its affinity for the D2 receptor (Buhusi & Meck, 2005).

Feedback from self-initiated actions can come from both internal monitoring of transition controller outputs through efference copy signals and from action-generated sensory feedback (reafference). This hypothesis is supported by work from Xin Jin and colleagues, who showed that disrupting auditory sensory feedback during a lever-pressing timing task disrupted timing performance. A major source of the auditory feedback (sound of the lever press) appears to come from the secondary auditory cortex, which sends projections to the striatum to regulate performance (Bakhurin & Yin, 2022; Cook et al., 2022).

While the striatonigral circuit is hypothesized to achieve velocity-to-position integration for simple proprioceptive transitions (Chapter 7), it may not be sufficient for the integration needed for interval timing. The completion of this simple transition can generate an efference copy signal through collateral projections from the BG output neurons to the thalamus. Each simple transition, lasting a few hundred milliseconds, thus represents a single "pulse" that enters the higher-order transition controller and accumulates in the integrator.

This hypothesis predicts that disrupting the thalamocortical targets of the BG outputs should also impair timing. Indeed, inactivation of the mediodorsal thalamus, a major target of BG output that projects to the prefrontal cortex, produced a rightward shift in peak time on probe trials and impaired timing precision (Lusk, Meck, & Yin, 2020). This observation suggests that collaterals of SNr projections may send a copy of the ongoing motor commands to the mediodorsal thalamus, which in turn projects to the prefrontal cortex. In support of this idea, inactivation of the prefrontal cortex also impaired peak time precision (Buhusi, Reyes, Gathers, Oprisan, & Buhusi, 2018).

8.4.3 Direct and Indirect Pathways

Using optogenetic stimulation, it is possible to pause the timing mechanism by activating the nigrocollicular pathway (Toda et al., 2017). Toda et al. used a peak-interval procedure in which sucrose solution was delivered to head-fixed mice every 10 seconds (Figure 8.5). The most immediate effect of stimulation is suppression of licking. This is to be expected given the inhibitory nigral projection to the orofacial regions of the lateral superior colliculus (Rossi et al., 2016). In addition, on peak trials, stimulation also produced a rightward shift in the onset of the next bout of licking. This shift is proportional to the duration of stimulation, suggesting that stimulation paused the timer. These results suggest that the BG not only generates high-level commands for licking but also efference copies that enter some accumulator, similar to a countdown mechanism. On the other hand, activation of the direct pathway appeared to restart the timer (Bakhurin et al., 2020). Mice showed a peak time roughly at the trained interval of 10 seconds. When dSPNs were stimulated after reward delivery, the next peak would appear approximately 10 seconds following the onset of stimulation (Figure 8.7). Without altering the representation of the time interval, direct pathway stimulation can reset the timer.

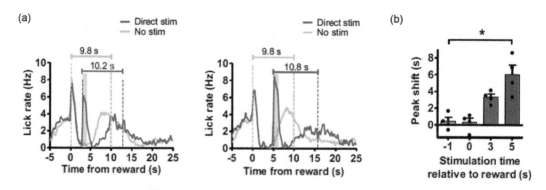

FIGURE 8.7 Direct pathway stimulation resets the internal clock. (a) Laser stimulation (25 Hz) was delivered to activate dSPNs. Left, mean licking rate during probe trials at 3 seconds post reward. Scale bars reflect the population's mean peak times for probe trials. Right, mean licking rate at 5 seconds following reward. (b) Magnitudes of shifts in peak time as a function of the duration of direct pathway stimulation. From Bakhurin et al. (2020) with permission.

If the striatonigral projection represents the accumulator, then striatopallidal activation could stop or reverse accumulation by discharging the integrator. Increasing indirect pathway activity, whether by increasing striatopallidal output or by increasing SNr output, is expected to slow down or stop the timer (Bakhurin et al., 2020; Toda et al., 2017).

8.5 SERIAL ORDER

The BG have long been implicated in the ordering and sequencing of behavior (Mushiake & Strick, 1995). Neural activity related to behavioral sequences has been reported in multiple areas in the frontal cortex and the striatum (Kermadi & Joseph, 1995; Tanji & Mushiake, 1996; Tanji & Shima, 1996). Graybiel argues that the anatomy of the BG allows them to combine smaller behavioral units into larger sequences, a process she calls chunking, inspired by the concept of chunking in the study of short-term memory (Graybiel, 1998; Miller, 1956). Chunking is thus similar to the concatenation of simple behavioral units described above.

In simple transitions, e.g. to turn one's head to the left, the order of the positions reached is not arbitrary. Due to the continuous nature of space, it is impossible to jump from one point to another. It is necessary to traverse all the intermediate points. On the other hand, this constraint is not present in higher-order transitions. Elements in a behavioral sequence can be arbitrary, as each element is represented as an event. Serial order is a property of higher-order transitions, like knocking and then turning the knob. The order can be arbitrary. As Lashley puts it (Lashley, 1951): "Pronunciation of the word 'right' consists first of retraction and elevation of the tongue, expiration of air and activation of the vocal cords; second, depression of the tongue and jaw; third, elevation of the tongue to touch the dental ridge, stopping of vocalization, and forceful expiration of air with depression of the tongue and jaw. These movements have no intrinsic order of association. Pronunciation of the word 'tire' involves the same motor elements in reverse order. Such movements occur in all permutations. The order must therefore be imposed upon the motor elements by some organization other than direct associative connections between them."

To control serial order, it is necessary to detect order independently of the individual elements. Such a detector, as the input function of this controller, is not activated until the elements have occurred in the right order. For example, a detector for AB is activated whenever AB is shown, but not AA, BB, or BA. Its output thus stands for a unique sequence. Similar starting and ending elements can also be shared with other sequence detectors. For example, the letter B may be shared by boy, blank, bay, etc.

In a ramp-to-threshold mechanism, a constant input to multiple elements, each with a different threshold, will result in sequential activation. Sequences with a fixed order can be explained by successively higher thresholds for later elements in the sequence (McKellar et al., 2019). The lower the threshold, the earlier the activation. Such a mechanism may work for very short sequences, but other mechanisms are needed for longer learned sequences to couple distinct transitions. For longer and more complex sequences, it is probably necessary to use recurrent circuits. For example, one may use a circuit in which merely detecting the first element A will not activate the sequence detector but result in reverberation of A. This transient memory mechanism maintains the input to the detector for some time until the second input B arrives. The B input connection has a weight that is strong enough to turn on the sequence detector for AB (Powers, 1973). Thus, completion of each of the elements in the right order would activate the input function to reduce error in this controller, and the completion of the final unit would turn off the sequence detector. As long as there is an error signal, the whole sequence will be generated.

Because the elements to be combined and ordered are often different, simple repetition is not sufficient for controlling serial order. Purely reentrant projections from the BG outputs may implement repetition of a reference signal, but they are not sufficient for the serial activation of different action elements in a sequence. However, as reviewed in Chapter 4, BG outputs may also reach thalamic regions that project to a different cortical region. After the completion of one action, such projections make it possible to activate a different reference signal for a different action, thus binding simple transitions in a sequence.

There is abundant evidence implicating the BG in the learning and production of sequential behaviors, though the underlying circuit mechanisms remain obscure (Graybiel, 1998; Hikosaka, Nakamura, Sakai, & Nakahara, 2002). Below, we shall review some empirical results that shed light on the contributions of the BG.

8.6 DORSOLATERAL STRIATUM AND GROOMING

One innate behavioral sequence that has been extensively studied is grooming in rodents (Fentress & Stilwell, 1973). Rats show a syntactic chain of up to 25 distinct grooming movements, with four sequential phases in a fixed serial order. Each phase lasts a few seconds, starting with small rapid strokes by both paws around the nose, followed by single strokes made by one paw, then by large strokes by both paws over the entire face, and then by body licking (Berridge & Fentress, 1987b; Cromwell & Berridge, 1996; Kalueff et al., 2016). The pattern is intact even after removing facial tactile inputs by cutting the trigeminal afferent fibers (Berridge & Fentress, 1987a). In decerebrate rats, transected either above the superior colliculus or above the pons and cerebellum, syntactic chains can still be produced, indicating that pattern generation does not require the cortico-BG networks (Berridge, 1989a). As in licking behavior, the role of the BG output in grooming is to provide top-down regulation of the innate sequences. Descending BG outputs to the brainstem is necessary and sufficient for generating complex syntactic chains. Deficits in sequence completion are observed after striatal lesions or nigrostriatal DA depletion. In particular, lesions of the anterior dorsolateral striatum (DLS) reduced completion of the grooming sequence, whereas lesions of the ventral pallidum and globus pallidus reduced grooming duration but did not significantly affect grooming syntax (Aldridge & Berridge, 1998; Berridge & Whishaw, 1992; Cromwell & Berridge, 1996). The anterior DLS receives extensive projections from the somatosensory cortex as well as the primary motor cortex, including regions typically associated with whisker and forepaw representations. It is in a position to receive convergent corticostriatal inputs from M1 and S1 projections (Flaherty & Graybiel, 1994). Single-unit activity in the striatum can be correlated with particular grooming movements such as forelimb strokes and body licks, though the experiments lacked precise behavioral quantification to determine the exact relationship (Aldridge & Berridge, 1998). Interestingly, the emergence of the detailed grooming sequences also parallels striatal maturation (Stallman, Berridge, & Colonnese, 1996).

Striatal DA also contributes to the performance of the grooming sequence. DA depletion or deletion of D1 receptors impairs grooming (Berridge, 1989b; Cromwell, Berridge, Drago, & Levine, 1998). On the other hand, D1 receptor activation not only increases grooming but also increases the probability of completing all four phases of a syntactic chain in a specific serial order (Taylor, Rajbhandari, Berridge, & Aldridge, 2010). In hyperdopaminergic mice with reduced expression of DA transporters (DAT knockdown), there are also more rigid syntactic grooming chain patterns, with more stereotyped and predictable syntactic grooming sequences (Berridge & Aldridge, 2000a, b; Berridge, Aldridge, Houchard, & Zhuang, 2005). These findings suggest that increased DA signaling increases sequence stereotypy, possibly by increasing the gain of the sequence repetition controller.

8.7 SEQUENCE LEARNING

The acquisition of new sequences has also been shown to depend on the striatum. Hikosaka and colleagues trained monkeys on a sequence learning task and examined the role of the striatum. They trained monkeys to press 10 button presses with many different variations. Two stimuli were presented as a set, and the monkey must press the two in a particular order. After performing five consecutive sets, it must complete a "hyperset" with 10 button presses (Hikosaka, Rand, Miyachi, & Miyashita, 1995). The monkey had to discover the correct sequence by trial and error. Once well-trained, performance of this task was effector-specific; that is, performance depended on the hand used for the learning. Performance was much worse when the other hand was used (Rand, Hikosaka,

Miyachi, Lu, & Miyashita, 1998). Muscimol inactivation of the anterior striatum (both caudate and putamen) impaired learning of new sequences, whereas inactivation of posterior putamen disrupted the execution of well-learned sequences (Miyachi, Hikosaka, Miyashita, Karadi, & Rand, 1997). DA depletion also impaired both learning and performance on this task (Matsumoto, Hanakawa, Maki, Graybiel, & Kimura, 1999). After unilateral DA depletion, a monkey could improve performance with its ipsilateral arm but not with its contralateral arm. Unlike the ipsilateral arm, the contralateral arm produced discrete movements which were not "chunked" together.

In rodents, the striatum is also critical for the acquisition of a simple sequence, in which mice had to press two levers in a particular order to earn a food pellet (Yin, 2010). As shown in Figure 8.8, in this task only one sequence was rewarded (LR, left lever press followed by right-lever press). RR, LL, and RL were not rewarded. Mice were first trained to press either lever separately, and pre-training lesions of the dorsomedial striatum (DMS) or DLS did not affect the acquisition of lever pressing on a single lever. However, when the mice were trained on the serial order task, the DLS-lesioned group showed a significant impairment. Damage to the DLS made the mice slower in initiating the sequence and in producing the second press in the correct sequence. Two types of errors are possible: repeats (LL, RR) or alternates (RL). The most common error made by DLS-lesioned mice was the production of repeats, especially the repeat of the action proximal to reward

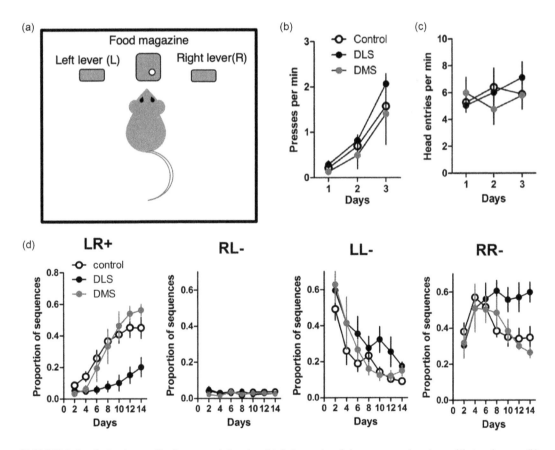

FIGURE 8.8 Striatal contribution to serial order. (a) Schematic of the operant chamber with two levers. (b) Acquisition of lever pressing on a single lever was not affected by striatal lesions. DLS, dorsolateral striatum; DMS, dorsomedial striatum. (c) Rate of head entries into the food magazine was not affected by striatal lesions. (d) Proportion of four possible sequences in lesioned and control mice. Only the LR sequence (a left lever press followed by a right-lever press) was rewarded (LR+). DLS lesions selectively impaired the acquisition of the LR sequence. The errors mostly consist of repetition of the RR sequence. From Yin (2010) with permission.

(RR). The alternative sequence of RL, in contrast, was almost never produced. Because the animal had learned to press the left lever during the preliminary lever press training, the LL sequence was often produced early in acquisition. With training, the LL sequence was extinguished. In contrast, there was a transient increase in the frequency of RR in all mice, along with an increase in LR. In DLS-lesioned mice, however, the RR error persisted even though it was never rewarded.

During learning, the proximal action was first selected. When LR was rewarded, the mouse initially repeated R, the action immediately preceding the reward. Only after repeated nonreinforcement of the RR did it stop performing the incorrect sequence. The effect of reward following LR, then, was initially a nonspecific increase in the frequency of all sequences ending with R (LR and RR). In the DLS lesion group, both LR and RR increased in frequency, though the incorrect RR remained the most common pattern produced. DLS lesions did not impair sequence discrimination; the lesioned mice did not produce RL and LL sequences any more frequently than did the control and DMS groups. Nor can the persistence of the incorrect RR sequence in the DLS group be explained by increased perseveration per se because the LL sequence was rapidly reduced during acquisition. However, the sequence with the same proximal action (R) as the correct sequence was repeated. As shown by the selective deficits in producing the correct LR sequence, DLS lesions impaired the binding of different actions to form a performance unit.

While the primary motor cortex does not seem to be critical for the acquisition of serial order, the secondary motor cortex and dorsomedial prefrontal cortex play a more important role, as lesions of these regions can significantly impair serial order learning (Ostlund, Winterbauer, & Balleine, 2009; Yin, 2009). In mice, the initiation of a simple sequence with two actions requires projections from the secondary motor cortex to the DLS. This glutamatergic synapse is strengthened by serial order learning (Rothwell et al., 2015).

In another study, Geddes et al. trained mice to press the left and right levers in a particular sequence (LLRR) (Geddes, Li, & Jin, 2018). As revealed by the order of acquisition, the mice appeared to combine actions into subsequences (LL and RR) first before incorporating these subsequences into the complete target sequence (LLRR). The RR subsequence was acquired first, followed by the slow acquisition of the LL subsequence.

Striatum-specific deletion of NMDA receptors drastically impaired this type of sequence learning. Without striatal NMDA receptors, mice developed a consistent right-lever bias. They could still increase the performance of R1 and R2, but the frequency of LL decreased rather than increased. When dSPNs were lesioned, mice had difficulty initiating the LL subsequence. When iSPNs were lesioned, mice were impaired in switching from the left to the right subsequences. Recording from identified dSPNs and iSPNs, Geddes et al. found that many dSPNs showed the start and end of a sequence, but this pattern is less common in iSPNs. Many iSPNs were more active during the transition from the left to the right subsequences, after the last press in the left subsequence and before the initiation of the first press in the right subsequence. Stimulation of dSPNs after the first or second left press often inserted an additional left press into the LL subsequence. This effect could not be explained as a restart of the whole sequence since stimulation on the second press of the RR subsequence also resulted in one additional right press. Rather, dSPN stimulation seems to "replay" the last action, as predicted by the repetition control model. Overall sequence length did not change significantly after dSPN stimulation; increasing the left subsequence by optogenetic stimulation was accompanied by a shortening of the right subsequence. Unlike dSPN stimulation, the right subsequence did not compensate to maintain the same overall sequence length after iSPN stimulation. Stimulation of iSPNs following the first right press instead resulted in entry into the food magazine to check for rewards.

These results suggest a hierarchical organization of different transitions within a sequence. The effects of dSPN and iSPN stimulation on performance support the predictions of the repetition control model: direct pathway activation is expected to repeat an action element, while indirect pathway activation is expected to switch to the next action element. Switching can be explained by discharging the integrator and accumulating the ongoing subsequence. When the integrator is discharged by iSPN stimulation, the next element is initiated (Fallon, Fernandez, & Yin, 2022). If the stimulation-elicited discharging occurs in the LL subsequence, this model predicts a switch to the RR sequence; if it occurs in the RR subsequence, then we would expect a switch to a food magazine entry, the next element in the sequence.

8.8 IMAGINATION

Transition control is not only necessary for generating complex action programs but also for so-called cognitive or higher functions, such as working memory, mental imagery, and attention. To understand the relationship between transition control and higher functions, we must introduce the concept of imagination. As mentioned in Chapter 1, Willis argued that imagination is one of the functions of the corpus striatum, though he did not define imagination or explain how it could be implemented by neural circuits (Willis, 1664). In the control hierarchy, however, imagination has a technical definition: according to Powers, a controller operates in imagination mode when it sends a copy of its output signal back to its own perceptual input function instead of commanding lower-level controllers to generate behavioral output (Powers, 1973). Imagination allows a higher-level controller to influence its own input function without going through the feedback function in the external environment. It can be considered a form of internal feedback shielded from environmental disturbances (Figure 8.9).

As perceptual signals at higher levels and signals generated by imagination share the same input channel, there can be competition between bottom-up perceptual input and imagined input. For example, it is difficult to imagine one piece of music while listening to a different piece. The imagination mode is also the means by which perceptual representations associated with the goal can be retrieved. For example, the desire for steak may either generate the instrumental action needed to acquire steak or simply to retrieve the steak representation—a perceptual memory that may also serve as a reference signal for goal-directed behavior. As discussed in Chapter 7, higher-level goal representations are content-addressable memories, which are presumably maintained by the cerebral cortex. These learned representations can be retrieved before action performance. A grocery list is an example of such reference memories, which can be used to retrieve the relevant goal representations to guide behavior.

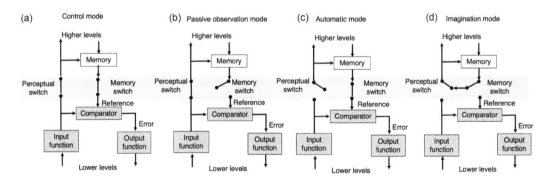

FIGURE 8.9 Different modes of operation at higher levels of the control hierarchy (Powers, 1973). (a) Control mode: both perceptual and memory switches are vertical. The control system functions normally in controlling its input, and the controlled variable is also sent to higher levels. (b) Passive observation mode: perceptual switch is vertical, but the memory is not. The control system is off as it does not receive a descending reference signal, but perceptual channels are still transmitting the input signals. (c) Automatic mode: memory switch is vertical, but perceptual switch is not. The control system operates normally, but no copy of the perceptual signal is sent to higher levels, and there is no conscious perception of feedback from the lower levels that perform the task. (d) Imagination mode: neither perception nor memory switch is vertical. A reference signal can activate the higher perceptual channels, resulting in perceptual experience in the imagination mode.

8.8.1 BG AND THE IMAGINATION MODE

Little is known about the neural mechanisms underlying imagination. The hypothesis advanced here is that the BG projections to the thalamocortical network are also critical for imagination by routing a reference command to a higher input channel. Some BG outputs only target brainstem or tectum, some only project to the thalamus, and some, perhaps the majority, send collaterals to multiple targets (Beckstead, 1983; McElvain et al., 2021; Parent & De Bellefeuille, 1982). As discussed earlier, BG outputs to the thalamus can provide an efference copy signal to coordinate the performance of a behavioral sequence in which different elements must be expressed in a particular order. This function requires axonal collateral projections that send descending commands to the colliculus or brainstem as well as a copy of the command to the thalamocortical circuit. On the other hand, some BG output channels may be used to activate the higher-order thalamic nuclei, which could in turn activate the relevant association cortical areas where the relevant perceptual memories are found. "Exclusive" BG projections to higher-order thalamic nuclei may be especially important for implementing imagination. But even if BG outputs to the thalamus have collaterals to other regions, the strength of the connectivity can be altered through learning so that the imagination mode can be used without strongly activating the downstream position controllers and generating overt movements. Whether this involves active suppression of lower levels in the action hierarchy remains unclear. It should be noted that imagination is often accompanied by subtle motor activation, e.g. galvanic skin responses, eye movements, muscle tension, or force generation (James, 1890).

Perceptual inputs from peripheral sensors and imagined signals may compete at the level of the thalamic nuclei. The competition is between bottom-up sensory inputs monitoring the state of the environment and BG outputs that "reenter" the thalamocortical network to activate specific perceptual representations or memories. There could be gating of the inputs to the thalamocortical system so that these distinct inputs are not in conflict. Through the corticothalamic projections, cortical drive can determine the mode of pallidothalamic action to gate the signals that reenter the thalamocortical network. In songbirds, it has been shown that, in the presence of strong cortical drive, BG output and thalamic activity could be entrained so that BG activity is propagated to the cortex (Goldberg et al., 2013). BG activation of the thalamocortical network may thus be gated by concurrent corticothalamic input. In principle, reentrant projections to the frontal cortex can activate specific goal representations and generate corresponding reference signals for action. Projections to the posterior cortex also target the inferotemporal and auditory cortex to activate specific perceptual memory representations (Middleton & Strick, 2000).

The imagination mode may be important for action planning and simulation, which require retrieval and coordination of reference signals. Note that this type of action planning is different from forward models of motor control discussed earlier in Chapter 5. Although imagination does involve predicting the consequences of actions, especially in retrieving memories associated with the feedback, it does not require a detailed model of environmental properties. There is no need to calculate the inverse kinematics or dynamics.

Some features of the controlled variables, however, are preserved in imagination. Regardless of the content of imagination, so long as it is a transition variable, similar control mechanisms will be needed, implemented by BG circuits that are also used for overt performance. For example, according to the present model, covert humming or internal replay of sequences should activate largely the same BG structures involved in the actual playing. Whether imagined or expressed in performance, the control process would occur at the same rate. For example, humming Bach's Goldberg Variations, whistling it, or playing it on the piano involve different effectors, but they require roughly the same amount of time as the tempo, a transition variable, is preserved in these different modes. What they share are the high-level reference signals. Likewise, other transition variables, like serial order, are also preserved.

The present model predicts that any manipulation that affects the rate of change in transition, such as DA depletion, can affect both overt movement and the content of imagination (or "thoughts"). In support of this prediction, Parkinson's patients often show both slowed movements and slowed verbal processing (Poliakoff, 2013). Interestingly, patients whose motor symptoms are asymmetric due to more DA depletion on one side also showed similar asymmetry in motor imagery. Not only do they show better performance on a finger sequencing task with their good hand, but their mental simulation of movement is also better with the good hand compared to the affected hand (Dominey, Decety, Broussolle, Chazot, & Jeannerod, 1995). There are thus overlapping neural circuits for action planning and execution (Decety & Grèzes, 2006). Likewise, a motor cortico-BG network (premotor cortex, bilateral putamen, and thalamus) is activated when the subject is imagining drawing figures or executing these movements (Castrop, Dresel, Hennenlotter, Zimmer, & Haslinger, 2012; Stephan et al., 1995). Areas engaged in mental simulation include bilateral premotor, prefrontal, supplementary motor, and left posterior parietal areas, and the caudate nuclei (Gerardin et al., 2000).

8.8.2 MENTAL ROTATION

Perhaps the clearest experimental demonstration of the imagination mode comes from studies of mental rotation. In classic experiments, Shepard and Metzler asked subjects to compare two objects with distinct orientations and determine if they were identical. This task required imagining one object being rotated into the same orientation as the other. The time it takes to determine whether two objects are identical is a linearly increasing function of the angular difference in their orientations (Shepard & Metzler, 1971).

Studies have reported that mental rotation and motor rotation share similar neural circuits, including the motor and premotor cortex, parietal cortex, and BG (Alivisatos & Petrides, 1997). Moreover, in support of the transition control model, the rate of rotation during mental rotation is comparable to that during overt rotation (Pellizzer & Georgopoulos, 1993; Wexler, Kosslyn, & Berthoz, 1998). The rate of rotation, a transition variable, is what overt and imagined processes have in common.

In some Parkinson's patients with bradykinesia, there is also slowed mental rotation (Amick, Schendan, Ganis, & Cronin-Golomb, 2006; Lee, Harris, & Calvert, 1998). Harris et al. reported a case of a woman with a lesion in the right BG and severe mental rotation impairments (Harris, Harris, & Caine, 2002). She had no difficulty recognizing the objects being rotated, but she could not map the left and right sides of objects to her egocentric reference frame. In Huntington's disease, there is a significant reduction in the speed of mental rotation (Lineweaver, Salmon, Bondi, & Corey-Bloom, 2005). Mental rotation of perceptual images requires manipulation of the corresponding cortical representations, which can be achieved using the BG output to the thalamocortical network. As already mentioned above, it is possible that a subset of BG output can selectively activate the thalamocortical circuit without affecting the midbrain or brainstem position controllers to generate overt movement. This explains how there could be bradykinesia without slowing mental rotation (Duncombe, Bradshaw, Iansek, & Phillips, 1994).

The present model suggests that the same BG circuits that are involved in overt movement also play a critical role in mental simulation or imagination. The control mechanisms required for transition variables such as tempo and serial order are preserved in imagination mode. One prediction of this account is that any manipulation that affects the rate of change in transition, such as DA depletion, should affect both overt movement and imagination. This is consistent with findings from studies of Parkinson's patients, who may show impairments in overt movement as well as mental rotation. The rate of rotation during mental rotation is comparable to that during overt rotation, supporting the idea that the rate of transition is a key feature preserved in the imagination mode.

8.9 WORKING MEMORY

Working memory is here defined as the sustained activation of an internal representation in the absence of corresponding perceptual inputs. For example, when pursuing a prey, the input does not seize completely when it is transiently obscured from view. Working memory has both retrospective and prospective functions (Konorski, 1967). Retrospectively, it functions to bridge a gap in sensory input or a lag in perceptual feedback. Prospectively, it can be used to activate relevant reference representations. For example, the same mechanism is used for maintaining an online representation of an apple just observed or desiring the same apple.

Working memory is traditionally thought to require a recurrent excitatory network in the cortex (Robbins & Arnsten, 2009; Williams & Goldman-Rakic, 1995). But studies have also implicated the BG. For example, lesions to components of the associative cortico-BG circuit, e.g. either the prefrontal cortex or its striatal target in the caudate, can produce significant working memory deficits (Aarts, van Holstein, & Cools, 2011; Divac, Rosvold, & Szwarcbart, 1967; Levy, Friedman, Davachi, & Goldman-Rakic, 1997).

Baddeley's original model of working memory includes three components: a central executive, an articulatory or phonological loop, and a visuospatial sketch pad (Baddeley, 1992). The central executive is a selector that determines what enters the other systems. The articulatory loop is based on internal rehearsal using imagined auditory feedback, for example as one articulates a phone number using the imagination mode before dialing it. The visuospatial sketch pad is similar to the mental imagery discussed above. Recent work has shown that Baddeley's account is far too restricted, as working memory could be extended to other modalities as well (e.g. tactile) (Jaffe & Constantinidis, 2021). The phonological loop and visuospatial sketchpad can both be explained by transition control. The phonological loop relies on the reentry to the auditory and visual cortical regions, and the visuospatial sketchpad can be explained by the mechanism for mental rotation discussed above. According to the transition control model, the manipulation of working memory representations in the absence of overt performance uses the same control mechanisms as the repetition of current action (e.g. lever press) and switching to a different action.

Studies in humans have suggested a role for the BG in gating and manipulating working memory, whether to maintain the ongoing representation or to switch to a different representation. PD patients are impaired at shifting in verbal expression as well as movement, as well as in cognitive set switching (Cools et al., 1984; Van den Bercken & Cools, 1982). This deficit in thinking or shifting of mental representations is analogous to deficits in adjusting the velocity and tempo of movements. The mechanism for holding signals online using reverberation is the same as repetition control, whereas the mechanism for switching requires the termination of the ongoing program, just as in behavioral switching. At one extreme, there is distractibility, and at the other, perseverance.

Some argue that striatal DA mediates switching, whereas PFC DA contributes to the persistence of working memory representations (Cools & D'Esposito, 2011; Gruber, Dayan, Gutkin, & Solla, 2006; Williams & Goldman-Rakic, 1995). In switching internal representations, distinct channels are used to access the thalamocortical network via reentrant projections for maintenance of existing representations and for switching to different representations once the ongoing representation is terminated. Just like the coordination of behavioral sequences, the direct and indirect pathways are thought to play opposite roles in this process. The direct pathway could be responsible for repeating the same routine, whereas the indirect pathway contributes to updating or switching to the next element. There is some evidence in support of this hypothesis. For example, overexpression of striatal D2 receptors, which presumably reduces iSPN output, results in impaired set shifting in mice (Kellendonk et al., 2006).

8.10 SUMMARY

Higher-order transition variables can be built using simple transition variables. One cycle of a repeating pattern is a simple transition or "event". Tempo, the rate of change in events, can be adjusted by changing the interval between events. One example of a repetitive behavioral sequence is licking in rodents, which is characterized by stereotyped bouts. While the pattern generators are found in the brainstem, BG output to the tectum and brainstem nuclei can directly regulate licking by adjusting the duty cycle, or the "on time" of pattern generation. It is hypothesized that simple transitions can be concatenated to form more complex sequences of behavior, by coupling transition controllers in a nested network. The output of one transition controller can be registered as individual pulses at the input of another controller. One cycle of a simple transition can be counted as a single pulse being accumulated in a coupled integrator that generates a more abstract and longer-lasting transition.

An emergent property of transition control is time. There is no dedicated sensory organ for time, but the sense of time emerges at the higher levels, where the derivatives of perceptual configurations are computed and the rate of change is explicitly represented. Timing is therefore a by-product of transition control. Not surprisingly, the BG have been implicated in interval timing in the seconds-to-minutes range.

Serial order is the ordering of arbitrary event-level transitions. Innate sequences like grooming in rodents can be genetically programmed, but often learning is required to produce different behaviors in a specific order. Such learning also requires the sensorimotor cortico-BG network. The efference copy from BG output collaterals to the thalamocortical network can provide feedback representing distinct components of the sequence.

Imagination is a key functional mode of the control hierarchy, especially at the transition control level. In imagination mode, a transition controller may send output to its own input function, making it possible to activate higher-level perceptual representations without generating overt behavior. It can be implemented by reentrant signaling from BG outputs that mainly change the state of the thalamo-cortical network without commanding overt behaviors via descending projections to the brainstem.

REFERENCES

Aarts, E., van Holstein, M., & Cools, R. (2011). Striatal dopamine and the interface between motivation and cognition. *Frontiers in Psychology*, 2, 163.

Aldridge, J. W., & Berridge, K. C. (1998). Coding of serial order by neostriatal neurons: A "natural action" approach to movement sequence. *Jouranl of Neuroscience*, 18(7), 2777–2787.

Alivisatos, B., & Petrides, M. (1997). Functional activation of the human brain during mental rotation. *Neuropsychologia*, 35(2), 111–118.

Amick, M. M., Schendan, H. E., Ganis, G., & Cronin-Golomb, A. (2006). Frontostriatal circuits are necessary for visuomotor transformation: Mental rotation in parkinson's disease. *Neuropsychologia*, 44(3), 339–349. doi: 10.1016/j.neuropsychologia.2005.06.002.

Baddeley, A. (1992). Working memory. *Science*, 255(5044), 556–559.

Bakhurin, K. I., Li, X., Friedman, A. D., Lusk, N. A., Watson, G. D., Kim, N., & Yin, H. H. (2020). Opponent regulation of action performance and timing by striatonigral and striatopallidal pathways. *Elife*, 9, e54831.

Bakhurin, K. I., & Yin, H. H. (2022). Closing the loop on models of interval timing. *Nature Neuroscience*, 25(3), 270–271.

Beckstead, R. (1983). Long collateral branches of substantia nigra pars reticulata axons to thalamus, superior colliculus and reticular formation in monkey and cat. Multiple retrograde neuronal labeling with fluorescent dyes. *Neuroscience*, 10(3), 767–779.

Benecke, R., Rothwell, J., Dick, J., Day, B., & Marsden, C. (1987). Disturbance of sequential movements in patients with parkinson's disease. *Brain*, 110(2), 361–379.

Berridge, K. C. (1989a). Progressive degradation of serial grooming chains by descending decerebration. *Behavioural Brain Research*, 33(3), 241–253.

Berridge, K. C. (1989b). Substantia nigra 6-ohda lesions mimic striatopallidal disruption of syntactic grooming chains: A neural systems analysis of sequence control. *Psychobiology*, 17(4), 377–385.

Berridge, K. C., & Aldridge, J. W. (2000a). Super-stereotypy i: Enhancement of a complex movement sequence by systemic dopamine d1 agonists. *Synapse*, 37(3), 194–204. doi: 10.1002/1098-2396(20000901)37:3<194::AID-SYN3>3.0.CO;2-A.

Berridge, K. C., & Aldridge, J. W. (2000b). Super-stereotypy ii: Enhancement of a complex movement sequence by intraventricular dopamine d1 agonists. *Synapse*, 37(3), 205–215. doi: 10.1002/1098-2396(20000901)37:3<205::AID-SYN4>3.0.CO;2-A.

Berridge, K. C., Aldridge, J. W., Houchard, K. R., & Zhuang, X. (2005). Sequential super-stereotypy of an instinctive fixed action pattern in hyper-dopaminergic mutant mice: A model of obsessive compulsive disorder and tourette's. *BMC Biology*, 3, 4. doi: 10.1186/1741-7007-3-4.

Berridge, K. C., & Fentress, J. C. (1987a). Deafferentation does not disrupt natural rules of action syntax. *Behavioural Brain Research*, 23(1), 69–76.

Berridge, K. C., & Fentress, J. C. (1987b). Disruption of natural grooming chains after striatopallidal lesions. *Psychobiology*, *15*(4), 336–342.

Berridge, K. C., & Robinson, T. E. (1998). What is the role of dopamine in reward: Hedonic impact, reward learning, or incentive salience? *Brain Research Review*, *28*(3), 309–369.

Berridge, K. C., & Whishaw, I. Q. (1992). Cortex, striatum and cerebellum: Control of serial order in a grooming sequence. *Experimental Brain Research*, *90*(2), 275–290.

Bignall, K. E., & Schramm, L. (1974). Behavior of chronically decerebrated kittens. *Experimental Neurology*, *42*(3), 519–531. doi: 10.1016/0014-4886(74)90075-2.

Buhusi, C. V., & Meck, W. H. (2005). What makes us tick? Functional and neural mechanisms of interval timing. *Nature Review Neuroscience*, *6*(10), 755–765.

Buhusi, C. V., Reyes, M. B., Gathers, C.-A., Oprisan, S. A., & Buhusi, M. (2018). Inactivation of the medial-prefrontal cortex impairs interval timing precision, but not timing accuracy or scalar timing in a peak-interval procedure in rats. *Frontiers in Integrative Neuroscience*, *12*, 20.

Castrop, F., Dresel, C., Hennenlotter, A., Zimmer, C., & Haslinger, B. (2012). Basal ganglia–premotor dysfunction during movement imagination in writer's cramp. *Movement Disorders*, *27*(11), 1432–1439.

Chandler, S. H., & Goldberg, L. J. (1988). Effects of pontomedullary reticular formation stimulation on the neuronal networks responsible for rhythmical jaw movements in the guinea pig. *Journal of Neurophysiology*, *59*(3), 819–832.

Chen, Z., Travers, S. P., & Travers, J. B. (2001). Muscimol infusions in the brain stem reticular formation reversibly block ingestion in the awake rat. *American Journal of Physiology-Regulatory, Integrative and Comparative Physiology*, *280*(4), R1085–R1094.

Chen, Z., Zhang, Z.-Y., Zhang, W., Xie, T., Li, Y., Xu, X.-H., & Yao, H. (2021). Direct and indirect pathway neurons in ventrolateral striatum differentially regulate licking movement and nigral responses. *Cell Reports*, *37*(3), 109847.

Cook, J. R., Li, H., Nguyen, B., Huang, H.-H., Mahdavian, P., Kirchgessner, M. A., ... Jin, X. (2022). Secondary auditory cortex mediates a sensorimotor mechanism for action timing. *Nature Neuroscience*, *25*(3), 330–344.

Cools, A. R. (1981). Physiological significance of the striatal system: New light on an old concept. In: Szentagothai, J., Hamori, J., Palkovits, M. (Eds.), *Regulatory Functions of the CNS Subsystems* (pp. 227–230). New York: Elsevier.

Cools, A. R., & D'Esposito, M. (2011). Inverted-u-shaped dopamine actions on human working memory and cognitive control. *Biological Psychiatry*, *69*(12), e113–e125.

Cools, A. R., van den Bercken, J. H., Horstink, M. W., van Spaendonck, K. P., & Berger, H. J. (1984). Cognitive and motor shifting aptitude disorder in parkinson's disease. *Journal of Neurology, Neurosurgery & Psychiatry*, *47*(5), 443–453.

Coull, J. T., Vidal, F., Nazarian, B., & Macar, F. (2004). Functional anatomy of the attentional modulation of time estimation. *Science*, *303*(5663), 1506–1508. doi: 10.1126/science.1091573.

Cromwell, H. C., & Berridge, K. C. (1996). Implementation of action sequences by a neostriatal site: A lesion mapping study of grooming syntax. *Journal of Neuroscience*, *16*(10), 3444–3458.

Cromwell, H. C., Berridge, K. C., Drago, J., & Levine, M. S. (1998). Action sequencing is impaired in d1a-deficient mutant mice. *European Journal of Neuroscience*, *10*(7), 2426–2432.

Decety, J., & Grèzes, J. (2006). The power of simulation: Imagining one's own and other's behavior. *Brain Research*, *1079*(1), 4–14.

Delfs, J., & Kelley, A. (1990). The role of d 1 and d 2 dopamine receptors in oral stereotypy induced by dopaminergic stimulation of the ventrolateral striatum. *Neuroscience*, *39*(1), 59–67.

Divac, I., Rosvold, H. E., & Szwarcbart, M. K. (1967). Behavioral effects of selective ablation of the caudate nucleus. *Journal of Comparative and Physiological Psychology*, *63*(2), 184–190.

Dominey, P., Decety, J., Broussolle, E., Chazot, G., & Jeannerod, M. (1995). Motor imagery of a lateralized sequential task is asymmetrically slowed in hemi-parkinson's patients. *Neuropsychologia*, *33*(6), 727–741.

Duncombe, M. E., Bradshaw, J. L., Iansek, R., & Phillips, J. G. (1994). Parkinsonian patients without dementia or depression do not suffer from bradyphrenia as indexed by performance in mental rotation tasks with and without advance information. *Neuropsychologia*, *32*(11), 1383–1396.

Fallon, I. P., Fernandez, S., & Yin, H. H. (2022). The striatal indirect pathway resets the neural representation of numerosity. *Paper Presented at the Annual Meeting of the Society for Neuroscience*, San Diego.

Faull, R., & Mehler, W. (1978). The cells of origin of nigrotectal, nigrothalamic and nigrostriatal projections in the rat. *Neuroscience*, *3*(11), 989–1002.

Fentress, J. C., & Stilwell, F. P. (1973). Grammar of a movement sequence in inbred mice. *Nature*, *244*(5410), 52–53.
Flaherty, A. W., & Graybiel, A. M. (1994). Input-output organization of the sensorimotor striatum in the squirrel monkey. *Journal of Neuroscience*, *14*(2), 599–610.
Fowler, S. C., & Mortell, C. (1992). Low doses of haloperidol interfere with rat tongue extensions during licking: A quantitative analysis. *Behavioral Neuroscience*, *106*(2), 386.
Geddes, C. E., Li, H., & Jin, X. (2018). Optogenetic editing reveals the hierarchical organization of learned action sequences. *Cell*, *174*(1), 32–43. e15.
Gerardin, E., Sirigu, A., Lehericy, S., Poline, J. B., Gaymard, B., Marsault, C., ... Le Bihan, D. (2000). Partially overlapping neural networks for real and imagined hand movements. *Cerebral Cortex*, *10*(11), 1093–1104.
Gibbon, J. (1977). Scalar expectancy-theory and webers law in animal timing. *Psychological Review*, *84*(3), 279–325. doi: 10.1037//0033-295x.84.3.279.
Gibbon, J., & Church, R. M. (1990). Representation of time. *Cognition*, *37*(1–2), 23–54.
Goldberg, J. H., Farries, M. A., & Fee, M. S. (2013). Basal ganglia output to the thalamus: Still a paradox. *Trends in Neurosciences*, *36*(12), 695–705.
Graybiel, A. M. (1998). The basal ganglia and chunking of action repertoires. *Neurobiology of Learning and Memory*, *70*(1–2), 119–136.
Grill, H. J., & Norgren, R. (1978). The taste reactivity test. Ii. Mimetic responses to gustatory stimuli in chronic thalamic and chronic decerebrate rats. *Brain Research*, *143*(2), 281–297. doi: 10.1016/0006-8993(78)90569-3.
Gruber, A. J., Dayan, P., Gutkin, B. S., & Solla, S. A. (2006). Dopamine modulation in the basal ganglia locks the gate to working memory. *Journal of Computational Neuroscience*, *20*(2), 153.
Harris, I. M., Harris, J. A., & Caine, D. (2002). Mental-rotation deficits following damage to the right basal ganglia. *Neuropsychology*, *16*(4), 524–537.
Hikosaka, O., Nakamura, K., Sakai, K., & Nakahara, H. (2002). Central mechanisms of motor skill learning. *Current Opinion in Neurobiology*, *12*(2), 217–222.
Hikosaka, O., Rand, M. K., Miyachi, S., & Miyashita, K. (1995). Learning of sequential movements in the monkey: Process of learning and retention of memory. *Journal of Neurophysiology*, *74*(4), 1652–1661.
Hintiryan, H., Foster, N. N., Bowman, I., Bay, M., Song, M. Y., Gou, L., ... Dong, H. W. (2016). The mouse cortico-striatal projectome. *Nature Neuroscience*, *19*(8), 1100–1114. doi: 10.1038/nn.4332.
Inchul, P., Amano, N., Satoda, T., Murata, T., Kawagishi, S., Yoshino, K., & Tanaka, K. (2005). Control of orofacio-lingual movements by the substantia nigra pars reticulata: High-frequency electrical microstimulation and gaba microinjection findings in rats. *Neuroscience*, *134*(2), 677–689.
Jaffe, R., & Constantinidis, C. (2021). Working memory: From neural activity to the sentient mind. *Comprehensive Physiology*, *11*(4), 1–41.
James, W. (1890). *The Principles of Psychology* (vol. 1): New York: Henry Holt.
Kalueff, A. V., Stewart, A. M., Song, C., Berridge, K. C., Graybiel, A. M., & Fentress, J. C. (2016). Neurobiology of rodent self-grooming and its value for translational neuroscience. *Nature Reviews Neuroscience*, *17*(1), 45.
Kase, D., Uta, D., Ishihara, H., & Imoto, K. (2015). Inhibitory synaptic transmission from the substantia nigra pars reticulata to the ventral medial thalamus in mice. *Neuroscience Research*, *97*, 26–35.
Kellendonk, C., Simpson, E. H., Polan, H. J., Malleret, G., Vronskaya, S., Winiger, V., Moore, H., & Kandel, E. R. (2006). Transient and selective overexpression of dopamine D2 receptors in the striatum causes persistent abnormalities in prefrontal cortex functioning. *Neuron*, *49*, 603–615.
Kelley, A. E., Lang, C. G., & Gauthier, A. M. (1988). Induction of oral stereotypy following amphetamine microinjection into a discrete subregion of the striatum. *Psychopharmacology (Berl)*, *95*(4), 556–559.
Kermadi, I., & Joseph, J. P. (1995). Activity in the caudate nucleus of monkey during spatial sequencing. *Journal of Neurophysiology*, *74*(3), 911–933.
Kiehn, O. (2016). Decoding the organization of spinal circuits that control locomotion. *Nature Reviews Neuroscience*, *17*(4), 224.
Konorski, J. (1967). *Integrative Activity of the Brain*: Chicago, IL: University of Chicago Press.
Krosigk, M., & Smith, A. (1991). Descending projections from the substantia nigra and retrorubral field to the medullary and pontomedullary reticular formation. *European Journal of Neuroscience*, *3*(3), 260–273.
Lashley, K. S. (1951). The problem of serial order in behavior. In: Jeffress, L. A. (Ed.), *Cerebral Mechanisms in Behavior: The Hixon Symposium* (pp. 112–146). New York: Wiley.
Lee, A., Harris, J., & Calvert, J. (1998). Impairments of mental rotation in parkinson's disease. *Neuropsychologia*, *36*(1), 109–114.

Lee, J., Wang, W., & Sabatini, B. L. (2020). Anatomically segregated basal ganglia pathways allow parallel behavioral modulation. *Nature Neuroscience*, *23*(11), 1388–1398.

Levy, R., Friedman, H. R., Davachi, L., & Goldman-Rakic, P. S. (1997). Differential activation of the caudate nucleus in primates performing spatial and nonspatial working memory tasks. *Journal of Neuroscience*, *17*(10), 3870–3882.

Lineweaver, T. T., Salmon, D. P., Bondi, M. W., & Corey-Bloom, J. (2005). Differential effects of alzheimer's disease and huntington's disease on the performance of mental rotation. *Journal of the International Neuropsychological Society*, *11*(1), 30–39.

Lowe, A. A. (1984). Tongue movements–brainstem mechanisms and clinical postulates. *Brain, Behavior and Evolution*, *25*(2–3), 128–137.

Lusk, N., Meck, W. H., & Yin, H. H. (2020). Mediodorsal thalamus contributes to the timing of instrumental actions. *Journal of Neuroscience*, *40*(33), 6379–6388.

Maricq, A. V., Roberts, S., & Church, R. M. (1981). Methamphetamine and time estimation. *Journal of Experimental Psychology: Animal Behavior Processes*, *7*(1), 18.

Matsumoto, N., Hanakawa, T., Maki, S., Graybiel, A. M., & Kimura, M. (1999). Role of nigrostriatal dopamine system in learning to perform sequential motor tasks in a predictive manner. *Journal of Neurophysiology*, *82*(2), 978–998.

McElvain, L. E., Chen, Y., Moore, J. D., Brigidi, G. S., Bloodgood, B. L., Lim, B. K., ... Kleinfeld, D. (2021). Specific populations of basal ganglia output neurons target distinct brain stem areas while collateralizing throughout the diencephalon. *Neuron*, *109*, 1–18.

McElvain, L. E., Friedman, B., Karten, H. J., Svoboda, K., Wang, F., Deschenes, M., & Kleinfeld, D. (2018). Circuits in the rodent brainstem that control whisking in concert with other orofacial motor actions. *Neuroscience*, *368*, 152–170. doi: 10.1016/j.neuroscience.2017.08.034.

McKellar, C. E., Lillvis, J. L., Bath, D. E., Fitzgerald, J. E., Cannon, J. G., Simpson, J. H., & Dickson, B. J. (2019). Threshold-based ordering of sequential actions during drosophila courtship. *Current Biology*, *29*(3), 426–434. e426.

Meck, W. H. (1983). Selective adjustment of the speed of internal clock and memory processes. *Journal of Experimental Psychology: Animal Behavior Processes*, *9*(2), 171.

Meck, W. H. (1996). Neuropharmacology of timing and time perception. *Cognitive Brain Research*, *3*(3–4), 227–242.

Meck, W. H., & Church, R. M. (1983). A mode control model of counting and timing processes. *Journal of Experimental Psychology: Animal Behavior Processes*, *9*(3), 320.

Mello, G. B., Soares, S., & Paton, J. J. (2015). A scalable population code for time in the striatum. *Current Biology*, *25*(9), 1113–1122. doi: 10.1016/j.cub.2015.02.036.

Merchant, H., Grahn, J., Trainor, L., Rohrmeier, M., & Fitch, W. T. (2015). Finding the beat: A neural perspective across humans and non-human primates. *Philosophical Transactions of the Royal Society B: Biological Sciences*, *370*(1664), 20140093.

Merchant, H., Harrington, D. L., & Meck, W. H. (2013). Neural basis of the perception and estimation of time. *Annual Review of Neuroscience*, *36*, 313–336. doi: 10.1146/annurev-neuro-062012-170349.

Middleton, F. A., & Strick, P. L. (2000). Basal ganglia and cerebellar loops: Motor and cognitive circuits. *Brain Research Review*, *31*(2–3), 236–250.

Miller, G. A. (1956). The magical number seven, plus or minus two: Some limits on our capacity for processing information. *Psychological Review*, *63*(2), 81.

Miyachi, S., Hikosaka, O., Miyashita, K., Karadi, Z., & Rand, M. K. (1997). Differential roles of monkey striatum in learning of sequential hand movement. *Experimental Brain Research*, *115*(1), 1–5.

Moore, J. D., Kleinfeld, D., & Wang, F. (2014). How the brainstem controls orofacial behaviors comprised of rhythmic actions. *Trends in Neurosciences*, *37*(7), 370–380.

Mushiake, H., & Strick, P. L. (1995). Pallidal neuron activity during sequential arm movements. *Journal of Neurophysiology*, *74*(6), 2754–2758.

Nakamura, Y., & Katakura, N. (1995). Generation of masticatory rhythm in the brainstem. *Neuroscience Research*, *23*(1), 1–19.

Ostlund, S. B., Winterbauer, N. E., & Balleine, B. W. (2009). Evidence of action sequence chunking in goal-directed instrumental conditioning and its dependence on the dorsomedial prefrontal cortex. *Journal of Neuroscience*, *29*(25), 8280–8287. doi: 10.1523/JNEUROSCI.1176-09.2009.

Parent, A., & De Bellefeuille, L. (1982). Organization of efferent projections from the internal segment of globus pallidus in primate as revealed by flourescence retrograde labeling method. *Brain Research*, *245*(2), 201–213.

Pellizzer, G., & Georgopoulos, A. P. (1993). Common processing constraints for visuomotor and visual mental rotations. *Experimental Brain Research*, *93*(1), 165–172.

Poliakoff, E. (2013). Representation of action in parkinson's disease: Imagining, observing, and naming actions. *Journal of Neuropsychology*, *7*(2), 241–254.

Powers, W. T. (1973). *Behavior: Control of Perception*: New Canaan, CT: Benchmark Publications.

Rand, M. K., Hikosaka, O., Miyachi, S., Lu, X., & Miyashita, K. (1998). Characteristics of a long-term procedural skill in the monkey. *Experimental Brain Research*, *118*(3), 293–297.

Redgrave, P., Marrow, L., & Dean, P. (1992). Topographical organization of the nigrotectal projection in rat: Evidence for segregated channels. *Neuroscience*, *50*(3), 571–595.

Robbins, T. W., & Arnsten, A. F. (2009). The neuropsychopharmacology of fronto-executive function: Monoaminergic modulation. *Annual Review of Neuroscience*, *32*, 267–287.

Roitman, M. F., Stuber, G. D., Phillips, P. E., Wightman, R. M., & Carelli, R. M. (2004). Dopamine operates as a subsecond modulator of food seeking. *Journal of Neuroscience*, *24*(6), 1265–1271.

Rossi, M. A., Li, H. E., Lu, D., Kim, I. H., Bartholomew, R. A., Gaidis, E., ... Yin, H. H. (2016). A gabaergic nigrotectal pathway for coordination of drinking behavior. *Nature Neuroscience*, *19*(5), 742–748. doi: 10.1038/nn.4285.

Rossi, M. A., & Yin, H. H. (2015). Elevated dopamine alters consummatory pattern generation and increases behavioral variability during learning. *Frontiers in Integrative Neuroscience*, *9*, 37. doi: 10.3389/fnint.2015.00037.

Rothwell, P. E., Hayton, S. J., Sun, G. L., Fuccillo, M. V., Lim, B. K., & Malenka, R. C. (2015). Input- and output-specific regulation of serial order performance by corticostriatal circuits. *Neuron*, *88*(2), 345–356. doi: 10.1016/j.neuron.2015.09.035.

Sacks, O. (1991). *Awakenings*: London: Duckworth & Co.

Salamone, J. D., Cousins, M. S., Maio, C., Champion, M., Turski, T., & Kovach, J. (1996). Different behavioral effects of haloperidol, clozapine and thioridazine in a concurrent lever pressing and feeding procedure. *Psychopharmacology (Berl)*, *125*(2), 105–112.

Sasamoto, K., Zhang, G., & Iwasaki, M. (1990). Two types of rhythmical jaw movements evoked by stimulation of the rat cortex. *Japanese Journal of Oral Biology*, *32*(1), 57–68.

Satoda, T., Amano, N., Masuda, Y., Uchida, T., Yasui, Y., & Mizuno, N. (2002). The sites of origin of projection fibers from the cerebral cortex to the jaw region of the striatum of the rat. *Neuroscience Letters*, *332*(1), 9–12.

Schultz, W., Dayan, P., & Montague, P. R. (1997). A neural substrate of prediction and reward. *Science*, *275*(5306), 1593–1599.

Shepard, R. N., & Metzler, J. (1971). Mental rotation of three-dimensional objects. *Science*, *171*(3972), 701–703.

Stallman, E. L., Berridge, K. C., & Colonnese, M. T. (1996). Ontogeny of action syntax in altricial and precocial rodents: Grooming sequences of rat and guinea pig pups. *Behaviour*, *133*(15–16), 1165–1195.

Stephan, K. M., Fink, G. R., Passingham, R. E., Silbersweig, D., Ceballos-Baumann, A. O., Frith, C. D., & Frackowiak, R. S. (1995). Functional anatomy of the mental representation of upper extremity movements in healthy subjects. *Journal of Neurophysiology*, *73*(1), 373–386.

Tanji, J., & Mushiake, H. (1996). Comparison of neuronal activity in the supplementary motor area and primary motor cortex. *Cognitive Brain Research*, *3*(2), 143–150.

Tanji, J., & Shima, K. (1996). Supplementary motor cortex in organization of movement. *European Neurology*, *36*(Suppl 1), 13–19.

Taylor, J. L., Rajbhandari, A. K., Berridge, K. C., & Aldridge, J. W. (2010). Dopamine receptor modulation of repetitive grooming actions in the rat: Potential relevance for tourette syndrome. *Brain Research*, *1322*, 92–101.

Toda, K., Lusk, N. A., Watson, G. D. R., Kim, N., Lu, D., Li, H. E., ... Yin, H. H. (2017). Nigrotectal stimulation stops interval timing in mice. *Current Biology*, *27*(24), 3763–3770 e3763. doi: 10.1016/j.cub.2017.11.003.

Travers, J. B., Dinardo, L. A., & Karimnamazi, H. (1997). Motor and premotor mechanisms of licking. *Neuroscience & Biobehavioral Reviews*, *21*(5), 631–647.

Travers, J. B., & Norgren, R. (1983). Afferent projections to the oral motor nuclei in the rat. *Journal of Comparative Neurology*, *220*(3), 280–298.

Treisman, M. (1963). Temporal discrimination and the indifference interval: Implications for a model of the "internal clock". *Psychological Monographs: General and Applied*, *77*(13), 1.

Van den Bercken, J. H., & Cools, A. R. (1982). Evidence for a role of the caudate nucleus in the sequential organization of behavior. *Behavioural Brain Research*, *4*(4), 319–327.

von Krosigk, M., Smith, Y., Bolam, J. P., & Smith, A. D. (1992). Synaptic organization of gabaergic inputs from the striatum and the globus pallidus onto neurons in the substantia nigra and retrorubral field which project to the medullary reticular formation. *Neuroscience, 50*(3), 531–549.

Wexler, M., Kosslyn, S. M., & Berthoz, A. (1998). Motor processes in mental rotation. *Cognition, 68*(1), 77–94.

Williams, G. V., & Goldman-Rakic, P. S. (1995). Modulation of memory fields by dopamine dl receptors in prefrontal cortex. *Nature, 376*(6541), 572–575.

Willis, T. (1664). *Cerebri Anatome*: Amsterdam: Joannes de Someren.

Yasui, Y., Nakano, K., Nakagawa, Y., Kayahara, T., Shiroyama, T., & Mizuno, N. (1992). Non-dopaminergic neurons in the substantia nigra project to the reticular formation around the trigeminal motor nucleus in the rat. *Brain Research, 585*(1–2), 361–366.

Yasui, Y., Tsumori, T., Ando, A., & Domoto, T. (1995). Demonstration of axon collateral projections from the substantia nigra pars reticulata to the superior colliculus and the parvicellular reticular formation in the rat. *Brain Research, 674*(1), 122–126.

Yasui, Y., Tsumori, T., Ono, K., & Kishi, T. (1997). Nigral axon terminals are in contact with parvicellular reticular neurons which project to the motor trigeminal nucleus in the rat. *Brain Research, 775*(1–2), 219–224.

Yin, H. H. (2009). The role of the murine motor cortex in action duration and order. *Frontiers in Integrative Neuroscience, 3*, 23. doi: 10.3389/neuro.07.023.2009.

Yin, H. H. (2010). The sensorimotor striatum is necessary for serial order learning. *Journal of Neuroscience, 30*(44), 14719–14723. doi: 10.1523/JNEUROSCI.3989–10.2010.

Yin, H. H. (2014). Action, time and the basal ganglia. *Philosophical Transactions of the Royal Society B: Biological Sciences, 369*(1637), 20120473.

9 Motivation

So far, we have discussed *how* the basal ganglia (BG) output can specify actions. In the action control hierarchy, each level controls a distinct variable, and a given descending reference signal is proportional to the error signal from the immediately higher level. The current reference must be proportional to the error at a higher-level controller. But where does the higher reference signal come from? This is a question about *why* any action is performed, or what is commonly called "motivation."

An enormous literature has implicated the BG in motivation (Ikemoto & Panksepp, 1999; Ikemoto, Yang, & Tan, 2015; Smith, Tindell, Aldridge, & Berridge, 2009). Damage to the limbic BG often results in abulia and apathy, which are distinct from motor symptoms like akinesia. A patient with akinesia may attempt to move but is unable to do so, but a patient with abulia has no desire to move and makes no attempt to initiate action (Berrios & Grli, 1995). On the other hand, increasing dopamine (DA) signaling often increases motivated behaviors, and many parts of the BG are also effective sites for self-stimulation. For example, animals will perform arbitrary actions in order to activate the mesolimbic DA pathway (Wise, 2004).

Such observations suggest that a major function of the BG is to convert some motivation signal to an appropriately scaled command for action, but it is unclear how this can be achieved (Schmidt et al., 2008). In this chapter, we shall review research on BG contributions to motivated behaviors and consider how the transition control model sheds light on the neural basis of motivation.

9.1 ASPECTS OF MOTIVATION

Motivation usually refers to four aspects of behavior.

1. *Activation*: General activation is related to arousal and manifests in behaviors like exploration and locomotion. It is usually measured in terms of overall activity level.
2. *Valence*: Motivated behavior is directed toward or away from specific locations and objects. Such attraction or repulsion is often explained using the concept of valence. Valence is distinct from activation as it attempts to account for the direction of behavior. Positive valence is associated with seeking or attraction; negative valence, with avoidance or rejection.
3. *Selection*: Since distinct motivational systems must share the same set of effectors, the organism is presented with the choice of which goal to pursue. The definition of the goal constitutes the "reason for action", the answer to the *why* question.
4. *Effort*: Given the desire for some goal, one might vary how much effort one exerts in pursuit of it. Often this would require some form of cost/benefit analysis and regulation of energy, expenditure.

Motivational deficits after damage to the BG typically involve one or more of the above aspects.

Symptoms include lack of spontaneous movement and speech, withdrawal, and loss of initiative, even though movements are relatively normal (Caplan et al., 1990; Mega & Cohenour, 1997). A striking set of symptoms, known as an autoactivation deficit, has been reported after damage to the associative cortico-BG network (Laplane & Dubois, 2001). Patients report an "empty mind," despite intact conscious awareness, and above all lack of any desire to move. For example, a patient stays in bed for half an hour with an unlit cigarette in his mouth and explains his behavior by saying, "I am waiting for a light." When explicitly instructed to move, however, such patients can still initiate action normally. Such symptoms are not usually observed after sensorimotor striatal lesions.

9.2 LIMBIC BG AND REWARD

Historically, the hypothalamus has been the structure most often associated with motivated behaviors. Many behaviors, such as attacking and copulation, can be elicited by stimulating the hypothalamus (Hess & Akert, 1955). Brain areas that are highly connected with the hypothalamus are collectively known as the limbic system (MacLean, 1952). The term limbic is often used to describe a wide variety of functions, especially emotion, motivation, and regulation of interoceptive states (Risold, Thompson, & Swanson, 1997; Swanson, 2000).

The limbic BG circuits send major projections to the hypothalamus (Groenewegen, Berendse, & Haber, 1993; Haber, Groenewegen, Grove, & Nauta, 1985; Maldonado-Irizarry, Swanson, & Kelley, 1995; Mogenson, Swanson, & Wu, 1983; Nauta, Smith, Faull, & Domesick, 1978). Limbic BG target the anterior and lateral hypothalamus as well as the anterior midbrain, which targets more posterior and medial hypothalamic regions. These areas have been implicated in the regulation of arousal and energy expenditure, digestion, and fight or flight behaviors (Hess, 1957). Attack, locomotion, and feeding behaviors can be elicited by ventral tegmental area (VTA) stimulation (Kim et al., 2015; Sheard & Flynn, 1967; Wang, Tan, Zhang, & Luo, 2013; Wyrwicka & Doty, 1966). Thus, the limbic BG can be viewed as a higher level for the regulation of motivated behaviors.

The part of the limbic BG that has attracted the most attention is the nucleus accumbens, which has long been implicated in locomotor and feeding behavior (Herrick, 1926). Mogenson and colleagues proposed the concept of the "limbic-motor interface" to describe accumbens function, emphasizing the transformation of incentive motivation into movement, as manifested in a behavior like a rat approaching a food reward (Mogenson, Jones, & Yim, 1980).

VTA DA neurons innervate the nucleus accumbens and other related regions, like the prefrontal cortex, via the mesolimbic pathway. To the lay public, this pathway is known as the brain's reward pathway. According to a popular view, DA represents reward value, whether the reward in question is food, sex, drugs of abuse, social interaction, or brain stimulation. This view is based on interpretations of early experimental results on DA. For example, injection of DA antagonists reduced lever pressing for food rewards—a finding that was initially interpreted as producing a state of anhedonia or lack of pleasure and that DA signals a hedonically positive reward (Wise, 1982; Wise, Spindler, DeWit, & Gerberg, 1978). But later work questioned this interpretation. DA neurons are not only activated by rewards or reward-predicting stimuli but also by salient stimuli regardless of valence, such as aversive stimuli like foot shocks, loud noise, and air puffs (Barter et al., 2015; Horvitz, 2000; Lammel et al., 2012; McCullough, Sokolowski, & Salamone, 1993). The idea that DA reflects positive valence is also contradicted by studies using facial expressions as measures of valence (Berridge & Robinson, 1998). Taste stimuli generate a stereotyped sequence of behavioral responses (Grill & Norgren, 1978a, b). Sweet tastes elicit "liking" expressions, which include rhythmic tongue protrusions and rhythmic lateral lip licking. In contrast, bitter solutions elicit gaping eyes, head shaking, and face washing. These responses, which can be elicited soon after birth, are observed across a number of species, including humans (Figure 9.1). Berridge and Robinson found that DA depletion did not disrupt oral movements associated with palatable food (Berridge & Robinson, 1998; Berridge, Venier, & Robinson, 1989). Rats still showed normal liking reactions to reward even in the absence of DA signaling. Moreover, DA-depleted rats could still discriminate between sucrose and quinine, showing that they could still exhibit normal reactions to stimuli typically associated with positive and negative valence (Berridge & Robinson, 1998; Berridge et al., 1989). On the other hand, dopamine transporter (DAT) knockdown mice, with enhanced DA signaling, did not show increased liking facial expressions in response to sucrose (Pecina, Cagniard, Berridge, Aldridge, & Zhuang, 2003).

To explain these findings, Berridge and Robinson draw a distinction between "liking" and "wanting." Wanting refers to the appetite and desire to eat, whereas liking refers to the palatability

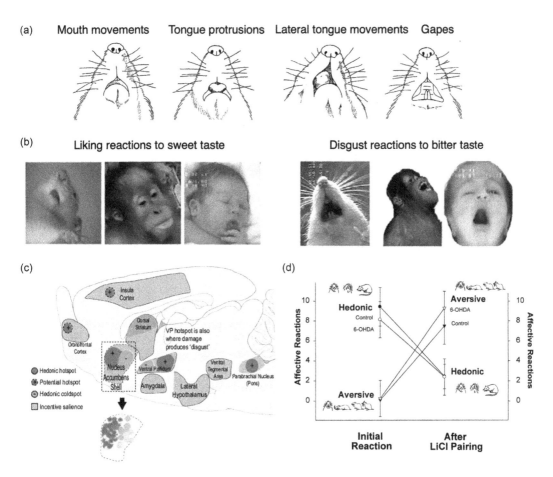

FIGURE 9.1 Affective reactions and their neural substrates. (a) Taste reactivity responses in rats. From Grill and Norgren (1978a) with permission. Hedonic reactions include tongue protrusions, lateral tongue movements, and paw licks. Aversive reactions include gapes, headshakes, and face/paw wipes. (b) Liking and disgust reactions are highly conserved. The same reactions are found in rodents and primates. (c) Illustration of the neural circuits implicated in affective valence. Hot spots (red) are sites where opioid stimulation enhances "liking" reactions elicited by sucrose. Cold spots (blue) are sites where the same opioid stimulation suppresses "liking" reactions to sucrose. (b and c) From Berridge and Kringelbach (2015) with permission. (d) Dopamine depletion with 6-hydroxydopamine (6-OHDA) did not influence affective reactions. Nor did it affect the change in affective reactions after induction of taste aversion (when food is paired with lithium chloride injection). From Berridge and Robinson (1998) with permission.

or pleasure associated with eating. They argue that DA signals "incentive salience," which is used to invigorate approach behavior to salient stimuli. As such, DA manipulations affect the wanting, rather than liking, component of motivated behaviors. According to them, since standard behavioral measures like lever pressing or reward port entry measure wanting rather than liking, manipulations of DA are expected to affect these measures. In contrast, measures like taste reactivity reflect the liking component, so they are not affected by DA manipulations.

In contrast to the lack of any effect of DA depletion on taste reactivity, lesions or pharmacological manipulations of areas in the limbic BG dramatically alter taste reactivity. For example, lesions of the caudal ventral pallidum (VP) result in the loss of "'liking'" reactions to sweet taste (Berridge & Kringelbach, 2015). Instead, the liking reactions are replaced by disgust reactions such as gapes and headshakes (Cromwell & Berridge, 1993; Ho & Berridge, 2014). Berridge and colleagues discovered a gradient of hedonic reactions by manipulating neural activity in the accumbens shell and VP, in regions they called "hedonic hotspots" (Castro & Berridge, 2014; Pecina, Smith, & Berridge, 2006). These areas appear to be critical for modulating unconditional hedonic responses to rewards (Figure 9.1). Although DA signaling is not required for generating liking responses or learning about the value of reward, some studies have implicated opioid signaling in these processes. In the accumbens shell, there are regions where different liking and disgust reactions may be elicited using opioid signaling. For example, mu opioid stimulation in specific shell regions can increase liking reactions to sucrose (Berridge & Kringelbach, 2015; Pecina et al., 2006).

9.3 VALENCE AND BIDIRECTIONAL CONTROL

While the distinction between wanting and liking is useful, it is important to note that they are normally correlated, even if they are mediated by distinct neural substrates. This distinction is similar to the traditional distinction between preparatory and consummatory behaviors. Preparatory behaviors include running to approach a goal, as in measures of wanting, whereas consummatory behaviors include eating, drinking, and copulating, as in measures of liking (Konorski, 1967). When a rat approaches a piece of food and consumes it, wanting precedes liking. Common experimental measures like lever pressing or locomotion measure preparatory behaviors, whereas taste reactivity tests measure orofacial components of consummatory behaviors.

In both preparatory and consummatory behaviors, there is bidirectional control, in which behavior can increase or decrease the value of a specific input variable. They differ mainly in the perceptual input variable being controlled. For example, in approaching a reward, visual and perhaps olfactory inputs are controlled. When consuming the same reward, taste and tactile inputs are more important. In a normal sequence of motivated behavior, there is a particular order in which different behavioral elements are expressed.

What is commonly called valence reflects the operation of opponent systems for bidirectional control. There are antagonistic neural systems for opponent output functions at every level, manifested in both preparatory and consummatory behaviors (Konorski, 1967). If the reference signal is set to a high level and the perception is too low, then we act to acquire the perception. If the reference level is low, then we act to reduce that same perception. For example, we may ingest food or spit it out. The sign of the error signal determines the valence, and different sets of effectors are activated depending on the sign of the error. For ingestive behaviors, the relevant perceptual variables reflect more or less of the food, and the outputs are generated by error signals reflecting "not enough" or "too much." "Not enough" error signals might generate outputs such as tongue protrusion to acquire flavor, and "too much" error signals can produce disgust reactions like spitting it out or face washing. For "preparatory" behaviors like approach and retreat, the errors reflect perceptual variables such as "farther" or "closer" in relation to the food. It is not the absolute value of the input variable (i.e. distance) that determines behavioral output, but the difference between this input and the reference state. The sign of this error signal determines whether to approach or retreat.

Valence or value is not inherent in the reward itself but is determined by the magnitude and sign of the error signal in specific control systems (Figure 9.2). Only by comparison with the reference can something be too much or not enough. Consequently, the same food can be rewarding when the animal is hungry but aversive when it is sated. When deprived, the reference signal for food reward rate control increases, and when compared with the low food, input generates an "not enough" error signal. This error signal is converted to descending reference signals that ultimately activate the appetitive system and suppress the disgust system. For example, very salty food normally elicits aversive reactions, but after sodium depletion, the same food can elicit ingestive behaviors (Cabanac, 1971; Cabanac & Lafrance, 1990).

Motivation

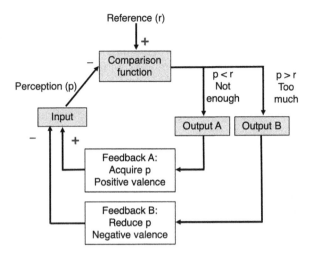

FIGURE 9.2 Bidirectional control and valence—Errors can be positive or negative, but neural signals can only be positive. For bidirectional control, it is necessary to have a pair of antagonistic systems, both using positive neural signals but commanding antagonistic sets of effectors. Thus, "not enough" errors generate seeking and approach behaviors, while "too much" errors generate avoidance and disgust behaviors. This type of organization is found in many controllers involved in both preparatory and consummatory behaviors.

9.4 DISTINCT ACCUMBENS OUTPUTS REGULATE REPARATORY AND CONSUMMATORY BEHAVIORS

Available evidence suggests a key functional dissociation between the accumbens core and shell: the core is critical for preparatory behaviors and the shell for consummatory behaviors. They belong to largely distinct cortico-BG circuits with distinct patterns of connectivity. As already mentioned, the shell and downstream VP contain hedonic hotspots and cold spots that are critical for opponent consummatory behaviors as measured by taste reactivity, with distinct areas mediating aversive and appetitive behaviors such as tongue protrusion and gaping (Berridge & Kringelbach, 2015; Pecina & Berridge, 2000). DA signaling in the accumbens shell has also been implicated in conditioned taste aversion learning (Fenu, Bassareo, & Di Chiara, 2001; Reynolds & Berridge, 2002). On the other hand, the core appears to be involved in regulation of approach and avoidance behavior. Inactivation of the caudal core in rats can promote avoidance behavior (Hamel, Thangarasa, Samadi, & Ito, 2017). Anterior cingulate cortical projections to the accumbens core are necessary for learned approach behavior (Parkinson, Willoughby, Robbins, & Everitt, 2000). Local infusions of a α-amino-3-hydroxy-5-methyl-4-isoxazolepropionic acid (AMPA)/kainate glutamate receptor antagonist disrupted approach behavior toward reward-associated but did not affect the acquisition. Infusions of the N-methyl-D-aspartate receptor antagonist AP-5 impaired only the acquisition, but not performance. In addition, DA antagonism decreased responses to the conditioned stimulus (Di Ciano, Cardinal, Cowell, Little, & Everitt, 2001).

Together, these results show that both core and shell regions are involved in bidirectional control of specific transition variables. The accumbens core and its corticostriatal inputs from the medial prefrontal cortex appear to be critical for locomotor approach and retreat, where the controlled variable might be distance to target or rate of progression. On the other hand, the nucleus accumbens shell and associated circuits are critical for the regulation of ingestive behaviors, such as licking and chewing, that are generated to control transitions in taste and other related sensory inputs (Maldonado-Irizarry et al., 1995; Prinssen, Balestra, Bemelmans, & Cools, 1994; Yin, 2016).

Outputs from limbic BG circuits can reach specific pattern generators in consummatory behavior, like chewing, sniffing, and licking, as well as in preparatory behaviors like locomotion, approach, and avoidance (Ikemoto & Panksepp, 1999; Mogenson et al., 1980). These are used to control input variables related to chemical senses like taste and smell. The olfactory tubercle, also located in the ventral striatum, seems to be a part of a cortico-BG circuit that controls odor transitions by coordinating approach, sniffing, and other related behaviors. This can be distinguished from the nearby ventrolateral striatum, which generates command signals that regulate inputs from the orofacial region. The limbic output flow could also influence additional controlled variables, as yet undefined, by commanding a variety of effectors, including skeletomotor, autonomic, and neuroendocrine.

9.5 REINFORCEMENT

Another influential concept often used to explain motivation is reinforcement. Many argue that DA serves as a reinforcement signal, which may be hedonically neutral or even associated with negative valence. The concept of reinforcement has traditionally played two roles: one in performance and the other in learning. In Chapter 11, we shall examine the role of reinforcement in learning, but for now we will focus on its role in performance.

Reinforcement is traditionally defined as an event that, when following some behavior, increases the frequency of that behavior in the future (Olds, 1977; Thorndike, 1911). In positive reinforcement, a rat may press a lever to obtain food, whereas in negative reinforcement, it may press a lever to avoid a painful shock. In both cases, the behavior of lever pressing is increased. The reinforcing properties of food, drugs, and other rewards refer to their ability to cause a repetition of some behavior, independent of subjective pleasure. In this sense, reinforcement does not need to be hedonically positive, as is implied by the term "reward."

The circularity of the conventional definition of reinforcement has long been noted. It is apparent if rephrased thus: a behavior is repeated because it is followed by something that causes the repetition of behavior. To avoid circularity, Hull attempted to define reinforcement more narrowly as drive reduction. Drives are uncomfortable internal stimulations that can be reduced through behavior. For example, food deprivation is supposed to increase the hunger drive, and the deprived animal is more likely to press a lever to reduce this drive (Hull, 1943). This formulation appears to be less circular. If drive reduction fails to reinforce a behavior, then the hypothesis is falsified. In drive reduction theory, however, there is no explanation of how drives arise or when they are reduced. A drive is treated just like any external stimulus in a linear causation system. The implicit assumption is that behavior simply maximizes reinforcement. A drive like hunger really refers to some error signal in a control system. To explain how it works, one must define the controlled variable and the reference state. The phenomenon that Hull was attempting to explain is error reduction in control systems for essential variables (Cannon, 1932). These variables are usually under homeostatic control; that is to say, they have tonic and stable reference signals, and significant deviations from such setpoints threaten survival. But the Hullian concept of reinforcement does not take into account negative feedback. It is an open loop concept, as reinforcement cannot be too much or not enough. There is no reference value by which the current value can be compared. This assumption is also found in modern RL, in which the concept of the reward function plays a similar role as reinforcement (Sutton & Barto, 2018). But just like Hullian reinforcement theory, RL also lacks an adequate definition of reward (Keramati & Gutkin, 2014).

9.6 SELF-STIMULATION

The basic phenomenon that the concept of reinforcement attempts to explain is the repetition of behavior given some desirable consequence. DA is often thought to be a reinforcement signal because DA pathways are very effective sites for self-stimulation. Olds and Milner first discovered that rats could learn to press a lever to deliver electrical stimulation through an implanted electrode (Olds & Milner, 1954). This phenomenon is now known as intracranial self-stimulation. In extreme cases, rats would press the lever at high rates until exhaustion, even at the expense of starving themselves (Olds, 1977). Self-stimulation results suggest that, whenever a large DA signal follows a behavior, the probability of that behavior will increase in the future.

While many brain sites were found to be effective sites for self-stimulation, activation of the mesolimbic pathway probably produces the most robust effects, resulting in high rates of lever pressing. DA depletion in the ventral striatum or VTA reduces self-stimulation (Fibiger, LePiane, Jakubovic, & Phillips, 1987). Animals will also self-administer drugs that increase DA signaling, such as DA agonists, cocaine, or amphetamine (Roberts & Koob, 1982). Action-contingent delivery of DA agonists or DA uptake inhibitors directly into the nucleus accumbens also serves as effective reinforcement (Carlezon, Devine, & Wise, 1995; Ikemoto, Glazier, Murphy, & McBride, 1997).

According to an early interpretation, self-stimulation produces an artificial reward (Fibiger, 1978; Wise, 1978). Although there are sites where pleasurable sensations can be generated, in most cases, including stimulation of DA pathways, this does not appear to be the case (Berridge & Kringelbach, 2015; Olds, 1977). Unlike activation of other regions, such as the lateral septum (Heath, 1963), activation of the DA pathway does not produce any obvious subjective pleasure. For example, animals do not prolong the stimulation when given the chance to do so (Olds, 1977; Rossi, Sukharnikova, Hayrapetyan, Yang, & Yin, 2013). Consequently, self-stimulation is often interpreted as the artificial activation of a neural reinforcement signal. Thus, independent of any sensation of pleasure, food may trigger DA release, which increases food-seeking behavior. Artificial reinforcers like drugs are assumed to have a similar effect to the extent that they elicit DA release.

Self-stimulation of DA pathways and their targets can be explained by repetition control. Artificially injecting a signal into the reference pathway for repetition control would generate repetition, and a larger signal would produce a larger error signal and more repetitions. Whether the stimulation activates the reference signal or the error signal, the result would be repetition of the last action.

As described in previous chapters, the BG output nuclei, such as VTA and substantia nigra pars reticulata (SNr) GABA neurons, send axon collaterals to DA neurons, which in turn project to the corticostriatal circuit where the action command originates. Such projections could reactivate the action command so that the reference signal is replayed via a reverberating cortico-BG loop. Since DA only changes the gain, by itself, it may not be sufficient to generate actions. Some concurrent excitatory input is needed, and DA can amplify this descending command to generate transition reference signals. In self-stimulation, the artificially elicited DA signal can promote repetition by acting on D1 receptors. What allows successful repetition of the action is the organization of the entire action hierarchy; what is repeated is a higher-level transition reference.

According to the present perspective, the so-called reinforcement effect of DA is due to repetition control. By increasing the gain of repetition control, increased DA can promote the "replay" of a reference signal. Its immediate effect is to adjust performance online, which should be distinguished from learning. Exactly what is repeated during DA self stimulation depends on the transition variable that is being controlled by the circuit being stimulated. It could be a simple transition or a short behavioral sequence. According to the present model, DA is not a teaching signal, though it could have an indirect role in learning. We shall return to this topic in later chapters, when we consider the concept of reward prediction error.

The adaptive gain hypothesis explains a few puzzling observations. First, stimulation frequency can determine the rate, or latency of future action (Ikemoto & Panksepp, 1999). This is to be expected if the DA signal contributes to the repetition and persistence of the ongoing action command. Increasing DA stimulation produces the opposite effect as that produced by a larger natural reward, which can produce a longer "post-reinforcement pause." Moreover, compared to natural rewards, mesolimbic self-stimulation is more rapidly extinguished. When the stimulation is no longer delivered upon lever pressing, animals quickly stop pressing. By contrast, normal extinction of instrumental behavior trained with food rewards is much slower, especially when partial reinforcement schedules are used. During extinction, behavior is at first characterized by frustration and higher than usual rate of lever pressing. Following a long delay, there can also be spontaneous recovery, a return to a high level of pressing even in the absence of retraining (Amsel, 1962). On the other hand, extinction of DA pathway self-stimulation behavior does not seem to involve active suppression of learned action as in normal extinction (Bouton, 2004). Rather, it appears to be immediate because the repetition reference is no longer boosted by the artificially injected DA signal. The role of DA in learning and plasticity will be considered in Chapters 12 and 13.

9.7 EFFORT EXERTION

In earlier chapters, we have seen that DA depletion in the dorsal striatum can reduce gain in simple transition (velocity) as well as higher-order transitions (repetition and sequence), resulting in reduced peak velocity as well as reduced tempo. This suggests that the nigrostriatal DA pathway mainly adjusts the adaptive gain of velocity and repetition control. Whereas the nigrostriatal pathway mainly targets the associative and sensorimotor BG networks, the mesolimbic DA pathway mainly targets the limbic BG. If mesolimbic DA also acts as adaptive gain, then what are the relevant controlled variables in the limbic BG circuits? Here we shall focus on the control of reward rate in operant (instrumental) conditioning. This does not mean that the limbic BG are only concerned with food rewards. Rather, we shall focus on reward rate control because it is better understood than the control of other variables.

In operant conditioning, the controlled variable is reward rate, the number of rewards in some time window, and the rate of pressing is the means by which reward rate can be controlled. Reward rate reference is assumed to be proportional to the error in energy homeostasis. Animals are first food-deprived, so that the homeostatic error (experienced as hunger) is reflected in a high reward rate reference. The actual reward can be monitored using specific perceptual inputs, such as those associated with consumption. Such feedback is independent of delayed feedback from the digestive system. The latter also plays a key role in energy homeostasis, but feedback from consummatory behaviors is sufficient to reduce the reward rate error, which temporarily stops the food-seeking behavior.

The monitored reward rate is compared to the reward rate reference to generate a reward rate error. For example, a hungry rat entering an operant chamber starts with a high reward rate reference (and error), which requests a large amount of food within some time window. The reward rate error simply reflects desire or wanting. It is also similar to the concept of marginal utility in economics, though it does not assume that animals maximize rewards (Alchian & Allen, 1977). With a high reward rate error, the animal will attempt to obtain and consume rewards as quickly as possible.

In standard operant conditioning, the feedback function can be defined as the rate of reward as a function of the rate of lever pressing (Baum, 1973). It can be manipulated by changing the ratio on a fixed ratio (FR) schedule. The ratio is the number of presses required to earn a piece of food. FR schedules are convenient for analysis because they have a linear feedback function. To obtain reward, the rat must convert the reward rate error into a reference signal for action repetition, i.e. a bout of lever pressing proportional to the ratio requirement: press once on FR1 or ten times on FR10. The conversion is determined by the gain in the output function of reward rate controller. The output of this reward rate controller can adjust the reference signal of the repetition controller to dictate press run length and tempo.

What is commonly called effort can be defined as the gain of reward rate control—the amplification of the reward rate error when generating the press rate reference. Such output gain or error sensitivity should not be equated with the behavioral output itself. It refers to how much the rat will press for a specific unit of desired reward. Effort cannot be measured by the rate of lever pressing but by the rate of lever pressing divided by the reward rate error. A hungry rat is more willing to exert effort to defend a given rate of reward input (Ettinger & Staddon, 1983; Staddon, 1983).

When reward rate is controlled, increasing the ratio requirement presents a disturbance that must be countered by increasing the rate of pressing. To achieve successful control, the behavior must mirror the effort requirement. Otherwise, there would have been a reduction in reward rate. This requires a high gain in translating desire for food into a longer bout of lever pressing. If the desired reward rate is 5 pellets per minute, it is approximated by the actual reward rate earned when the ratio requirement is very low. For example, at FR1 the rat presses 5 times per minute, earning 5 pellets, so we infer that the rat wants roughly 5 pellets per minute on average. If the ratio requirement increased from FR1 to FR5, the same behavioral output of 5 presses per minute would have resulted in a drastic reduction in reward rate, from 5 pellets to 1 pellet. A reward rate controller with a high gain is able to generate the desired rate of 5 pellets per minute perfectly by increasing the rate of lever pressing to 25 presses per minute. Indeed, this pattern is observed, though the quality of control declines as the ratio requirement increases (i.e. as the schedule gets leaner). There is obviously a physical limit to how quickly the animal can press, but aside from this limit, the cost of pressing, in terms of energy consumption and other variables like fatigue, also becomes significant at higher ratios. To model behavior at high ratios, one would need to perform a cost–benefit analysis, essentially subtracting the cost from the reward rate. It requires a description of conflicting controllers, for example those that minimize fatigue or energy expenditure.

So long as the action of lever pressing is effective, the animal can attempt to control the reward rate, but there is no reason to waste effort in an environment where the action has little effect on reward. This is the case when the schedule of reinforcement is too lean and the cost of pressing becomes too high. At very high ratios, the animal will simply give up. The reference reward rate signal is still sent, and a large error signal is still experienced since there is no reward input to reduce it, but the animal stops pressing as the gain of the reward controller is either lower or the reward rate error must recruit alternative means of reducing the error (Figure 9.3).

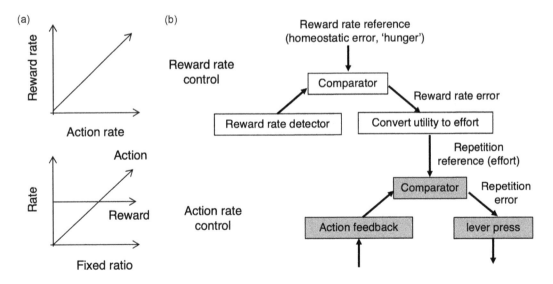

FIGURE 9.3 Effort and the control hierarchy. (a) Top, reward rate is proportional to press rate on a fixed ratio schedule. On this schedule, the ratio is the number of presses required to earn one reward. Bottom, illustration of a perfect controller of reward rate. As the ratio requirement increases, the desired reward rate is maintained (indicated by the flat line), while the lever press rate increases, assuming that the ratio is not too high and the cost is negligible. (b) The reward rate controller is higher on the motivational hierarchy, and recruits an action rate controller by altering the repetition reference rate.

9.7.1 Limbic BG and Effort Regulation

Previous work has implicated the limbic cortico-BG network and mesolimbic pathway in effort regulation. For example, lesions of the medial prefrontal cortex (including the anterior cingulate cortex), which projects to the ventromedial striatum, reduced how much effort rats were willing to exert for rewards (Walton, Bannerman, Alterescu, & Rushworth, 2003; Walton, Bannerman, & Rushworth, 2002). Rats were trained to climb a barrier to obtain a large reward in one arm of a T-maze or to obtain a small reward in the other arm, which has no barrier. Normally, most rats preferred the large reward arm with the barrier, but after excitotoxic lesions to the medial prefrontal cortex, the rats showed a nearly exclusive preference for the low reward arm with no barrier. Human neuroimaging studies have also implicated the anterior cingulate cortex in task difficulty or mental effort (Botvinick, Cohen, & Carter, 2004; Naccache et al., 2005).

In rats, the activation of A2A receptors in the nucleus accumbens core reduces effort exertion as measured by operant lever pressing on a FR task. A2A activation can increase indirect pathway activity (Mingote et al., 2008). Adenosine signaling can thus regulate effort by modulating the activity of the ventral striatopallidal pathway. These observations suggest that the activation of the indirect pathway can reduce effort exertion on a FR task (Mingote et al., 2008; Salamone, Correa, Farrar, Nunes, & Pardo, 2009).

In a human neuroimaging study, the subjects were trained to press a button in response to cues in order to earn money (Kohli et al., 2018). In the easy condition, there was plenty of time to press and a high reward probability. In the hard condition, the subject was given very little time to press; it was necessary to press quickly to obtain a reward and the reward probability was also far lower. Ventral striatum activity was much greater in the hard condition. This finding supports the idea that ventral striatal activity reflects effort exertion rather than expected value or reward probability.

9.7.2 DA Contribution to Effort

Salamone and colleagues have shown that accumbens DA contributes to effort regulation (Salamone et al., 2009). Accumbens DA depletion dramatically reduced lever pressing on FR schedules with high ratio requirements but had a smaller impact on performance when the ratio was low. Importantly, it did not reduce consumption of food rewards (Aberman & Salamone, 1999; John D Salamone, Correa, Mingote, & Weber, 2005). When allowed to choose between regular chow on the floor or a more palatable food by pressing on a lever, normal rats pressed the lever to earn palatable foods and neglected less desirable home chow, but rats with accumbens DA depletion rarely pressed the lever, though they still ate the home chow (Sokolowski & Salamone, 1998). This finding suggests that DA depletion reduces the gain of reward rate control: without affecting the desire for food, it drastically reduces the willingness to exert effort to obtain it. Only when the effort requirement is low can DA-depleted animals achieve adequate control.

In contrast, increasing mesolimbic DA signaling can increase effort exertion. Hyperdopaminergic mice (DAT knockdown) show higher rates of lever pressing compared to controls (Yin, Zhuang, & Balleine, 2006). When given a choice, hyperdopaminergic mice pressed more on high-cost levers compared to wild-type mice (Beeler, Daw, Frazier, & Zhuang, 2010). This is what is expected with a higher gain in reward rate control. In general, higher DA promotes energy expenditure and exploration, while lower DA favors energy conservation and exploitation (Salamone, Wisniecki, Carlson, & Correa, 2001).

A recent study in humans also supported the role of DA in "cognitive" effort (Westbrook et al., 2020). Participants had to decide whether to engage in demanding cognitive tasks. Higher DA synthesis capacity in the caudate nucleus was associated with greater willingness to exert cognitive effort, as measured by willingness to perform a difficult version of the N-BACK working memory task. In addition, in people with low DA synthesis capacity, methylphenidate (which is expected to increase DA signaling) increased motivation to work.

Activation of the mesolimbic pathway can increase impulsivity and delay discounting. The main effect of increasing reward rate gain is to increase the rate at which the reward error is corrected. Increasing gain or increasing reward rate error is thus expected to increase preference for immediate gratification (faster error reduction), which is associated with delay discounting. Indeed, hungry animals show delay discounting and risk aversion (preference for certain rather than probabilistic reward), whereas sated animals do not (Leblond, Fan, Brynildsen, & Yin, 2011; Mona Leblond, Sukharnikova, Yu, Rossi, & Yin, 2014; Li et al., 2021).

9.7.3 Conflating Reward Rate and Performance

A common misunderstanding is to assume that reward rate is the independent variable and behavior is the dependent variable. In operant behavior, reward rate is not an independent variable; it depends on the rate of pressing. As the animal actively controls the reward rate specified by the internal reference, to model operant behavior, two parallel and simultaneous functions are needed: the effect of food on behavior and the effect of behavior on food.

Previous attempts to explain the contribution of DA to motivation neglect the feedback function (Niv, Daw, Joel, & Dayan, 2007). For example, according to the response rigor model, tonic DA represents opportunity cost, which is defined as the amount of reward that would have been collected had there been action. It is assumed that a higher reward rate "invigorates" performance, as it causes a higher rate of response, and that the animal must "choose actions and latencies that maximize the accumulated rewards." This assumption is contradicted by a number of well-known findings. As already discussed, leaner schedules, in which the amount of reward for a given amount of effort is smaller, result in a higher rate of pressing. When the cost of action is negligible, animals will control a constant overall reward rate while varying the rate of pressing according to the ratio requirements (Yin, 2013). A higher reward rate does not generate a higher action rate. Increasing the rate of reward per action actually reduces the rate of action produced.

Reward rate is often conflated with performance. Performance vigor is often used to infer a higher reward rate or higher reward value (Hamid et al., 2016). In the current model, performance is proportional to reward rate error, which requires an internal comparison process and an amplification factor (gain) that converts error into a reference for a lower-level action repetition controller. Without the hierarchical organization relating reward rate and action rate, it is not possible to explain the observations on different schedules of reinforcement and manipulations of feedback functions.

The hypothesis advanced here is that the cortico-BG network associated with the nucleus accumbens core and its DA innervation from the mesolimbic pathway can implement reward rate control. This transition control system compares reference inputs from homeostatic controllers (e.g. hunger) and generates outputs that determine effort exertion in preparatory behaviors. In the case of operant conditioning, reward rate error is converted to a reference for action repetition, so that the desire for one more reward becomes a command for a specific run of lever pressing. The same analysis can be applied to avoidance behavior, as long as the controlled variable is defined precisely (Powers, 1971). Regardless of hedonic valence, mesolimbic DA is needed to convert the higher error into more effort.

9.8 DA AND FORCE GENERATION

Recent results provide further support for the role of DA in effort generation (Figure 9.4). Hughes et al. examined the relationship between VTA DA neurons and force in mice (Hughes et al., 2020). The mice were prevented from moving by head fixation but can still exert force on the sensors placed on the head fixation setup. Mice were trained on a fixed-time task, in which a sucrose reward was delivered every 10 seconds. After training, the mice showed anticipatory behavior, as detected by force generated in the forward direction, which would have resulted in approach behavior if the mice were free to move. They found a striking relationship between the firing rate of VTA DA neurons and the force generated. In particular, different populations of DA neurons were found with

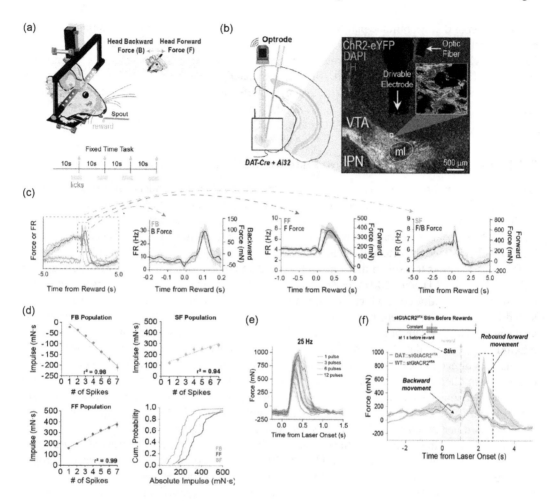

FIGURE 9.4 Dopamine (DA) and force exertion. (a) Illustration of a fixed-time reward task. (b) Placement of the optrode in the ventral tegmental area (VTA) and expression of ChR2 in DA neurons (tyrosine hydroxylase-positive, TH+). (c) Three populations of DA neurons were found that represent force in different directions. FB, fast backward. FF, fast forward. SF, slow forward. (d) Summary of the relationship between impulse and number of spikes in different populations of DA neurons. (e) Optogenetic stimulation of DA neurons with ChR2 increased forward force generation. (f) Optogenetic inhibition with stGtACR2 produced backward force generation.

different direction preferences: some neurons increased firing during forward force exertion, while others increased firing during backward force exertion. The bust firing of DA neurons strongly predicted the production of force (Figure 9.4). The amount of force generated over time (impulse) is determined quantitively by the number of spikes in the DA burst.

Optogenetic excitation generates force in the forward direction, and optogenetic inhibition produces backward force. VTA DA neurons can regulate the magnitude, direction, and duration of force used to move toward or away from any motivationally relevant stimuli. Every pulse of light exciting VTA DA neurons produced a proportional increase in impulse.

Impulse, or a change in momentum, is necessary for the initiation of any movement. According to the impulse-momentum theorem:

$$J = \Delta p \tag{9.1}$$

where J is impulse and p is momentum. Since momentum is defined as:

$$p = mv \tag{9.2}$$

where m is mass and v is velocity, the relationship between impulse and velocity is:

$$J = m\Delta v \quad (9.3)$$

Assuming a constant mass, equation (9.3) shows that more impulse generates more velocity. This explains why DA activity could be correlated with velocity in freely moving animals (Barter et al., 2015). Selective stimulation of SNc DA neurons can directly initiate movements, and selective inhibition of SNc DA neurons can suppress movement initiation (da Silva, Tecuapetla, Paixão, & Costa, 2018).

Earlier, we discussed the example of operant conditioning on an FR schedule as an example of effort exertion for the sake of convenience. On a different behavioral task, effort can be manifested differently depending on how the effective behavior or operant is defined, whether it is the number of presses, speed of locomotion, or force exerted. On the fixed-time task, even though reward delivery is not contingent upon force exertion, the generation of approach behavior (and force) is necessary for collecting the reward. This observation suggests that the reward rate error can be converted into any action, depending on prior experience.

One prediction of the adaptive gain model is that DA activity could be especially high when animals are prevented from reaching their goals. Without sufficient negative feedback, the error accumulates, generating persistent references (e.g. for repetition of lever pressing), and further increasing adaptive gain.

9.9 A LABILE MOTIVATIONAL HIERARCHY

By definition, the highest level of the control hierarchy is one whose reference signal is not altered by any other level. It could be intrinsic homeostatic control systems, in which essential variables are genetically programmed to ensure survival. We can have control over the lower reference settings like those for joint angles or body configurations, but we cannot arbitrarily choose to be hungry or thirsty, cold or hot, as such states reflect errors in controllers for essential variables, which cannot be altered voluntarily and arbitrarily. As Schopenhauer puts it: "A man can do what he wills but he cannot will what he wills" (Schopenhauer, 1978).

The relationship between the higher level and lower level in the motivational hierarchy can be described as "in order to," as in "he runs to the fountain in order to drink water." What precedes "in order to" is the means; and what follows it, the end. However, the means–end relationship is different from the hierarchical relationship that characterizes the action hierarchy. In the action hierarchy, the lower and higher levels are often innately coupled, and the variables they control are also often related measures of the same physical variables. Thus, muscle length control is necessary for body configuration control, which is necessary for movement velocity control. To change a joint angle, it is necessary to change muscle tension. On the other hand, in the motivational hierarchy, the relationship between the variables is more arbitrary and labile. The coupling between means and ends is not fixed.

Although homeostatic controllers have innately coupled output functions, which include autonomic and neuroendocrine outputs, they cannot command arbitrary actions. The genome may include a program for temperature regulation that is already effective at birth, using outputs like blood vessel dilation and constriction, sweating, panting, or shivering. Yet it does not contain any program for behaviors like putting on a sweater or starting a fire. Likewise, the taste reactivity measures reviewed earlier reflect innately coupled outputs, but behaviors like tongue protrusion and gaping are not sufficient to control food rewards in any natural environment.

In the motivational hierarchy, the level where the error is first initiated is the lead level. Error can be generated either by a learned incentive (e.g. sight of chocolate) or a homeostatic error (hunger). The lead controller can command different actions that could reduce errors. In the natural environment, what is desired is seldom readily available. Through learning, it is possible to establish a labile motivational hierarchy that coordinates means and ends in a flexible manner.

The limbic cortico-BG networks generate reward errors, similar to desires in ordinary language. These errors can directly command a number of controllers using innately specified projections to the hypothalamus, mesencephalic locomotor region, periaqueductal gray, and orofacial pattern generators in the brainstem. They can generate specific behavioral outputs, like approach behavior or consummatory behaviors. However, often these behaviors may not be sufficient to satisfy the desires, and sometimes they may even be inappropriate. Instead, one must learn to perform new actions to reduce the higher-level error. This type of learning modifies the motivational hierarchy so that reward rate errors can recruit specific transition controllers for the production of specific actions. As the higher level can alter the reference signal for the lower level, it can also suppress the innately coupled consummatory system. For example, salivation in anticipation of food reward can be suppressed during the performance of learned instrumental actions toward food (Ellison & Konorski, 1964). We will return to the question of instrumental learning in later chapters.

9.10 PARALLEL BG NETWORKS AND MOTIVATED BEHAVIORS

As already mentioned, the parallel and partially overlapping BG circuit allows coupling between different transition variables. They can thus implement the motivational hierarchy, in which one transition controller (e.g. reward rate) alters the reference for another (e.g. action rate). This is the basic mechanism underlying operant conditioning or instrumental learning. Figure 9.5 illustrates possible interactions between different cortico-BG networks.

One possibility is that the anatomical connections between different cortico-BG networks, such as the spiraling striato-nigro-striatal projections, allow unidirectional propagation of signals from limbic to sensorimotor networks (Haber, 2003; Joel & Weiner, 2000). Striatal neurons may activate their dopaminergic afferents by disinhibiting the neighboring loop. The adaptive gain mechanism can be used not only to sustain a particular action via reverberation or reselection but also to potentiate and recruit a related transition controller.

Suppose a hungry animal desires a specific food. The representation of the desired food can recruit the specific action representation via corticostriatal projections. The error will begin to accumulate at a higher level, in the limbic BG. In inexperienced animals, this could be expressed

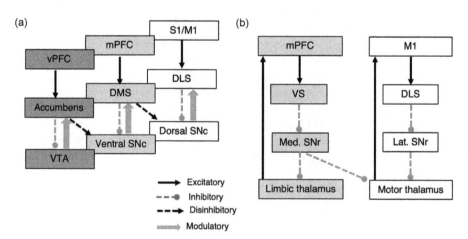

FIGURE 9.5 Interaction between basal ganglia networks. (a) Spiraling projections that allow feedforward interactions from the limbic network to the associative network and finally to the sensorimotor network. A given striatal area inhibits a region in the midbrain and also projects to nonreciprocal cortico-thalamic pathways. (b) Unidirectional connection between limbic and sensorimotor networks. Using disinhibition, the limbic striatum can ultimately increase activity in the motor cortex. Based on Aoki et al. (2019), with permission. DMS, dorsomedial striatum; DLS, dorsolateral striatum; PFC, prefrontal cortex, SNc, substantia nigra pars compacta, VS, ventral striatum; VTA, ventral tegmental area.

in innately coupled outputs such as the production of approach or consummatory behaviors. But following instrumental learning, the same error signal can influence the reference input of other transition control systems, for example those for repetition of specific actions. As adaptive gain, DA signaling could affect the next available controller lower on the motivational hierarchy, possibly using the spiraling striato-nigro-striatal projections (Haber, Fudge, & McFarland, 2000). This account predicts a wave of DA activity from ventromedial limbic regions to dorsolateral sensorimotor regions in the striatum during the performance of instrumental actions. Recent work has found DA waves in a medial to lateral direction at the time of reward-guided actions (Hamid, Frank, & Moore, 2021).

The spiraling striatonigral projections are not the only possible mechanism. The limbic network can exert a unidirectional influence on the sensorimotor network via nigrothalamic projections (Aoki et al., 2019). Activation of direct pathway spiny projection neurons (dSPNs) in the limbic striatum can ultimately influence the medial prefrontal cortex, whereas activation of dSPNs in the sensorimotor striatum can only change motor cortical activity, as predicted by the classic loop model. Moreover, limbic striatal projections also reach SNr neurons that target the motor thalamus, allowing the limbic BG output to influence the sensorimotor loop. In contrast, the sensorimotor striatal projections do not affect the limbic thalamus or prefrontal cortex.

A labile motivational hierarchy can be implemented by the parallel cortico-BG networks controlling different transition variables. Each network could have multiple functional divisions, but at present such classification is not yet possible, so we can only provide a crude sketch of the neural implementation of the motivational hierarchy.

Through learning, the limbic networks can recruit the associative and sensorimotor networks to generate appropriate actions. The associative network is linked to exteroceptive perception, including the perception of space, related to physical changes that are farther away from the animal, whole-field motion, and complex configurations of objects. It orients the organism toward any exteroceptive perceptual input it desires and to pursue those inputs. The sensorimotor network is mainly concerned with controlling proprioceptive transitions needed to perform specific action sequences. In most motivated behaviors, all three of these networks are needed, but their activation may be coordinated dynamically.

An experimental observation that sheds light on the motivational hierarchy is Pavlovian-instrumental transfer (PIT). In PIT, animals receive separate Pavlovian and instrumental training phases, in which they learn independently to associate a cue (CS) with food (unconditioned stimulus, US) and to press a lever for the same food. Then, on probe trials, the cue is presented with the lever available. During the presentation of the CS, which is usually a long and sustained stimulus lasting a few minutes, the rate of lever pressing is significantly increased. This is known as the transfer effect. A CS that predicts food can increase instrumental actions for the same food, even though the CS is never directly paired with instrumental training (Trapold & Overmier, 1972). Two forms of PIT have been identified. In general transfer, any reward-related stimulus could produce an arousing effect and potentiate instrumental performance. In outcome-specific transfer, the anticipation of a specific reward selectively enhances performance of the action associated with that reward (Balleine & O'Doherty, 2010; Corbit, Janak, & Balleine, 2007). According to the present model, the cue can activate the reward rate controller, acting as a proxy for homeostatic error and increasing the reward rate reference signal transiently. Moreover, PIT can also increase the gain of the reward rate controller. The latter possibility would predict more prominent transfer effects when the effort requirement is low but reduced transfer when the effort requirement is high. These predictions remain to be tested.

Lesion studies have shown that the nucleus accumbens is critical for PIT, even though it is not essential for instrumental learning (Corbit, Muir, & Balleine, 2001). In particular, the accumbens shell is necessary for the outcome-specific form of PIT but not for general PIT. In contrast, lesions of the accumbens core reduce general PIT but spare outcome-specific PIT. Lesions of the striatum-like central nucleus of the amygdala abolish general transfer but spare outcome-specific transfer. In contrast, lesions of the basolateral amygdala, a cortex-like structure that strongly projects to the nucleus accumbens, impair outcome-specific transfer but spare general transfer (Corbit & Balleine, 2005). Thus, distinct neural circuits mediate the action selection and general activating influences of predictive stimuli on instrumental performance. The basolateral amygdala and nucleus accumbens shell are involved in outcome-specific representations that recruit a specific action module associated with that outcome (Balleine & Dickinson, 1998; Johnson, Gallagher, & Holland, 2009). The central amygdala and the nucleus accumbens core appear to be more critical for general arousal effects of any reward predictor. Inactivation of the VTA reduced both outcome-specific and general PIT (Corbit et al., 2007). These results suggest that the cortico-acumbens core circuit and mesolimbic DA projections from the VTA are not only critical for translating error in reward rate control into instrumental lever pressing but also responsible for the PIT effect.

9.11 SUMMARY

The limbic BG network is implicated in motivated behavior, and damage to this network produces symptoms like abulia. There are distinct cortico-BG circuits implicated in consummatory and preparatory components of motivated behavior. For example, the nucleus accumbens shell appears to be a key node in a control system for gustatory or consummatory transitions, whereas the nucleus accumbens core is a key node in a controller for preparatory behavior and effort regulation.

Mesolimbic DA depletion reduces gain in reward rate control, thus decreasing the error to reference conversion and the repetition reference signal. Increasing DA increases willingness to work for a given amount of reward without increasing the amount of reward desired. Depending on the environmental contingency, the output of the reward rate controller can manifest in different behaviors, whether approaching reward or pressing a lever.

The available evidence does not support the interpretation of mesolimbic self-stimulation as simply generating subjective pleasure. But it can be explained by the repetition control model and the role of mesolimbic DA as adaptive gain in repetition control. It is hypothesized that mesolimbic DA is critical for effort exertion in general, regardless of the outcome valence. DA depletion in the nucleus accumbens does not disrupt food consumption but dramatically reduces effort exertion. According to the present model, effort is determined largely by the gain of reward rate control, as it recruits any number of effective actions. One transition variable is linked with another through a feedback function, defining reward rate as a function of press rate.

A labile motivational hierarchy can explain the contributions of parallel cortico-BG circuits in motivated behavior. The motivational hierarchy coordinates means and ends, selecting specific instrumental actions to satisfy specific needs and desires. This organization can be implemented by parallel but overlapping cortico-BG networks, which allow limbic BG outputs to exert a unidirectional influence on the sensorimotor network. Through learning, homeostatic error signals for essential physiological variables and predictors of error reduction in these variables can recruit the action hierarchy for the production of specific actions.

REFERENCES

Aberman, J., & Salamone, J. D. (1999). Nucleus accumbens dopamine depletions make rats more sensitive to high ratio requirements but do not impair primary food reinforcement. *Neuroscience, 92*(2), 545–552.

Alchian, A. A., & Allen, W. R. (1977). *Exchange and Production: Competition, Coordination, and Control*: Belmont, CA: Wadsworth Publishing Company Belmont.

Amsel, A. (1962). Frustrative nonreward in partial reinforcement and discrimination learning: Some recent history and a theoretical extension. *Psychological Review, 69*, 306–328.

Aoki, S., Smith, J. B., Li, H., Yan, X., Igarashi, M., Coulon, P., ... Jin, X. (2019). An open cortico-basal ganglia loop allows limbic control over motor output via the nigrothalamic pathway. *Elife, 8*, e49995.

Balleine, B. W., & Dickinson, A. (1998). Goal-directed instrumental action: Contingency and incentive learning and their cortical substrates. *Neuropharmacology, 37*(4–5), 407–419.

Balleine, B. W., & O'Doherty, J. P. (2010). Human and rodent homologies in action control: Corticostriatal determinants of goal-directed and habitual action. *Neuropsychopharmacology, 35*(1), 48–69. doi: 10.1038/npp.2009.131.

Barter, J., Li, S., Lu, D., Rossi, M., Bartholomew, R., Shoemaker, C. T., ... Yin, H. H. (2015). Beyond reward prediction errors: The role of dopamine in movement kinematics. *Frontiers in Integrative Neuroscience, 9*, 39.

Baum, W. M. (1973). The correlation-based law of effect. *Journal of the Experimental Analysis of Behavior, 20*(1), 137–153.

Beeler, J. A., Daw, N. D., Frazier, C. R., & Zhuang, X. (2010). Tonic dopamine modulates exploitation of reward learning. *Frontiers in Behavioral Neuroscience, 4*, 170.

Berridge, K. C., & Kringelbach, M. L. (2015). Pleasure systems in the brain. *Neuron, 86*(3), 646–664.

Berridge, K. C., & Robinson, T. E. (1998). What is the role of dopamine in reward: Hedonic impact, reward learning, or incentive salience? *Brain Research Review, 28*(3), 309–369.

Berridge, K. C., Venier, I. L., & Robinson, T. E. (1989). Taste reactivity analysis of 6-hydroxydopamine-induced aphagia: Implications for arousal and anhedonia hypotheses of dopamine function. *Behavioral Neuroscience, 103*(1), 36–45.

Berrios, G., & Grli, M. (1995). Abulia and impulsiveness revisited: A conceptual history. *Acta Psychiatrica Scandinavica, 92*(3), 161–167.

Botvinick, M. M., Cohen, J. D., & Carter, C. S. (2004). Conflict monitoring and anterior cingulate cortex: An update. *Trends in Cognitive Sciences, 8*(12), 539–546.

Bouton, M. E. (2004). Context and behavioral processes in extinction. *Learning & Memory, 11*, 485–494.

Cabanac, M. (1971). Physiological role of pleasure. *Science, 173*(4002), 1103–1107.

Cabanac, M., & Lafrance, L. (1990). Postingestive alliesthesia: The rat tells the same story. *Physiology & Behavior, 47*(3), 539–543.

Cannon, W. (1932). *The Wisdom of the Body*: New York: W. W. Norton.

Caplan, L. R., Schmahmann, J. D., Kase, C. S., Feldmann, E., Baquis, G., Greenberg, J. P., ... Hier, D. B. (1990). Caudate infarcts. *Archives of Neurology, 47*(2), 133–143.

Carlezon, W. A., Devine, D. P., & Wise, R. A. (1995). Habit-forming actions of nomifensine in nucleus accumbens. *Psychopharmacology (Berl), 122*(2), 194–197.

Castro, D. C., & Berridge, K. C. (2014). Opioid hedonic hotspot in nucleus accumbens shell: Mu, delta, and kappa maps for enhancement of sweetness "liking" and "wanting". *Journal of Neuroscience, 34*(12), 4239–4250.

Corbit, L. H., & Balleine, B. W. (2005). Double dissociation of basolateral and central amygdala lesions on the general and outcome-specific forms of pavlovian-instrumental transfer. *Journal of Neuroscience, 25*(4), 962–970.

Corbit, L. H., Janak, P. H., & Balleine, B. W. (2007). General and outcome-specific forms of pavlovian-instrumental transfer: The effect of shifts in motivational state and inactivation of the ventral tegmental area. *European Journal of Neuroscience, 26*(11), 3141–3149.

Corbit, L. H., Muir, J. L., & Balleine, B. W. (2001). The role of the nucleus accumbens in instrumental conditioning: Evidence of a functional dissociation between accumbens core and shell. *Journal of Neuroscience, 21*(9), 3251–3260.

Cromwell, H. C., & Berridge, K. C. (1993). Where does damage lead to enhanced food aversion: The ventral pallidum/substantia innominata or lateral hypothalamus? *Brain Research, 624*(1–2), 1–10.

da Silva, J. A., Tecuapetla, F., Paixão, V., & Costa, R. M. (2018). Dopamine neuron activity before action initiation gates and invigorates future movements. *Nature, 554*(7691), 244.

Di Ciano, P., Cardinal, R. N., Cowell, R. A., Little, S. J., & Everitt, B. J. (2001). Differential involvement of nmda, ampa/kainate, and dopamine receptors in the nucleus accumbens core in the acquisition and performance of pavlovian approach behavior. *Journal of Neuroscience, 21*(23), 9471–9477.

Ellison, G. D., & Konorski, J. (1964). Separation of the salivary and motor responses in instrumental conditioning. *Science, 146*, 1071–1072.

Ettinger, R. H., & Staddon, J. E. (1983). Operant regulation of feeding: A static analysis. *Behavioral Neuroscience, 97*(4), 639–653.

Fenu, S., Bassareo, V., & Di Chiara, G. (2001). A role for dopamine d1 receptors of the nucleus accumbens shell in conditioned taste aversion learning. *Journal of Neuroscience, 21*(17), 6897–6904.

Fibiger, H. C. (1978). Drugs and reinforcement mechanisms: A critical review of the catecholamine theory. *Annual Review of Pharmacology and Toxicology, 18*(1), 37–56.

Fibiger, H. C., LePiane, F., Jakubovic, A., & Phillips, A. (1987). The role of dopamine in intracranial self-stimulation of the ventral tegmental area. *Journal of Neuroscience, 7*(12), 3888–3896.

Grill, H. J., & Norgren, R. (1978a). The taste reactivity test. I. Mimetic responses to gustatory stimuli in neurologically normal rats. *Brain Research*, *143*(2), 263–279.

Grill, H. J., & Norgren, R. (1978b). The taste reactivity test. Ii. Mimetic responses to gustatory stimuli in chronic thalamic and chronic decerebrate rats. *Brain Research*, *143*(2), 281–297.

Groenewegen, H. J., Berendse, H. W., & Haber, S. N. (1993). Organization of the output of the ventral striatopallidal system in the rat: Ventral pallidal efferents. *Neuroscience*, *57*(1), 113–142.

Haber, S. N. (2003). The primate basal ganglia: Parallel and integrative networks. *Journal of Chemical Neuroanatomy*, *26*(4), 317–330.

Haber, S. N., Fudge, J. L., & McFarland, N. R. (2000). Striatonigrostriatal pathways in primates form an ascending spiral from the shell to the dorsolateral striatum. *Journal of Neuroscience*, *20*(6), 2369–2382.

Haber, S. N., Groenewegen, H. J., Grove, E. A., & Nauta, W. J. (1985). Efferent connections of the ventral pallidum: Evidence of a dual striato pallidofugal pathway. *Journal of Comparative Neurology*, *235*(3), 322–335.

Hamel, L., Thangarasa, T., Samadi, O., & Ito, R. (2017). Caudal nucleus accumbens core is critical in the regulation of cue-elicited approach-avoidance decisions. *Eneuro*, *4*(1).

Hamid, A. A., Frank, M. J., & Moore, C. I. (2021). Wave-like dopamine dynamics as a mechanism for spatiotemporal credit assignment. *Cell*, *184*(10), 2733–2749. e2716.

Hamid, A. A., Pettibone, J. R., Mabrouk, O. S., Hetrick, V. L., Schmidt, R., Vander Weele, C. M., … Berke, J. D. (2016). Mesolimbic dopamine signals the value of work. *Nature Neuroscience*, *19*(1), 117–126. doi: 10.1038/nn.4173.

Heath, R. G. (1963). Electrical self-stimulation of the brain in man. *American Journal of Psychiatry*, *120*(6), 571–577.

Herrick, C. J. (1926). *Brains of Rats and Men*: Chicago: University of Chicago Press.

Hess, W. R. (1957). *The Functional Organization of the Diencephalon*: New York: Grune & Stratton.

Hess, W. R., & Akert, C. (1955). Experimental data on role of hypothalamus in mechanism of emotional behavior. *AMA Archives of Neurology & Psychiatry*, *73*(2), 127–129.

Ho, C. Y., & Berridge, K. C. (2014). Excessive disgust caused by brain lesions or temporary inactivations: Mapping hotspots of the nucleus accumbens and ventral pallidum. *European Journal of Neuroscience*, *40*(10), 3556–3572.

Horvitz, J. C. (2000). Mesolimbocortical and nigrostriatal dopamine responses to salient non-reward events. *Neuroscience*, *96*(4), 651–656.

Hughes, R. N., Bakhurin, K. I., Petter, E. A., Watson, G. D., Kim, N., Friedman, A. D., & Yin, H. H. (2020). Ventral tegmental dopamine neurons control the impulse vector during motivated behavior. *Current Biology*, *30*, 1–14.

Hull, C. (1943). *Principles of Behavior*: New York: Appleton-Century-Crofts.

Ikemoto, S., Glazier, B. S., Murphy, J. M., & McBride, W. J. (1997). Role of dopamine d1 and d2 receptors in the nucleus accumbens in mediating reward. *Journal of Neuroscience*, *17*(21), 8580–8587.

Ikemoto, S., & Panksepp, J. (1999). The role of nucleus accumbens dopamine in motivated behavior: A unifying interpretation with special reference to reward-seeking. *Brain Research Reviews*, *31*(1), 6–41.

Ikemoto, S., Yang, C., & Tan, A. (2015). Basal ganglia circuit loops, dopamine and motivation: A review and enquiry. *Behavioural Brain Research*, *290*, 17–31.

Joel, D., & Weiner, I. (2000). The connections of the dopaminergic system with the striatum in rats and primates: An analysis with respect to the functional and compartmental organization of the striatum. *Neuroscience*, *96*(3), 451–474.

Johnson, A. W., Gallagher, M., & Holland, P. C. (2009). The basolateral amygdala is critical to the expression of pavlovian and instrumental outcome-specific reinforcer devaluation effects. *Journal of Neuroscience*, *29*(3), 696–704.

Keramati, M., & Gutkin, B. (2014). Homeostatic reinforcement learning for integrating reward collection and physiological stability. *Elife*, *3*. doi: 10.7554/eLife.04811.

Kim, I. H., Rossi, M. A., Aryal, D. K., Racz, B., Kim, N., Uezu, A., … Soderling, S. H. (2015). Spine pruning drives antipsychotic-sensitive locomotion via circuit control of striatal dopamine. *Nature Neuroscience*, *18*(6), 883–891. doi: 10.1038/nn.4015.

Kohli, A., Blitzer, D. N., Lefco, R. W., Barter, J. W., Haynes, M. R., Colalillo, S. A., Ly, M., & Zink, C. F. (2018). Using Expectancy Theory to quantitatively dissociate the neural representation of motivation from its influential factors in the human brain: An fMRI study. *Neuroimage*, *178*, 552–561.

Konorski, J. (1967). *Integrative Activity of the Brain*: Chicago, IL: University of Chicago Press.

Lammel, S., Lim, B. K., Ran, C., Huang, K. W., Betley, M. J., Tye, K. M., … Malenka, R. C. (2012). Input-specific control of reward and aversion in the ventral tegmental area. *Nature*, *491*(7423), 212–217.

Laplane, D., & Dubois, B. (2001). Auto-activation deficit: A basal ganglia related syndrome. *Movement Disorders*, *16*(5), 810–814.

Leblond, M., Fan, D., Brynildsen, J. K., & Yin, H. H. (2011). Motivational state and reward content determine choice behavior under risk in mice. *PLoS One*, *6*(9), e25342. doi: 10.1371/journal.pone.0025342.

Leblond, M., Sukharnikova, T., Yu, C., Rossi, M. A., & Yin, H. H. (2014). The role of pedunculopontine nucleus in choice behavior under risk. *European Journal of Neuroscience*, *39*(10), 1664–1670.

Li, H. E., Rossi, M. A., Watson, G. D., Moore, H. G., Cai, M. T., Kim, N., ... Hughes, R. N. (2021). Hypothalamic-extended amygdala circuit regulates temporal discounting. *Journal of Neuroscience*, *41*(9), 1928–1940.

MacLean, P. D. (1952). Some psychiatric implications of physiological studies on frontotemporal portion of limbic system (visceral brain). *Electroencephalography & Clinical Neurophysiology*, *4*(4), 407–418.

Maldonado-Irizarry, C. S., Swanson, C. J., & Kelley, A. E. (1995). Glutamate receptors in the nucleus accumbens shell control feeding behavior via the lateral hypothalamus. *Journal of Neuroscience*, *15*(10), 6779–6788.

McCullough, L. D., Sokolowski, J. D., & Salamone, J. D. (1993). A neurochemical and behavioral investigation of the involvement of nucleus accumbens dopamine in instrumental avoidance. *Neuroscience*, *52*(4), 919–925.

Mega, M. S., & Cohenour, R. C. (1997). Akinetic mutism: Disconnection of frontal-subcortical circuits. *Cognitive and Behavioral Neurology*, *10*(4), 254–259.

Mingote, S., Font, L., Farrar, A. M., Vontell, R., Worden, L. T., Stopper, C. M., ... Chrobak, J. J. (2008). Nucleus accumbens adenosine a2a receptors regulate exertion of effort by acting on the ventral striatopallidal pathway. *Journal of Neuroscience*, *28*(36), 9037–9046.

Mogenson, G. J., Jones, D. L., & Yim, C. Y. (1980). From motivation to action: Functional interface between the limbic system and the motor system. *Progress in Neurobiology*, *14*(2–3), 69–97.

Mogenson, G. J., Swanson, L., & Wu, M. (1983). Neural projections from nucleus accumbens to globus pallidus, substantia innominata, and lateral preoptic-lateral hypothalamic area: An anatomical and electrophysiological investigation in the rat. *Journal of Neuroscience*, *3*(1), 189–202.

Naccache, L., Dehaene, S., Cohen, L., Habert, M.-O., Guichart-Gomez, E., Galanaud, D., & Willer, J.-C. (2005). Effortless control: Executive attention and conscious feeling of mental effort are dissociable. *Neuropsychologia*, *43*(9), 1318–1328.

Nauta, W. J., Smith, G. P., Faull, R. L., & Domesick, V. B. (1978). Efferent connections and nigral afferents of the nucleus accumbens septi in the rat. *Neuroscience*, *3*(4–5), 385–401.

Niv, Y., Daw, N. D., Joel, D., & Dayan, P. (2007). Tonic dopamine: Opportunity costs and the control of response vigor. *Psychopharmacology (Berl)*, *191*(3), 507–520.

Olds, J. (1977). *Drives and Reinforcements: Behavioral Studies of Hypothalamic Functions*: New York: Raven Press.

Olds, J., & Milner, P. (1954). Positive reinforcement produced by electrical stimulation of septal area and other regions of rat brain. *Journal of Comparative and Physiological Psychology*, *47*(6), 419.

Parkinson, J. A., Willoughby, P. J., Robbins, T. W., & Everitt, B. J. (2000). Disconnection of the anterior cingulate cortex and nucleus accumbens core impairs pavlovian approach behavior: Further evidence for limbic cortical-ventral striatopallidal systems. *Behavioral Neuroscience*, *114*(1), 42–63.

Pecina, S., & Berridge, K. C. (2000). Opioid site in nucleus accumbens shell mediates eating and hedonic 'liking' for food: Map based on microinjection fos plumes. *Brain Research*, *863*(1–2), 71–86.

Pecina, S., Cagniard, B., Berridge, K. C., Aldridge, J. W., & Zhuang, X. (2003). Hyperdopaminergic mutant mice have higher "wanting" but not "liking" for sweet rewards. *Journal of Neuroscience*, *23*(28), 9395–9402.

Pecina, S., Smith, K. S., & Berridge, K. C. (2006). Hedonic hot spots in the brain. *Neuroscientist*, *12*(6), 500–511.

Pfaus, J. G., & Phillips, A. G. (1991). Role of dopamine in anticipatory and consummatory aspects of sexual behavior in the male rat. *Behavioral Neuroscience*, *105*(5), 727.

Powers, W. T. (1971). A feedback model for behavior: Application to a rat experiment. *Behavioral Science*, *16*(6), 558–563.

Prinssen, E. P., Balestra, W., Bemelmans, F. F., & Cools, A. R. (1994). Evidence for a role of the shell of the nucleus accumbens in oral behavior of freely moving rats. *Journal of Neuroscience*, *14*(3 Pt 2), 1555–1562.

Reynolds, S. M., & Berridge, K. C. (2002). Positive and negative motivation in nucleus accumbens shell: Bivalent rostrocaudal gradients for gaba-elicited eating, taste "liking"/"disliking" reactions, place preference/avoidance, and fear. *Journal of Neuroscience*, *22*(16), 7308–7320.

Risold, P. Y., Thompson, R. H., & Swanson, L. W. (1997). The structural organization of connections between hypothalamus and cerebral cortex. *Brain Research Review*, *24*(2–3), 197–254.

Roberts, D. C., & Koob, G. F. (1982). Disruption of cocaine self-administration following 6-hydroxydopamine lesions of the ventral tegmental area in rats. *Pharmacology Biochemistry and Behavior, 17*(5), 901–904.

Rossi, M. A., Sukharnikova, T., Hayrapetyan, V. Y., Yang, L., & Yin, H. H. (2013). Operant self-stimulation of dopamine neurons in the substantia nigra. *PLoS One, 8*(6), e65799.

Salamone, J. D., Correa, M., Farrar, A. M., Nunes, E. J., & Pardo, M. (2009). Dopamine, behavioral economics, and effort. *Frontiers in Behavioral, 3*, 13. doi: 10.3389/neuro.08.013.2009.

Salamone, J. D., Correa, M., Mingote, S. M., & Weber, S. M. (2005). Beyond the reward hypothesis: Alternative functions of nucleus accumbens dopamine. *Current Opinion in Pharmacology, 5*(1), 34–41.

Salamone, J. D., Wisniecki, A., Carlson, B., & Correa, M. (2001). Nucleus accumbens dopamine depletions make animals highly sensitive to high fixed ratio requirements but do not impair primary food reinforcement. *Neuroscience, 105*(4), 863–870.

Schmidt, L., d'Arc, B. F., Lafargue, G., Galanaud, D., Czernecki, V., Grabli, D., ... Pessiglione, M. (2008). Disconnecting force from money: Effects of basal ganglia damage on incentive motivation. *Brain, 131*(5), 1303–1310.

Schopenhauer, A. (1978). *Preisschrift Über Die Freiheit des Willens* (vol. 305): Berlin: Felix Meiner Verlag.

Sheard, M. H., & Flynn, J. P. (1967). Facilitation of attack behavior by stimulation of the midbrain of cats. *Brain Research, 4*(4), 324–333.

Smith, K. S., Tindell, A. J., Aldridge, J. W., & Berridge, K. C. (2009). Ventral pallidum roles in reward and motivation. *Behavioural Brain Research, 196*(2), 155–167.

Sokolowski, J., & Salamone, J. (1998). The role of accumbens dopamine in lever pressing and response allocation: Effects of 6-ohda injected into core and dorsomedial shell. *Pharmacology Biochemistry and Behavior, 59*(3), 557–566.

Staddon, J. E. R. (1983). *Adaptive Behavior and Learning*: Cambridge, MA: Cambridge University Press.

Sutton, R. S., & Barto, A. G. (2018). *Reinforcement Learning: An Introduction*: Cambridge, MA: MIT Press.

Swanson, L. W. (2000). Cerebral hemisphere regulation of motivated behavior. *Brain Research, 886*(1–2), 113–164.

Thorndike, E. L. (1911). *Animal Intelligence: Experimental Studies*: New York: Macmillan.

Trapold, M. A., & Overmier, J. B. (1972). The second learning process in instrumental learning. In: Black, A. A., & Prokasy, W. F. (Eds.), *Classical Conditioning ii: Current Research and Theory* (pp. 427–452). New York: Appleton-Century-Crofts.

Walton, M. E., Bannerman, D. M., Alterescu, K., & Rushworth, M. F. (2003). Functional specialization within medial frontal cortex of the anterior cingulate for evaluating effort-related decisions. *Journal of Neuroscience, 23*(16), 6475–6479.

Walton, M. E., Bannerman, D. M., & Rushworth, M. F. (2002). The role of rat medial frontal cortex in effort-based decision making. *Journal of Neuroscience, 22*(24), 10996–11003.

Wang, S., Tan, Y., Zhang, J., & Luo, M. (2013). Pharmacogenetic activation of midbrain dopamine neurons produces hyperactivity. *Neuroscience Bulletin, 29*(5), 1–8.

Westbrook, A., Van Den Bosch, R., Määttä, J., Hofmans, L., Papadopetraki, D., Cools, R., & Frank, M. (2020). Dopamine promotes cognitive effort by biasing the benefits versus costs of cognitive work. *Science, 367*(6484), 1362–1366.

Wise, R. A. (1978). Catecholamine theories of reward: A critical review. *Brain Research, 152*(2), 215–247.

Wise, R. A. (1982). Neuroleptics and operant beahvior: The anhedonia hypothesis. *Behavioral Brain Sciences, 5*, 39–87.

Wise, R. A. (2004). Dopamine, learning and motivation. *Nature Reviews Neuroscience, 5*(6), 483–494. doi: 10.1038/nrn1406.

Wise, R. A., Spindler, J., DeWit, H., & Gerberg, G. J. (1978). Neuroleptic-induced" anhedonia" in rats: Pimozide blocks reward quality of food. *Science, 201*(4352), 262–264.

Wyrwicka, W., & Doty, R. W. (1966). Feeding induced in cats by electrical stimulation of the brain stem. *Experimental Brain Research, 1*(2), 152–160.

Yin, H. H. (2013). Restoring purpose in behavior. In: *Computational and Robotic Models of the Hierarchical Organization of Behavior* (pp. 319–347). Berlin: Springer.

Yin, H. H. (2016). The basal ganglia and hierarchical control in voluntary behavior. In: Soghomonian, J.-J. (Ed.), *The Basal Ganglia-Novel Perspectives on Motor and Cognitive Functions* (vol. in press, pp. 513–566). Berlin: Springer.

Yin, H. H., Zhuang, X., & Balleine, B. W. (2006). Instrumental learning in hyperdopaminergic mice. *Neurobiology of Learning and Memory, 85*(3), 283–288.

10 Actions and Goals

Different goals may require the use of a particular action, just as different actions may be used to reach a particular goal. Learning not only adds new goals or objects of desire but also new means of reaching a particular goal. Once a goal is selected, the production of behavior must be coordinated precisely in space and time to reach it.

This chapter considers the role of the basal ganglia (BG) in goal-directed behavior. We shall start by examining how the BG contribute to the control of simple spatial relationships, in particular providing guidance needed to target specific goals in space. We shall then consider control of more complex relationships, such as contingencies and causal relationships, and the contributions of the cortico-BG networks in the learning and expression of goal-directed actions.

10.1 APPROACHING A GOAL

As goals like mates or food are spatially localized, to reach them, behavior must be guided precisely and continuously in space. How is such guidance achieved? A spatial location can be defined egocentrically, with self as the origin (e.g. forward, backward), or allocentrically, with some arbitrary location as the origin (e.g. east, west). In motivated behavior, direction is also defined relative to the external goal object. "Closer and farther" reflect the value of a spatial relationship variable. Getting closer to the goal could involve movement in any direction.

As discussed in Chapter 6, the midbrain orientation controllers are critical for orienting movements toward some external object (Watanabe & Munoz, 2010). For example, in the case of gaze shifts, a salient stimulus in the peripheral visual field triggers eye and head movements to capture a new target. Such behavior, traditionally labeled "reflex," reflects the operation of lower-level position controllers just below the BG. Relying on these controllers only, one cannot deliberately look in the other direction, away from the salient stimulus, or at arbitrary targets. Volitional orienting requires descending projections from the brain to set arbitrary reference states for the position and orientation controllers. The BG are a major source of such descending commands.

In primates, a distinction is made between smooth pursuit, in which the eyes follow a moving visual target continuously, and saccades, which are rapid eye movements to change the point of fixation (Robinson, 1968). There is a dedicated oculomotor circuit, including the frontal eye field, caudate nucleus, and SNr. In both smooth pursuit and saccades, the BG neurons are strongly modulated (Basso, Pokorny, & Liu, 2005; Hikosaka & Wurtz, 1983; Yoshida & Tanaka, 2009). In humans, striatal lesions produce deficits in smooth pursuit (Lekwuwa & Barnes, 1996a, b) and even abolish voluntary saccades and smooth pursuits (Chung, Moon, Song, & Kim, 2006).

Moreover, the BG have also been associated with impaired tracking performance using other effectors. Caudate lesions, for example, can impair the ability to follow a moving light on the screen with a hand (Bowen, 1969). Lesioned monkeys showed mis-reaching above the target. When tracking a moving target with the hand, patients with Parkinson's disease also show greater lag and higher error (Flowers, 1978).

10.2 COMPULSORY APPROACH

A striking symptom after striatal (caudate) lesions is that animals walk incessantly and, when encountering a barrier, push continuously against it, even causing self-injury. This observation was made in classic studies by Magendie and Nothnagel in the 19th century, and later known as "obstinate progression" or "cursive hyperkinesia" (Butcher & Fox, 1968; Mettler & Mettler, 1942). However, Villablanca and colleagues showed that it is due to extensive forebrain lesions beyond the striatum. More selective lesions limited to the striatum (mainly caudate) in cats produced a somewhat different set of symptoms, which they called compulsory approaching syndrome (Villablanca, 2010; Villablanca, Marcus, & Olmstead, 1976; Villablanca, Marcus, Olmstead, & Avery, 1976). In compulsory approach, cats with caudate lesions follow any moving target. Unlike obstinate progression, the behavior is directed at a specific target, yet the cats would avoid obstacles and adjust their walking or running speed according to changes in their target speed. If pushed away, they would return immediately, only disengaging when a more salient new target was introduced. The approach behavior is not only guided by vision but also by other sensory modalities. For example, tactile stimulus would result in cats "sticking" to the target. If the snout is touched, the cat will push forward, maintaining contact with a moving hand.

To reach any spatially localized goal, the controller must update its representation of the current distance to the goal to generate the appropriate action. In obstinate progression, there appears to be a loss of distance feedback. As a result, the error persists because it is not reduced by negative feedback. Hence the persistent approach and pushing against the target or obstacle. But the loss of feedback is not sufficient to explain the compulsory approaching syndrome because the lesioned cat does stop forward movement once the target is reached. Some form of negative feedback, perhaps conveyed via the tactile inputs, can still reduce error and terminate approach behavior.

Obstinate progression and compulsory approaching suggest a malfunctioning spatial relationship control system that requires the associative BG network. Bidirectional control of distance to the target is achieved by approach or retreat. If the distance reference has a negative sign, then approach is driven by a positive distance error, which represents "too far" (actual distance minus reference distance). Eliminating this signal, the system will no longer approach targets, as no distance is too far. On the other hand, avoidance is driven by a negative distance error signaling "too close" (actual distance is less than reference distance). Eliminating this signal could result in the inability to retreat or disengage during pursuit. The distance relationship can be monitored using multiple senses, including visual, tactile, and auditory. Rooting, touching, rearing, running, and jumping are the corresponding outputs that act in parallel to reduce error. Moreover, if the reference distance is fixed at zero or a low value due to a lesion, then any perceptual input to the distance controller can trigger a positive distance error (too far), which is converted into pursuit movement. This could be the case in compulsory approach. The distance reference, fixed at a low level, corresponds to a state of being close to and clinging to the target. Approach behavior is constantly generated by the "too far" error.

10.3 A STRIATAL CIRCUIT FOR RELATIONSHIP CONTROL AND CONTINUOUS PURSUIT

To follow a moving target, self-velocity must match the target velocity, though this matching is a byproduct of maintaining a certain distance between self and target. Target velocity is not perceived directly during self-movement. For example, when driving a car to follow another, one may match the velocity of the other car by maintaining a certain distance, which is the controlled variable. Direct perception of the target velocity is not possible, given self movement. Deviation from reference (desired distance) is an error signal, which can be used to regulate the behavior of stepping on the gas pedal.

How can the distance error be converted into a velocity reference? To answer this question, Kim et al. trained mice to follow a continuously moving target with a sucrose spout (Kim et al., 2019). As shown in Figure 10.1, the target moves along the horizontal (x) axis for small distances, so that no locomotion is required, and the mouse is able to move its head and body to track the spout. Self-velocity lags distance to target, as revealed by cross correlation analysis of these two variables, suggesting that the mouse uses some perception of distance to modulate self-velocity. The distance perception could represent an error when it deviates from the reference state. Such distance control is learned and intermittent.

Actions and Goals

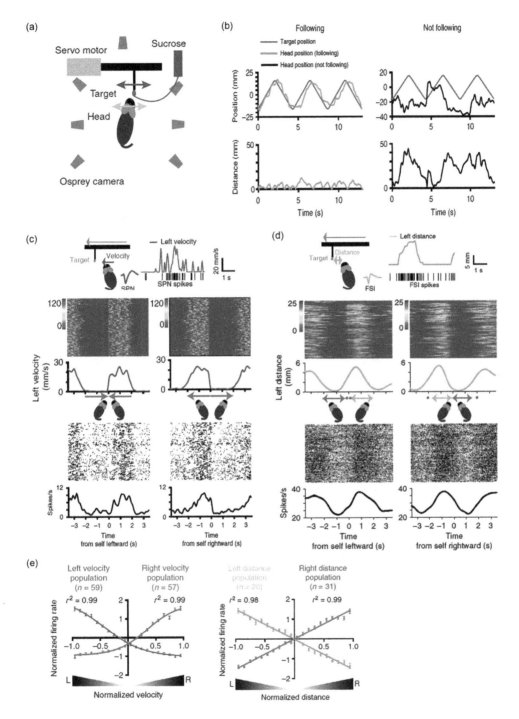

FIGURE 10.1 Fast-spiking interneuron–spiny projection neuron (FSI–SPN) circuit and control of distance to target. Proximity to reward is controlled in a continuous reward pursuit task. (a) Illustration of the behavioral task. Three-dimensional motion capture was used to measure the position of the target sucrose spout and mouse head. (b) Behavioral performance on the task. When the mouse is following the target, the distance error is maintained at a low level continuously. When the mouse is not following, errors are not reduced. (c) Representative SPN with firing rate correlated with leftward velocity. (d) Representative FSI with firing rate correlated with left distance. (e) Population summary of velocity-representing SPNs and distance-representing FSIs. From Kim et al. (2019) with permission.

Using this behavioral task, Kim et al. compared single-unit activity of fast-spiking interneurons (FSIs) and spiny projection neurons (SPNs) during continuous pursuit. Both FSIs and SPNs receive cortical inputs, and the FSIs project to the SPNs, forming a feedforward inhibition circuit (Chapter 3). During pursuit, at any time the horizontally moving target could be on either side of the mouse, and the distance is a measure of the error (Figure 10.1). Many SPNs were correlated with movement velocity: some increased firing with leftward movement and others with rightward movement. On the other hand, many FSIs represented distance from self to target. Some FSIs increased firing when the target moved to the left side of the animal and decreased firing when it moved to the right. Other FSIs showed the opposite pattern. When the target was directly in front of the mouse, the firing rate was close to the middle of the dynamic range. Different types of errors (left vs. right) were indicated by the sign of the signal in relation to the effective zero.

As described in Chapter 2, the FSIs (PV+) are often tonically active with high firing rates. These neurons receive excitatory inputs from the cortex, especially sensory cortices, and project to the SPNs, each synapsing on many SPNs (Lee et al., 2019; Planert, Szydlowski, Hjorth, Grillner, & Silberberg, 2010). The sparsely firing SPNs are suitable for representing movement velocity. Their firing rate is close to zero at rest, with short bursts of activity. On the other hand, tonically active neurons are suitable for the representation of distance as a time-varying analog signal. Inputs from sensory cortical areas may send distance-related signals to FSIs. A distance error signaling "too much to the left" would activate leftward SPNs and suppress rightward neurons, whereas "too much to the right" would activate rightward SPNs and suppress leftward neurons.

These results suggest that the FSI–SPN circuit can be used to transform the output of the distance controller into the reference for the velocity controller. The distance error is used to update the velocity reference signal. As shown in Figure 10.2, if the change in distance is positive (farther),

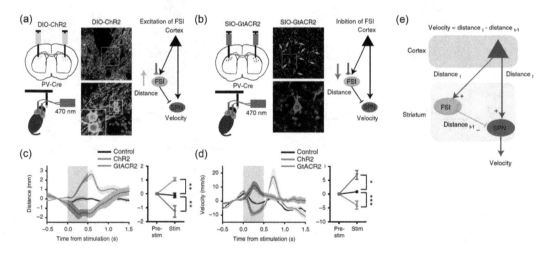

FIGURE 10.2 Distance-to-velocity transformation. (a) Illustration of channelrhodopsin (ChR2) experiments for exciting striatal FSIs. Cre-dependent ChR2 virus was injected into parvalbumin (PV)-Cre mice to express ChR2 in PV+ neurons. (b) Illustration of GtACR2 (Guillardia theta anion-conducting channelrhodopsins, an inhibitory opsin activated by blue light) experiments for inhibiting FSIs. (c) Summary of stimulation effects on distance to target. (d) Summary of stimulation effects on self-velocity. (e) Proposed circuit model for converting distance representation to velocity command. The fast-spiking interneuron–spiny projection neuron (FSI–SPN) circuit can act as a differentiator if FSIs and their SPN targets share common cortical inputs that represent distance. Left distance error is converted to a leftward velocity command, and right distance is converted to a rightward velocity command. See the text for a more detailed explanation. From Kim et al. (2019) with permission.

then the differentiator circuit generates a net positive output to drive the leftward velocity controller, causing movement toward the target. If it is decreasing, the circuit will generate a net negative output to the leftward velocity controller, causing deceleration. As one approaches the target, velocity decreases as distance decreases. When pursuit is initiated, distance is acquired by the control system and generates the velocity command. Since there is no subtraction of the past distance signal, the initiation of pursuit produces a large net excitatory drive to the SPN. Once the animal starts to track the target, the last distance signal is now subtracted from the current distance signal to update the velocity command at each time step. As the left distance decreases, the leftward velocity drive is reduced.

The control of the distance to target is intermittent. Distance does not become a controlled variable until the target has been selected and the controller is engaged. Initially, there is no distance error, as the system is not activated. Once the distance controller is turned on, the distance input is compared with the reference distance value, generating the error signal required for the instantaneous velocity reference. In support of this argument, when mice were not actively following the target, the FSI representation of self-target distance was dramatically degraded.

Striatal FSIs not only represent target distance but are also necessary for successful pursuit performance. Inhibition of these neurons using chemogenetics or optogenetics reliably disrupted pursuit behavior, by reducing distance and increasing velocity. In contrast, selective excitation of FSIs increased distance and decreased velocity. Inhibition of dSPNs also reduced velocity during pursuit, suggesting that the distance-to-velocity conversion is achieved mainly by FSI projections to the direct pathway.

10.4 APPROACH BEHAVIOR AND FEEDBACK

In approach behavior, the feedback function defines change in distance to target as a function of behavioral output. The *polarity* of feedback must be negative for approach behavior to be successful, e.g. running forward must reduce distance to target. There appear to be innate settings that convert distance error into approach behavior which reduces this error. However, a key limitation of such controllers is revealed when the feedback polarity is reversed, e.g. when negative feedback turns into positive feedback. This is illustrated by a study of approach behavior by Hershberger. He tested the ability of young chicks with little locomotor experience to adapt to a reversal in the feedback function (Hershberger, 1986). Chicks were trained to approach a food dish on a runway. The feedback function was then reversed in polarity, such that as they approached the dish, the visible environment moved away from them. Consequently, instead of reducing the distance as in a normal environment, the approach behavior increased distance. The chicks are then predicted to chase the food cup away, exhibiting the behavior of a control system when the feedback is positive. Hershberger found that many chicks showed such behavior. At least in young chicks, then, the default setting converts distance errors to descending commands for forward locomotion. It is therefore difficult for them to stop moving or to move backward to reach the food. The chick might stop, only to reengage in the futile approach behavior as the distance error is generated again. Figure 10.3 illustrates the feedback polarity reversal in this task.

Since polarity reversal turns negative feedback into positive feedback, persisting in the control effort would increase error. To adapt to a reversal in the feedback polarity, one must learn to move away from the goal in order to acquire it. By itself, the control system for simple approach behavior is not capable of adapting to a reversal in the feedback function. Higher levels are needed to recruit the opposite retreat system while suppressing the default connection to the approach system.

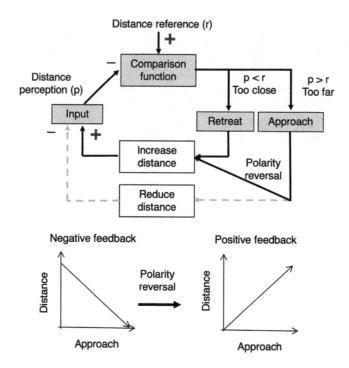

FIGURE 10.3 Feedback polarity in approach behavior. Schematic showing feedback polarity in a bidirectional controller for distance and the effect of feedback polarity reversal.

10.5 LEARNING TO APPROACH

In the type of appetitive Pavlovian conditioning commonly studied, a stimulus (e.g. an auditory cue) predicts food reward, and with learning, the animal generates an anticipatory approach to the reward-predicting cue. What changes during learning is primarily the spatial target of the anticipatory approach behavior and the timing of the behavior (Day & Carelli, 2007). In other words, learning is required to determine what to approach and when.

Zener described significant individual differences in the conditioned response (CR) elicited by a reward-predicting conditional stimulus (CS, sound of a bell). Some dogs approached the source of the bell CS and then backed up to the reward location, a food pan just below the animal's nose. Others only glanced at the CS and immediately fixed their gaze on the food, the unconditioned stimulus (US). Still others looked back and forth between the source of the CS and food in the US. This observation suggests that there are two distinct targets for approach behavior, CS source, and food pan, but there could be conflict in the target selection process (Zener, 1937).

Approach behavior has been extensively analyzed in studies of autoshaping or sign tracking (Brown & Jenkins, 1968; Buzsáki, 1982; Schwartz & Gamzu, 1977; Tomie, Brooks, & Zito, 2014). Prior to learning, rats approach the reward itself. After learning, they are more likely to approach the food-predictive CS before returning to the food hopper to retrieve the reward (Tomie, Grimes, & Pohorecky, 2008). This is called "sign tracking." Not all animals will show sign tracking. Some will simply approach the reward directly after CS presentation. Such behavior is known as "goal tracking" (Clark, Hollon, & Phillips, 2012; Flagel et al., 2010b).

Sign tracking relies on innate approach systems but only adds the predictor as a new target. In the natural environment, such behavior would usually lead one closer to the reward. However, as shown by Hershberger's results, such a system alone cannot handle polarity reversal in the feedback function. Sign tracking often persists even when the contingency between action and outcome has been reversed (Hearst & Jenkins, 1974; Meyer et al., 2012; Williams & Williams, 1969).

Both the acquisition and performance of approach require the limbic corticostriatal circuit, including the anterior cingulate cortex and its target in the accumbens core (Brog, Salyapongse, Deutch, & Zahm, 1993; Bussey, Everitt, & Robbins, 1997; Cardinal et al., 2002; Day, Roitman, Wightman, & Carelli, 2007; Day, Wheeler, Roitman, & Carelli, 2006; Parkinson, Willoughby, Robbins, & Everitt, 2000). Lesions of the core or of the anterior cingulate or a disconnection between these two structures impair the acquisition of Pavlovian approach behavior (Parkinson et al., 2000). N-methyl-D-aspartate (NMDA) activation in the accumbens core is necessary for learning approach behaviors (Kelley, Smith-Roe, & Holahan, 1997). Blockade of accumbens DA receptors can increase the latency to initiate an approach (McGinty, Lardeux, Taha, Kim, & Nicola, 2013). Local infusion of a D1-like DA receptor antagonist or an NMDA receptor antagonist immediately after training also impaired this form of learning (Dalley et al., 2005).

Neurons in the accumbens core can change their activity systematically during stimulus–reward learning (Day et al., 2006). Nicola and colleagues examined the activity of accumbens neurons in the acquisition of approach behavior (Vega-Villar, Horvitz, & Nicola, 2019). In their study, a discriminative stimulus (auditory or visual S+) signaled the insertion of a lever and reward delivery (no lever press was required). After animals learned to approach the S+ selectively, more neurons were excited than inhibited by the cue, and the cue-evoked excitation was stronger for discriminative stimuli compared with the unrewarded cue (S−). Such activity preceded locomotion and predicted the speed and latency of subsequent approach behavior. Blocking NMDA receptors reduced S+ approaches in moderately trained mice. In the drug-free extinction test, despite exposure to a comparable number of cue-reward pairings during training, animals that had been treated with APV during training had a reduced cue-elicited approach. The proportion of neurons excited by S+ also increased, while the proportion of neurons inhibited by S+ decreased. N-methyl-D-aspartate receptors (NMDAR) antagonist injections into the accumbens core prevented cued approach learning and the learning-related changes in neural activity. These results suggest that glutamatergic inputs to the acumbens core and NMDA-dependent synaptic plasticity is required for this type of learning. *In vitro* studies in brain slices have shown that accumbens NMDARs also play a permissive role in long-term potentiation (LTP) (Hunt & Castillo, 2012). A major source of glutamatergic projections to the accumbens is the basolateral amygdala, a cortex-like structure. These projections are potentiated by NMDA and D1 activation (Floresco, Blaha, Yang, & Phillips, 2001) and have also been implicated in anticipatory approach behavior (Stuber et al., 2011). Hebbian plasticity could modify a set of glutamatergic synapses on accumbens core neurons so that approach behavior could be generated to predict rewards. Such plasticity can redefine the input function of the control system. Consequently, the initial target for the approach behavior is altered from the goal to the cue, an example of stimulus substitution in Pavlovian conditioning.

DA antagonism in the accumbens core impaired sign tracking rather than goal tracking (Flagel et al., 2010a; Saunders & Robinson, 2012). Using fast-scan cyclic voltammetry to measure DA signaling in the accumbens, Flagel et al. tested the contribution of DA to sign tracking and goal tracking (Figure 10.4). If CS-evoked dopamine release reflects reward prediction, it should increase in both sign trackers and goal trackers, but they found that the shift in phasic dopamine from the reward to the cue occured only in sign trackers. In goal trackers, the DA response did not differ significantly between cue and reward presentations over the course of learning (Flagel et al., 2010a). Systemic administration of flupenthixol, a DA antagonist, reduced CR performance in both groups. Importantly, when tested later without flupenthixol, sign trackers still failed to show sign tracking, indicating that dopamine antagonism blocked acquisition of this behavior. These results suggest that DA does not act as a general teaching signal for stimulus–reward learning but is more important for the acquisition and expression of approach behavior as measured by sign tracking. We shall return to this type of learning in Chapter 12.

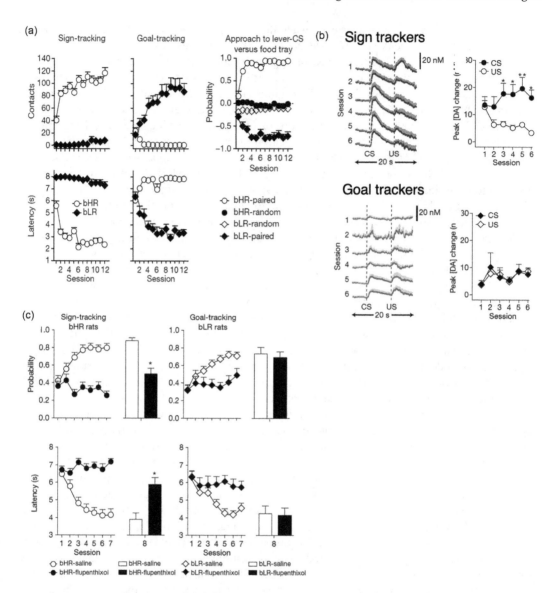

FIGURE 10.4 Sign tracking and goal tracking. (a) Development of sign tracking versus goal-tracking CRs in rats. bHR, high responder to novelty. bLR, low responder to novelty. (b) Phasic DA signaling in the nucleus accumbens core as measured by fast-scan cyclic voltammetry during stimulus–reward learning. (c) Systematic injection of the DA antagonist flupenthixol blocked expression of sign tracking or goal tracking. Probability and latency of CR expression are shown during seven sessions of learning. However, when rats were tested without any drug on a probe test during session 8, bLR rats (goal trackers) learned the CS–US association, as indicated by the expression of the goal-tracking CR, even though the drug prevented goal tracking during training. bHR rats (sign trackers) showed less goal tracking compared to the saline group on session 8, suggesting that they failed to acquire the CS–US association. From Flagel et al. (2010a) with permission.

10.6 CONTINGENCY, ASSOCIATIVE STRUCTURES, AND ANALYSIS OF CONDITIONING EXPERIMENTS

Simple approach behavior requires continuous control of distance to target, which is perhaps the prototype of all relationship variables. For example, we say we are *closer* to our goal or *closer* in time (Lakoff & Johnson, 2008; Shepard, 1981). In addition, there are aslo complex contingencies

ns and Goals

that incorporate both spatial and temporal requirements. For example, in stimulus–reward learning, what is required is to approach a particular target at a particular time, which is defined in relation to some other event.

Historically, temporal contiguity or simple pairing of two events (e.g. CS and US) was considered the key determinant of Pavlovian conditioning. However, often mere pairing of CS and US is not sufficient (Rescorla, 1968, 1988). For example, when the probability of the US in the absence of CS is the same as probability in the presence of the CS, acquisition of the CR is reduced. Contingency can be quantified as the difference between two probabilities (often called delta p): probability of reward given stimulus minus probability of reward in the absence of stimulus (Hammond, 1980; Rescorla, 1968).

Contingency can also be viewed as a higher-level relationship variable critical for guiding many behaviors. The experienced contingency is proportional to the correlation between two time-varying rate variables: in Pavlovian conditioning, the correlation between the rate of CS (stimulus) and the rate of US (outcome), and in instrumental conditioning, the rate of lever pressing (action) and the rate of reward (outcome). Unlike temporal contiguity, the rates can be integrated over a longer time window (Gallistel, 1990; Gallistel & Gibbon, 2000). Contingency and contiguity are not mutually exclusive but rather involve distinct controlled variables at different levels.

Such relationships are traditionally analyzed using associative structures. Three basic elements can be identified in a typical behavioral experiment: a cue, an action, and an outcome (Bolles, 1972). In associationist theories of learning, these elements and the associations among them form the content of learning (Colwill & Rescorla, 1990; Dickinson, 1997; Mackintosh, 1974; Rescorla, 1980). For example, the formation of stimulus––outcome (S–O) associations is thought to be responsible for Pavlovian learning, whereas the formation of action–outcome (A–O) associations is thought to be the basis for instrumental learning.

Associationist models, however, lack any working model of performance. An association, being one-dimensional, cannot explain the timing, duration, direction, and other features of behavior. Nevertheless, associative structures are still useful as a heuristic for classifying behavioral observations. Figure 10.5 summarizes the traditional classification of associative structures and their relationship to behavioral feedback functions.

A key prediction from the associationist analysis is that independent manipulation of outcome (O) value should affect performance guided by S–O or A–O associations (Dickinson, 1997). To test which of the associative structures is responsible for performance, a common method is post-training outcome devaluation, induced either by induction of taste aversion to the food reward during training or by specific satiety of the food. Both will immediately reduce desire for the reward, as reflected in consumption data. To test whether the outcome representation is acquired during learning and used in guiding behavioral performance, a probe test can be conducted under extinction conditions in the absence of reward feedback.

To assess which relationship is being controlled, contingencies between the specific elements of associative structures can be manipulated. In Pavlovian contingency (S–O) degradation, free rewards are introduced independently of the CS presentation. In instrumental contingency (A–O) degradation, free rewards are presented independently of lever pressing. A–O contingency computation requires monitoring of volitional actions to compute the delta p-value, the difference between the probability of an outcome given action and the probability of an outcome given no action.

The performance of Pavlovian CRs is usually reduced by outcome devaluation or S–O contingency degradation but is less affected by manipulation of the A–O contingency. For example, Sheffield tested whether salivation in Pavlovian conditioning was controlled by the A–O contingency or by the S–O contingency (Sheffield, 1965). In his experiment, dogs received pairings between a tone and a food reward. But the production of the salivation CR during the tone canceled the food delivery on that trial. While maintaining the tone-food (S–O) contingency, the salivation-food (A–O) contingency was abolished. If salivation was controlled by its relationship with food, then the dogs should not salivate to the tone. The dogs acquired and maintained salivation to the

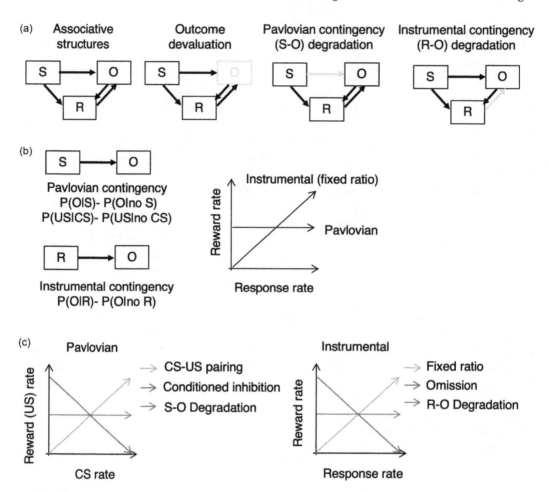

FIGURE 10.5 Feedback polarity and contingency. (a) Standard associative structures in the analysis of conditioning experiments. S, stimulus; R, response; O, outcome. (b) Defining Pavlovian and instrumental contingencies in terms of probabilities. On the right, the feedback function is shown for these two types of conditioning. Note that reward rate is here defined as the rate of delivery, not the rate of reward collection. (c) Relationship between associative structures and feedback function. Left, Pavlovian contingency can be defined as the correlation between the rate of reward and the rate of the conditioned stimulus. This correlation is altered by manipulations of the Pavlovian (S–O or CS-US) contingency. Right, feedback function associated with different manipulations of the instrumental (R–O) contingency. Only instrumental conditioning sets up a feedback function relating response rate and the rate of reward delivery, but some feedback function (not shown here) still exists even in Pavlovian conditioning, as behavior is still required for obtaining or consuming reward, even if it is not required for causing reward delivery.

tone, even though salivation canceled most of the food they could have obtained by not salivating. These results are therefore similar to those discussed earlier on reversal in feedback polarity, since the A–O contingency is an example of a feedback function.

10.7 GOAL-DIRECTED ACTIONS

In instrumental conditioning, if an action is mediated by the A–O association, degradation of the A–O contingency or outcome devaluation is expected to reduce performance (Colwill & Rescorla, 1986; Dickinson, 1989, 1994). Colwill and Rescorla trained rats to perform two different actions for two distinct food rewards separately. After training, one of the rewards was devalued by prefeeding, and the subject was then given a probe test, during which both levers were available, but no reward was delivered. If performance is governed by the A–O representation, then we would expect a reduction in the action earning the devalued reward compared to the action earning the non-devalued reward. Indeed, they found that rats selectively reduced lever pressing, leading to the devalued food reward.

As discussed in Chapter 9, instrumental conditioning can be modeled using a reward rate controller on top of an action repetition controller. Devaluation by specific satiety reduces the homeostatic error that is the source of reference signals to the reward rate controller, reducing reward rate reference (demand) and rate of pressing. On the other hand, degradation is a manipulation of the feedback function. Free background rewards reduce the rewards that must be earned by action (Chapter 9). Under this condition, one would also expect a reduction in performance that is proportional to the background reward rate. Consequently, the rate of lever pressing will be proportional to the difference between desired reward rate and free reward rate multiplied by the ratio requirement. Of course this model only applies to cases where the effort or cost of pressing is negligible, but it sheds light on the effect of manipulations like instrumental contingency degradation.

According to Dickinson, for an action to be considered goal-directed, it must meet two criteria (Dickinson, 1989). First, the subject must have knowledge of the relationship between action and outcome (belief criterion). This could be tested using contingency manipulations like contingency degradation. Second, the subject must have desire for the outcome in question (desire criterion). This can be tested using outcome devaluation. These two criteria exclude target-directed behavior, such as sign tracking or Pavlovian approach behavior. Approach behavior certainly satisfies the desire criterion, as performance is reduced by devaluation. Yet it does not meet the belief criterion, as performance is relatively insensitive to A–O manipulations. It is perhaps more informative to rename the belief criterion "agency criterion." In humans, judgments of action efficacy as well as rate of operant performance strongly depend on the A–O contingency (Chatlosh, Neunaber, & Wasserman, 1985).

According to Dickinson, goal-directed action selection is mediated by bidirectional associations between action and outcome. The outcome–action (O–A) mechanism resembles the classic ideomotor theory (James, 1890), in which action is initiated by an internal representation of the desired outcome. For example, when craving chocolate, the chocolate representation appears first and activates the specific action that will earn chocolate. In the forward association between action and outcome (A–O), one can retrieve the expected outcome given the performance of some action, predicting the consequences of one's own actions.

A–O and O–A associations are similar to the forward and inverse models in theories of motor control, which we discussed in Chapter 5. But unlike these models, in the control hierarchy, there is no need for detailed predictions of environmental properties. The representations of actions and outcomes are abstract because they are reference signals from higher levels of the control hierarchy. They initiate multiple comparisons at multiple levels to generate the needed signals at lower levels to generate actions.

10.8 NEURAL BASIS OF ACTION–OUTCOME LEARNING

In human neuroimaging work, learning about actions and their reward consequences involves the associative striatum (O'Doherty et al., 2004; Tricomi, Delgado, & Fiez, 2004). For example, Tricomi and colleagues presented reward and punishment either after a predictive cue or following a button press by the subject. The caudate nucleus was activated when the subjects believed their action (button press) caused the outcome, regardless of the outcome valence (Tricomi et al., 2004). Studies also found associative striatal activity related to action–contingent rewards (Delgado, Nystrom, Fissell, Noll, & Fiez, 2000; Shohamy et al., 2004). On the other hand, more passive forms of appetitive learning with only S–O contingencies appear to activate the ventral striatum (O'Doherty et al., 2004). Medial prefrontal cortex tracked local changes in action–outcome correlations as well as subjective ratings of the causal efficacy of actions (Tanaka, Balleine, & O'Doherty, 2008).

Activity in the ventromedial prefrontal cortex and the right anterior caudate was correlated with the probability of reward for some action. Activity in the inferior frontal gyrus and the left posterior caudate, on the other hand, varied with the probability of noncontingent rewards. Thus, distinct corticostriatal circuits may support estimation of contingent and noncontingent rewards. A–O contingency (delta p) was also negatively correlated with activity in the parietal association cortex (inferior and superior parietal lobules) and the frontal cortex (left middle frontal gyrus) (Liljeholm, Tricomi, O'Doherty, & Balleine, 2011). Recent work found representation of the action–contingent outcome in the medial prefrontal cortex and representation of the action-independent outcome in the anterior cingulate cortex (Morris, Dezfouli, Griffiths, Le Pelley, & Balleine, 2022). These results suggest that the associative cortico-BG network detects and segregates the unique causal effects of actions from their context. The segregation of outcomes due to self-action (reafference) or other causes in the environment (exafference) is critical for determining agency (von Holst & Mittelstaedt, 1950). A comparison between these two variables can generate a signal proportional to the contingency, which can then be converted to a reference signal for action repetition.

10.8.1 Striatal Activity Modulated by Reward Expectancy

Primate studies found that activity in the caudate nucleus was modulated by expected reward probability and magnitude (Hikosaka, Sakamoto, & Usui, 1989; Hollerman & Schultz, 1998). In experiments performed by Hikosaka and colleagues, monkeys had to attend to some visual cue, remember its location, and generate a saccade to that location when instructed to do so. In the one-direction rewarded condition, the monkeys were rewarded for correct saccades in one of four possible directions (left, right, up, down). Some caudate neurons showed sustained memory-related activity and phasic activity before saccades that was higher for rewarded directions. Many caudate neurons were preferentially active when one of the target positions (usually contralateral to the recording site) was rewarded (Kawagoe, Takikawa, & Hikosaka, 1998; Takikawa, Kawagoe, & Hikosaka, 2002). However, these studies did not examine behavioral changes related to manipulations of reward probability and magnitude.

Studies also found that the activity of reward-dependent caudate neurons were correlated with saccade velocity and latency (Itoh et al., 2003; Watanabe, Lauwereyns, & Hikosaka, 2003). Latencies were shorter and saccade velocity was higher on trials with large rewards compared to trials with small rewards (Takikawa, Kawagoe, Itoh, Nakahara, & Hikosaka, 2002). Some neurons fire before saccade execution, but their activity can predict saccade velocity and latency, especially for contralateral saccades. Firing rate was correlated with velocity not only in the peri-saccadic period but also before the cue presentation. In contrast, reward-depressed neurons did not show any correlation between firing rate and saccade velocity or latency. These results suggest that the outcome representations may activate these caudate neurons, which send velocity commands for eye movements.

The correlation between striatal activity and saccade velocity cannot be explained by reward expectancy since it is present for both rewarded and unrewarded saccades. One possibility is that the activity reflects reference input to saccade-generating modules, but saccades are prevented downstream through active inhibition. During the task, the monkey must wait until the fixation period is over before moving to the remembered target. The inhibition is only removed when the fixation period ends. Such activity may signal the maintenance of the desired target location representation and prime the relevant action modules in the striatum.

These results suggest that representations of future reward may potentiate specific action representations in the associative striatum, with selectivity for actions directed toward contralateral space (Hikosaka, Nakamura, & Nakahara, 2006). Inputs to the associative striatum can signal both discriminative stimulus and outcome representation. After learning, the presentation of discriminative stimuli can prime striatal action modules, producing anticipatory activity before reward delivery.

It should be noted that most studies on the neural correlates of reward expectancy fail to control the effort requirement. Nor did they explicitly manipulate the outcome value or A–O contingency.

Actions and Goals

To better understand the contributions of the associative cortico-BG network to the generation of goal-directed actions, we must turn to studies that directly manipulated outcome value and instrumental contingency.

10.8.2 ASSOCIATIVE CORTICO-BG NETWORK AND A–O LEARNING

Balleine and Dickinson showed that goal-directed actions, as defined by sensitivity to outcome devaluation and contingency degradation, were abolished by lesions of the prelimbic area in the medial prefrontal cortex (Balleine & Dickinson, 1998). Rats were trained to perform two distinct actions for two different reward outcomes on ratio schedules. Prelimbic cortex lesions before training did not abolish the acquisition of lever pressing. After training, devaluation was achieved by inducing satiety as one of the outcomes and testing performance on a probe test in which both actions were available but no reward was presented. During the probe test, rats with prelimbic cortical lesions were not sensitive to outcome devaluation. Whereas control rats selectively reduced performance of the action earning the devalued reward, lesioned rats did not. Likewise, lesioned rats also failed to reduce performance of the action leading to the degraded outcome.

The prelimbic region is a major source of projections to the limbic and associative striatal areas. Lesions of the mediodorsal thalamus, which strongly projects to the prelimbic cortex, can also produce insensitivity to outcome devaluation and contingency degradation (Corbit & Balleine, 2003; Corbit, Muir, & Balleine, 2003; Killcross & Coutureau, 2003). On the other hand, the nucleus accumbens, which receives strong prelimbic projections, is not necessary for A–O learning (Corbit, Muir, & Balleine, 2001). Lesions of the accumbens shell do not alter sensitivity to outcome devaluation or to instrumental contingency degradation. Lesions of the accumbens core reduce sensitivity to devaluation without impairing the sensitivity to contingency degradation.

Moreover, lesions of the nucleus accumbens often resulted in reduced lever pressing, yet they did not significantly impair the ability to select actions based on outcome value or action-outcome contingency. Such results suggest a dissociation between the motivational influence by which effort is exerted through the reward rate control system and the recruitment of the appropriate action repetition controller (Chapter 9). The output of the reward rate controller may be reduced due to accumbens lesions, but it is still able to recruit the appropriate instrumental action previously learned to reduce the specific reward rate error.

10.8.3 POSTERIOR DORSOMEDIAL STRIATUM

The key striatal node that is critical for A–O learning is the posterior dorsomedial striatum (pDMS), another major target of prelimbic cortical projections. Pre-training or post-training lesions of the pDMS rendered performance insensitive to instrumental contingency degradation or outcome devaluation (Yin, Ostlund, Knowlton, & Balleine, 2005). As shown in Figure 10.6, rats with pDMS, anterior dorsomedial striatum (aDMS), or sham lesions were trained to perform two actions for two distinct outcomes (press one lever for food pellets and a different lever for sucrose solution). One of the rewards was devalued by pre-feeding. On the probe test, sham control rats and aDMS rats selectively reduced pressing for the devalued outcome, but rats with pDMS lesions pressed both levers equally, indicating that their performance was not regulated by anticipation of a specific outcome. Inactivation of the pDMS abolished sensitivity to contingency degradation, while inactivation of the anterior DMS had no effect. When the contingency was selectively degraded by presentation of noncontingent rewards, the pDMS-lesioned rats were not able to selectively reduce pressing on the degraded lever.

In these experiments, the key probe test was conducted without rewards in a short extinction session. When rewards are omitted, rats are forced to rely on prior learning and their memory of the contingency. When rewards are delivered, performance can be driven by either prior learning or new learning, conflating learning and performance. It is therefore informative to compare

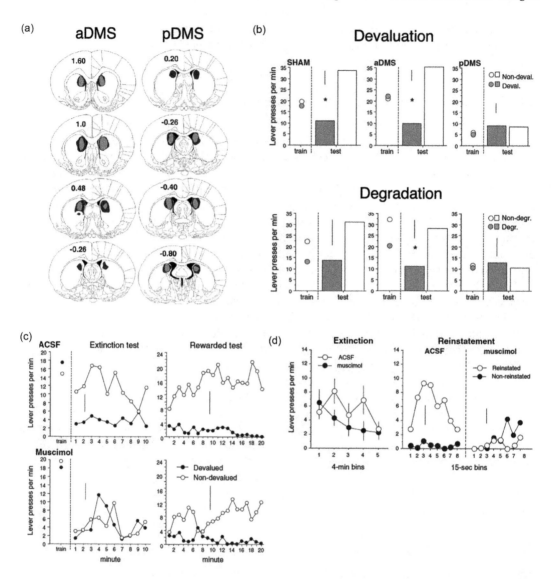

FIGURE 10.6 Dorsomedial striatum and action–outcome (A–O) learning. (a) Schematic illustration of excitotoxic (*N*-methyl-D-aspartate, NMDA) lesion placement. Numbers indicated mm from the bregma (anterior–posterior). (b) Results from outcome devaluation and instrumental contingency degradation tests. The rate of lever pressing on the last day of training is shown on the left and during the extinction test immediately after devaluation treatment on the right. SHAM, sham-lesioned control group; aDMS, anterior dorsomedial striatum; pDMS, posterior dorsomedial striatum. (c) Muscimol inactivation of pDMS on the devaluation test abolished sensitivity to outcome devaluation on the probe test (extinction) but not on the rewarded test. (d) pDMS inactivation also abolished outcome-selective reinstatement. After extinction of lever pressing, a single reward previously earned by one of the levers was presented, resulting in a selective increase in lever pressing on that lever in the control group (ACSF, artificial cerebrospinal fluid infusion into pDMS). However, such reinstatement of instrumental performance was not shown when the pDMS was inactivated with muscimol. Each vertical line represents 1 standard error of the difference of the means, a measure of within-subject variance.

performance on the probe test with that on a separate reward test. When given reward feedback, rats with pDMS lesions were still able to reduce pressing of the lever, earning the devalued reward just like controls. They were still able to learn to discriminate between the consequences of the two actions. This observation shows that lesions of the pDMS prevented rats from retrieving the A–O contingency from memory.

Actions and Goals

To test whether specific outcomes can recruit the actions associated with them in lesioned rats, a reinstatement test was performed. Once lever pressing has stopped completely after extinction, a single presentation of the outcome can selectively reinstate the extinguished action. This effect is known as outcome-specific reinstatement, showing that animals can use the outcome to retrieve the specific action associated with it. Prior to the reinstatement test, the level of performance on the two levers during the extinction phase was close to zero and did not differ between the pDMS inactivation group (muscimol) and vehicle control group (Yin, Ostlund, et al., 2005). After the introduction of a single reward, control rats selectively increased their pressing of the lever originally earning that outcome. In contrast, the pDMS inactivation group failed to show reinstatement of instrumental performance.

Yin et al. also tested whether acquisition of new A–O contingencies requires striatal NMDA receptor activation (Yin, Knowlton, & Balleine, 2005). They first trained rats to press two levers (left or right) for sucrose solution. After this pre-training phase, they infused the NMDA antagonist APV into the pDMS just before a new learning session, during which two distinct new outcomes were associated with the two levers: pressing one lever delivered food pellets and pressing the other delivered fruit punch. Thus, under the influence of NMDAR blockade in the pDMS, rats were exposed to two new A–O contingencies. Learning during this session was tested the next day by outcome devaluation, sating the animals on either the pellets or fruit punch before assessing their performance on the two levers of extinction. During this probe session, no drug was injected, so the rats were tested on their memory of the contingencies based on what they had learned previously during the training session. Rats in the control group, which received artificial cerebrospinal fluid injections during learning, selectively reduced pressing on the lever that earned the now-devalued outcome, indicating that they had learned the new A–O relationship. In contrast, rats that received NMDAR blockade did not show sensitivity to devaluation. Blocking NMDA receptors in the pDMS during new A–O contingency learning selectively disrupted sensitivity to devaluation of this outcome during the probe test. Blockade of NMDA receptors 1 hour after the acquisition session or just before the probe test had no significant effect. NMDARs in the pDMS therefore appear to be critical for new action–outcome learning.

The prelimbic cortex–pDMS–SNr–medial dorsal thalamus circuit appears to be critical for the acquisition of A–O contingencies. Lesions of different components of this associative cortico-BG circuit can render behavior less sensitive to devaluation and degradation (Balleine & Dickinson, 1998; Corbit & Balleine, 2003; Corbit et al., 2003; Yin, Ostlund, et al., 2005). The medial prefrontal cortex (including the prelimbic cortex) projects to both the limbic and associative regions of the striatum, but it is the corticostriatal projections to the pDMS that are critical for A–O learning. Disconnection studies showed that such learning depends on prelimbic projections to the pDMS, but not on prelimbic projections to the accumbens (Hart, Bradfield, & Balleine, 2018; Hart, Bradfield, Fok, Chieng, & Balleine, 2018). Thus, A–O learning appears to require a specific subset of corticostriatal projections from the medial prefrontal cortex.

The prelimbic–pDMS pathway originates mostly in Layer 5 of the prelimbic cortex. This layer contains both intratelencephalic tract (IT) neurons that project bilaterally to the striatum and pyramidal track (PT) neurons that project ipsilaterally to the brain stem and spinal cord with collaterals to the striatum (see Chapter 2 for a discussion of these two corticostriatal pathways) (Shepherd, 2013). Hart et al. showed that the prelimbic IT neurons are critical for A–O learning (Figure 10.7). Disconnecting the PT connections but preserving the crossed connections from IT neurons has no effect on goal-directed learning, as assessed using an outcome devaluation test. However, selectively inhibiting the IT prelimbic-pDMS projection blocked goal-directed action.

10.8.4 Direct and Indirect Pathways

To examine the contributions of the direct and indirect pathways in the pDMS to instrumental learning, Balleine and colleagues used chemogenetics to inhibit the activity of dSPNs and iSPNs in the pDMS during acquisition and then assessed the content of learning using devaluation. Inhibiting

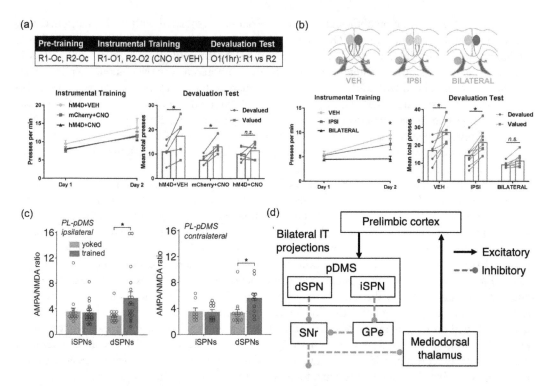

FIGURE 10.7 Dorsomedial striatum (DMS) is a key site for goal-directed learning and plasticity. (a) Inactivating the prelimbic-posterior DMS (pDMS) projection using chemogenetics blocks A–O (A–O) learning. Top, experimental design. R1, action 1; R2, action 2; Oc, common outcome earned by either action. CNO, clozapine-N-oxide. When the prelimbic inputs to pDMS are inactivated with hM4D, there is no effect on the rate of acquisition, but when tested after devaluation without inhibition (no CNO), rats with PL–pDMS inactivation did not reduce pressing on the lever, earning the devalued outcome. (b) Selective inactivation of IT neurons in the prelimbic cortex during training. From Hart, Bradfield, Fok, Chieng, and Balleine (2018) with permission. (c) With instrumental learning, projections from prelimbic cortex (PL) to pDMS show significant increases in the AMPA/NMDA ratio (a measure of glutamatergic synaptic strength) in direct pathway spiny projection neurons (dSPNs), but not in indirect pathway spiny projection neuron (iSPNs). From Fisher, Ferguson, Bertran-Gonzalez, and Balleine (2020) with permission. (d) Schematic showing the key circuit necessary for A-O learning.

dSPN activity during training prevented acquisition of the A–O contingency, as indicated by reduced sensitivity to devaluation on the choice test, yet blocking the indirect pathway had no clear effect (Peak, Chieng, Hart, & Balleine, 2020). However, while iSPN inhibition failed to prevent initial learning, it impaired the updating of the instrumental contingency. Rats were exposed to the same A–O contingencies as during initial training, but with only one of the two levers rewarded at a time, and these contingencies were reversed frequently within a session. Whereas choice behavior of control rats closely followed the new contingencies, inhibition of iSPNs in the pDMS impaired the ability to switch to the new rewarded lever following a contingency change. Compared to control rats, these rats persisted in pressing the old rewarded lever. These results suggest that dSPNs in the pDMS are critical for goal-directed learning and for initiation of the learned action, whereas iSPNs contribute to flexible updating of the learned contingency. Further supporting this conclusion, lever press training induced an increase in activity-related gene expression (immediate-early gene Zif268) in dSPNs, suggesting that dSPNs were preferentially activated by A–O learning. Increased activity was found to be lateralized with respect to the position of the lever. More activation was found on the side contralateral to the lever. Unilateral inhibition of dSPNs reduced pressing on the lever located on the contralateral side of the inhibited striatum. This observation suggests that A–O learning preferentially recruits dSPNs for actions in contralateral space.

10.8.5 Synaptic Plasticity in pDMS

The behavioral findings just reviewed are in agreement with *in vitro* brain slice results on synaptic plasticity in the striatum. As reviewed in Chapter 4, in brain slices, striatal LTP is most easily found in the dorsomedial striatum and usually depends on NMDA receptor and D1 receptor activation (Calabresi, Pisani, Mercuri, & Bernardi, 1992; Dang et al., 2006; Kerr & Wickens, 2001; Yin, Park, Adermark, & Lovinger, 2007). *Ex vivo* experiments also implicated corticostriatal plasticity in instrumental learning. After limited instrumental training, brain slices were taken from rats, and training-induced changes in plasticity were assessed based on light-evoked activity in both dSPNs and iSPNs. Axon terminals of prelimbic cortical projections to the pDMS were activated using optogenetic stimulation, and the SPN response was recorded using voltage clamp whole-cell patch-clamp recording. AMPA/NMDA ratios, a measure of glutamatergic synaptic strength, increased on dSPNs bilaterally following instrumental training (Fisher et al., 2020). These results suggest that goal-directed learning potentiates prelimbic inputs to dSPNs in the pDMS.

Corticostriatal LTP requires activation of extracellular signal-regulated kinase (ERK), a member of the mitogen-activated protein kinase pathway (Calabresi et al., 2000). The ERK pathway, which can be activated by dopamine and glutamate receptor signaling, is known to contribute to synaptic plasticity (Hawes, Gillani, Evans, Benkert, & Blackwell, 2013; Sgambato, Pages, Rogard, Besson, & Caboche, 1998; Valjent, Caboche, & Vanhoutte, 2001). Activation of D1 DA receptors and NMDA glutamate receptors, which is critical for striatal LTP, can initiate ERK signaling. ERK activity is increased in dSPNs early in instrumental training. Pharmacological blockade of ERK phosphorylation in the pDMS also impaired A–O learning, rendering subsequent probe test performance insensitive to devaluation (Shiflett & Balleine, 2011; Shiflett, Brown, & Balleine, 2010).

10.8.6 The Role of DA in Goal-Directed Actions

Recently, Jin and colleagues examined DA release in the mouse dorsomedial striatum evoked by direct optogenetic stimulation of the SNc DA neurons (Hollon et al., 2021). They trained mice to press a lever on a fixed ratio schedule for optogenetic stimulation of SNc DA neurons. They then used a within-subject design to compare DA evoked under self-stimulation and passive stimulation conditions. In the self-stimulation phase, mice earned optogenetic stimulation by pressing a lever. In the subsequent passive playback phase, the levers were retracted, and the mice received noncontingent stimulations. Using fast-scan cyclic voltammetry to measure DA in the DMS, they showed that self-stimulated DA release was reduced compared with the DA evoked by passive playback stimulation, suggesting that lever pressing can reduce DA signaling. The activity of DA neurons was also reduced in the contingent condition, showing that the reduction in DA signaling as measured by striatal voltammetry was at least partly due to reduced firing rates of SNc DA neurons. Action-induced suppression was not observed when the mouse pressed the inactive lever, so inhibition appeared to be specific to the reinforced action.

It remains unclear how self-generated action may suppress DA firing. One possibility is direct inhibitory projections from the striosome compartment to DA neurons, but at present there is no data on the functional role of these projections. Another candidate is the indirect pathway, which can have a net inhibitory effect on DA neurons (iSPN–GPe–SNr–SNc).

One caveat is that, since voltammetry measures overall DA signaling in a given area, it can only reveal net changes in DA concentration. It lacks the spatial resolution to see if there are differences in DA signaling within a particular striatal area. Consequently, these results cannot exclude the possibility that the efference copy of the instrumental action may suppress DA signaling to other channels while promoting signaling to the recently active channel. This possibility remains to be tested.

In mice performing a reward-guided task on a running wheel, Hamid and colleagues showed that striatal DA signaling shows distinct wave-like patterns of activation across large striatal regions following reward delivery (Hamid, Frank, & Moore, 2021). Interestingly, when the task was running-contingent (instrumental), the wave traveled from limbic to sensorimotor regions. Even before reward delivery, there was significant ramping of DA activity in the dorsomedial striatum. On the other hand, during the "Pavlovian" task, when the reward was not contingent upon running behavior, the DA wave traveled in the opposite direction from DLS to DMS. When the contingency was shifted during a session, the direction of DA wave propagation was also shifted accordingly. In principle, this pattern of DA signaling could be achieved using disinhibition via striato-nigro-striatal projection (Haber, Fudge, & McFarland, 2000). As described in Chapter 9, the limbic network is in a position to influence DA signaling in the associative and sensorimotor networks. When generating instrumental actions, the limbic network controlling reward rate is expected to be the lead level in the motivational hierarchy, with outputs that affect the associative and sensorimotor networks. In contrast, when reward is delivered noncontingently, the sensorimotor network may be activated by the salient sensory inputs first. The pattern of DA wave propagation is therefore in agreement with this model.

10.9 BG CONTRIBUTIONS TO NEUROPROSTHETIC CONTROL

So far, our examination of instrumental learning has focused on studies using instrumental conditioning, in which overt lever presses are rewarded. However, similar learning could occur even when there is no explicit action requirement. Carmena, Costa, and colleagues studied the role of the BG in the instrumental control of neural activity. Animals received food rewards when they altered neural activity in specific brain regions, a procedure that was first introduced to study feedback modulation of motor cortical activity (Fetz, 1969). This approach has often been used in brain–machine interface research, with applications in neuroprosthetic control. For example, reward was delivered when animals changed neural activity to move a cursor to one of two targets. Each target was associated with a distinct food reward. No overt instrumental action was required, yet animals could readily learn to alter neural activity systematically. Their "neural performance" was also highly sensitive to instrumental contingency degradation. Moreover, striatal neurons systematically changed their activity with learning, and activity of neurons in motor cortex and the striatum also became more coherent. This pattern suggests that corticostriatal plasticity makes it possible for cortical inputs to activate the relevant striatal neurons more reliably with training. Learning to control motor cortical activity also requires striatal plasticity.Genetic deletion of striatal NMDARs disrupted learning on this task, similar to findings on standard instrumental learning (Koralek, Jin, Long II, Costa, & Carmena, 2012).

In addition, both mice and rats could learn to modulate spike activity in the primary visual cortex in a goal-directed manner. Even in the absence of visual stimuli presentation, visual cortex neurons can be volitionally modulated, and this effect depends on the associative BG circuit. Inhibition of the DMS impaired such learning, but not production of learned cortical patterns (Neely, Koralek, Athalye, Costa, & Carmena, 2018). Neural activity in the DMS that receives input from the primary visual cortex also changed with learning.

These results suggest that instrumental training can alter the state of a cortical region via the associative cortico-BG network, even without requiring overt actions. Although no overt movements were detected in this task, it is unclear whether there are subtle behavioral effects associated with neuroprosthetic training. It is possible that subtle movements may play a role in altering visual cortical activity, but sensitive behavioral measures will be needed to detect these changes.

Visual cortical inputs that are associated with error reduction or goal achievement can somehow repeat themselves via BG activation, achieving the desired perceptual state. This mechanism is similar to the imagination mode discussed in Chapter 8. According to the present model, it requires BG projections to the thalamocortical network to alter cortical activity. Whether the input in question is auditory or visual, this form of learning also requires the DMS and presumably, plasticity at the corticostriatal synapses.

10.10 SUMMARY

Relationships are higher-level representations that guide behavior. In the control of spatial proximity, the simplest possible spatial relationship, representation of the distance between self and target, can be used to command self-velocity. Classic studies showed that lesions to the caudate nucleus produce compulsive approach behavior, implicating the caudate nucleus in spatial proximity control. In mice, the striatal FSI–SPN feedforward inhibition circuit is critical for continuous pursuit behavior, by converting the distance relationship variable as represented by FSIs to instantaneous velocity commands from dSPNs.

Animals can learn to approach any arbitrary goal, including neutral stimuli that are predictors of reward. Such behavior is guided by the stimulus–reward contingency, which can be acquired through learning and depends on synaptic plasticity in the nucleus accumbens core. More broadly, the limbic cortico-BG network mediates a conditioned approach to predictors of reward. The development of such an anticipatory approach requires stimulus–reward learning, but not knowledge of the instrumental contingency. On the other hand, goal-directed instrumental actions are governed by the desire for a specific outcome as well as the belief that a particular action will lead to that outcome. These criteria can be experimentally tested using instrumental contingency degradation and outcome devaluation. Using such behavioral assays, studies have shown that the associative cortico-BG network, especially the pDMS and associated circuits, is critical for learning action–outcome contingencies.

REFERENCES

Balleine, B. W., & Dickinson, A. (1998). Goal-directed instrumental action: Contingency and incentive learning and their cortical substrates. *Neuropharmacology, 37*(4–5), 407–419.

Basso, M. A., Pokorny, J. J., & Liu, P. (2005). Activity of substantia nigra pars reticulata neurons during smooth pursuit eye movements in monkeys. *European Journal of Neuroscience, 22*(2), 448–464.

Bolles, R. (1972). Reinforcement, expectancy, and learning. *Psychological Review, 79*, 394–409.

Bowen, F. (1969). Visuomotor deficits produced by cryogenic lesions of the caudate. *Neuropsychologia, 7*(1), 59–65.

Brog, J. S., Salyapongse, A., Deutch, A. Y., & Zahm, D. S. (1993). The patterns of afferent innervation of the core and shell in the "accumbens" part of the rat ventral striatum: Immunohistochemical detection of retrogradely transported fluoro-gold. *Journal of Comparative Neurology, 338*(2), 255–278.

Brown, P. L., & Jenkins, H. M. (1968). Auto-shaping the pigeon's key peck. *Journal of the Experimental Analysis of Behavior, 11*, 1–8.

Bussey, T. J., Everitt, B. J., & Robbins, T. W. (1997). Dissociable effects of cingulate and medial frontal cortex lesions on stimulus-reward learning using a novel pavlovian autoshaping procedure for the rat: Implications for the neurobiology of emotion. *Behavioral Neuroscience, 111*(5), 908.

Butcher, L. L., & Fox, S. S. (1968). Motor effects of copper in the caudate nucleus: Reversible lesions with ion-exchange resin beads. *Science, 160*(3833), 1237–1239.

Buzsáki, G. (1982). The "where is it?" reflex: Autoshaping the orienting response. *Journal of the Experimental Analysis of Behavior, 37*(3), 461–484.

Calabresi, P., Gubellini, P., Centonze, D., Picconi, B., Bernardi, G., Chergui, K., … Greengard, P. (2000). Dopamine and camp-regulated phosphoprotein 32 kda controls both striatal long-term depression and long-term potentiation, opposing forms of synaptic plasticity. *Journal of Neuroscience, 20*(22), 8443–8451.

Calabresi, P., Pisani, A., Mercuri, N. B., & Bernardi, G. (1992). Long-term potentiation in the striatum is unmasked by removing the voltage-dependent magnesium block of nmda receptor channels. *European Journal of Neuroscience, 4*(10), 929–935.

Cardinal, R. N., Parkinson, J. A., Lachenal, G., Halkerston, K. M., Rudarakanchana, N., Hall, J., … Everitt, B. J. (2002). Effects of selective excitotoxic lesions of the nucleus accumbens core, anterior cingulate cortex, and central nucleus of the amygdala on autoshaping performance in rats. *Behavioural Neuroscience, 116*(4), 553–567.

Chatlosh, D., Neunaber, D., & Wasserman, E. (1985). Response-outcome contingency: Behavioral and judgmental effects of appetitive and aversive outcomes with college students. *Learning and Motivation*, *16*(1), 1–34.

Chung, P.-W., Moon, H.-S., Song, H. S., & Kim, Y. B. (2006). Ocular motor apraxia after sequential bilateral striatal infarctions. *Journal of Clinical Neurology*, *2*(2), 134–136.

Clark, J. J., Hollon, N. G., & Phillips, P. E. (2012). Pavlovian valuation systems in learning and decision making. *Current Opinion in Neurobiology*, *22*(6), 1054–1061.

Colwill, R. M., & Rescorla, R. A. (1986). Associative structures in instrumental learning. In: Bower, G. (Ed.), *The Psychology of Learning and Motivation* (vol. 20, pp. 55–104). New York: Academic Press.

Colwill, R. M., & Rescorla, R. A. (1990). Evidence for the hierarchical structure of instrumental learning. *Animal Learning & Behavior*, *18*(1), 71–82.

Corbit, L. H., & Balleine, B. W. (2003). The role of prelimbic cortex in instrumental conditioning. *Behavioural Brain Research*, *146*(1–2), 145–157.

Corbit, L. H., Muir, J. L., & Balleine, B. W. (2001). The role of the nucleus accumbens in instrumental conditioning: Evidence of a functional dissociation between accumbens core and shell. *Journal of Neuroscience*, *21*(9), 3251–3260.

Corbit, L. H., Muir, J. L., & Balleine, B. W. (2003). Lesions of mediodorsal thalamus and anterior thalamic nuclei produce dissociable effects on instrumental conditioning in rats. *European Journal of Neuroscience*, *18*(5), 1286–1294.

Dalley, J. W., Laane, K., Theobald, D. E., Armstrong, H. C., Corlett, P. R., Chudasama, Y., & Robbins, T. W. (2005). Time-limited modulation of appetitive pavlovian memory by d1 and nmda receptors in the nucleus accumbens. *Proceedings of the National Academy of Sciences*, *102*(17), 6189–6194.

Dang, M. T., Yokoi, F., Yin, H. H., Lovinger, D. M., Wang, Y., & Li, Y. (2006). Disrupted motor learning and long-term synaptic plasticity in mice lacking nmdar1 in the striatum. *Proceedings of the National Academy of Sciences*, *103*(41), 15254–15259.

Day, J. J., & Carelli, R. M. (2007). The nucleus accumbens and pavlovian reward learning. *Neuroscientist*, *13*(2), 148–159.

Day, J. J., Roitman, M. F., Wightman, R. M., & Carelli, R. M. (2007). Associative learning mediates dynamic shifts in dopamine signaling in the nucleus accumbens. *Nature Neuroscience*, *10*(8), 1020–1028.

Day, J. J., Wheeler, R. A., Roitman, M. F., & Carelli, R. M. (2006). Nucleus accumbens neurons encode pavlovian approach behaviors: Evidence from an autoshaping paradigm. *European Journal of Neuroscience*, *23*(5), 1341–1351.

Delgado, M. R., Nystrom, L. E., Fissell, C., Noll, D., & Fiez, J. A. (2000). Tracking the hemodynamic responses to reward and punishment in the striatum. *Journal of Neurophysiology*, *84*(6), 3072–3077.

Dickinson, A. (1989). Expectancy theory in animal conditioning. In: Klein, S. B., & Mowrer, R. R. (Eds.), *Contemporary Learning Theories* (pp. 279–308). Hillsdale, NJ: Lawrence Erlbaum Associates.

Dickinson, A. (1994). Instrumental conditioning. In: Mackintosh, N. J. (Ed.), *Animal Learning and Cognition* (pp. 45–79). Orlando, FL: Academic.

Dickinson, A. (1997). Bolles's psychological syllogism. In: Bouton, M. E., & Fanselow, M. S. (Eds.), *Learning, Motivation, and Cognition* (pp. 345–367). Washington, DC: APA.

Fetz, E. E. (1969). Operant conditioning of cortical unit activity. *Science*, *163*(3870), 955–958.

Fisher, S. D., Ferguson, L. A., Bertran-Gonzalez, J., & Balleine, B. W. (2020). Amygdala-cortical control of striatal plasticity drives the acquisition of goal-directed action. *Current Biology*, *30*(22), 4541–4546. e4545.

Flagel, S. B., Clark, J. J., Robinson, T. E., Mayo, L., Czuj, A., Willuhn, I., … Akil, H. (2010a). A selective role for dopamine in stimulus-reward learning. *Nature*, *469*(7328), 53–57. doi: 10.1038/nature09588.

Flagel, S. B., Robinson, T. E., Clark, J. J., Clinton, S. M., Watson, S. J., Seeman, P., … Akil, H. (2010b). An animal model of genetic vulnerability to behavioral disinhibition and responsiveness to reward-related cues: Implications for addiction. *Neuropsychopharmacology*, *35*(2), 388–400.

Floresco, S. B., Blaha, C. D., Yang, C. R., & Phillips, A. G. (2001). Dopamine d1 and nmda receptors mediate potentiation of basolateral amygdala-evoked firing of nucleus accumbens neurons. *Journal of Neuroscience*, *21*(16), 6370–6376.

Flowers, K. (1978). Lack of prediction in the motor behaviour of parkinsonism. *Brain*, *101*(1), 35–52.

Gallistel, C. R. (1990). *The Organization of Learning*: Cambridge, MA: MIT Press.

Gallistel, C. R., & Gibbon, J. (2000). Time, rate, and conditioning. *Psychological Review*, *107*(2), 289–344.

Haber, S. N., Fudge, J. L., & McFarland, N. R. (2000). Striatonigrostriatal pathways in primates form an ascending spiral from the shell to the dorsolateral striatum. *Journal of Neuroscience*, *20*(6), 2369–2382.

Hamid, A. A., Frank, M. J., & Moore, C. I. (2021). Wave-like dopamine dynamics as a mechanism for spatio-temporal credit assignment. *Cell, 184*(10), 2733–2749. e2716.

Hammond, L. J. (1980). The effect of contingency upon the appetitive conditioning of free-operant behavior. *Journal of the Experimental Analysis of Behavior, 34*(3), 297–304.

Hart, G., Bradfield, L. A., & Balleine, B. W. (2018). Prefrontal corticostriatal disconnection blocks the acquisition of goal-directed action. *Journal of Neuroscience, 38*(5), 1311–1322. doi: 10.1523/JNEUROSCI.2850-17.2017.

Hart, G., Bradfield, L. A., Fok, S. Y., Chieng, B., & Balleine, B. W. (2018). The bilateral prefronto-striatal pathway is necessary for learning new goal-directed actions. *Current Biology, 28*(14), 2218–2229 e2217. doi: 10.1016/j.cub.2018.05.028.

Hawes, S. L., Gillani, F., Evans, R. C., Benkert, E. A., & Blackwell, K. T. (2013). Sensitivity to theta-burst timing permits ltp in dorsal striatal adult brain slice. *Journal of Neurophysiology, 110*(9), 2027–2036.

Hearst, E., & Jenkins, H. M. (1974). *Sign-Tracking: The Stimulus-Reinforcer Relation and Directed Action*: Austin, TX: Psychonomic Society.

Hershberger, W. A. (1986). An approach through the looking glass. *Animal Learning & Behavior, 14*, 443–451.

Hikosaka, O., Nakamura, K., & Nakahara, H. (2006). Basal ganglia orient eyes to reward. *Journal of Neurophysiology, 95*(2), 567–584.

Hikosaka, O., Sakamoto, M., & Usui, S. (1989). Functional properties of monkey caudate neurons. Iii. Activities related to expectation of target and reward. *Journal of Neurophysiology, 61*(4), 814–832.

Hikosaka, O., & Wurtz, R. H. (1983). Visual and oculomotor functions of monkey substantia nigra pars reticulata. Iii. Memory-contingent visual and saccade responses. *Journal of Neurophysiology, 49*(5), 1268–1284.

Hollerman, J. R., & Schultz, W. (1998). Dopamine neurons report an error in the temporal prediction of reward during learning. *Nature Neuroscience, 1*(4), 304–309.

Hollon, N. G., Williams, E. W., Howard, C. D., Li, H., Traut, T. I., & Jin, X. (2021). Nigrostriatal dopamine signals sequence-specific action-outcome prediction errors. *Current Biology, 31*(23), 5350–5363. e5355.

Hunt, D. L., & Castillo, P. E. (2012). Synaptic plasticity of nmda receptors: Mechanisms and functional implications. *Current Opinion in Neurobiology, 22*(3), 496–508.

Itoh, H., Nakahara, H., Hikosaka, O., Kawagoe, R., Takikawa, Y., & Aihara, K. (2003). Correlation of primate caudate neural activity and saccade parameters in reward-oriented behavior. *Journal of Neurophysiology, 89*(4), 1774–1783.

James, W. (1890). *The Principles of Psychology* (vol. 1): New York: Henry Holt.

Kawagoe, R., Takikawa, Y., & Hikosaka, O. (1998). Expectation of reward modulates cognitive signals in the basal ganglia. *Nature Neuroscience, 1*(5), 411–416.

Kelley, A. E., Smith-Roe, S. L., & Holahan, M. R. (1997). Response-reinforcement learning is dependent on n-methyl-d-aspartate receptor activation in the nucleus accumbens core. *Proceedings of the National Academy of Sciences, 94*(22), 12174–12179.

Kerr, J. N., & Wickens, J. R. (2001). Dopamine d-1/d-5 receptor activation is required for long-term potentiation in the rat neostriatum in vitro. *Journal of Neurophysiology, 85*(1), 117–124.

Killcross, S., & Coutureau, E. (2003). Coordination of actions and habits in the medial prefrontal cortex of rats. *Cerebral Cortex, 13*(4), 400–408.

Kim, N., Li, H. E., Hughes, R. N., Watson, G. D. R., Gallegos, D., West, A. E., … Yin, H. H. (2019). A striatal interneuron circuit for continuous target pursuit. *Nature Communications, 10*(1), 2715. doi: 10.1038/s41467-019-10716-w.

Koralek, A. C., Jin, X., Long II, J. D., Costa, R. M., & Carmena, J. M. (2012). Corticostriatal plasticity is necessary for learning intentional neuroprosthetic skills. *Nature, 483*(7389), 331–335.

Lakoff, G., & Johnson, M. (2008). *Metaphors We Live By*: Chicago: University of Chicago Press.

Lee, C. R., Yonk, A. J., Wiskerke, J., Paradiso, K. G., Tepper, J. M., & Margolis, D. J. (2019). Opposing influence of sensory and motor cortical input on striatal circuitry and choice behavior. *Current Biology, 29*(8), 1313–1323. e1315.

Lekwuwa, G., & Barnes, G. (1996a). Cerebral control of eye movements ii. Timing of anticipatory eye movements, predictive pursuit and phase errors in focal cerebral lesions. *Brain, 119*(2), 491–505.

Lekwuwa, G., & Barnes, G. (1996b). Cerebral control of eye movements: I. The relationship between cerebral lesion sites and smooth pursuit deficits. *Brain, 119*(2), 473–490.

Liljeholm, M., Tricomi, E., O'Doherty, J. P., & Balleine, B. W. (2011). Neural correlates of instrumental contingency learning: Differential effects of action-reward conjunction and disjunction. *Journal of Neuroscience, 31*(7), 2474–2480. doi: 10.1523/JNEUROSCI.3354-10.2011.

Mackintosh, N. J. (1974). *The Psychology of Animal Learning*: London: Academic Press.

McGinty, V. B., Lardeux, S., Taha, S. A., Kim, J. J., & Nicola, S. M. (2013). Invigoration of reward seeking by cue and proximity encoding in the nucleus accumbens. *Neuron, 78*(5), 910–922.

Mettler, F. A., & Mettler, C. C. (1942). The effects of striatal injury. *Brain: A Journal of Neurology, 65*(3), 242–255.

Meyer, P. J., Lovic, V., Saunders, B. T., Yager, L. M., Flagel, S. B., Morrow, J. D., & Robinson, T. E. (2012). Quantifying individual variation in the propensity to attribute incentive salience to reward cues. *PLoS One, 7*(6), e38987. doi: 10.1371/journal.pone.0038987.

Morris, R. W., Dezfouli, A., Griffiths, K. R., Le Pelley, M. E., & Balleine, B. W. (2022). The neural bases of action-outcome learning in humans. *Journal of Neuroscience, 42*(17), 3636–3647.

Neely, R. M., Koralek, A. C., Athalye, V. R., Costa, R. M., & Carmena, J. M. (2018). Volitional modulation of primary visual cortex activity requires the basal ganglia. *Neuron, 97*(6), 1356–1368. e1354.

O'Doherty, J., Dayan, P., Schultz, J., Deichmann, R., Friston, K., & Dolan, R. J. (2004). Dissociable roles of ventral and dorsal striatum in instrumental conditioning. *Science, 304*(5669), 452–454.

Parkinson, J. A., Willoughby, P. J., Robbins, T. W., & Everitt, B. J. (2000). Disconnection of the anterior cingulate cortex and nucleus accumbens core impairs pavlovian approach behavior: Further evidence for limbic cortical-ventral striatopallidal systems. *Behavioral Neuroscience, 114*(1), 42–63.

Peak, J., Chieng, B., Hart, G., & Balleine, B. W. (2020). Striatal direct and indirect pathway neurons differentially control the encoding and updating of goal-directed learning. *Elife, 9*, e58544.

Planert, H., Szydlowski, S. N., Hjorth, J. J., Grillner, S., & Silberberg, G. (2010). Dynamics of synaptic transmission between fast-spiking interneurons and striatal projection neurons of the direct and indirect pathways. *Journal of Neuroscience, 30*(9), 3499–3507.

Rescorla, R. A. (1968). Probability of shock in the presence and absence of cs in fear conditioning. *Journal of Comparative and Physiological Psychology, 66*(1), 1–5.

Rescorla, R. A. (1980). *Pavlovian Second-Order Conditioning: Studies in Associative Learning*: New York: Psychology Press.

Rescorla, R. A. (1988). Pavlovian conditioning: It's not what you think it is. *American Psychologist, 43*(3), 151.

Robinson, D. A. (1968). Eye movement control in primates. *Science, 161*(3847), 1219–1224.

Saunders, B. T., & Robinson, T. E. (2012). The role of dopamine in the accumbens core in the expression of pavlovian-conditioned responses. *European Journal of Neuroscience, 36*(4), 2521–2532. doi: 10.1111/j.1460-9568.2012.08217.x.

Schwartz, B., & Gamzu, E. (1977). Pavlovian control of operant behavior. In: Honig, W., & Staddon, J. E. R. (Eds.), *Handbook of Operant Behavior* (pp. 53–97). Englewood Cliffs, NJ: Prentice Hall.

Sgambato, V., Pages, C., Rogard, M., Besson, M.-J., & Caboche, J. (1998). Extracellular signal-regulated kinase (erk) controls immediate early gene induction on corticostriatal stimulation. *Journal of Neuroscience, 18*(21), 8814–8825.

Sheffield, F. D. (1965). Relation between classical and instrumental conditioning. In: Prokasy, W. F. (Ed.), *Classical Conditioning* (pp. 302–322). New York: Appleton-Century-Crofts.

Shepard, R. N. (1981). Psychophysical complementarity. In: Kubovy, M. (Ed.), *Perceptual Organization* (pp. 279–341). Mahwah, NJ: Lawrence Erlbaum.

Shepherd, G. M. (2013). Corticostriatal connectivity and its role in disease. *Nature Reviews Neuroscience, 14*(4), 278–291.

Shiflett, M. W., & Balleine, B. W. (2011). Contributions of erk signaling in the striatum to instrumental learning and performance. *Behavioural Brain Research, 218*(1), 240–247. doi: 10.1016/j.bbr.2010.12.010.

Shiflett, M. W., Brown, R. A., & Balleine, B. W. (2010). Acquisition and performance of goal-directed instrumental actions depends on erk signaling in distinct regions of dorsal striatum in rats. *Journal of Neuroscience, 30*(8), 2951–2959. doi: 10.1523/JNEUROSCI.1778-09.2010.

Shohamy, D., Myers, C. E., Grossman, S., Sage, J., Gluck, M. A., & Poldrack, R. A. (2004). Cortico-striatal contributions to feedback-based learning: Converging data from neuroimaging and neuropsychology. *Brain, 127*(Pt 4), 851–859.

Stuber, G. D., Sparta, D. R., Stamatakis, A. M., van Leeuwen, W. A., Hardjoprajitno, J. E., Cho, S., … Deisseroth, K. (2011). Excitatory transmission from the amygdala to nucleus accumbens facilitates reward seeking. *Nature, 475*(7356), 377.

Takikawa, Y., Kawagoe, R., & Hikosaka, O. (2002). Reward-dependent spatial selectivity of anticipatory activity in monkey caudate neurons. *Journal of Neurophysiology, 87*(1), 508–515.

Takikawa, Y., Kawagoe, R., Itoh, H., Nakahara, H., & Hikosaka, O. (2002). Modulation of saccadic eye movements by predicted reward outcome. *Experimental Brain Research, 142*(2), 284–291.

Tanaka, S. C., Balleine, B. W., & O'Doherty, J. P. (2008). Calculating consequences: Brain systems that encode the causal effects of actions. *Journal of Neuroscience, 28*(26), 6750–6755. doi: 10.1523/JNEUROSCI.1808-08.2008.

Tomie, A., Brooks, W., & Zito, B. (2014). Sign-tracking: The search for reward. In: Klein, S. B., & Mowrer, R. R. (Ed.), *Contemporary Learning Theories: Volume II: Instrumental Conditioning Theory and the Impact of Biological Constraints on Learning* (p. 108). New York: Taylor and Francis.

Tomie, A., Grimes, K. L., & Pohorecky, L. A. (2008). Behavioral characteristics and neurobiological substrates shared by pavlovian sign-tracking and drug abuse. *Brain Research Reviews*, *58*(1), 121–135.

Tricomi, E. M., Delgado, M. R., & Fiez, J. A. (2004). Modulation of caudate activity by action contingency. *Neuron*, *41*(2), 281–292.

Valjent, E., Caboche, J., & Vanhoutte, P. (2001). Mitogen-activated protein kinase/extracellular signal-regulated kinase induced gene regulation in brain. *Molecular Neurobiology*, *23*(2), 83–99.

Vega-Villar, M., Horvitz, J. C., & Nicola, S. M. (2019). Nmda receptor-dependent plasticity in the nucleus accumbens connects reward-predictive cues to approach responses. *Nature Communications*, *10*(1), 1–16.

Villablanca, J. R. (2010). Why do we have a caudate nucleus. *Acta Neurobiol Exp (Wars)*, *70*(1), 95–105.

Villablanca, J. R., Marcus, R. J., & Olmstead, C. E. (1976). Effects of caudate nuclei or frontal cortical ablations in cats. I. Neurology and gross behavior. *Experimental Neurology*, *52*(3), 389–420.

Villablanca, J. R., Marcus, R. J., Olmstead, C. E., & Avery, D. L. (1976). Effects of caudate nuclei or frontal cortex ablations in cats: Iii. Recovery of limb placing reactions, including observations in hemispherectomized animals. *Experimental Neurology*, *53*(2), 289–303.

von Holst, E., & Mittelstaedt, H. (1950). The reafference principle. In: Martin, R. (Ed.), *The Collected Papers of Erich Von Holst* (pp. 139–173). Coral Gables, FL: University of Miami Press.

Watanabe, K., Lauwereyns, J., & Hikosaka, O. (2003). Neural correlates of rewarded and unrewarded eye movements in the primate caudate nucleus. *Journal of Neuroscience*, *23*(31), 10052–10057.

Watanabe, M., & Munoz, D. P. (2010). Presetting basal ganglia for volitional actions. *Journal of Neuroscience*, *30*(30), 10144–10157.

Williams, D. R., & Williams, H. (1969). Automaintenance in the pigeon: Sustained pecking despite contingent non-reinforcement. *Journal of the Experimental Analysis of Behavior*, *12*, 511–520.

Yin, H. H., Knowlton, B. J., & Balleine, B. W. (2005). Blockade of nmda receptors in the dorsomedial striatum prevents action-outcome learning in instrumental conditioning. *European Journal of Neuroscience*, *22*, 505–512.

Yin, H. H., Ostlund, S. B., Knowlton, B. J., & Balleine, B. W. (2005). The role of the dorsomedial striatum in instrumental conditioning. *European Journal of Neuroscience*, *22*, 513–523.

Yin, H. H., Park, B. S., Adermark, L., & Lovinger, D. M. (2007). Ethanol reverses the direction of long-term synaptic plasticity in the dorsomedial striatum. *European Journal of Neuroscience*, *25*(11), 3226–3232.

Yoshida, A., & Tanaka, M. (2009). Neuronal activity in the primate globus pallidus during smooth pursuit eye movements. *Neuroreport*, *20*(2), 121–125.

Zener, K. (1937). The significance of behavior accompanying conditioned salivary secretion for theories of the conditioned response. *The American Journal of Psychology*, *50*(1/4), 384–403.

11 Corticostriatal Contributions to Habits and Behavioral Automaticity

A central question in the study of learning is concerned with its content: what is learned? Studies have demonstrated different types of learning, which can be inferred using behavioral assays. This chapter further examines the contributions of different basal ganglia (BG) networks to learning. We shall begin with a review of the multiple memory systems framework, according to which the BG mediate habit learning or procedural learning in contrast to declarative learning mediated by the medial temporal lobe. We shall then review studies on the shift in the mode of behavioral control during habit formation and skill learning and the underlying neural substrates in the associative and sensorimotor cortico-BG networks. Finally, we shall consider how the control hierarchy may shed light on the development of behavioral automaticity.

11.1 MULTIPLE MEMORY SYSTEMS

According to Thorndike and Hull, learning consists of modifications of S–R associations, the equivalent of state-policy mappings in modern reinforcement learning. The animal is assumed to be passively responding to a succession of external and internal stimuli impinging upon its sensory system, and its tendency to respond is modified by a reinforcement signal. However, this view was challenged by Tolman (Tolman, 1932; Tolman, Ritchie, & Kalish, 1946a, b). According to Tolman, what is learned is not a specific behavior but a set of contingencies (Tolman et al., 1946b). The stimuli are not connected to the outgoing responses, but elaborated into a cognitive map, an internal representation that indicates various routes and environmental relationships. Having formed such a map, an animal may flexibly choose routes and responses even if they have never been explicitly reinforced (McNamara, Long, & Wike, 1956).

An experimental test of these ideas is known as the place/response task. In a typical version of this task, a rat has to find food at either end of a T-shaped maze. After it is trained to turn west (left), occasional probe trials are introduced, on which the orientation of the maze is reversed, so that the starting arm, instead of facing north, now faces south, with an inverted T (Figure 11.1). To reach reward, the rat has to use the spatial landmarks rather than remembered direction of turning. In other words, if the rat simply repeated the left-turning behavior at the choice point, which was repeatedly reinforced during the training phase, it would not have succeeded. Hull's theory predicts that the rat would turn left, as it was trained to do. Tolman, however, predicts that the rat would turn right and reach the food reward because its decision at the choice point is based on knowledge of the cognitive map. Rather than simply repeating the reinforced behavior of a left turn, the rat learns where the food is located in relation to landmarks and generates whatever behavior is needed to reach that location. This is known as the "place" strategy.

Results from studies using the place/response task turned out to be ambiguous, as rats were found to use both place and response strategies, depending on the cues available (Restle, 1957). Consequently, they failed to resolve the debate between Hull and Tolman, though they had a major impact on the subsequent work on the neurobiology of learning.

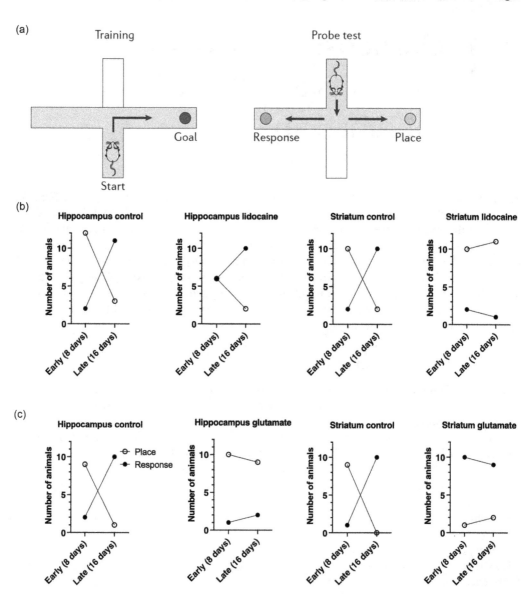

FIGURE 11.1 Dissociable neural systems for place and response strategies. (a) Illustration of the place/response task. In this example, a rat is trained to turn right to obtain reward from the east arm. On the probe test, the maze is inverted, so that the rat starts from the north arm. If the rat chooses the east arm by turning left (a response never reinforced during training), it is using the 'place' strategy. If it simply repeats the right turn response and goes to the west arm, then it is using the 'response' strategy. From Yin and Knowlton (2006) with permission. (b) Lidocaine inactivation of the hippocampus reduced the use of place strategy on the early test. Lidocaine inactivation of the striatum reduced the use of the response strategy on the late test. Based on Packard and McGaugh (1996). (c) Glutamate infusion into the hippocampus promoted use of place strategy even after extended training (16 days). Glutamate infusion into the striatum promoted the use of response strategy even after limited training. Based on Packard (1999).

11.2 PLACE AND RESPONSE

One way to resolve the debate between Hull and Tolman is to posit the existence of two dissociable learning systems. In an early proposal, Hirsh made a distinction between two types of learning, S–R and stimulus–stimulus (S–S), which are mediated by distinct memory systems (Hirsh, 1974). S–S refers to the relationship between two events, e.g. cloud-rain. Hirsh proposed that the hippocampus is responsible for such learning. If the animal is transported in a maze, it may acquire the configuration of the maze using surrounding cues and later generate the appropriate behavior flexibly to reach the goal, even if the behavior produced was never reinforced during training (Redish, 2016; Tolman, 1932). O'Keefe and Nadel proposed that, more broadly, Tolman's cognitive map is implemented by the hippocampus (O'Keefe & Nadel, 1978). On the other hand, others have argued that the S–R system, often equated with procedural memory or nondeclarative memory, is implemented by the BG (Mishkin, 1984; White, 1997). This system transforms specific sensory inputs into motor outputs, without the need to retrieve any cognitive map.

Experimental evidence for the dissociation just described comes from lesion studies in rats. Whishaw found that lesions of the dorsomedial striatum (DMS) (but not the dorsolateral striatum [DLS]) impaired the use of spatial navigation in a hidden platform water maze task (Whishaw, Mittleman, Bunch, & Dunnett, 1987). There are two versions of this task. In the place version, a transparent platform is located below the surface of the water but is invisible from the water level. To find the platform from any starting point, it is necessary to form a cognitive map of the environment, including the location of the invisible platform. For the cued version, the platform is made visible by a piece of styrofoam attached to it. Rats can simply swim toward it without an internal representation of the platform's location. DMS lesions impair both the acquisition and retention of both cue and place tasks. Lesioned rats also relied more on a cue-based strategy (Devan, McDonald, & White, 1999; Devan & White, 1999).

According to the modern multiple memory systems model, the BG and hippocampus implement the S–R and S–S systems, respectively. This prediction was tested by Packard and McGaugh using the place/response task (Packard & McGaugh, 1996). They trained rats to use a cross maze for 14 days, 4 trials a day. On days 8 and 16, a single probe trial was given, in which rats were placed in the start box opposite that used in training. Just before the probe trials, rats received local infusions of either saline or lidocaine (a sodium channel blocker) to inactivate the hippocampus or DLS. The results are shown in Figure 11.1b. Control rats showed place strategy on the early (day 8) probe trial and response strategy on the later (day 16) probe trial, indicating that extended training promoted a shift in the behavioral strategy. Inactivating the hippocampus with lidocaine showed no preference for place or response on the first probe trial, but a response strategy on the day 16 trial. However, rats with dorsolateral striatal inactivation displayed a place preference on the first probe trial and a response strategy on the second probe trial.

Packard observed the opposite pattern when glutamate was infused into the hippocampus and striatum (Packard, 1999). On days 4–6, rats received injections of either glutamate or saline into the dorsal hippocampus or the dorsal striatum (Figure 11.1c). On days 8 and 16, a probe trial was conducted. As expected, saline-treated control rats used the place strategy on day 8 and the response strategy on day 16. Rats receiving hippocampal glutamate injections used the place strategy when tested early or late. On the other hand, rats receiving glutamate in the striatum displayed response learning on days 8 and 16, indicating an accelerated shift in the use of the response strategy. These results suggest that hippocampal activation promotes the use of the place strategy, whereas dorsal striatal activation promotes the use of the response strategy.

11.3 BG AND PROCEDURAL LEARNING

The BG have also been implicated in procedural learning, which refers to acquisition of skills through repetition. Patients with damage to the medial temporal lobe, including the hippocampus, showed severe amnesia for recent events and facts, yet they could still learn new skills like mirror drawing, even though they could not recall the experience of learning (Corkin, 1968; Milner, 1970; Scoville & Milner, 1957; Weiskrantz & Warrington, 1979). On the other hand, patients with Parkinson's disease were impaired in procedural learning without showing impairment in recall and recognition tests of declarative memory (Saint-Cyr, Taylor, & Lang, 1988). A similar dissociation has been observed in a discrimination task in which pairs of objects are presented and the subject must choose the rewarded item in each pair. Normal subjects could learn these discriminations quickly. Amnesics with damage to the medial temporal lobe (including the hippocampus) could still learn these discriminations, but their performance improved more slowly compared to controls and Parkinson's patients, and they failed to show explicit knowledge of these associations. Similarly, in the "weather prediction" task, participants learn to predict categorical outcomes (sun or rain) based on some combinations of cues. Amnesics gradually improved their performance on this task, even though they had no explicit memory for the testing episode. By contrast, Parkinsonian patients were impaired at learning but had intact explicit memory (Knowlton, Mangels, & Squire, 1996). This double dissociation suggests that the BG play a role in procedural learning that is gradually acquired and dissociable from the role of the hippocampus (Saint-Cyr et al., 1988).

However, BG contributions to implicit learning have been questioned. Some studies have shown that the BG are critical for learning from action-contingent feedback, regardless of whether learning is implicit or explicit (Foerde & Shohamy, 2011a, b). In one study, Foerde and Shohamy required participants to learn to associate cues with outcomes. The participants had to choose which flower (outcome) is preferred by a given butterfly (cue). The butterfly-flower association was probabilistic, so the subject had to learn through trial and error which outcome was most probable given the cue. The feedback on whether they chose correctly was given either immediately or with a 6-second delay. Whereas age-matched controls showed comparable performance when feedback was immediate or delayed, Parkinson's patients were impaired when learning from immediate feedback but performed just as well as the controls when the feedback as delayed. Thus, when feedback is delayed, learning may require an alternative neural system that is not affected by DA depletion. Using fMRI, Foerde and Shohamy showed that in healthy participants, this alternative neural system included the hippocampus, which was selectively sensitive to delayed feedback but not to immediate feedback. In contrast, the dorsal striatum was preferentially activated during learning from immediate feedback, but not when the feedback was delayed.

These results suggest that immediate action-contingent feedback plays a key role in the type of learning mediated by the BG. The distinction between action-contingent and noncontingent feedback or between immediate and delayed feedback may be more useful than the traditional distinction between declarative and procedural learning.

11.4 LIMITATIONS OF MULTIPLE MEMORY SYSTEMS FRAMEWORK

According to the multiple memory systems model, learning can take place in one system or another, or simultaneously, and different neural systems may compete for expression during performance. Although these conclusions are supported by considerable evidence, a few assumptions of the model have been questioned.

In the neural control hierarchy, there is no need to have separate components for performance, perception, and memory. There is no system that is exclusively dedicated to memory, which is a key emergent property at the higher levels. As such, it is distributed widely where the higher-order perceptual input functions are found. The common belief that the hippocampus is specifically dedicated to memory is misleading (Hassabis & Maguire, 2007; Murray, Wise, & Graham, 2018). Although lesion studies have dissociated the contributions of the hippocampus and striatum, it is somewhat misleading to compare the functions of these two structures at the level of behavior. The hippocampus, which contains glutamatergic projection neurons, can be considered a cortical component that projects to the ventral striatum. Contrasting hippocampal and striatal function at the level of behavior is therefore similar to the traditional tendency to contrast cortical and BG function discussed in Chapter 1. At the level of behavior, a cortical area and its corresponding BG circuit work together to achieve some function, so it would be more appropriate to consider their contributions in conjunction (Yin & Knowlton, 2006).

There is considerable functional heterogeneity within the striatum. In the original studies by Packard and colleagues (Packard, 1999; Packard & McGaugh, 1996), the striatal manipulations were mostly confined to the DLS. More restricted lesions showed that the DLS is critical for the response strategy, whereas the DMS (associative striatum) contributes to the use of the place strategy (Yin & Knowlton, 2004). Despite extensive training that promoted the use of the response strategy in control rats, most rats with DLS lesions still used the place strategy. In contrast, most rats with DMS lesions used the response strategy regardless of the amount of training, similar to rats with hippocampal lesions. These observations suggest that the idea that the dorsal striatum as a whole is responsible for response learning is too simplistic. The hippocampal formation and parts of the associative striatum appear to be components of the same circuit for navigation guided by distal cues.

As discussed in Chapter 10, studies have demonstrated distinct modes of behavioral control, as defined by specific relationship variables. Performance of goal-directed actions is sensitive to changes in outcome value or instrumental contingency (Colwill & Rescorla, 1986; Dickinson, 1994). The associative network, with posterior dorsomedial striatum (pDMS) as a key node, is critical for learning the action–outcome (A–O) contingency. The explicit representation of the A–O contingency is a type of declarative knowledge, yet the hippocampus (and associated medial temporal lobe), which is traditionally considered the center for declarative memory, is not necessary for action–outcome learning (Corbit, Ostlund, & Balleine, 2002). These observations suggest that the traditional category of implicit, nondeclarative, or procedural learning fails to capture the contribution of the BG adequately.

11.5 FROM ACTIONS TO HABITS

A variety of evidence has shown that, with training, behavior becomes more stereotyped and efficient, requiring less attention and cognitive resources. Such performance is often considered "habitual" (Dickinson, 1985). Traditionally, habits are defined as well-learned behavioral routines triggered by specific sensory cues. As Darwin noted, "some actions, which were at first performed consciously, have become through habit and association converted into reflex actions" (Darwin, 1872). Habits are characterized by reduced conscious awareness and attentional demand, as well as more stereotyped and efficient performance. In his chapter on habits, James describes the example of a discharged veteran who would immediately drop his dinner when he heard "Attention!" and bring his hands down to assume a military posture (James 1890). Thus, habitual performance can be produced despite undesirable consequences.

According to Thorndike's law of effect, behavior is generated by a sensorimotor transformation via an S–R connection, which can be modified by learning (Thorndike, 1907). This account predicts that performance should not be affected by devaluation or instrumental contingency manipulations, as the outcome representation is not a part of the S–R associative structure and thus plays no role in performance (Dickinson, 1989). As a general account of instrumental performance, this model is falsified by the demonstration of goal-directed actions, which are dramatically reduced by devaluation or contingency degradation. Nevertheless, there is evidence for habitual control of instrumental performance. For example, as reviewed earlier, pDMS lesions do not abolish lever pressing per se but merely render performance insensitive to devaluation and degradation (Yin, Ostlund, Knowlton, & Balleine, 2005). Such performance is what the S–R account predicts.

Adams and Dickinson found that lever pressing performance is sensitive to devaluation early on, after limited training, but extended training produced devaluation-insensitive behavior (Adams, 1982; Adams & Dickinson, 1981). Based on these results, Dickinson proposed a distinction between goal-directed and habitual actions (Dickinson, 1985). This distinction is based on operational definitions of these two types of behavior. The performance of goal-directed actions is sensitive to devaluation and changes in the instrumental contingency. Habitual actions, on the other hand, are less affected by these experimental manipulations. For example, on the omission test, pressing a lever could cancel the delivery of a pre-scheduled pellet, so that reward is only earned when one refrains from pressing (Davis & Bitterman, 1971; Dickinson, Squire, Varga, & Smith, 1998;

Holland, 1979). This is a complete reversal in the instrumental contingency, changing the polarity of the feedback function from negative to positive. Now the action rate is negatively correlated with reward rate. Rats with extended training showed reduced sensitivity to the imposition of the omission contingency: they did not reduce lever pressing as readily when the contingency was reversed as rats with limited training. Recall that insensitivity to the omission contingency is also found in Pavlovian approach responses (Brown & Jenkins, 1968; Hershberger, 1986). But such insensitivity is attributed to the fact that approach behavior is mediated by representation of the stimulus–outcome (S–O) contingency rather than the action–outcome (A–O) contingency.

11.6 SENSORIMOTOR STRIATUM AND HABIT FORMATION

The shift from a goal-directed mode to a habitual mode parallels the shift from a place strategy to a response strategy (Packard & McGaugh, 1996). Such observations therefore suggest the existence of parallel systems for generating instrumental actions. They raise the question of how these functional systems are implemented by the brain, in particular by different striatal regions like the DMS and DLS.

The BG have often been associated with habit formation and the development of automaticity, based on most neuropsychological studies that did not use behavioral assays like outcome devaluation (Knowlton et al., 1996; Mishkin, 1984; Partiot et al., 1996; Saint-Cyr et al., 1988). To study striatal contribution to habit formation, Yin et al. trained rats under a random interval schedule of reinforcement (RI, a type of variable interval [VI] schedule), which has been shown to promote habit formation (Yin, Knowlton, & Balleine, 2004). On such a schedule, food rewards become available after a variable time interval, and lever pressing is required to deliver the reward. As shown in Figure 11.2, when conditioned taste aversion was induced to the food pellet reward, rats stopped consuming it when tested, showing no desire for the food. Yet when these rats were allowed to press the lever on a short probe test conducted in extinction, they still pressed the lever as often as control rats that did not receive the devaluation treatment. This pattern of behavior is what is expected from habitual performance. However, a different pattern was observed in rats with pretraining excitotoxic lesions of the DLS. Lesioned rats did not show any deficit in learning to press the lever. Nor did they show any deficit in taste aversion, as measured by reduced consumption of the reward. The key difference was revealed on the probe test after devaluation. Whereas control rats showed devaluation-insensitive performance on the probe test, the lesioned animals reduced their lever pressing more readily. This pattern suggests that their performance was based on action–outcome representations (Yin et al., 2004).

In addition to outcome devaluation, sensitivity to instrumental contingency manipulations has also been used as an assay for habit formation. If DLS lesions make rats rely more on the alternative goal-directed system, then they may also render performance more sensitive to changes in the instrumental contingency. The most drastic contingency manipulation is omission, in which the action is arranged to be negatively correlated with the outcome. Yin et al. tested this prediction by inactivating the DLS while imposing an omission contingency and found that rats with DLS inactivation showed higher sensitivity to the omission contingency (Yin, Knowlton, & Balleine, 2006).

Neuroimaging experiments in humans also found sensorimotor striatum activity during habitual performance. While in the scanner, participants were trained to press a button in the scanner for a food reward to be consumed later (Tricomi, Balleine, & O'Doherty, 2009). To promote habit formation, a VI schedule was used, just as in rodent studies. One group of participants received limited training, and another group received extensive training. Following training, the reward outcome was devalued, and participants were tested in the fMRI scanner for a 3-minute extinction test. This test showed that the overtrained group, but not the limited training group, showed reduced sensitivity to outcome devaluation, replicating the behavioral results from rats. In addition, there was little activation in the posterior putamen (analogous to the DLS) on the first day of training, when performance was shown

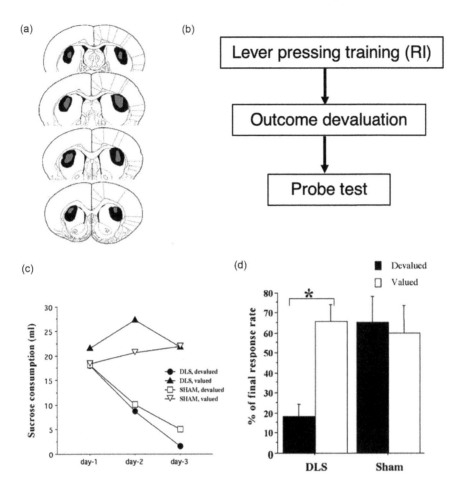

FIGURE 11.2 Lesions of sensorimotor striatum impair habit formation. (a) Coronal sections showing placement of lesions in the rat dorsolateral striatum (DLS). (b) Experimental design. (c) Consumption of sucrose reward after devaluation using lithium chloride injections paired with sucrose, which induces taste aversion to sucrose. Pretraining DLS lesions did not impair the rate of lever pressing. (d) Lever pressing on the probe test after devaluation. Rats with DLS lesions remained sensitive to outcome devaluation despite extended training that induced habitual performance in control rats: they reduced pressing after outcome devaluation. From Yin, Knowlton, and Balleine (2004) with permission.

to be goal-directed and sensitive to devaluation. During extended training, however, activity in the posterior putamen increased, especially on the third day, when performance was shown to be habitual.

Studies have also suggested that stress can promote habit formation and increase engagement of the sensorimotor cortico-BG network (Schwabe, Dickinson, & Wolf, 2011). In one study, rats were trained on a ratio schedule for food rewards. Whereas control rats showed highly goal-directed performance, as indicated by sensitivity to devaluation and instrumental contingency degradation, rats exposed to chronic stress showed reduced sensitivity to these manipulations as well as significant structural changes in the corticostriatal circuits (Dias-Ferreira et al., 2009). The changes in the associative and sensorimotor networks were strikingly different: there was atrophy of the medial prefrontal cortex and the associative striatum but hypertrophy of the sensorimotor striatum. Examination of the morphology of SPNs in the DLS showed increased dendritic arbors in rats exposed to chronic stress. Thus, stress promotes habit formation, possibly due to reduced excitatory corticostriatal transmission in the associative network and enhanced corticostriatal transmission in the sensorimotor network.

11.6.1 PLASTICITY MECHANISMS UNDERLYING HABITS

The results reviewed above suggest that DLS activity may become more responsive to cortical inputs as a result of habit formation. So far, only a few studies have examined corticostriatal synaptic plasticity mechanisms underlying habit formation. In one study, O'Hare et al. used an *ex vivo* approach to study changes in the DLS after training. They trained mice on VI schedules to generate habitual performance, as confirmed by devaluation (O'Hare et al., 2016). After training, their brains were removed, and synaptic strength was measured by stimulating corticostriatal afferents and recording evoked responses from direct pathway spiny projection neurons (dSPNs) and indirect pathway spiny projection neuron (iSPNs) in the DLS using 2-photon calcium imaging. Training altered the corticostriatal activation function in both dSPNs and iSPNs. Insensitivity to outcome devaluation in individual animals was used to quantify the degree of habit formation. Reduced sensitivity to devaluation was correlated with larger evoked output from both dSPNs and iSPNs, as measured by amplitudes of calcium transients, a measure of corticostriatal synaptic strength. Since large calcium transients indicate burst firing, the corticostriatal inputs from the primary somatosensory and motor cortices to the sensorimotor striatum appear to be more effective at eliciting burst firing in SPNs following habit formation. Weakening of inputs to dSPNs was correlated with suppression of performance during omission (lever pressing canceled reward delivery). On the other hand, iSPN output predicted habit expression but was unrelated to suppression during omission. These data suggest that reduced performance in habitual animals is not a simple reversal of the habit formation process. Weakened corticostriatal inputs to the direct pathway could be responsible for suppression of habitual actions, which is needed for both extinction and omission.

In addition, a shift in the relative timing of fire between dSPNs and iSPNs also strongly correlated with the degree of habitual performance: dSPNs more frequently fire before iSPNs in habitual mice. This observation suggests that habit formation is not only associated with increased glutamatergic transmission at both direct and indirect pathways but also with a shift in their relative timing. However, how these neuronal populations change in behaving animals remains unclear.

Habit formation has also been associated with a long-lasting increase in fast-spiking interneuron (FSI) excitability in the DLS (O'Hare et al., 2017). FSIs from habitual mice showed more sustained firing in response to current injections, compared with FSIs from goal-directed mice. This suggests that corticostriatal projections can more effectively recruit FSIs as a result of training. Selective inhibition of FSIs using chemogenetics prevented the expression of habitual performance. While FSIs exert a strong inhibitory effect on many SPNs, surprisingly, they appear to have an excitatory effect on some SPNs that show high burst firing. Highly active SPNs may be facilitated while less active SPNs are inhibited by the FSIs. This may allow the activation of the appropriate combination of SPNs while suppressing others SPNs. Consequently, SPNs that are strongly driven by cortical inputs may be potentiated further while the less active SPNs may be silenced, increasing the signal-to-noise ratio in corticostriatal recruitment of the appropriate SPNs.

11.7 DEVELOPMENT OF AUTOMATICITY

The results reviewed so far suggest a shift from DMS-dependent goal-directed action control to DLS-dependent habitual action control. There are similarities between this shift and the shift observed during the learning of a sequential button-pressing task in human subjects. Learning of new sequences preferentially activates the associative cortico-BG network (Jueptner, Frith, Brooks, Frackowiak, & Passingham, 1997; Jueptner, Stephan, et al., 1997). In particular, the caudate nucleus is differentially active, along with the anterior prefrontal cortex and anterior cingulate gyrus. On the other hand, during the performance of well-learned sequences, more posterior regions, including the putamen and premotor and motor cortical areas, became more active. Interestingly, when subjects were instructed to pay attention again during performance, the anterior regions again became more active.

There are similarities between habit formation with extended training and the development of automaticity in human behavior. In both, there appears to be a transfer of the main locus of activity from the associative striatum to the sensorimotor striatum as a result of extensive training. While these studies did not use devaluation to assess whether behavior was habitual, they showed that sequence generation was insensitive to dual-task interference, a common measure of automaticity in human behavior (Shiffrin & Schneider, 1977).

In mice, the sensorimotor corticostriatal pathway has also been implicated in skill learning (Costa, Cohen, & Nicolelis, 2004). Costa et al. recorded single-unit activity from the mouse motor cortex and sensorimotor striatum during a rotarod task. On this task, animals learned to maintain balance on a rotating rod at an increasing speed of rotation. Mice exhibited fast improvement in the task during the initial session and more gradual improvement across days. Many neurons in both the motor cortex and DLS changed their firing rates while mice were running on the rotarod as they improved their performance. The performance measures (time spent on the rod) suggest two phases in learning on this task: an initial phase of fast learning, characterized by rapid improvement, and a slower, more gradual phase, characterized by gradual refinement and more incremental improvements in performance. In both striatum and motor cortex, there was a dramatic increase in the number of task-related neurons recruited during the first session as performance rapidly improved. Additional training did not significantly increase the proportion of related neurons, which remained stable for the rest of the training days. The activity of many neurons was correlated with the speed of rotation, which was proportional to the speed of running as long as the mice stayed on the rotarod. The sensorimotor corticostriatal projections appear to rapidly recruit SPNs during this initial learning phase, during which running-related neurons in both motor cortex and DLS increased rapidly. However, during the later phase, the changes were much slower and more gradual, and showed different patterns in the cortex and striatum. Whereas speed-correlated neurons increased in the striatum, they decreased in the motor cortex. These results suggest a critical role for corticostriatal plasticity *in vivo* in skill learning.

Experiments also showed that extensive rotarod training was accompanied by long-lasting potentiation of glutamatergic transmission in the striatum (Yin et al., 2009). Synaptic strength was measured *ex vivo* after rotarod training. In the DMS, potentiation of excitatory transmission was only observed early in training; with extended training, synaptic strength decreased to naive levels. In the DLS (sensorimotor striatum), by contrast, no substantial potentiation was observed during the early learning phase but long-lasting potentiation of glutamatergic transmission developed with extended training. This pattern suggests that, early in acquisition, glutamatergic inputs to the DMS are preferentially strengthened, whereas the inputs to the DLS are gradually potentiated with extended training. This potentiation occurred mostly in iSPNs. Although synaptic strength in the DMS was higher early in training, extensive training induced potentiation in the DLS. In addition, *in vivo* electrophysiological recording from the DLS also showed that firing rate modulation (running compared with resting) also increased significantly in the DLS, but not in the DMS. This pattern is consistent with *ex vivo* observations of corticostriatal plasticity, suggesting that increased corticostriatal transmission contributes to the selective recruitment of DLS neurons during task performance.

During early learning of a skill like rotarod running, the associative circuit appears to be more critical, whereas with extensive training, the sensorimotor circuit becomes more engaged. However, this does not imply that only one of these circuits is active at a time. Both DMS and DLS could be engaged during performance, but their relative contributions may change with training. The associative striatal output may modulate the influence of the sensorimotor striatal output on effectors, but currently little is known about the neural dynamics in these areas or how they interact (Thorn, Atallah, Howe, & Graybiel, 2010).

11.8 HABITS AND SKILLS

The studies discussed above suggest some common underlying neural mechanisms in diverse phenomena like the shift from the place strategy to response strategy on a T-maze, habit formation in instrumental conditioning, and the development of automaticity in sequence and skill learning (Yin & Knowlton, 2006). For example, DMS lesions reduce the use of the place strategy and impair action–outcome learning, whereas DLS lesions reduce the use of the response strategy and impair habit formation. When the goal-directed system mediated by the associative BG circuit is damaged, an alternative habitual system mediated by the sensorimotor circuit appears to be responsible for performance. Skill learning and the development of automaticity appear to require the sensorimotor cortico-BG network, including the DLS.

These observations raise a number of questions. What is responsible for the shift from goal-directed actions to habitual actions or the development of automaticity with extended training? How can these findings be related to the representations of kinematic variables in the BG discussed in earlier chapters? To address these questions, let us first consider two features that habits and skills have in common: reduced attentional demand and effector specificity.

11.8.1 Reduced Attentional Demand

Resistance to interference and reduced attentional demand are common properties of behavioral automaticity (Shiffrin & Schneider, 1977). Reduced attentional demand often allows successful performance of a secondary task. For example, a skilled driver can perform some other task successfully while driving.

During initial instrumental learning, the key controlled variable is reward rate. Such control requires considerable attentional resources, as one must simultaneously monitor reward presentations, self-generated actions, and how these variables covary. In habitual performance, such monitoring is no longer necessary. Attentional demand is high during initial training due to the creation of a forward model, namely the anticipation of the outcome, including some of its sensory properties, when generating the action. This could be achieved using the imagination mode, without creating a detailed internal model of the environment.

11.8.2 Effector Specificity

Successful performance of a skill is often limited to the set of effectors (e.g. hand) with which it was originally trained. Studies in monkeys have demonstrated that extended training makes performance more specific to the trained hand (Rand, Hikosaka, Miyachi, Lu, & Miyashita, 1998; Rand et al., 2000).

In the control hierarchy, behavioral flexibility is a property of the higher level that has multiple means at its disposal (motor equivalence). These means are equivalent with respect to the higher goal in that they can all provide 'satisfaction' or error reduction. But they may differ in other respects, such as efficiency and effort required, which can be independently controlled variables. The degree of effector specificity in performance therefore reflects the lead level that is responsible for initiating the action. In the action control hierarchy, the higher the lead level, the more functionally equivalent actions there are, with more flexibility in which effectors are used for performance. For example, if the lead level is to stay warm, there are many learned actions that can achieve this aim in an adult human. But if the lead level is to reduce the distance between the right hand and the cup, there is far less flexibility in the effectors available to reduce the distance error.

The development of effector specificity suggests that, in both habits and skills, the lead level shifts from a higher level of functional integration to a lower level (Yin & Knowlton, 2006). This is in accord with the idea that the associative cortico-BG network represents a higher level in the motivational hierarchy than the sensorimotor network.

11.9 NATURE OF FEEDBACK

Related to the question of effector specificity is the type of feedback involved in each type of behavioral performance. For example, the animal performs a sequence of actions, and only at the end is a reward delivered. This type of feedback, which is intermittent and updated occasionally, reaches the highest level of the hierarchy, where it affects slowly changing controlled variables like the rate of reward. In addition, there are also intermediate types of feedback that reflect subgoals like the completion of a single lever press in a press sequence, sometimes called secondary reinforcement. Each subgoal could be associated with a specific sensory state (e.g. the sound that indicates the completion of press) that is generated by the behavior output. Completion of the subgoal reduces a local error signal. Finally, there is rapid and continuous feedback from virtually all sensory modalities during behavior, and continuous control of these input variables.

In initial instrumental learning, what is learned is the action–outcome contingency, in which the action is actually a complex sequence of behaviors and the outcome is the final outcome of that sequence. For example, the rat walks to the lever, presses it, moves to the reward port, and finally collects a piece of food. The associative network is critical for higher-order relationship control, based on the detection and control of action–outcome contingencies over time. This requires the ability to track recent actions as well as associated outcomes. The representation of relationship variables is not strongly associated with a specific program of action. It can take multiple actions. Hence the behavioral flexibility of the type that Tolman emphasized in his concept of the cognitive map.

With overtraining, as performance becomes habitual, it is no longer guided by the final feedback but by local feedback, especially feedback from each component of the sequence (Mowrer, 1960). The subgoals could be grouped together as a sequence of reference signals. Once initiated, the sequence can be generated purely based on local feedback from the completion of each subgoal, without the need to represent the final outcome (Figure 11.3). One prediction of this account is that very lean reinforcement schedules with rare "final outcome" feedback would promote habitual performance. Likewise, schedules with relatively weak action–outcome contingencies should also

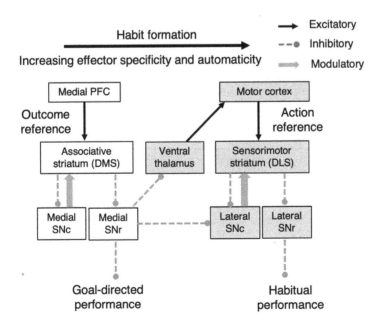

FIGURE 11.3 Shift from the associative to the sensorimotor cortico-basal ganglia network during habit formation. The illustration shows the hypothetical shift in the locus of control. DMS, dorsomedial striatum; SNc, substantia nigra pars compacta; SNr, substantia nigra pars reticulata.

promote reliance on local feedback, as higher-order perception of the relationship variable is weaker under such conditions. This appears to be the case in training procedures that promote habit formation (Dickinson, 1994; Perez & Dickinson, 2020).

Some have argued that the formation of behavioral sequences can explain insensitivity to devaluation and other traits of habitual performance (Dezfouli & Balleine, 2012, 2013; Amir Dezfouli, Lingawi, & Balleine, 2014). According to this account, with extended training, the basic units of actions are concatenated, so that devaluation of the final outcome has less impact on performance. This conclusion is supported by the finding that the sensorimotor striatum is also critical for learning arbitrary sequences (Yin, 2010). Standard tests like devaluation and degradation are conducted during extinction. In the absence of reward feedback, what initiates lever pressing could be the discriminative stimuli, and what sustains behavior could be local feedback from the actions themselves.

The type of perceptual feedback guiding performance can also change with training. For example, early in training, visual guidance is often needed to make the targeting movement correctly. There is often reliance on exteroceptive inputs, e.g. sight of the lever. These inputs largely reach the DMS from visual cortical regions. But with extended training, movement can be guided by proprioceptive or tactile feedback, which reaches the DLS from sensorimotor cortices. Thus, changes in experienced feedback can explain the differential engagement of different cortico-BG networks in A–O learning and habit formation.

On the other hand, while development of automaticity in skill learning resembles habits in reduced attentional demand and development of effector specificity, skill learning requires additional refinement of the control hierarchy beyond knowledge of the action–outcome contingency, serial order, or chunking of behavioral elements. It requires learned internal templates of reference signals for precise online control. The controlled variables allow detailed specification of action parameters like movement velocity and amplitude, which may be adjusted to improve efficiency of performance. With learning, performance of each element within the sequence can be refined, for example by specifying the exact proprioceptive or tactile feedback.

As reviewed earlier, in skill learning, the sensorimotor cortico-BG network also plays a critical role. Regardless of the amount of training, this network is crucial for generating the detailed movement patterns underlying the acquired motor skills (Dhawale, Wolff, Ko, & Ölveczky, 2021). The present account predicts that skilled learning should be accompanied by considerable changes in kinematic features of movement as well as representations of these kinematic variables in the sensorimotor BG. Exactly how such parameters change with specific types of skill development remains to be determined. Performance of skills may also require coordination with other regions, especially the cerebellum, which also projects to the thalamocortical network.

11.10 COMPULSIVE BEHAVIOR

The shift in the mode of action generation has often been associated with the development of compulsive behavior in animal models of addiction (Everitt & Robbins, 2005). Some believe that addiction involves a switch from controlled to compulsive drug seeking, which is characterized by persistent seeking despite aversive consequences. This is often measured using resistance to foot shock punishment in rodent models of addiction.

Everitt and colleagues have argued that neural mechanisms underlying the development of compulsive drug seeking are similar to those in habit formation (Everitt et al., 2008). This idea is sometimes called the "habit theory" of addiction. In rats, the development of cocaine seeking seems to resemble habit formation: at first, limbic BG circuitry (e.g. nucleus accumbens core) is involved, but with extended training, the DLS becomes critical for the maintenance of cocaine seeking (Belin & Everitt, 2008; Belin, Mar, Dalley, Robbins, & Everitt, 2008). Cocaine seeking is impaired by infusions of the dopamine (DA) receptor antagonist into the DLS (Belin & Everitt, 2008). Disconnecting the nucleus accumbens core from the dopaminergic innervation of the dorsal striatum also impairs cocaine seeking. Belin and Everitt therefore argued that this progression from ventral to more dorsal domains of the striatum is mediated by the spiraling connections with the midbrain DA neurons (see Chapter 4). Research on monkey cocaine self-administration also shows a similar pattern of progressive involvement of limbic, association, and sensorimotor striatal regions (Porrino, Lyons, Smith, Daunais, & Nader, 2004).

Others have questioned this interpretation of the habit theory of addiction (Vandaele & Ahmed, 2021). As pointed out by Vandaele and Ahmed, it is not sufficient to define habit as "not goal-directed"; nor is it adequate to rely exclusively on devaluation insensitivity as the operational definition of habit. They suggest that, on the contrary, overestimation of drug value may be responsible for devaluation-insensitive behavior. This argument is supported by subjective reports of addicts, who are usually aware of the aversive consequences of their actions but still perform such actions persistently. One prediction is that compulsive drug seeking should be accompanied by potentiation in the connection between outcome representation and the action rate control system. There is some evidence in support of this prediction. In one study on methamphetamine self-administration, as rats became drug-dependent, two frontostriatal circuits showed opposite patterns of change in resting-state functional connectivity (Hu et al., 2019). Following 20 days of self-administration, the circuit strength of an orbitofrontal-DMS projection increased, while prelimbic-ventral striatal circuit connectivity became weaker.

It is important to emphasize that experimental designs used in studies of addiction are different from those used to study habit formation. First, in distinguishing goal-directed and habitual performance, typically two tests are used—manipulations of the instrumental contingency, like degradation and omission, and manipulations of outcome value, like devaluation. By itself, devaluation sensitivity may not be sufficient. Secondly, the key probe tests are conducted in extinction without reward feedback. With outcome feedback after devaluation, habitual animals return to the goal-directed mode and reduce selection of the action leading to the devalued outcome. Final outcome feedback essentially serves as a reminder to interrupt their control of local sensory feedback. But studies of shock-resistant drug seeking rarely manipulate the instrumental contingency or use extinction probe tests; instead, they often use persistent responding despite electrical shocks paired with drug delivery as a measure of compulsive behavior (Belin & Everitt, 2008).

Devaluation insensitivity alone is not sufficient to distinguish between habits and compulsive seeking. Compulsive behavior is indicated by similar performance on rewarded and extinction probe tests, as the reward feedback (now aversive) is not expected to reduce performance. In this case, there is a conflict between two competing systems with different reference states. When the desire for the drug outweighs the reference signal from the shock avoidance system, the animal decides to pursue outcomes despite aversive consequences. Persistent desire and even obsession can be attributed to both enhanced activation of the drug-taking reference state and increased connectivity between this reference state and the action rate controller (Chapter 9). Drug seeking could be due to a high drug-taking reference, which is reflected in the corticostriatal recruitment of action modules. The drug-seeking action outweighs other options.

11.11 SUMMARY

According to the multiple learning and memory systems, the hippocampus is responsible for declarative learning, whereas the BG mediate procedural learning. While this distinction is a useful summary of neuropsychological observations, it fails to adequately classify different types of learning and their corresponding neural substrates. The BG do not merely implement procedural learning. Within the striatum, there is considerable functional heterogeneity. At least three major types of learning processes have been studied: learning to approach predictors of reward, learning to generate actions according to action–outcome relationships, and finally learning to execute behavioral sequences automatically.

During initial learning, performance is governed by explicit outcome representation and instrumental contingency, but with extended training, it can become habitual and no longer guided by outcome expectancy. In goal-directed actions, the outcome representation can guide action generation, whereas in habits, performance is initiated by the discriminative stimulus or feedback associated with the completion of behavioral elements. Habit formation requires the sensorimotor cortico-BG circuit, in particular the DLS. When the DLS is lesioned or inactivated, instrumental lever pressing becomes more sensitive to the devaluation of the reward or to changes in action–outcome contingency. Habit formation is also accompanied by plasticity in the glutamatergic corticostriatal projections of the DLS.

A similar shift in the locus of control from the associative to the sensorimotor BG network has been found during various types of skill learning. There are some similarities between habit formation in instrumental conditioning and skill learning and development of behavioral automaticity. In both, there is reduced attentional demand and increased effector specificity. It is hypothesized that such a shift reflects a shift from a higher level in the motivational hierarchy to a lower level. Each level is associated with distinct types of feedback and control distinct variables. The shift in the cortico-BG network governing performance during habit formation and skill learning is primarily due to changes in the type of feedback involved.

Compulsive drug-seeking behavior has been interpreted as an example of habits, characterized by persistent seeking despite aversive consequences. However, this interpretation has been questioned, as addicts are often fully aware of the negative consequences associated with drug taking. Overestimation of drug value may be responsible for devaluation-insensitive drug-seeking behavior, which is accompanied by potentiation in the associative corticostriatal projections.

REFERENCES

Adams, C. D. (1982). Variations in the sensitivity of instrumental responding to reinforcer devaluation. *Quarterly Journal of Experimental Psychology*, *33b*, 109–122.

Adams, C. D., & Dickinson, A. (1981). Instrumental responding following reinforcer devaluation. *Quarterly Journal of Experimental Psychology*, *33*, 109–122.

Belin, D., & Everitt, B. J. (2008). Cocaine seeking habits depend upon dopamine-dependent serial connectivity linking the ventral with the dorsal striatum. *Neuron*, *57*(3), 432–441.

Belin, D., Mar, A. C., Dalley, J. W., Robbins, T. W., & Everitt, B. J. (2008). High impulsivity predicts the switch to compulsive cocaine-taking. *Science*, *320*(5881), 1352–1355. doi: 10.1126/science.1158136.

Brown, P. L., & Jenkins, H. M. (1968). Auto-shaping the pigeon's key peck. *Journal of the Experimental Analysis of Behavior*, *11*, 1–8.

Colwill, R. M., & Rescorla, R. A. (1986). Associative structures in instrumental learning. In: Bower, G. (Ed.), *The Psychology of Learning and Motivation* (vol. 20, pp. 55–104). New York: Academic Press.

Corbit, L. H., Ostlund, S. B., & Balleine, B. W. (2002). Sensitivity to instrumental contingency degradation is mediated by the entorhinal cortex and its efferents via the dorsal hippocampus. *Journal of Neuroscience*, *22*(24), 10976–10984.

Corkin, S. (1968). Acquisition of motor skill after bilateral medial temporal-lobe excision. *Neuropsychologia*, *6*(3), 255–265.

Costa, R. M., Cohen, D., & Nicolelis, M. A. (2004). Differential corticostriatal plasticity during fast and slow motor skill learning in mice. *Current Biology*, *14*(13), 1124–1134.

Darwin, C. (1872). *The Expression of the Emotions in Man and Animals*: London: John Murray.

Davis, J., & Bitterman, M. E. (1971). Differential reinforcement of other behavior (dro): A yoked-control comparison. *Journal of the Experimental Analysis of Behavior*, *15*, 237–241.

Devan, B. D., McDonald, R. J., & White, N. M. (1999). Effects of medial and lateral caudate-putamen lesions on place- and cue- guided behaviors in the water maze: Relation to thigmotaxis. *Behavioural Brain Research*, *100*(1–2), 5–14.

Devan, B. D., & White, N. M. (1999). Parallel information processing in the dorsal striatum: Relation to hippocampal function. *Journal of Neuroscience*, *19*(7), 2789–2798.

Dezfouli, A., & Balleine, B. W. (2012). Habits, action sequences and reinforcement learning. *European Journal of Neuroscience*, *35*(7), 1036–1051. doi: 10.1111/j.1460-9568.2012.08050.x.

Dezfouli, A., & Balleine, B. W. (2013). Actions, action sequences and habits: Evidence that goal-directed and habitual action control are hierarchically organized. *PLoS Computational Biology*, *9*(12), e1003364. doi: 10.1371/journal.pcbi.1003364.

Dezfouli, A., Lingawi, N. W., & Balleine, B. W. (2014). Habits as action sequences: Hierarchical action control and changes in outcome value. *Philosophical Transactions of the Royal Society B: Biological Sciences*, *369*(1655), 20130482.

Dhawale, A. K., Wolff, S. B., Ko, R., & Ölveczky, B. P. (2021). The basal ganglia control the detailed kinematics of learned motor skills. *Nature Neuroscience*, *24*(9), 1256–1269.

Dias-Ferreira, E., Sousa, J. C., Melo, I., Morgado, P., Mesquita, A. R., Cerqueira, J. J., … Sousa, N. (2009). Chronic stress causes frontostriatal reorganization and affects decision-making. *Science*, *325*(5940), 621–625.

Dickinson, A. (1985). Actions and habits: The development of behavioural autonomy. *Philosophical Transactions of the Royal Society*, *B*, *308*, 67–78.

Dickinson, A. (1989). Expectancy theory in animal conditioning. In: Klein, S. B., & Mowrer, R. R. (Ed.), *Contemporary Learning Theories* (pp. 279–308). Hillsdale, NJ: Lawrence Erlbaum Associates.

Dickinson, A. (1994). Instrumental conditioning. In: Mackintosh, N. J. (Ed.), *Animal Learning and Cognition* (pp. 45–79). Orlando, FL: Academic Press.

Dickinson, A., Squire, S., Varga, Z., & Smith, J. W. (1998). Omission learning after instrumental pretraining. *Quarterly Journal of Experimental Psychology*, *51*(3), 271–286.

Everitt, B. J., Belin, D., Economidou, D., Pelloux, Y., Dalley, J. W., & Robbins, T. W. (2008). Neural mechanisms underlying the vulnerability to develop compulsive drug-seeking habits and addiction. *Philosophical Transactions of the Royal Society B: Biological Sciences*, *363*(1507), 3125–3135. doi: 10.1098/rstb.2008.0089.

Everitt, B. J., & Robbins, T. W. (2005). Neural systems of reinforcement for drug addiction: From actions to habits to compulsion. *Nature Neuroscience*, *8*(11), 1481–1489.

Foerde, K., & Shohamy, D. (2011a). Feedback timing modulates brain systems for learning in humans. *Journal of Neuroscience*, *31*(37), 13157–13167.

Foerde, K., & Shohamy, D. (2011b). The role of the basal ganglia in learning and memory: Insight from parkinson's disease. *Neurobiology of Learning and Memory*, *96*(4), 624–636.

Hassabis, D., & Maguire, E. A. (2007). Deconstructing episodic memory with construction. *Trends in Cognitive Sciences*, *11*(7), 299–306.

Hershberger, W. A. (1986). An approach through the looking glass. *Animal Learning & Behavior*, *14*, 443–451.

Hirsh, R. (1974). The hippocampus and contextual retrieval of information from memory: A theory. *Behavioral Biology*, *12*(4), 421–444.

Holland, P. C. (1979). Differential effects of omission contingencies on various components of pavlovian appetitive conditioned responding in rats. *Journal of Experimental Psychology: Animal Behavior Processes*, *5*(2), 178–193.

Hu, Y., Salmeron, B. J., Krasnova, I. N., Gu, H., Lu, H., Bonci, A., … Yang, Y. (2019). Compulsive drug use is associated with imbalance of orbitofrontal-and prelimbic-striatal circuits in punishment-resistant individuals. *Proceedings of the National Academy of Sciences*, *116*(18), 9066–9071.

James, W. (1890). *The Principles of Psychology*: New York: Henry Holt.

Jueptner, M., Frith, C. D., Brooks, D. J., Frackowiak, R. S., & Passingham, R. E. (1997). Anatomy of motor learning. Ii. Subcortical structures and learning by trial and error. *Journal of Neurophysiology*, *77*(3), 1325–1337.

Jueptner, M., Stephan, K. M., Frith, C. D., Brooks, D. J., Frackowiak, R. S., & Passingham, R. E. (1997). Anatomy of motor learning. I. Frontal cortex and attention to action. *Journal of Neurophysiology*, *77*(3), 1313–1324.

Knowlton, B. J., Mangels, J. A., & Squire, L. R. (1996). A neostriatal habit learning system in humans. *Science*, *273*(5280), 1399–1402.

McNamara, H. J., Long, J. B., & Wike, E. L. (1956). Learning without response under two conditions of external cues. *Journal of Comparative and Physiological Psychology*, *49*(5), 477.

Milner, B. (1970). Memory and the medial temporal regions of the brain. *Biology of Memory*, *23*, 31–59.

Mishkin, M., Malamut, B., & Bachevalier, J. (1984). Memories and habits: Two neural systems. In: Lynch, G., McGaugh, J. L., & Weinberger, N. (Eds.), *Neurobiology of Learning and Memory* (pp. 65–77). New York: Guilford Press.

Mowrer, O. (1960). *Learning Theory and Behavior*: New York: John Wiley & Sons.

Murray, E. A., Wise, S. P., & Graham, K. S. (2018). Representational specializations of the hippocampus in phylogenetic perspective. *Neuroscience Letters*, *680*, 4–12.

O'Hare, J. K., Ade, K. K., Sukharnikova, T., Van Hooser, S. D., Palmeri, M. L., Yin, H. H., & Calakos, N. (2016). Pathway-specific striatal substrates for habitual behavior. *Neuron*, *89*(3), 472–479. doi: 10.1016/j.neuron.2015.12.032.

O'Hare, J. K., Li, H., Kim, N., Gaidis, E., Ade, K., Beck, J., ... Calakos, N. (2017). Striatal fast-spiking interneurons selectively modulate circuit output and are required for habitual behavior. *Elife, 6*, e26231.

O'Keefe, J., & Nadel, L. (1978). *The Hippocampus as a Cognitive Map*: Oxford: Clarendon Press.

Packard, M. G. (1999). Glutamate infused posttraining into the hippocampus or caudate-putamen differentially strengthens place and response learning. *Proceedings of the National Academy of Sciences, 96*(22), 12881–12886.

Packard, M. G., & McGaugh, J. L. (1996). Inactivation of hippocampus or caudate nucleus with lidocaine differentially affects expression of place and response learning. *Neurobiology of Learning and Memory, 65*(1), 65–72.

Partiot, A., Verin, M., Pillon, B., Teixeira-Ferreira, C., Agid, Y., & Dubois, B. (1996). Delayed response tasks in basal ganglia lesions in man: Further evidence for a striato-frontal cooperation in behavioural adaptation. *Neuropsychologia, 34*(7), 709–721.

Perez, O., & Dickinson, A. (2020). A theory of actions and habits: The interaction of rate correlation and contiguity systems in free-operant behavior. *Psychological Review, 127*(6), 945.

Porrino, L. J., Lyons, D., Smith, H. R., Daunais, J. B., & Nader, M. A. (2004). Cocaine self-administration produces a progressive involvement of limbic, association, and sensorimotor striatal domains. *Journal of Neuroscience, 24*(14), 3554–3562.

Rand, M. K., Hikosaka, O., Miyachi, S., Lu, X., & Miyashita, K. (1998). Characteristics of a long-term procedural skill in the monkey. *Experimental Brain Research, 118*(3), 293–297.

Rand, M. K., Hikosaka, O., Miyachi, S., Lu, X., Nakamura, K., Kitaguchi, K., & Shimo, Y. (2000). Characteristics of sequential movements during early learning period in monkeys. *Experimental Brain Research, 131*(3), 293–304.

Redish, A. D. (2016). Vicarious trial and error. *Nature Reviews Neuroscience, 17*(3), 147.

Restle, F. (1957). Discrimination of cues in mazes: A resolution of the "place-vs.-response" question. *Psychological Review, 64*(4), 217.

Saint-Cyr, J., Taylor, A. E., & Lang, A. (1988). Procedural learning and neostriatal dysfunction in man. *Brain, 111*(4), 941–960.

Schwabe, L., Dickinson, A., & Wolf, O. T. (2011). Stress, habits, and drug addiction: A psychoneuroendocrinological perspective. *Experimental and Clinical Psychopharmacology, 19*(1), 53.

Scoville, W. B., & Milner, B. (1957). Loss of recent memory after bilateral hippocampal lesions. *Journal of Neurology, Neurosurgery, and Psychiatry, 20*(1), 11–21.

Shiffrin, W., & Schneider, R. M. (1977). Controlled and automatic human information processing: 11. Perceptual learning, automatic attending, and a general theory. *Psychological Review, 84*, 127–190.

Thorn, C. A., Atallah, H., Howe, M., & Graybiel, A. M. (2010). Differential dynamics of activity changes in dorsolateral and dorsomedial striatal loops during learning. *Neuron, 66*(5), 781–795. doi: 10.1016/j.neuron.2010.04.036.

Thorndike, E. L. (1907). The mental antecedents of voluntary movements. *The Journal of Philosophy, Psychology and Scientific Methods, 4*(2), 40–42.

Tolman, E. C. (1932). *Purposive Behavior in Animals and Man*: New York: Macmillan.

Tolman, E. C., Ritchie, B., & Kalish, D. (1946a). Studies in spatial learning. I. Orientation and the short-cut. *Journal of Experimental Psychology, 36*(1), 13.

Tolman, E. C., Ritchie, B., & Kalish, D. (1946b). Studies in spatial learning. Ii. Place learning versus response learning. *Journal of Experimental Psychology, 36*(3), 221.

Tricomi, E., Balleine, B. W., & O'Doherty, J. P. (2009). A specific role for posterior dorsolateral striatum in human habit learning. *European Journal of Neuroscience, 29*(11), 2225–2232. doi: 10.1111/j.1460-9568.2009.06796.x.

Vandaele, Y., & Ahmed, S. (2021). Habit, choice, and addiction. *Neuropsychopharmacology, 46*(4), 689–698.

Weiskrantz, L., & Warrington, E. K. (1979). Conditioning in amnesic patients. *Neuropsychologia, 17*(2), 187–194.

Whishaw, I. Q., Mittleman, G., Bunch, S. T., & Dunnett, S. B. (1987). Impairments in the acquisition, retention and selection of spatial navigation strategies after medial caudate-putamen lesions in rats. *Behavioural Brain Research, 24*(2), 125–138.

White, N. M. (1997). Mnemonic functions of the basal ganglia. *Current Opinion in Neurobiology, 7*(2), 164–169.

Yin, H. H. (2010). The sensorimotor striatum is necessary for serial order learning. *Journal of Neuroscience, 30*(44), 14719–14723. doi: 10.1523/JNEUROSCI.3989-10.2010.

Yin, H. H., & Knowlton, B. J. (2004). Contributions of striatal subregions to place and response learning. *Learning & Memory, 11*(4), 459–463.

Yin, H. H., & Knowlton, B. J. (2006). The role of the basal ganglia in habit formation. *Nature Reviews Neuroscience*, *7*(6), 464–476.

Yin, H. H., Knowlton, B. J., & Balleine, B. W. (2004). Lesions of dorsolateral striatum preserve outcome expectancy but disrupt habit formation in instrumental learning. *European Journal of Neuroscience*, *19*(1), 181–189.

Yin, H. H., Knowlton, B. J., & Balleine, B. W. (2006). Inactivation of dorsolateral striatum enhances sensitivity to changes in the action-outcome contingency in instrumental conditioning. *Behavioural Brain Research*, *166*(2), 189–196.

Yin, H. H., Mulcare, S. P., Hilario, M. R., Clouse, E., Holloway, T., Davis, M. I., ... Costa, R. M. (2009). Dynamic reorganization of striatal circuits during the acquisition and consolidation of a skill. *Nature Neuroscience*, *12*(3), 333–341. doi: 10.1038/nn.2261.

Yin, H. H., Ostlund, S. B., Knowlton, B. J., & Balleine, B. W. (2005). The role of the dorsomedial striatum in instrumental conditioning. *European Journal of Neuroscience*, *22*, 513–523.

12 Dopamine and Reinforcement Learning

Having considered the contributions of the BG to learning at the level of neural circuits and behavior, we are now ready to examine the algorithms governing learning in more detail. According to a popular interpretation, learning is achieved using a reward prediction error (RPE), which acts as a teaching signal used to update state–action associations. The RPE signal is thought to be implemented by phasic dopamine (DA), and the state–action associations are implemented by the weight of the corticostriatal synapse (Miller, 1981; Schultz, 1998).

This chapter provides a critical examination of the research on phasic DA signaling and reinforcement learning (RL). We shall begin by reviewing the development of the concept of prediction errors in theories of learning. We shall then review the original work on phasic DA activity that gave rise to the RPE hypothesis of DA function, as well as more recent research that challenges this hypothesis. Finally, we shall discuss key limitations of standard RL models used to explain the BG's role in learning, and alternative explanations for experimental findings commonly used to support the RPE model.

12.1 PREDICTION ERRORS

Prediction error is a key feature in modern theories of learning. Bush and Mosteller attempted to model Hull's idea that learning consists of a graduate increase in habit strength (sHr) as a function of reinforcement (Bush & Mosteller, 1951). According to them, the change in habit strength is proportional to the difference between the maximum possible strength (corresponding to a probability of 1 for the response) and the current strength. This difference is the effective reinforcement. Following each reinforcement, habit strength increases, but the amount of increase is offset by the current strength. During acquisition, the amount of learning on each trial decreases, creating a negatively accelerating learning curve. When the probability reaches 1, the associative strength will cease to grow as the learning curve asymptotes. The subtraction of prediction error limits the extent of learning, and the growth of habit strength is therefore capped at its maximum value. This model also generates a prediction on the probability of response on any trial (Bush & Mosteller, 1951). The implicit assumption is that the relevant behavioral measure is the probability of response during the trial. This measure is directly proportional to reward prediction.

12.1.1 Rescorla–Wagner Model

The next significant development in learning theory was the Rescorla–Wagner model. In this model, which focuses on Pavlovian conditioning, the effectiveness of the reinforcer (US, or reward) also decreases over the course of learning, similar to the Bush–Mosteller model. But a new feature was introduced: rather than the associative strength accruing to the cue only, learning would depend on the "total associative strength of the compound in which that stimulus appears, whereas for Hull only the strength of the component in question was relevant" (Rescorla & Wagner, 1972, P. 74). The prediction error is the difference between the actual outcome of a trial and the amount predicted by all existing cues. This small change allows the model to explain the so-called cue competition effects like blocking and relative validity (Dickinson, 1980; Kamin, 1967; Rescorla, 1968; Wagner, Logan, & Haberlandt, 1968). For example, in blocking, the presence of another previously learned predictor impairs learning of a new predictor. As effective reinforcement is reduced by the sum of all current predictors of the US (Vsum), if there is already a predictor present, then the new predictor will not 'absorb' learning. Consequently, learning of the association between the new predictor and the outcome will be 'blocked.'

The key intuition in the Rescorla–Wagner model is surprise, as the prediction error indicates how surprising the US is. If it is already predicted by another cue, it will not be surprising, so its ability to change the association is also reduced. When the reward is more than predicted, a positive RPE will increase associative strength. When the reward is less than expected (e.g. when it is omitted as in extinction), a negative RPE will reduce associative strength. This model is also known as a US processing model because the prediction term is subtracted from the maximum reinforcement signal, thus reducing the processing of the US or reinforcement.

12.1.2 Temporal Difference Algorithm

The next stage in the evolution of prediction error was the development of the temporal difference (TD) algorithm, which can be viewed as a real-time extension of the Rescorla–Wagner model (Sutton, 1988; Sutton & Barto, 1987). The prediction error is known as the TD error, which reflects the difference between the current prediction and the last prediction one time step ago. It is used to improve the estimates of value function, the sum of all future rewards. Both Bush–Mostellar and Rescorla–Wagner use a maximum value as the target. On the other hand, the TD algorithm uses bootstrapping, in which the current estimate serves as the target value. Compared to traditional supervised methods in neural networks that compare the final outcome with some observation (for example the Widrow–Hoff delta rule), TD error can be computed incrementally at each time step. TD learning also incorporates temporal discounting, so that greater changes are made to more recent predictions. The primary reinforcement signal decreases as it becomes predicted and is gradually replaced by acquired predictions. The reinforcement signal also moves backward in time from the primary reinforcer to the earliest predictor. Figure 12.1 provides a summary of the different models that use prediction errors.

12.2 PRINCIPLES OF RL

RL models have components that are similar to the associative structures discussed in Chapter 10. This is hardly surprising since they are all influenced by Hull's conceptual framework. The state is similar to the traditional notion of a stimulus, which could be internal or external. The policy refers to the behavioral output, similar to the response. Finally, the reward function is similar to the traditional concept of reinforcement. Each state or state–action pair is assigned a value indicating the desirability of that state, i.e. its reward function. What is modified in RL is assumed to be the mapping from states to actions, expressed as the probability of taking a particular action in a given state.

In supervised learning, the target response of the system is known. Feedback tells the system the deviation from the target, and this deviation is used to update the weights (as implemented by synaptic strength). With training, the system's output will eventually converge on the target. RL is usually considered unsupervised learning because there is no explicit target before learning. The learning system does not get instructions on what the correct response is. Instead, the feedback tells the system whether the current state is better or worse than before. This is based on a reward function, which assigns some value to each state. In computing the reward function, there is usually temporal discounting, so that immediate rewards are weighted more strongly than delayed rewards.

The most common model used to explain BG contribution to RL is the actor-critic model (Sutton & Barto, 2018). The actor contains an activation function that transforms value into probability of action selection. Its neural implementation is widely believed to be the dorsal striatum. The critic receives feedback and predicts future rewards based on the current state. This reward function is updated using the RPE. The critic also assigns a value to the actor's action, called the Q-value, according to the reward function (Sutton & Barto 1990). Without the critic, the actor can still perform actions, but it would not be able to select the best action to maximize reward because it would not know which action is better.

Bush-Mostellar (1955)

$\Delta p_n = \beta (\lambda - p_n)$

P_n is the probability of response on trial n.

λ is the asymptote of learning (maximum possible).

$\lambda - p_n$ is the prediction error.

β is the learning rate parameter (between 0 and 1).

Rescorla-Wagner (1972)

$\Delta V_{cs} = \alpha\beta (\lambda - V_{sum})$

ΔV_{CS} is the change in associative strength of CS

V_{sum} is the total associative strength of all stimuli present

λ is the asymptote of learning

$\lambda - V_{sum}$ is the prediction error.

α is the CS-dependent rate parameter.

β is the US-dependent rate parameter.

Temporal difference (1988)

$V_t \leftarrow V_t + \alpha(R_{t+1} + \gamma V_{t+1} - V_t)$

Or

$V_t \leftarrow (1-\alpha)V_t + \alpha(R_{t+1} + \gamma V_{t+1})$

V_t is the value function at time t

R_t is the reward at time t

$R_{t+1} + \gamma V_{t+1}$ is the TD target

$R_{t+1} + \gamma V_{t+1} - V_t$ is the TD prediction error

α is the learning rate parameter

γ is the temporal discount rate

FIGURE 12.1 Evolution of prediction error models of learning.

The actor selects actions based on the Q-value, which can be updated using RPE (δ):

$$\delta = r - Q$$

where δ is the prediction error and r is the reward. Note that Q is the value of the chosen action, not all available actions. A learning rate parameter α is used to determine step size and scale the impact of the RPE on subsequent Q-value according to the following equation:

$$Q_{t+1} = Q_t + \alpha * \delta$$

The critic starts out by predicting a zero reward, so the TD error is zero until the first reward. As soon as a reward is received, however, the TD error increases to 1. It updates the activation function in the actor and increases the tendency to repeat the last action.

The Q-value is typically interpreted as the weight of the activation function of the actor, which is implemented by the striatum (Barto, 1995; Houk, Adams, & Barto, 1995). The corticostriatal input is plastic, reflecting the weight that can be modified by RPE, which is thought to be implemented by the phasic DA signal. For example, if a lever press results in more reward than predicted, then a positive RPE will be generated, and the Q-value will be increased by strengthening the corticostriatal synapse to that particular unit.

12.3 PHASIC DA AND RPE

The development of models using prediction errors proceeded independently of work on neuroscience, but influential experiments discovered a striking parallel between the phasic activity of DA neurons and RPE (Hollerman & Schultz, 1998; Tomas Ljungberg, Apicella, & Schultz, 1991; Ljungberg, Apicella, & Schultz, 1992; Mirenowicz & Schultz, 1994; Romo & Schultz, 1990; Tobler, Fiorillo, & Schultz, 2005; Waelti, Dickinson, & Schultz, 2001). In these experiments, Schultz and colleagues recorded single unit activity from DA neurons in behaving monkeys. Because substantia nigra pars compacta (SNc) and ventral tegmental area (VTA) DA neurons appear to show similar properties, they are usually lumped together in the analysis. Because it had already been well established that DA depletion results in akinesia, Schultz and colleagues first attempted to examine the relationship between DA activity and motor function (Ungerstedt, 1971). Yet despite multiple attempts, they did not find a clear relationship between DA activity and motor output (Delong, Crutcher, & Georgopoulos, 1983; Romo & Schultz, 1990; Schultz, Ruffieux, & Aebischer, 1983). Instead, many DA neurons showed burst firing after the presentation of rewards (Ljungberg et al., 1992; Romo & Schultz, 1990; Schultz, 1986). Such responses were less common with aversive stimuli (Hollerman & Schultz, 1998; Schultz & Romo, 1987).

In early studies, instrumental conditioning methods were used. For example, when a pair of pictures appeared, the monkey had to touch the lever below the rewarded picture to receive a drop of juice (Schultz, Apicella, & Ljungberg, 1993). Many DA neurons showed phasic activity when monkeys touched the hidden food or when the picture appeared (Romo & Schultz, 1990). Later studies used Pavlovian conditioning. For example, monkey viewed visual stimuli that predicted reward delivery, and licking was used as the main behavioral measure (Fiorillo, Tobler, & Schultz, 2003).

Regardless of the task used, a common pattern was reported. First, DA shows a short burst of activity immediately after reward delivery. Later, this phasic DA response is generated earlier in time, in response to cues that predict reward or at the time of action initiation (Schultz, Dayan, & Montague, 1997; Suri & Schultz, 2001). The response to the primary reward only occurs when the reward is not predicted. But if a predicted reward is omitted (negative prediction error), there is a dip in DA activity at the time of expected reward (Tomas Ljungberg et al., 1991). Presenting the reward earlier than expected is also sufficient to elicit the full phasic response found with unpredicted rewards (Hollerman & Schultz, 1998; Schultz et al., 1993). Together, these results suggest that the phasic DA response represents the RPE (Figure 12.2).

The TD algorithm can explain the shift in activation from reward to predictor during learning. Before learning, the presentation of the reward for the first time elicits a strong phasic response, as the predicted reward is zero and the current reward is maximal. The RPE, the difference between the two, is also high. After learning, the reward no longer elicits RPE, as it is fully predicted. Instead, the conditioned stimulus (CS), being the earliest predictor, elicits the RPE signal. When the reward is omitted, the RPE is negative, so that a dip in DA activity is expected at the time of reward delivery.

The key operation needed to compute the RPE is a subtraction, which can be implemented in multiple ways. In the simplest case, if a DA neuron receives two distinct inputs, one excitatory and the other inhibitory, the output of the DA neuron could be proportional to the difference between the two (Figure 12.2).

Some have proposed that distinct pathways mediate the CS-burst in DA neurons and the inhibition of response to the US (Contreras-Vidal & Schultz, 1999). Many prefrontal cortical neurons respond to primary rewards and reward-predicting stimuli (Schultz, Tremblay, & Hollerman, 1998). They are in a position to excite DA neurons directly and generate phasic activity at the time of the CS presentation (Kim et al., 2015; Overton & Clark, 1997). In addition, the prefrontal cortex also projects to the striosomes, which are a source of direct projections to DA neurons, which may produce delayed inhibition of DA neurons (Gerfen, 1992; Houk et al., 1995). When expected rewards

Dopamine and Reinforcement Learning

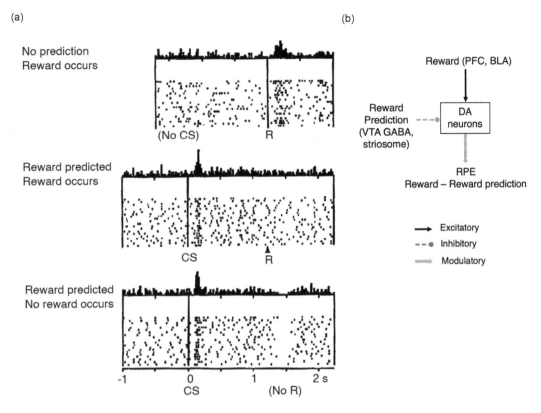

FIGURE 12.2 Phasic dopamine (DA) and reward prediction error (RPE). (a) Phasic DA activity that appears to support the RPE hypothesis. The phasic activity of DA neurons shifts to the time of the earliest sensory cue predicting reward. From Schultz, Dayan, and Montague (1997) with permission. (b) Possible neural circuit implementation of RPE, in which actual and predicted reward signals are compared, using excitatory and inhibitory inputs to DA neurons. PFC, prefrontal cortex; BLA, basolateral amygdala, GABA, glutamate decarboxylase; VTA, ventral tegmental area; CS, conditioned stimulus; R, reward.

are not received, striosomal inhibition of SNc is unopposed by excitation and reduces DA activity, thus producing the dip when the expected reward is omitted (Brown, Bullock, & Grossberg, 1999).

Some have proposed that the habenula is critical for computing negative prediction errors (Hikosaka, 2010; Matsumoto & Hikosaka, 2007). Many lateral habenula neurons often show the opposite pattern as DA neurons: they pause firing when the monkey receives a reward. These neurons are activated by predictors of aversive stimuli or reward omission. The habenula is believed to provide a precisely timed inhibitory input to DA neurons that reduces the effective reinforcement signal, but this projection is assumed to be indirect since habenula neurons are glutamatergic.

12.4 RESULTS THAT CHALLENGE THE RPE HYPOTHESIS

Despite its popularity, the RPE interpretation of DA signaling has been challenged repeatedly (Berridge, 2007; Redgrave, Prescott, & Gurney, 1999). First, DA neurons also respond to any novel, unexpected stimuli. For example, Horvitz et al. (1997) showed that unpredictable light flashes or tones could evoke phasic DA responses even when they did not predict any reward. This was also observed in early studies by Schultz (Ljungberg et al., 1992): DA neurons "...typically responded to a visual or auditory stimulus (outside the task) when it was presented unexpectedly, but stopped responding if the stimulus was repeated; a subtle sound outside the monkey's view was particularly effective." Redgrave argues that the latency (~70–100 ms) of the salience-related phasic DA response is too short for the signal to reflect recognition of reward (Redgrave et al., 1999). If the animals cannot even recognize what type of stimuli are present at the time of the short-latency response of DA, then DA neurons cannot adequately signal events that are 'better' or 'worse' than expected.

In response to these concerns, Schultz argues that there are two components of the DA response: the earlier component (<100 ms) reflects salience by detecting potential reward before fully processing it, and the later component (~150–200 ms) that represents reward value after identification of the object (Schultz, 2016). The earlier component could enhance attention to salient stimuli regardless of their value, but it is the later component that serves as the effective reinforcement or teaching signal. We shall return to this argument later.

Since aversive stimuli are viewed as the opposite of reward, the RPE hypothesis predicts the exact opposite response—a pause in DA neurons. Some have argued that DA neurons are generally less responsive to aversive stimuli or sometimes suppress firing (Fiorillo, 2013; Matsumoto & Hikosaka, 2009; Schultz, 1998, 2016). But other studies showed phasic DA activity in response to aversive stimuli. Other studies found similar phasic activity following aversive stimuli (Barter et al., 2015; Horvitz, 2000; Wang & Tsien, 2011). For example, as reviewed in Chapter 7, in SNc DA neurons, the representation of movement velocity is the same regardless of outcome valence (Barter et al., 2015). Neurons that show phasic activity during movement to collect sucrose rewards also show phasic activity during movement to avoid aversive air puffs. Their velocity tuning remained similar, even though the movements were very different.

12.5 VALUE, PERFORMANCE, AND RPE

In models based on prediction error, the key operation is a subtraction, but they differ in how the terms are defined and what is being subtracted from what. In the Bush–Mosteller model, the terms being compared are the maximum possible strength of the association between CS and response and the current strength. In the Rescorla–Wagner model, the total existing associative strength is subtracted from the maximum possible strength. The unconditioned stimulus (US) magnitude is assumed to be equal to the maximum, as the US is assumed to elicit unconditioned response (UR) in an S–R association. In TD, the last estimate is subtracted from the current estimate of reward value.

In prediction error models, the existing associative strength is equated with the probability of conditioned response (CR), and a change in associative strength or learning is therefore equated with a change in performance. If the maximum probability is 1, the error correction will result in convergence to 1. In RL, the equivalent of associative strength is the value function. In Pavlovian conditioning, values are assigned to arbitrary stimuli, whereas in instrumental conditioning, values are assigned to actions.

In the actor-critic model, the actor is often assumed to consist of a collection of striatal neurons representing different actions, like a giant switchboard with many units, each representing a distinct action (Suri, 2002). The activation function would largely reflect the connection weight at the corticostriatal synapse. Q-value might represent the strength of activation of a group of direct pathway spiny projection neurons (dSPNs), resulting in the selection of some action via disinhibition. If the corticostriatal synapses reflect an S–R connection, an increase in this weight should increase the probability of selecting the behavior, and decreasing the weight also reduces the probability of response (Barto, 1995; Houk et al., 1995; Joel, Niv, & Ruppin, 2002). However, RL does not explain how the action value generates any action.

12.6 PHASIC DA AND PERFORMANCE

More recent work has shown that the pattern of DA activity interpreted as reflecting RPE can be explained by changes in performance. Traditional experimental designs rely on the presentation and measurement of discrete events like cues and rewards. The implicit assumption is that neural activity around the time of these events is only related to them. What is ignored is what the animals are actually doing when stimuli are presented. Most studies that reported RPE signals in DA neurons used head-fixed animals, but such animals often move or attempt to move, but these movements were not measured.

As already reviewed in Chapter 7, SNc DA neurons signal kinematic variables, and opponent neuronal populations can be interpreted in relation to a specific direction of movement (Barter et al., 2015). There are distinct populations of SNc DA neurons with opponent phasic activity (Barter et al., 2015; Barter, Castro, Sukharnikova, Rossi, & Yin, 2014; Rossi, Fan, Barter, & Yin, 2013). Such activity is not predicted by the RPE model.

In head-restrained mice, recent work has directly tested the RPE hypothesis using precise measurements of force exerted (Bakhurin, Hughes, Barter, Zhang, & Yin, 2020; Hughes, Bakhurin, Barter, Zhang, & Yin, 2020, Bakhurin et al., 2023). As discussed in Chapter 9, VTA DA neurons represent force generated over time in a direction-specific manner (Hughes et al., 2020). On a stimulus–reward task with force measurements, opponent populations of VTA DA neurons were found. Many DA neurons showed phasic activity that predicted forward movement, which reflects anticipatory approach movement. The latency of forward movement after CS presentation was determined by the magnitude and latency of the phasic DA burst. Other DA neurons showed the opposite response; they were selectively active during backward force exertion.

During learning, there was a gradual shift in the timing of forward force exertion, from just after reward delivery to the time of CS presentation. This shift in phasic DA activity from the US to the CS can be explained by the shift in force exertion during acquisition (Figure 12.3). As force exertion became anticipatory, phasic DA activity increases as soon as the cue was delivered. These phasic force-related responses were found at the start of training. In fact, the relationship between force and DA activity remained the same throughout training, even though the force generated at the time of CS increased significantly. Similar force tuning was also found during the inter-trial interval when mice showed spontaneous movements.

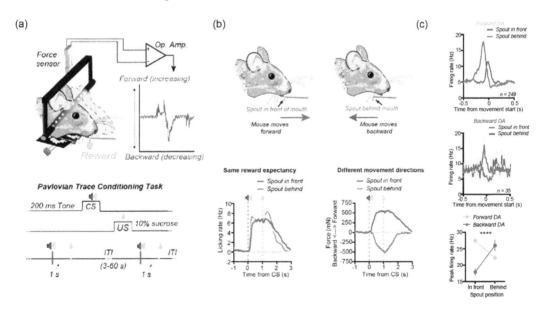

FIGURE 12.3 Patterns of phasic dopamine (DA) activity during stimulus–reward learning can be explained by changes in force exertion. (a) Illustration of the Pavlovian conditioning task. (b) Manipulation of reward spout location does not change reward prediction, as shown by anticipatory licking behavior, but changes the direction of force exertion. (c) There are two major types of ventral tegmental area DA neurons: one type that fires preferentially for forward force exertion and a second type that fires preferentially for backward force exertion. These neurons show distinct patterns of phasic activity when the location of reward spout is varied, even if the reward expectancy is constant. CS, conditioned stimulus; US, unconditioned stimulus.

If learning is accompanied by changes in performance, then how can learning and performance be dissociated? Bakhurin et al. attempted to dissociate the role of DA in performance and in RPE by manipulating the direction of force exertion on the stimulus–reward task while maintaining the same reward predictability. The reward spout was moved either in front of or behind the mouth. Because the mice received the same reward regardless of the direction of movement, there is no difference in reward predictability when the spout position is moved by a few millimeters. The RPE hypothesis therefore predicts that DA activity should be equivalent in these conditions. The change in spout location did not change reward prediction, as indicated by similar anticipatory licking (CR). But it dramatically changed the direction of force needed to reach the reward (Figure 12.3). This is also reflected in a change in DA activity when movement direction changes, reflecting the force tuning of each DA neuron recorded. Thus "forward" neurons fired more when the spout was in front and the movement was in the forward direction, whereas "backward" neurons fired more when the spout was moved slightly backward, so that backward force exertion was needed.

Previous studies showed that phasic DA signals were modulated by reward probability and magnitude (Eshel et al., 2015; Fiorillo et al., 2003), but these studies did not quantify the behavioral changes associated with these manipulations. When the force exerted was measured, Bakhurin et al. found that the changes in phasic DA could also be explained by changes in behavioral performance. For example, with larger rewards, the forward force exerted after the CS was also greater.

Another well-known observation often used to support the RPE hypothesis is pauses in DA activity at the time of predicted reward when the reward is omitted. This is interpreted as reflecting a negative RPE. However, Bakhurin et al. found that this pattern can also be explained by the changes in force exertion (Bakhurin et al., 2023). Normally, mice exerted force in the forward direction following reward delivery, but reward omission resulted in a pause or reduction in forward force exertion. This was reflected in a dip in the firing rates of DA neurons that preferentially fire during forward force exertion (Figure 12.4).

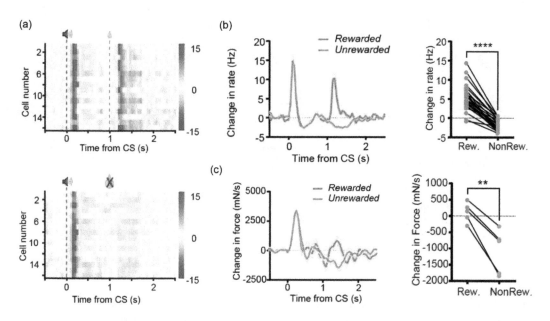

FIGURE 12.4 A dip in dopamine (DA) activity at the time of reward omission can be explained by the direction of force exertion. (a) Simultaneously recorded DA neurons that increase firing during forward force exertion during trials with expected reward delivery (top) and omission (bottom). Note the characteristic 'dip' in DA neurons when the reward is omitted. (b) Summary of DA activity on rewarded and unrewarded trials, showing the dip in activity. (c) Changes in force exertion show a similar pattern. CS, conditioned stimulus.

If the animal is not restrained, force generation is the equivalent of a change in movement velocity. These results are therefore in agreement with previous work on the relationship between DA activity and movement kinematics (Barter et al., 2015). In freely moving animals, DA activity represents variables like velocity. Force tuning in VTA DA neurons does not change in the course of learning. Nor is the relationship between SNc DA activity and kinematics altered by learning.

The behavioral changes observed during learning, such as greater and earlier anticipatory movement, can be explained, at least in part, by changes in the input to the DA neurons. Instead of driving learning in downstream neurons, plasticity in structures somewhere upstream of DA neurons or in the inputs to DA neurons may be responsible for the observed changes (Stuber et al., 2008).

DA neurons receive a wide variety of inputs (Watabe-Uchida, Zhu, Ogawa, Vamanrao, & Uchida, 2012). The correlation between the transition command (efference copy) and other inputs to the DA neurons could result in plasticity in the latter inputs, allowing them to activate DA directly as a result of learning. Consequently, predictive sensory inputs will be sufficient to activate DA neurons after learning. Whether the plasticity takes place at the glutamatergic synapses on DA neurons or elsewhere, the end result is anticipatory activation of DA neurons.

As mentioned earlier, Schultz argues that there are two components of DA signaling: an earlier component signaling salience and a later component signaling reward value. Neither component is related to movement (Schultz, 2016). According to him, large movements may activate DA neurons, but well-controlled movements involving a limited number of muscles do not (Schultz, 1986; Schultz et al., 1983). This argument fails to explain why DA neurons that show typical phasic activity relative to cues actually contribute to force generation. Nor can it explain the selectivity of DA neurons for different directions of movement or force exertion.

As reviewed in Chapter 6, the BG output nuclei send descending commands that specify specific body configuration changes, mainly through the reticulospinal pathway. Movements of the body, including head and torso movements, are the most basic and evolutionarily conserved movements, as opposed to restricted arm movements in a restrained monkey. This could explain why Schultz and colleagues failed to find a consistent relationship between DA activity and movement in their original studies, as their primary measure was arm electromyography in restrained monkeys sitting in a chair. While the monkey is expected to produce a variety of movements or attempts to move as measurable by force exerted, these movements were neglected in their experiments.

Some have attempted to reconcile the RPE hypothesis with other findings on DA. According to one view, there are multiple types of DA neurons that encode different variables: some encode reward, some aversion, and some movement (da Silva, Tecuapetla, Paixão, & Costa, 2018; Lammel, Ion, Roeper, & Malenka, 2011; Lammel et al., 2012; Menegas, Akiti, Amo, Uchida, & Watabe-Uchida, 2018). RPE is therefore considered only one of the functions performed by a subset of DA neurons. This argument is not supported by results showing that neurons related to kinematics satisfy the traditional criterion of RPE encoding. The results reviewed above suggest that the interpretation of phasic DA responses as RPE is mainly due to an experimental confound. All the patterns that appear to support the RPE hypothesis can be explained by subtle changes in performance, which were never measured in previous studies. Likewise, DA neurons that are considered to be related to aversion may also be explained by their role in specific types of movements, in particular backward movements.

There are unquestionably multiple types of DA neurons in the midbrain, but at least some of the heterogeneity can be explained by neurons that are preferentially active for movements in a particular direction. The arguments for multiplexing multiple variables in DA neurons suffer from the same limitations as the RPE hypothesis. Without precise behavioral measurements, it is impossible to rule out the possibility that the DA activity is related to the movement.

12.7 IS DA NECESSARY AND SUFFICIENT FOR LEARNING?

Experiments that manipulated DA signaling often failed to find significant effects on learning. For example, to test whether removing DA prevents learning, Palmiter and colleagues created DA-deficient mice that lack the enzyme, tyrosine hydroxylase, which is critical for DA synthesis

(Cannon & Palmiter, 2003). These mice lack DA and show akinesia, aphagia, and adipsia, just like rats with DA depletion in classic studies. Yet in spite of these drastic behavioral deficits, they were still able to exhibit normal learning (Robinson, Sandstrom, Denenberg, & Palmiter, 2005). For example, they could still learn to have a preference for a spout that delivered sucrose solution and to choose that sucrose spout over a spout that delivered water. DA-deficient mice also display a robust conditioned preference for morphine when given a DA precursor during the testing phases (Hnasko, Sotak, & Palmiter, 2005).

Other studies tested whether increasing DA signaling affects learning. An early attempt to increase DA signaling is to manipulate the DA transporter (DAT), which is critical for pumping the released DA in the synaptic cleft back into the cell. DAT knockdown mice have only ~10% of DA transporters compared to wild-type mice, with an elevated extracellular DA that is about 170% of control mice (Zhuang et al., 2001). In these mice, there was no enhancement of learning, though they did show enhanced performance (Cagniard, Balsam, Brunner, & Zhuang, 2006; Cagniard, Beeler et al., 2006; Yin, Zhuang, & Balleine, 2006). Chronically elevated DA enhances effort for a food reward without apparent effects on Pavlovian and operant learning. Using inducible DA knockdown mice, Cagniard et al. first trained mice before reducing DAT activity. Reducing DAT increased the tonic firing of DA neurons without changing phasic firing. By training mice prior to increasing DA tone, they could examine the impact of increasing dopaminergic transmission on the expression of a learned behavior. They showed that alterations in DA transmission can scale the performance of a previously learned behavior in the absence of new learning. The performance-scaling effects of DA depends on the motivational state. Increased response due to elevated DA was only observed in deprived animals. Optogenetic activation of DA neurons intended to mimic bursts of activity also increased movement without affecting learning (Barter et al., 2015; Hughes, Bakhurin, Petter, et al., 2020). Taken together, these results suggest that DA is neither necessary nor sufficient for learning.

Studies have used blocking to examine the causal role of DA in learning (Keiflin, Pribut, Shah, & Janak, 2019; Steinberg et al., 2013). Blocking is an example of the so-called cue competition effects that motivated the formulation of learning models using prediction errors (Rescorla & Wagner, 1972). In blocking, learning of a CS–US association is reduced (blocked) in the presence of an established predictor. For example, if the animal has already learned that some cue (CS1) predicts reward, then the presentation of a second predictor (CS2) will not produce much learning about the CS2-reward relationship. The presence of CS1 appears to "block" the learning of CS2 as a predictor. This phenomenon can be explained by prediction error models. According to the Rescorla–Wagner model, for example, the existing associative strength of the learned CS (Vsum) is simply subtracted from the actual reward (λ), yielding a reduced RPE. Consequently, learning of CS2 is reduced or 'blocked' because the US is not surprising (i.e. already predicted by CS1).

If blocking is due to reduced RPE (reduced by the existing predictor), then artificially restoring RPE by stimulation at the time of reward delivery should prevent blocking. Indeed, one study reported that blocking was prevented by stimulation of DA neurons at the time of reward (Steinberg et al., 2013). However, in their optogenetic experiments, there were no behavioral controls for stimulation-induced changes in attention to the CS or US, stimulus generalization, or CR performance. Nor were there detailed behavioral measures of what the rats were doing during the CS presentation. The only measure was time spent in the reward port during the CS, but it is unclear exactly what kind of behavior was detected by the beam break in the reward port or how it was affected by stimulation. DA activation at the time of omitted reward may also generate approach behavior, which makes the rat stay in the reward port. Without measuring the behavior, it is not possible to conclude that DA stimulation interfered with extinction learning.

A limitation of conventional blocking experiments is the lack of continuous behavioral measures or detailed characterization of how compound stimuli presentation affects performance. In fact, there has long been a debate on whether blocking actually affects learning or performance. Because blocking can be reversed by manipulations like spontaneous recovery, post-training reminders, and post-training extinction of the competing stimulus, some have argued that it is due to a performance deficit rather than a failure to learn (Blaisdell, Gunther, & Miller, 1999). Consequently, without the appropriate behavioral measures, it is difficult to rule out the role of performance in DA's observed effects on blocking.

12.8 LEARNING AND PERFORMANCE

Although historically the relationship between learning and performance has long been a source of contention (Kimble, 1961; Rescorla, 1980), many neuroscientists are not aware of the conceptual difficulties involved (Smith et al., 2004). In RL, it is often assumed that learning can be equated with a change in performance, but performance can change for many reasons independent of learning, such as changes in motivational state or fatigue (Kimble, 1961). On the other hand, learning can also occur without clear changes in performance. The relationship between learning and performance presented a major problem in traditional studies of learning, mainly due to the lack of any adequate model of performance (Rescorla, 1980).

A number of attempts have been made to reconcile the role of DA in performance and learning, but so far they have not been satisfactory because they have neglected key properties of the control hierarchy discussed in previous chapters. RL models equate reward prediction with the probability of action selection. They assume that reward prediction is directly translatable into performance, and every update in prediction is immediately reflected in a performance change. The higher the reward value, the higher the behavioral output (Glimcher, 2011; Hamid et al., 2016). However, as we have seen in the discussion of partial reinforcement and effort in Chapter 9, it is problematic to equate the value function with performance. Reward prediction can be dissociated from performance. More reward does not lead to more behavior, contrary to the basic assumption of RL. In operant conditioning, lowering the reward earned by pressing (i.e. partial reinforcement) can increase the rate of lever pressing, whereas increasing the reward earned by pressing can reduce lever pressing. These observations contradict the predictions of RL models.

To illustrate common confusions on the relationship between learning and performance, it is helpful to review the attempt to reconcile the RPE hypothesis with the incentive salience model of DA (McClure, Daw, & Montague, 2003). The incentive salience model focuses on DA contributions to performance rather than learning. It posits that DA encodes how attractive a particular target is in eliciting approach behavior and how hard an animal will work for reward (wanting, see Chapter 9) (Berridge & Robinson, 1998; Ikemoto, 2007; Salamone, Correa, Farrar, Nunes, & Pardo, 2009). However, McClure et al. argue that incentive salience is compatible with the RPE hypothesis as it is the equivalent of predicted future rewards (McClure et al., 2003). According to them, the RPE not only update reward prediction but also the probability of behavioral output. To support this argument, they discuss the observation of extinction mimicry, one of the original observations following injections of DA antagonists. Wise and colleagues reported that DA antagonism reduced performance gradually despite the delivery of rewards, just like extinction in normal animals (Wise, 1982, 2004; Wise, Spindler, DeWit, & Gerberg, 1978). The interpretation, then, is that DA antagonism mimicked the effect of natural extinction with reward omission, producing a gradual decline in performance.

According to the RPE model, reward omission after learning produces a negative RPE. McClure et al. argue that a similar process occurs when animals are given DA antagonists even when rewards are presented: DA antagonism results in a below-baseline rate of postsynaptic receptor activation, like a negative RPE upon reward delivery. Consequently, repeating the negative RPE results in a gradual reduction of the value estimates, as manifested in behavioral extinction. In other words, the decrease in performance following DA antagonism is interpreted as gradual unlearning (Figure 12.5).

With extinction mimicry, however, there is no demonstration of reinstatement, spontaneous recovery, or any of the other phenomena typically associated with natural extinction. In fact, Tombaugh and colleagues demonstrated that reduction of performance due to DA antagonism was in fact very different from that observed during extinction (Tombaugh, Anisman, & Tombaugh, 1980). Using operant conditioning with food rewards, they showed that the DA antagonist pimozide produces a much greater reduction in lever pressing than extinction. This difference was not apparent in the original studies by Wise and others because they used continuous reinforcement schedules (each lever press is followed by a food reward), which produced low rates of pressing and rapid extinction. However, the rate of pressing was much higher when partial reinforcement was used

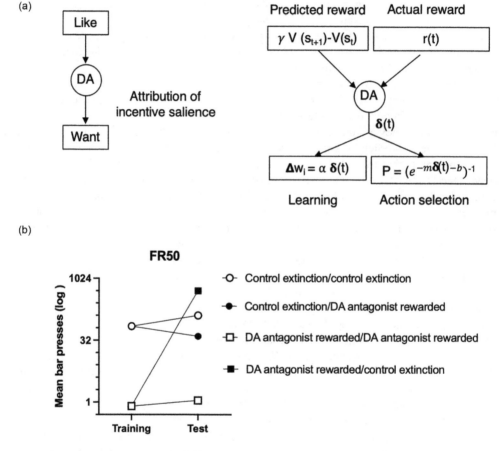

FIGURE 12.5 The effect of dopamine (DA) antagonism: learning or performance? (a) Equating incentive salience with reward prediction. Reward prediction error (RPE) is used to change both learning and performance. A positive error increases the value attributed to some state or action, making it more likely to be chosen in the future. Learning or change in the weight is proportional to the RPE. S_t, state; V, reward value function; r, reward; w_i, learned weights; RPE, $\delta(t)$; m, weighting factor or learning rate parameter; b, the effect of DA blockade. The RPE is used both as a learning signal to update reward value function and to bias action selection. The conversion of learned incentive value into a probability of action is equivalent to the incentive salience. Based on McClure, Daw, and Montague (2003). (b) Results that question the extinction mimicry account of DA antagonism. When rats were transferred from the pimozide treatment to extinction, the rate of pressing increased to levels seen in rats treated with pimozide for the first time or rats undergoing extinction for the first time. Modified from Tombaugh, Anisman, and Tombaugh (1980).

(for example fixed ratio [FR] or variable interval schedules) and remained high for some time even without reinforcement. Under these conditions, DA antagonism produced a much greater reduction in lever pressing compared to natural extinction.

The clearest result falsifying the negative RPE account of performance reduction came from experiments showing recovery of performance after DA antagonism. In one experiment, Tombaugh et al. trained rats on FR schedules to press a lever for food reward (Tombaugh, Tombaugh, & Anisman, 1979). They were then divided into two groups: one group received the DA antagonist pimozide but received normal FR training for 7 days; the other received vehicle injections for 7 days, but received the extinction contingency (vehicle extinction group). The group treated with DA antagonists during rewarded sessions reduced lever pressing over time, but the decrease in performance was not nearly as drastic as in the vehicle extinction group. During the second phase, each of the groups was further divided into two groups, one receiving vehicle extinction and the other pimozide in rewarded sessions. Pimozide always reduced performance, but when rats injected with pimozide during training were tested again without the drug (DA antagonist rewarded/control extinction group in Figure 12.5b), their performance immediately returned to a high level, close to that of control rats. This pattern cannot be explained by the RPE hypothesis. If DA blockade simply resulted in negative RPE and extinction, then we would expect lever pressing to remain low even when the animal is no longer under the influence of the drug. However, this is not what was observed. The immediate recovery of performance after DA antagonist showed that this treatment did not affect learning directly, but impaired performance. We must therefore reject the attempt by McClure et al. to reconcile learning and performance using RPE because their performance model of instrumental behavior is inadequate. On the other hand, these results support the adaptive gain hypothesis, according to which DA amplifies the reward rate error to generate a reference for action repetition.

12.9 ADAPTIVE GAIN AND REINFORCEMENT

In the transition control model, DA does not cause movements directly, but serves as a multiplicative gain that amplifies other inputs to the BG. In the simplest case, via D1 receptor activation, DA can increase excitability and response to excitatory inputs (Lahiri & Bevan, 2020). But the gain provided by DA is not fixed but adaptive—it can also vary according to ongoing demands. For example, the direct pathway can disinhibit DA neurons (dSPN–SNr–SNc DA). Such collateral projections can provide an efference copy of descending action commands, and in turn the DA neurons project back to the striatum to modulate its responsiveness.

Adaptive gain modulates the output function of a transition controller, transforming some higher errors into a specific reference for lower levels. The magnitude of gain determines how quickly the system reaches steady state as dictated by the reference value. The behavioral policy is not abstract and categorical, as assumed in RL; rather, it is a continuous and multidimensional vector that explicitly specifies action parameters. The behavioral significance of adaptive gain depends on the circuit being affected, what other inputs are arriving in the striatum at the same time, and the exact definition of the controlled variable. In the velocity control system associated with the sensorimotor network, the commands are integrated to yield reference position vectors (Chapter 7). This account predicts that DA stimulation can result in movement, so long as the relevant corticostriatal input is present. This is confirmed by optogenetic experiments showing direct effects of DA stimulation on movement (Barter et al., 2015; Hughes, Bakhurin, Petter et al., 2020). It also predicts bradykinesia and akinesia after DA depletion in the sensorimotor network (Chapter 7). In spatial relationship control, adaptive gain determines performance in a continuous pursuit task by determining how much distance error is amplified to generate velocity reference at each time step (Chapter 10). It can determine how well the animal follows a moving target by controlling the distance between itself and the target (Kim et al., 2019). This prediction remains to be tested. In the repetition controller associated with the associative and sensorimotor networks, adaptive gain can promote the repetition of an action reference (Chapter 8). In the reward rate controller associated with the limbic networks, adaptive gain can amplify the reward rate error to generate reference signals for a number of behaviors.

This account provides an explanation for results from Tombaugh et al. In their study the reward rate error is converted to a repetition reference, which serves as an action command specifying the number of presses in a bout (Tombaugh et al., 1980). DA could reactivate or sustain this command, so long as the cortical ensemble that provides the glutamatergic drive to the striatum is still active. This repetition function, especially in limbic and associative networks, is what is normally called reinforcement in DA self stimulation. DA antagonism disrupts this process. DA depletion in the accumbens core produced similar effects (Aberman & Salamone, 1999; Sokolowski & Salamone, 1998). This account also explains why lesions of the nucleus accumbens core reduces instrumental performance without affecting the content of learning (Corbit, Muir, & Balleine, 2001).

Adaptive gain for repetition control also explains the well-known self-stimulation of DA neurons. When the DA signal immediately follows some action, it promotes the repetition of that action, but this effect on performance can be transient. It cannot be equated with learning. Consequently, self-stimulation performance is often quickly extinguished since there is no repetition reference without stimulating DA neurons. It can only result in long-term changes under specific conditions, as we shall consider in Chapter 13.

Another key property of such self-stimulation behavior is lack of satiety, which is not surprising given the present framework. What is artificially injected is essentially a repetition reference. With natural rewards, this reference signal is proportional to the reward rate error, the difference between the current reward rate and the reward rate reference. It is set by the homeostatic error or hunger. With satiety, the reward rate reference declines along with the reward rate error, reducing the repetition reference. In self-stimulation, stimulation-induced DA signaling that coincides with excitatory corticostriatal reference input can reactivate the repetition reference indefinitely in the action repetition controller. This explains why, in some extreme cases, the animals will neglect all other activities, even eating and drinking (Olds, 1977).

The spiraling striato-nigro-striatal projections discussed in Chapter 9 make it possible for a given striatal region to influence DA neurons projecting to a different striatal region. Thus, adaptive gain is not limited to the striatal area that initiated the disinhibition, but can influence other regions, in particular regions that belong to lower levels of the motivational hierarchy. For example, the limbic striatum can influence DA neurons that project to the associative striatum. Modulating the output gain in the reward rate controller is equivalent to regulating effort exertion, e.g. reference for the rate of lever pressing. In support of this hypothesis, lesion or inactivation of the posterior dorsomedial striatum (pDMS) can also significantly reduce the rate of lever pressing (Yin, Ostlund, Knowlton, & Balleine, 2005). But unlike accumbens lesions, pDMS lesions also disrupted the learning of action–outcome contingencies (Corbit et al., 2001; Yin, Knowlton, & Balleine, 2005). This difference suggests that the accumbens-associated limbic BG circuits generate the reward rate error (e.g. more food pellets) that is not specific to a particular action, but the pDMS circuit is already selecting a specific action and specifying its repetition reference (e.g. keep pressing the left lever).

Negative feedback control of reward rate is responsible for the seemingly paradoxical observation that a higher reward rate reduces lever pressing. The press rate is determined by reward rate error and gain, not reward rate, contrary to common assumptions (Glimcher, 2011; Hamid et al., 2016). A higher reward rate can reduce the error more quickly, thus terminating lever pressing.

In operant conditioning on an FR schedule, the descending reference signal specifies the rate of repetition of some action. It is not a reward prediction, but a reward rate request that requires very different behavioral outputs depending on the feedback function. When the same amount of reward is requested (e.g. one pellet now), the actual rate of pressing is determined by the gain. The gain does not modulate the step size for learning, but the magnitude of the descending reference command for a transition variable. It determines whether the desire for one more piece of food is converted into a reference for a certain number of presses. Although higher errors (more deprivation) could result in a higher rate of pressing, the more important determinant of performance is the feedback function.

This account predicts that mesolimbic DA signaling is proportional to effort exerted, which should mirror the imposed schedule of reinforcement (feedback function) on an FR schedule (Chapter 9). Manipulation of the feedback function between action and outcome is expected to alter DA signaling systematically. Leaner schedules can produce higher DA signaling, but only when the animal is hungry (high reward rate error). The same reward rate error can be amplified by adaptive gain to generate a larger reference for press rate, force exerted, or whatever the operant may be.

Adaptive gain is not the only function of midbrain DA neurons. Adaptive gain is only enabled when the output of the BG affects the DA signaling back to the BG. While disinhibition through direct pathway output is a known mechanism for activating DA neurons (Lobb, Troyer, Wilson, & Paladini, 2011), it is not the only one. Various excitatory projections from both cortical and subcortical sources can also activate DA neurons (Lobb, Troyer, Wilson, & Paladini, 2011; Naito & Kita, 1994; Overton & Clark, 1997; Sesack & Pickel, 1992; Watabe-Uchida et al., 2012). In the absence of volitional actions, DA neurons can still be activated by salient inputs. Such short-latency activation may reflect efference copy signals from brainstem and collicular position controllers for orientation. The degree to which DA neurons are activated may depend on whether these inputs are surprising. All sensory channels are associated with some default reference states, which could be an intrinsic property of a lower-level controller or a result of top-down biases set by descending signals. Deviations from such states might trigger control action at the lower levels or modulate attention via gain modulation in the perceptual input function. For example, the superior colliculus, which generates head and eye movements to focus on salient stimuli, may activate DA neurons via direct projections (Comoli et al., 2003; Dommett et al., 2005). To the extent that perceptual inputs controlled by these lower-level orientation and position controllers deviate from their intrinsic reference signals, they can also activate DA neurons. That is to say, in general, more salient and surprising inputs can also produce greater activation of DA neurons. The role of this type of sensory surprise is not to strengthen state–action associations, but to prime the control hierarchy for action or to bias high level sensory processing.

12.10 REINFORCEMENT LEARNING VERSUS CONTROL

Prediction error algorithms resemble the error correction algorithm in negative feedback control. Subtraction is used in both. The main difference is in what the terms represent and how they are related to the rest of the system. This similarity at the algorithmic level is not accidental, as RL was also influenced by optimal control theory (Barto, Sutton, & Anderson, 1983). In optimal control, one attempts to find system parameters to minimize some arbitrary cost function. It is another example of the inside-out view of control criticized in Chapter 5.

The actor in the actor-critic model can be viewed as an optimal controller that acts on the plant (Figure 12.6). As Barto puts it: "The controller inputs labeled 'context' provide information pertinent to control task's objective. You might think of the context signals as specifying a 'motivational state' that implies certain control goals …. The critic also needs to know the motivational context of the task because its evaluations will be different depending on what the actor should be trying to do" (Barto, 1995). Where does this 'objective' come from? How does the controller know whether the consequence is good or bad? Barto does not address any of these questions. According to him, the critic can observe the consequences of the actor's behavior on the environment and specify the objectives, goals, and internal states, along with feedback from the actions. The critic is beginning to appear omniscient, but Barto does not explain how it can acquire all the relevant knowledge.

The TD target can be compared with the reference signal of a control system, the reward prediction with the perceptual input being controlled, the RPE with a change in control system output, and the learning rate parameter with the gain. But the functional roles of these terms are very different.

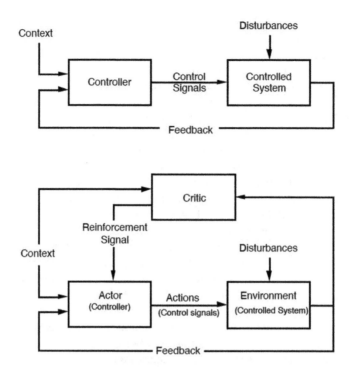

FIGURE 12.6 Relationship between feedback control and the actor-critic architecture. Top, standard illustration of a feedback controller. Bottom, illustration of the actor-critic model as an example of an optimal control system. Because this model assumes an inside-out view of control, the source of the reference signal is deliberately vague. Where reference signals are needed, the input is simply labeled context. The critic provides a teaching signal to adjust the connection between state (feedback plus context/reference) and actor in order to counter the environmental disturbance. From Barto (1995) with permission.

The actor-critic model is largely open loop, with feedback filtered through the perception of the observer or designer, and the error signal changes the state-policy mapping. There is no control of input, as the model assumes that what is controlled is the output, i.e. control signal sent to the plant. This adaptation process, which changes the so-called "control signals" to the plant, is misleadingly called learning. In transition control, the error signal is not a teaching signal. Its role in continuously adjusting performance can only be understood in the context of a control hierarchy. The change in performance cannot be equated with learning because it is possible to change controller output without any long-term changes in system parameters. In Chapter 13, we shall consider how learning can take place in the control hierarchy.

The actor-critic model, which essentially consists of an ideal observer coupled to a single controller, cannot find the right reference signals by itself. Instead of using a control hierarchy that compares input and reference to generate reference signals at every level, reference signals must be introduced indirectly using the reward function. One must tune a monolithic reward function for the controller so that the system can perform error reduction with a predetermined reference from the perspective of the ideal observer. Without this observer, the system will cease to function correctly. By defining the reward function, the observer is indirectly setting the reference state, just as a user sets the room temperature on a thermostat. Because the RL approach does not consider errors from the perspective of the controller, the system is not fully autonomous. Without the control hierarchy, the abstract descriptions cannot be successfully translated into actual behavioral output. Indeed, in practice, it has proven impossible to find the right reward function for even simple behaviors.

12.11 SUMMARY

The concept of prediction error has a long history in theories of learning. According to an influential hypothesis, phasic DA encodes a RPE. Schultz and colleagues showed that a short burst of DA activity follows reward presentation, but with repeated learning, this burst is shifted to the earliest predictor of the reward, as predicted by the RPE hypothesis. However, a number of experimental findings are incompatible with this hypothesis. RPE is supposed to be a scalar signal, but DA signals represent vectorial quantities. DA neurons are preferentially active for movements in different directions. They can represent kinematic variables or the equivalent in force vectors independent of reward prediction and in the absence of learning. Manipulations of DA signaling can alter performance without affecting learning. The pattern of phasic DA signaling commonly used to support the RPE hypothesis can be explained by changes in behavior during learning, but previous studies failed to measure these changes. For example, although anticipatory force exertion in response to reward, predicting cues increases with learning, DA neurons do not show significant changes in their force tuning. Such observations suggest that learning takes place upstream, resulting in increased excitatory drive to DA neurons when reward predictors are presented.

According to the transition control model, DA is not a teaching signal, but contributes to adaptive gain control, which continuously modifies performance online, whether to escape from harm or to approach reward. Despite the use of error correction in both RL models and transition control, their roles differ due to the distinct underlying models of behavior. The similarity in their algorithms can be attributed to the influence of optimal control theory in the development of RL. However, the actor-critic model, like modern control theory, is susceptible to the objectivist fallacy by externalizing the error correction process and imposing the observer's perspective. Rather than generating reference signals using multiple comparisons at various levels to control independent variables, the designer indirectly dictates the reference in RL models by defining the reward function. Since the actor-critic model is a compensator that is implicitly coupled to an ideal observer, it cannot function independently and requires the observer to close the loop.

REFERENCES

Aberman, J., & Salamone, J. D. (1999). Nucleus accumbens dopamine depletions make rats more sensitive to high ratio requirements but do not impair primary food reinforcement. *Neuroscience*, 92(2), 545–552.

Bakhurin, K. I., Hughes, R. N., Barter, J. W., Zhang, J., & Yin, H. H. (2020). Protocol for recording from ventral tegmental area dopamine neurons in mice while measuring force during head-fixation. *STAR Protocols*, 1(2), 100091.

Bakhurin, K. I., Hughes, R. N., Jiang, Q., Fallon, I. P., Yin, H. (2023). Force tuning explains changes in phasic dopamine signaling during stimulus-reward learning. bioRxiv:2023.2004. 2023.537994.

Barter, J. W., Castro, S., Sukharnikova, T., Rossi, M. A., & Yin, H. H. (2014). The role of the substantia nigra in posture control. *European Journal of Neuroscience*, 39 (9), 1465–1473.

Barter, J. W., Li, S., Lu, D., Rossi, M., Bartholomew, R., Shoemaker, C. T., ... Yin, H. H. (2015). Beyond reward prediction errors: The role of dopamine in movement kinematics. *Frontiers in Integrative Neuroscience*, 9, 39.

Barto, A. G. (1995). Adaptive critics and the basal ganglia. In: Houk, J. C, Davis, J. D. B. (Eds.), *Models of Information Processing in the Basal Ganglia* (pp. 215–232).

Barto, A. G., Sutton, R. S., & Anderson, C. W. (1983). Neuronlike adaptive elements that can solve difficult learning control problems. *IEEE Transactions on Systems, Man, and Cybernetics*, 5, 834–846.

Berridge, K. C. (2007). The debate over dopamine's role in reward: The case for incentive salience. *Psychopharmacology (Berl)*, 191(3), 391–431.

Berridge, K. C., & Robinson, T. E. (1998). What is the role of dopamine in reward: Hedonic impact, reward learning, or incentive salience? *Brain Research Review*, 28(3), 309–369.

Blaisdell, A. P., Gunther, L. M., & Miller, R. R. (1999). Recovery from blocking achieved by extinguishing the blocking cs. *Animal Learning & Behavior*, 27(1), 63–76.

Brown, J., Bullock, D., & Grossberg, S. (1999). How the basal ganglia use parallel excitatory and inhibitory learning pathways to selectively respond to unexpected rewarding cues. *Journal of Neuroscience*, 19(23), 10502–10511.

Bush, R. R., & Mosteller, F. (1951). A mathematical model for simple learning. *Psychological Review*, *58*(5), 313.
Cagniard, B., Balsam, P. D., Brunner, D., & Zhuang, X. (2006). Mice with chronically elevated dopamine exhibit enhanced motivation, but not learning, for a food reward. *Neuropsychopharmacology*, *31*(7), 1362–1370. doi: 10.1038/sj.npp.1300966.
Cagniard, B., Beeler, J. A., Britt, J. P., McGehee, D. S., Marinelli, M., & Zhuang, X. (2006). Dopamine scales performance in the absence of new learning. *Neuron*, *51*(5), 541–547.
Cannon, C. M., & Palmiter, R. D. (2003). Reward without dopamine. *Journal of Neuroscience*, *23*(34), 10827–10831.
Comoli, E., Coizet, V., Boyes, J., Bolam, J. P., Canteras, N. S., Quirk, R. H., ... Redgrave, P. (2003). A direct projection from superior colliculus to substantia nigra for detecting salient visual events. *Nature Neuroscience*, *6*(9), 974–980. doi: 10.1038/nn1113.
Contreras-Vidal, J. L., & Schultz, W. (1999). A predictive reinforcement model of dopamine neurons for learning approach behavior. *Journal of Computational Neuroscience*, *6*(3), 191–214.
Corbit, L. H., Muir, J. L., & Balleine, B. W. (2001). The role of the nucleus accumbens in instrumental conditioning: Evidence of a functional dissociation between accumbens core and shell. *Journal of Neuroscience*, *21*(9), 3251–3260.
da Silva, J. A., Tecuapetla, F., Paixão, V., & Costa, R. M. (2018). Dopamine neuron activity before action initiation gates and invigorates future movements. *Nature*, *554*(7691), 244.
Delong, M. R., Crutcher, M. D., & Georgopoulos, A. P. (1983). Relations between movement and single cell discharge in the substantia nigra of the behaving monkey. *Journal of Neuroscience*, *3*(8), 1599–1606.
Dickinson, A. (1980). *Contemporary Animal Learning Theory* (vol. 1): Cambridge: Cambridge University Press.
Dommett, E., Coizet, V., Blaha, C. D., Martindale, J., Lefebvre, V., Walton, N., ... Redgrave, P. (2005). How visual stimuli activate dopaminergic neurons at short latency. *Science*, *307*(5714), 1476–1479.
Eshel, N., Bukwich, M., Rao, V., Hemmelder, V., Tian, J., & Uchida, N. (2015). Arithmetic and local circuitry underlying dopamine prediction errors. *Nature*, *525*(7568), 243–246. doi: 10.1038/nature14855.
Fiorillo, C. D. (2013). Two dimensions of value: Dopamine neurons represent reward but not aversiveness. *Science*, *341*(6145), 546–549.
Fiorillo, C. D., Tobler, P. N., & Schultz, W. (2003). Discrete coding of reward probability and uncertainty by dopamine neurons. *Science*, *299*(5614), 1898–1902.
Gerfen, C. R. (1992). The neostriatal mosaic: Multiple levels of compartmental organization in the basal ganglia. *Annual Review of Neuroscience*, *15*, 285–320.
Glimcher, P. W. (2011). Understanding dopamine and reinforcement learning: The dopamine reward prediction error hypothesis. *Proceedings of the National Academy of Sciences*, *108*(Suppl. 3), 15647–15654.
Hamid, A. A., Pettibone, J. R., Mabrouk, O. S., Hetrick, V. L., Schmidt, R., Vander Weele, C. M., ... Berke, J. D. (2016). Mesolimbic dopamine signals the value of work. *Nature Neuroscience*, *19*(1), 117–126. doi: 10.1038/nn.4173.
Hikosaka, O. (2010). The habenula: From stress evasion to value-based decision-making. *Nature Reviews Neuroscience*, *11*(7), 503–513.
Hnasko, T. S., Sotak, B. N., & Palmiter, R. D. (2005). Morphine reward in dopamine-deficient mice. *Nature*, *438*(7069), 854–857.
Hollerman, J. R., & Schultz, W. (1998). Dopamine neurons report an error in the temporal prediction of reward during learning. *Nature Neuroscience*, *1*(4), 304–309.
Horvitz, J. C. (2000). Mesolimbocortical and nigrostriatal dopamine responses to salient non-reward events. *Neuroscience*, *96*(4), 651–656.
Horvitz, J. C., Stewart, T., Jacobs, B.L. (1997). Burst activity of ventral tegmental dopamine neurons is elicited by sensory stimuli in the awake cat. *Brain Research*, *759*, 251–258.
Houk, J. C., Adams, J. L., & Barto, A. G. (1995). A model of how the basal ganglia generates and uses neural signals that predict reinforcement. In: Houk, J. C., Davis, J. L., & Beiser, D. G. (Eds.), *Models of Information Processing in the Basal Ganglia* (pp. 249–270). Cambridge, MA: MIT Press.
Hughes, R. N., Bakhurin, K. I., Barter, J. W., Zhang, J., & Yin, H. H. (2020). A head-fixation system for continuous monitoring of force generated during behavior. *Frontiers in Integrative Neuroscience*, *14*, 11.
Hughes, R. N., Bakhurin, K. I., Petter, E. A., Watson, G. D., Kim, N., Friedman, A. D., & Yin, H. H. (2020). Ventral tegmental dopamine neurons control the impulse vector during motivated behavior. *Current Biology*, *30*, 1–14.
Ikemoto, S. (2007). Dopamine reward circuitry: Two projection systems from the ventral midbrain to the nucleus accumbens-olfactory tubercle complex. *Brain Research Review*, *56*(1), 27–78. doi: 10.1016/j.brainresrev.2007.05.004.

Joel, D., Niv, Y., & Ruppin, E. (2002). Actor–critic models of the basal ganglia: New anatomical and computational perspectives. *Neural Networks, 15*(4–6), 535–547.

Kamin, L. J. (1969). *Predictability, Surprise, Attention, and Conditioning*. In: Campbell, B. A., Church, R. M. (Eds.), *Punishment and Aversive Behavior* (pp. 279–296). New York: Appleton-Century-Crofts.

Keiflin, R., Pribut, H. J., Shah, N. B., & Janak, P. H. (2019). Ventral tegmental dopamine neurons participate in reward identity predictions. *Current Biology, 29*(1), 93–103. e103.

Kim, I. H., Rossi, M. A., Aryal, D. K., Racz, B., Kim, N., Uezu, A., ... Soderling, S. H. (2015). Spine pruning drives antipsychotic-sensitive locomotion via circuit control of striatal dopamine. *Nature Neuroscience, 18*(6), 883–891. doi: 10.1038/nn.4015.

Kim, N., Li, H. E., Hughes, R. N., Watson, G. D. R., Gallegos, D., West, A. E., ... Yin, H. H. (2019). A striatal interneuron circuit for continuous target pursuit. *Nature Communications, 10*(1), 2715. doi: 10.1038/s41467-019-10716-w.

Kimble, G. A. (1961). *Hilgard and Marquis' Conditioning and Learning* (2nd edition). New York: Appleton-Century-Crofts.

Lahiri, A. K., & Bevan, M. D. (2020). Dopaminergic transmission rapidly and persistently enhances excitability of d1 receptor-expressing striatal projection neurons. *Neuron, 106*(2), 277–290.

Lammel, S., Ion, D. I., Roeper, J., & Malenka, R. C. (2011). Projection-specific modulation of dopamine neuron synapses by aversive and rewarding stimuli. *Neuron, 70*(5), 855–862.

Lammel, S., Lim, B. K., Ran, C., Huang, K. W., Betley, M. J., Tye, K. M., ... Malenka, R. C. (2012). Input-specific control of reward and aversion in the ventral tegmental area. *Nature, 491*(7423), 212–217.

Ljungberg, T., Apicella, P., & Schultz, W. (1991). Responses of monkey midbrain dopamine neurons during delayed alternation performance. *Brain Research, 567*(2), 337–341.

Ljungberg, T., Apicella, P., & Schultz, W. (1992). Responses of monkey dopamine neurons during learning of behavioral reactions. *Journal of Neurophysiology, 67*(1), 145–163.

Lobb, C. J., Troyer, T. W., Wilson, C. J., & Paladini, C. A. (2011). Disinhibition bursting of dopaminergic neurons. *Frontiers in Systems Neuroscience, 5*, 25.

Matsumoto, M., & Hikosaka, O. (2007). Lateral habenula as a source of negative reward signals in dopamine neurons. *Nature, 447*(7148), 1111–1115.

Matsumoto, M., & Hikosaka, O. (2009). Two types of dopamine neuron distinctly convey positive and negative motivational signals. *Nature, 459*(7248), 837–841.

McClure, S. M., Daw, N. D., & Read Montague, P. (2003). A computational substrate for incentive salience. *Trends in Neurosciences, 26*(8), 423–428.

Menegas, W., Akiti, K., Amo, R., Uchida, N., & Watabe-Uchida, M. (2018). Dopamine neurons projecting to the posterior striatum reinforce avoidance of threatening stimuli. *Nature Neuroscience, 21*(10), 1421.

Miller, R. (1981). *Meaning and Purpose in the Intact Brain*: New York: Oxford University Press.

Mirenowicz, J., & Schultz, W. (1994). Importance of unpredictability for reward responses in primate dopamine neurons. *Journal of Neurophysiology, 72*(2), 1024–1027.

Naito, A., & Kita, H. (1994). The cortico-nigral projection in the rat: An anterograde tracing study with biotinylated dextran amine. *Brain Research, 637*(1–2), 317–322.

Olds, J. (1977). *Drives and Reinforcements: Behavioral Studies of Hypothalamic Functions*: New York: Raven Press.

Overton, P., & Clark, D. (1997). Burst firing in midbrain dopaminergic neurons. *Brain Research Reviews, 25*(3), 312–334.

Redgrave, P., Prescott, T. J., & Gurney, K. (1999). Is the short-latency dopamine response too short to signal reward error? *Trends in Neuroscience, 22*(4), 146–151.

Rescorla, R. A. (1968). Probability of shock in the presence and absence of cs in fear conditioning. *Journal of Comparative and Physiological Psychology, 66*(1), 1–5.

Rescorla, R. A. (1980). *Pavlovian Second-Order Conditioning: Studies in Associative* Learning: Mahwah, NJ: Lawrence Erlbaum.

Rescorla, R. A., & Wagner, A. R. (1972). A theory of pavlovian conditioning: Variations in the effectiveness of reinforcement and nonreinforcement. *Classical Conditioning II: Current Research and Theory, 2*, 64–99.

Robinson, S., Sandstrom, S. M., Denenberg, V. H., & Palmiter, R. D. (2005). Distinguishing whether dopamine regulates liking, wanting, and/or learning about rewards. *Behavioral Neuroscience, 119*(1), 5.

Romo, R., & Schultz, W. (1990). Dopamine neurons of the monkey midbrain: Contingencies of responses to active touch during self-initiated arm movements. *Journal of Neurophysiology, 63*(3), 592–606.

Rossi, M. A., Fan, D., Barter, J. W., & Yin, H. H. (2013). Bidirectional modulation of substantia nigra activity by motivational state. *PLoS One, 8*(8), e71598.

Salamone, J. D., Correa, M., Farrar, A. M., Nunes, E. J., & Pardo, M. (2009). Dopamine, behavioral economics, and effort. *Frontiers in Behavioral Neuroscience, 3*, 13. doi: 10.3389/neuro.08.013.2009.

Schultz, W. (1986). Responses of midbrain dopamine neurons to behavioral trigger stimuli in the monkey. *Journal of Neurophysiology, 56*(5), 1439–1461.

Schultz, W. (1998). Predictive reward signal of dopamine neurons. *Journal of Neurophysiology, 80*(1), 1–27.

Schultz, W. (2016). Dopamine reward prediction-error signalling: A two-component response. *Nature Reviews Neuroscience, 17*(3), 183–195. doi: 10.1038/nrn.2015.26.

Schultz, W., Apicella, P., & Ljungberg, T. (1993). Responses of monkey dopamine neurons to reward and conditioned stimuli during successive steps of learning a delayed response task. *Journal of Neuroscience, 13*(3), 900–913.

Schultz, W., Dayan, P., & Montague, P. R. (1997). A neural substrate of prediction and reward. *Science, 275*(5306), 1593–1599.

Schultz, W., & Romo, R. (1987). Responses of nigrostriatal dopamine neurons to high-intensity somatosensory stimulation in the anesthetized monkey. *Journal of Neurophysiology, 57*(1), 201–217.

Schultz, W., Ruffieux, A., & Aebischer, P. (1983). The activity of pars compacta neurons of the monkey substantia nigra in relation to motor activation. *Experimental Brain Research, 51*(3), 377–387.

Schultz, W., Tremblay, L., & Hollerman, J. R. (1998). Reward prediction in primate basal ganglia and frontal cortex. *Neuropharmacology, 37*(4–5), 421–429.

Sesack, S. R., & Pickel, V. M. (1992). Prefrontal cortical efferents in the rat synapse on unlabeled neuronal targets of catecholamine terminals in the nucleus accumbens septi and on dopamine neurons in the ventral tegmental area. *Journal of Comparative Neurology, 320*(2), 145–160.

Smith, A. C., Frank, L. M., Wirth, S., Yanike, M., Hu, D., Kubota, Y., ... Brown, E. N. (2004). Dynamic analysis of learning in behavioral experiments. *Journal of Neuroscience, 24*(2), 447–461.

Sokolowski, J., & Salamone, J. (1998). The role of accumbens dopamine in lever pressing and response allocation: Effects of 6-ohda injected into core and dorsomedial shell. *Pharmacology Biochemistry and Behavior, 59*(3), 557–566.

Steinberg, E. E., Keiflin, R., Boivin, J. R., Witten, I. B., Deisseroth, K., & Janak, P. H. (2013). A causal link between prediction errors, dopamine neurons and learning. *Nature Neuroscience, 16*(7), 966–973.

Stuber, G. D., Klanker, M., de Ridder, B., Bowers, M. S., Joosten, R. N., Feenstra, M. G., & Bonci, A. (2008). Reward-predictive cues enhance excitatory synaptic strength onto midbrain dopamine neurons. *Science, 321*(5896), 1690–1692. doi: 10.1126/science.1160873.

Suri, R. E. (2002). Td models of reward predictive responses in dopamine neurons. *Neural Networks, 15*(4–6), 523–533.

Suri, R. E., & Schultz, W. (2001). Temporal difference model reproduces anticipatory neural activity. *Neural Computation, 13*(4), 841–862.

Sutton, R. S. (1988). Learning to predict by the methods of temporal differences. *Machine Learning, 3*(1), 9–44.

Sutton, R. S., & Barto, A. G. (1990). Time-Derivative Models of Pavlovian Reinforcement. In: Gabriel, M., & Moore, J. W. (Eds.), *Learning and Computational Neuroscience: Foundations of Adaptive Networks* (pp. 497–537). Cambridge, MA: MIT Press.

Sutton, R. S., & Barto, A. G. (2018). *Reinforcement Learning: An Introduction*: Cambridge, MA: MIT Press.

Tobler, P. N., Fiorillo, C. D., & Schultz, W. (2005). Adaptive coding of reward value by dopamine neurons. *Science, 307*(5715), 1642–1645.

Tombaugh, T. N., Anisman, H., & Tombaugh, J. (1980). Extinction and dopamine receptor blockade after intermittent reinforcement training: Failure to observe functional equivalence. *Psychopharmacology (Berl), 70*(1), 19–28.

Tombaugh, T. N., Tombaugh, J., & Anisman, H. (1979). Effects of dopamine receptor blockade on alimentary behaviors: Home cage food consumption, magazine training, operant acquisition, and performance. *Psychopharmacology (Berl), 66*(3), 219–225.

Ungerstedt, U. (1971). Adipsia and aphagia after 6-hydroxydopamine induced degeneration of the nigro-striatal dopamine system. *Acta Physiologica Scandinavica, 367*, 95–122.

Waelti, P., Dickinson, A., & Schultz, W. (2001). Dopamine responses comply with basic assumptions of formal learning theory. *Nature, 412*(6842), 43–48.

Wagner, A. R., Logan, F. A., & Haberlandt, K. (1968). Stimulus selection in animal discrimination learning. *Journal of Experimental Psychology, 76*(2p1), 171.

Wang, D. V., & Tsien, J. Z. (2011). Convergent processing of both positive and negative motivational signals by the vta dopamine neuronal populations. *PLoS One, 6*(2), e17047.

Watabe-Uchida, M., Zhu, L., Ogawa, S. K., Vamanrao, A., & Uchida, N. (2012). Whole-brain mapping of direct inputs to midbrain dopamine neurons. *Neuron, 74*(5), 858–873. doi: 10.1016/j.neuron.2012.03.017.

Wise, R. A. (1982). Neuroleptics and operant beahvior: The anhedonia hypothesis. *Behavioral Brain Sciences, 5*, 39–87.

Wise, R. A. (2004). Dopamine, learning and motivation. *Nature Reviews Neuroscience*, *5*(6), 483–494. doi: 10.1038/nrn1406.

Wise, R. A., Spindler, J., DeWit, H., & Gerberg, G. J. (1978). Neuroleptic-induced" anhedonia" in rats: Pimozide blocks reward quality of food. *Science*, *201*(4352), 262–264.

Yin, H. H., Knowlton, B. J., & Balleine, B. W. (2005). Blockade of nmda receptors in the dorsomedial striatum prevents action-outcome learning in instrumental conditioning. *European Journal of Neuroscience*, *22*, 505–512.

Yin, H. H., Ostlund, S. B., Knowlton, B. J., & Balleine, B. W. (2005). The role of the dorsomedial striatum in instrumental conditioning. *European Journal of Neuroscience*, *22*, 513–523.

Yin, H. H., Zhuang, X., & Balleine, B. W. (2006). Instrumental learning in hyperdopaminergic mice. *Neurobiology of Learning and Memory*, *85*(3), 283–288.

Zhuang, X., Oosting, R. S., Jones, S. R., Gainetdinov, R. R., Miller, G. W., Caron, M. G., & Hen, R. (2001). Hyperactivity and impaired response habituation in hyperdopaminergic mice. *Proceedings of the National Academy of Sciences*, *98*(4), 1982–1987.

13 Reorganization, Exploration, and Plasticity

Having questioned the reward prediction error (RPE) hypothesis and the actor-critic model, we are now ready to consider alternative explanations of basal ganglia (BG) contributions to learning. This chapter addresses how cortico-BG networks can be reorganized through learning and how such reorganization is manifested in the types of learning traditionally associated with the BG. We shall review the role of exploratory behavior and behavioral variability in learning, focusing on BG contributions to bird song learning and performance. We shall then consider the role of neuromodulation and synaptic plasticity in BG-dependent learning and how short-term adaptive gain control can enable long-term plasticity needed for learning.

13.1 LEARNING FROM A CONTROL PERSPECTIVE

A successful control system automatically varies output to maintain the controlled variable near its reference level. It can be effective over a wide range of environmental conditions without changing its parameters. A change in performance in the control system does not necessarily indicate learning. For example, as one drinks water from a cup, the weight of the cup decreases, and the amount of torque needed to hold it also decreases. Negative feedback control automatically adjusts the torque output without requiring any learning. In fact, many phenomena traditionally called learning are just error correction or short-term adaptation, as they do not require long-term changes in the control system. In traditional experiments testing the RPE hypothesis, changes in the reward probability or reward size do not necessarily entail long-term learning.

Learning can be defined as long-term changes in system parameters. A variable can take on different values, but regardless of its value, the functions remain the same. A parameter, on the other hand, is a term that determines the form of a function. A change in parameter changes how one variable is related to another. Its physical implementation requires long-term changes in the connections between system components.

In a control hierarchy, how performance is altered by learning depends on exactly which parameters are changed. Within each control system, there are two major possibilities, as illustrated in Figure 13.1.

1. Changing the input function redefines the controlled variable. For example, suppose one wishes to avoid the rain and stay dry. To achieve anticipatory control, it is possible to redefine the input variable to incorporate predictors of rain, like clouds. Consequently, clouds can act as a proxy for rain to the relevant controllers responsible for rain avoidance, which are turned on as soon as the cloud appears. This is the basis for associative learning, especially Pavlovian conditioning.
2. Changing the output could generate a different output for a given error signal. This could alter the amount and pattern of reference signals sent to lower levels. In addition, learning can also alter the connection between the output of one controller and the comparator of a lower-level controller, recruiting controllers to achieve a higher-level reference state. As we have seen in earlier chapters, these changes correspond to what is normally called instrumental or operant learning.

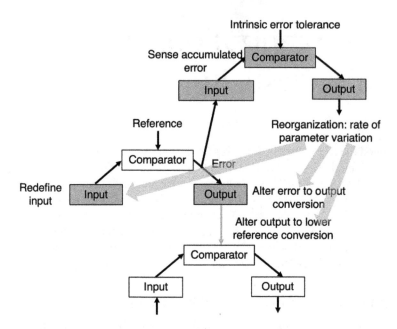

FIGURE 13.1 Reorganization of control hierarchy. The reorganization system can detect accumulated error, a measure of current system failure, and generate output once the error exceeds some threshold. The output increases the rate of variation in system parameters, which is normally very low. Gray arrows indicate places in the controller where parameters could be altered by experience. When error is reduced, the rate of parameter variation decreases, thus saving the latest parameters.

Analysis of learning in terms of associative structures (Chapter 10) suggests some of the controlled variables that can be modified by learning. For example, in Pavlov's original studies, salivation is the 'unconditioned response' (UR). Although salivation does not affect food delivery, it still affects how food is received by moistening the food (meat powder), making it easier to swallow and digest (Domjan, 2005; Dworkin, 1993; Pavlov, 1927). The effective US for salivation conditioning, such as dry meat powder or acid solution, generate a disturbance to an existing control system. By moistening the meat powder or diluting the acid solution, the UR of salivation reduces the error. Assuming that the controlled variable is moisture in the mouth, the dry meat powder US introduces a disturbance to the controller, generating an error signal that is converted into the UR output. The incorporation of the CS (e.g. sound of a metronome) into the definition of the perceptual controlled variable means that the CS can have the same effect as the actual disturbance presented by the US. Even though it is a previously 'neutral' stimulus that does not present a disturbance to this particular controller, learning endows the CS with the new capacity to act as a proxy disturbance. Although Pavlovian conditioning often involves much more than stimulus substitution, the refinement of the input function to incorporate a predictor is usually necessary for generating anticipatory control.

On the other hand, in instrumental learning, what is reorganized is mainly the output function. The error signal from a higher controller, when transformed into a reference signal, can recruit lower-level controllers. Consequently, a higher-level reference can be achieved by commanding lower-level controllers. As discussed in Chapter 9, this is what happens when desire for food rewards is converted to a command for pressing a lever. This learning process is traditionally known as response substitution, i.e. a new response is added to a given stimulus (Kimble, 1961). Instrumental learning can establish or modify the connections between reward rate controllers (outcome) and action rate controllers (action).

13.2 EXPLORATION AND REORGANIZATION

In the actor-critic model, the actor must compare the action values of all possible actions in a state before selecting the highest-valued action. But if action values are low to begin with, how can the right action be generated in the first place before any reinforcement? A common solution is to use a more permissive activation function linking state and policy, allowing the selection of actions with low values and some probability (Sutton & Barto, 2018). But this process can be too slow. The higher the number of possible actions, the longer it will take to explore, and the number of possible actions is determined by how each action is defined. Is pushing the lever with the snout a distinct action from, say, pushing with both the snout and the right paw? Is pushing a lever slightly harder a distinct action? In practice, effective exploration is difficult to achieve in reinforcement learning (RL), given the enormous action space. It could take many millions of trials for even the simplest behaviors, and such inefficiency is only masked by the computational power used in running RL in simulation.

Conventional RL models also fail to specify exactly when exploration is needed. In the control hierarchy, learning is needed when the existing organization and set of parameters prove to be inadequate for controlling some important variable (Powers, 1973). Modifications of the functions or relationships between variables in control systems make it possible to control new variables or to control existing variables better. Unlike RL, the conditions for reorganization are clearly defined: once the accumulated error in essential variables crosses a threshold, reorganization is initiated in the form of random variation in system parameters. The rate of variation is proportional to the accumulated error. Once error is reduced, the rate of variation will also be reduced. Stopping parameter variation is the equivalent of saving the latest set of parameters. Reorganization can occur for virtually all system parameters.

Persistent errors in essential control systems can promote exploration. When placed in an operant chamber, at first a sated rat may explore a little, but it is far more likely to groom or take a nap. In contrast, a hungry rat is very active. It will visit different locations, stopping often to investigate, and sniffing and touching constantly to explore the environment, especially if it expects to find food there. At the transition level, random changes in parameters are manifested in behavioral variability. Reorganization could result in variations in reference for relationships, repetition, or velocity controllers, depending on the system affected. Behavioral sequences for exploration reflect variations in such transition variables. For example, in foraging, the key variation in parameter is at the relationship control level, where distinct target references can be injected. This type of variation could produce approach behavior toward novel targets.

During initial instrumental learning, for example, the rat might bump into the lever, use its paws, its teeth, or its head to "press the lever" (Skinner, 1963). In mice, variability in lever pressing duration is initially high, but decreases with learning, reflecting the production of more stereotyped lever presses (Yin, 2009). In rats, behavioral variability on an operant task is also regulated as a graded function of recent reward rate (Dhawale, Miyamoto, Smith, & Ölveczky, 2019). When reward is plentiful, behavioral variability is reduced; when reward is scarce, variability is increased.

13.3 BG AND EXPLORATION

The limbic cortico-BG networks have been associated with exploratory behavior and novelty-seeking (Bardo, Donohew, & Harrington, 1996). In mammals, exploration uses the same systems for preparatory and approach behaviors as reviewed in Chapter 9. Dopamine (DA) agonists can increase exploration, but the expression of exploratory behavior depends on the context (Ikemoto & Panksepp, 1999). In the absence of specific objects, behaviors like rearing, digging, or forward locomotion are exhibited (Carr & White, 1987; Swanson, Heath, Stratford, & Kelley, 1997). When there are objects in the environment, there can be an approach or manipulation (Cador, Taylor, & Robbins, 1991; Kelley & Delfs, 1991).

DA has been hypothesized to regulate exploration and exploitation (Frank, Doll, Oas-Terpstra, & Moreno, 2009; Humphries, Khamassi, & Gurney, 2012). Increasing DA is thought to promote exploitation, whereas decreasing DA is thought to promote exploration (Humphries et al., 2012). This effect can be explained by the role of DA as an adaptive gain. As discussed in earlier chapters, activation of D1 receptors may increase repetition and persistence of ongoing behavior by adjusting the gain in controllers for higher-order transition variables like proximity relationships and repetition. Indeed, manipulations that increase DA signaling often produce stereotypical behavioral sequences (Berridge, Aldridge, Houchard, & Zhuang, 2005; Saka, Goodrich, Harlan, Madras, & Graybiel, 2004). On the other hand, when the overall DA level is reduced, there is less behavioral persistence. DA acting on the repetition controller can transiently stop the exploration and permit further reduction of the error. This is similar to so-called exploitation in RL terminology.

13.4 BIRD SONG LEARNING

The study of bird song has shed light on the role of behavioral exploration and variability in learning. Male songbirds learn to sing to attract females, and there is a sexually dimorphic cortico-BG circuit (Figure 13.2) dedicated to song learning and performance (Brainard & Doupe, 2002; Marler & Hamilton, 1966; Mooney, 2009). Bird song consists of easily distinguishable units called syllables, each lasting about 100–200 ms and characterized by a distinct pitch. A short sequence of 2–7 syllables is called a motif (Figure 13.2).

Bird song learning can be divided into two stages. During the first stage, the juvenile bird simply listens to the songs of its tutor and learns to memorize the tutor's song. Practicing alone, it gradually learns to match its song to the memorized model (Konishi, 1965). This "sensorimotor learning" stage is characterized by high-performance variability, which is gradually reduced until a match to a model song is achieved (Tchernichovski, Mitra, Lints, & Nottebohm, 2001). Variability observed during practice reflects exploration that is critical for matching self-generated songs to tutor songs.

Brainard and colleagues tested the role of auditory feedback in singing by distorting the pitch that the birds could hear using a headset. When the feedback pitch heard is different from the 'intended' pitch, birds shift the pitch of their vocalizations to compensate (Sakata & Brainard, 2006, 2008; Sober & Brainard, 2009; Tumer & Brainard, 2007). This adjustment is limited to a single targeted syllable. If a specific syllable is replaced by white noise, the bird will also change its pitch to avoid the noise. Within hours of receiving distorted auditory feedback, birds selectively reduced the production of those syllable variants to avoid the pitch disruption (Tumer & Brainard, 2007). Normally, white noise is delivered within 16–30 ms of the syllable. When an additional 100 ms was added to the delay, there was no longer a compensatory change in pitch. What is therefore avoided is pitch deviation within a very short time period after pitch production. Outside of this window, the pitch input is no longer controlled, so white noise would no longer constitute a disturbance.

The pitch, or fundamental frequency, of a syllable depends on the configuration of the upper vocal tract. Birds adjust pitch through cyclical movements of the hyoid skeleton attached to the larynx and an expansion of the esophagus. At the beginning of each syllable, the cavity expands and extends into the cranial end of the esophagus. At the end of a syllable, the esophagus collapses, reducing the cavity volume (Riede, Suthers, Fletcher, & Blevins, 2006). The volume of the oropharyngeal and esophagus cavities determines the fundamental frequency of each syllable. To change pitch, it is necessary to change the vocal tract configuration, similar to changing posture. The production of each syllable thus involves velocity control of the vocal tract. The feedback function defines a specific relationship between the rate of change in the vocal tract configuration and auditory feedback. Pitch distortion changes this feedback function, as the same commands to muscles now result in different auditory feedback. Deviation from the reference pitch is actively resisted, as birds vary behavioral output in order to control the perceived pitch.

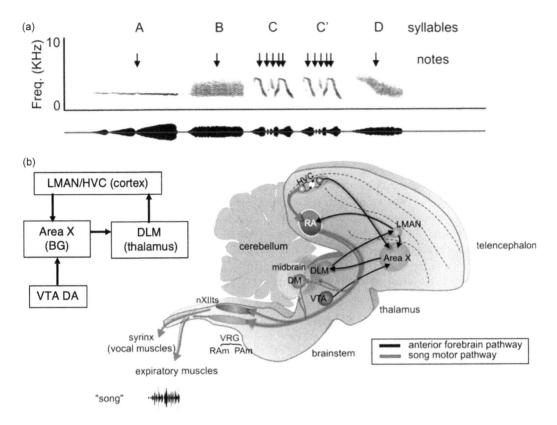

FIGURE 13.2 Bird song. (a) A spectrograph shows the fundamental frequency of bird song. Notes are the basic units, which can be grouped to form syllables. A motif is a sequence of syllables. (b) Illustration of the basal ganglia (BG) circuit involved in bird song. The anterior forebrain pathway is the equivalent of a cortico-BG–thalamocortical pathway. It is a key pathway dedicated to bird song. Area X is the equivalent of striatum and pallidum. DLM is a thalamic nucleus that receives BG output and, in turn, projects back to the prefrontal analog (LMAN). From Mooney (2009) with permission.

The bird first learns to form a template for tutor song through passive listening. During the so-called sensorimotor learning stage, the juvenile bird, the bird learns a set of vocal tract reference signals corresponding to the auditory inputs varies pitch in order to match the auditory input with the acquired template. Auditory feedback is critical during this stage. It is needed to adjust the connection between a specific set of reference signals (reflecting desired pitch, volume, duration, etc.) and the transition reference signals for vocal tract transitions that generate the syllable so that the vocal tract changes would produce the desired sounds. The relevant input variables being controlled are not only auditory but also include additional proprioceptive inputs from the vocal tract changes. These proprioceptive inputs must be controlled in order to achieve particular auditory inputs. In other words, the song template is more than just an auditory representation. Rather, it involves the tuning of virtually the entire action hierarchy that ultimately commands the vocal effectors. There are multiple templates, corresponding to acquired reference states at multiple levels. Consequently, the bird learns to tune the hierarchy so that a given pitch reference would generate a specific vocal tract transition command requesting a specific proprioceptive input.

Once the action hierarchy is established, auditory feedback is no longer necessary because the lower levels, e.g. proprioceptive transition control, can produce the needed sounds. The feedback function that defines the pitch change as a function of vocal tract transition command is stable. It is not subject to many environmental disturbances, as the generation of pitch comes from a protected part of the body, and the medium by which the sound travels to the auditory system is also stable. Consequently, the descending reference to the vocal tract configuration can reliably generate the desired pitch. In adult birds that have learned to sing, then, it is only necessary to have proprioceptive feedback monitoring the change in vocal tract configuration. The control of this variable is the equivalent of the control of pitch. Auditory feedback is no longer necessary, as long as there is a stable feedback function between vocal tract configuration reference signals and the sound produced. The same is true for a trained singer who cannot hear her own voice. Removing auditory feedback does not disrupt singing immediately, though the performance may degrade over time without auditory feedback. On the other hand, introducing some disturbance in the proximal action feedback, for example by manipulating the vocal tract, would immediately disrupt performance.

In addition, it is possible that the proprioceptive commands can also generate 'imaginary' auditory inputs. As discussed in Chapter 8, the imagination mode allows the efference copy of action commands to activate specific sensory representations without necessarily generating overt movement. This form of internal feedback could range from imagining music in one's head to humming music. Such a mechanism is probably what allowed Beethoven to compose music in spite of his deafness.

13.5 BG CONTRIBUTIONS TO BIRD SONG

Songbirds have a dedicated cortico-BG–thalamocortical circuit, known as the anterior forebrain pathway (AFP), that is critical for song learning (Johnson, Sablan, & Bottjer, 1995; Kao, Doupe, & Brainard, 2005). As shown in Figure 13.2, it contains the equivalent of the striatum and pallidum (area X), thalamus (DLM), premotor cortex (HVC), and prefrontal cortex (lateral magnocellular nucleus of the anterior nidopallium or LMAN) as well as projections from DA neurons (Gale, Person, & Perkel, 2008; Goldberg, Farries, & Fee, 2013; Luo, Ding, & Perkel, 2001; Perkel, 2004). It is similar to the mammalian BG–thalamocortical network, though the different components like striatum and pallidum are not as spatially segregated. The outputs of the AFP reach the RA, a nucleus that projects topographically to motor neurons innervating the syrinx, the organ for song production.

The prefrontal analogue LMAN projects to the RA as well as the BG. It is also a target of BG outputs, thus forming a reentrant circuit. Its output neurons often send one axonal branch to the RA and another to the BG (Luo et al., 2001; Nixdorf-Bergweiler, Lips, & Heinemann, 1995). This organization is similar to the pyramidal tract (PT) pathway in the mammalian brain. LMAN lesions do not significantly impair song performance in adult birds, but produce profound learning deficits in juvenile birds (Bottjer, Miesner, & Arnold, 1984). LMAN is critical for vocal exploration in both juvenile birds and adult birds. During the sensorimotor learning stage, it is critical for generation of variability or vocal babbling. Although its contribution to performance is reduced as the song becomes crystalized, even in adults it contributes to trial-to-trial variability in pitch (Andalman & Fee, 2009; Kao et al., 2005; Olveczky, Andalman, & Fee, 2005).

Variability is also found in adult birds, though it is less common. In adult zebra finches, there is modulation in the variability of syllable pitch by social context. When singing alone (undirected singing), pitch variability is high, but it is reduced when singing in front of a female bird. LMAN lesions reduced cross-rendition variation in pitch during undirected song to a level found during directed song. In other words, LMAN appears to be critical for social context-dependent modulation of variability (Kao & Brainard, 2006). Likewise, variability in AFP activity is also associated with variability in song structure (Hessler & Doupe, 1999; Kao et al., 2005).

While lesions of the AFP abolish normal song learning in juveniles, they do not abolish song performance in adult birds (Scharff & Nottebohm, 1991; Sohrabji, Nordeen, & Nordeen, 1990). Rather, lesions impair adaptive modification of song in adults, especially in the generation of variability.

There are different types of variability in bird song, such as in pitch and in syllable duration. There can also be variability in higher-order features like syllable sequences (Derégnaucourt, Mitra, Fehér, Pytte, & Tchernichovski, 2005; Sossinka & Böhner, 1980; Tchernichovski et al., 2001). These are comparable to different types of transition variables discussed earlier, such as velocity, repetition, and serial order.

The songbird BG appear to be critical for variability in pitch. BG lesions eliminate rapid within-syllable-pitch variations (Ali et al., 2013; Kojima, Kao, Doupe, & Brainard, 2018). Ali and his colleagues delivered loud noise bursts whenever the syllable pitch or duration was below or above a threshold value. They tested whether birds can learn to avoid such aversive feedback by varying either syllable pitch or duration. When the operant was defined as pitch variation, bilateral lesions of Area X largely abolished learning. In contrast, when the operant was syllable duration, lesions of the premotor cortex analog area (HVC) impaired learning. This result suggests that the BG are critical for pitch variation.

In adult birds, Mooney and colleagues recently studied the role of the striatum in song variability (Singh Alvarado et al., 2021). They recorded neural activity from the striatum during singing. As the male bird switched from undirected to female-directed song performance, striatal activity decreased. This suppression was unrelated to the male's body movement but reflected a significant decrease in pitch variability. Neither HVC cells nor their axon terminals in the BG showed significant changes in singing-related activity when alternating between undirected and directed singing, suggesting that the striatum may contribute to the modulation of pitch variability. In addition, noradrenergic signaling in the BG reduced vocal variability by suppressing striatal output. Alpha-adrenergic antagonism selectively increased the variability of directed songs, but D1 antagonism did not. These results therefore suggest that singing-related striatal activity could contribute to vocal exploration during practice. Inhibition of striatal activity reduced the variability across renditions in the pitch of targeted syllables.

The studies reviewed above demonstrate a critical role for the BG in song learning and performance. Bird song can be analyzed as a set of vocal transitions using both proprioceptive and auditory feedback. In learning, the BG are necessary for generating pitch variability, which is similar to movement velocity control, and for associating reference signals for specific vocal tract configurations with auditory feedback. In adult birds, the BG are still critical for performance if there are significant environmental disturbances that alter the feedback function between behavioral output and auditory feedback or for context-dependent modulation of song variability.

13.6 DA AND BIRD SONG

The bird's BG also receive a dopaminergic projection from the ventral tegmental area (VTA). In adult Bengalese finches, lesions of DA inputs to the BG impaired vocal learning (Hoffmann, Saravanan, Wood, He, & Sober, 2016). When the pitch of a target syllable was either above or below a threshold, white noise was presented. This procedure produced compensatory changes in vocal pitch to avoid white noise. DA depletion impaired the ability to alter pitch to avoid white noise, but did not change number, quality, or variability of song performances. Thus, DA appears to be especially critical for pitch-adaptation learning.

Roberts and colleagues found that pitch-contingent stimulation of VTA dopaminergic terminals in Area X could gradually change the pitch of the target syllable without affecting other syllables in the motif (Xiao et al., 2018). Photo-stimulation was delivered within 25 ms of the syllable-pitch measurement and persisted for 100 ms, overlapping with syllable production. When the stimulation was contingent upon the highest third of all pitch variants, the pitch of the targeted syllable increased. When the stimulation was contingent upon the lowest third, there was a significant decrease in pitch. In other words, the activation of VTA terminals promoted the repetition of the most recently performed variant. In contrast, optogenetic inhibition following higher pitch renditions resulted in decreased pitch, and inhibition following low pitch renditions resulted in increased pitch of the targeted syllable. Excitation and inhibition of dopaminergic terminals therefore have opponent effects on future performances of targeted song syllables.

Pitch adaptation induced by optogenetic stimulation of VTA DA terminals in area X depends on D1Rs (Hisey, Kearney, & Mooney, 2018). When a D1 antagonist was infused into Area X, optogenetic stimulation of VTA terminals induced little or no pitch learning. In adult birds, pitch adaptation to white noise presentation also depends on D1R activation. In juvenile zebra finches, lesions of VTA DA neurons that project to the BG disrupted copying of the tutor song during the sensorimotor learning stage (Hisey et al., 2018).

13.7 LESSONS FROM BIRD SONG

The cortico-BG circuit is thought to compare sensory feedback with the song template acquired (Troyer & Doupe, 2000a, b). Work on bird song supports the idea, elaborated in Chapter 7, that reference signals reach the BG via corticostriatal projections. The efference copy from the BG output to the thalamocortical circuit may coordinate the serial activation of different elements in a sequence.

In actor-critic models of bird song learning, syllable quality is thought to reflect action value (Chen & Goldberg, 2020; Fee & Goldberg, 2011). DA-modulated plasticity in the actor module adjusts the weight of each state–action pair according to its predicted quality. But syllable quality is not a property of any auditory input. It depends on deviation from the stored value of the desired input; higher quality is closer to the template value. For any neural signal to reflect performance error, there must be some internal comparison process, so that deviation from reference produces compensatory changes.

Feedback is not limited to auditory feedback but must also include proprioceptive and other sensory inputs generated during singing. There can be multiple templates at multiple levels of the hierarchy, implemented by reference signals. They represent desired states for the sequence of vocal tract configurations as well as desired pitch. Learning to sing is difficult and time-consuming, given the number of parameters that must be tuned at multiple levels. During practice, in the sensorimotor stage, variability is needed to find the appropriate combination of parameters in transition control, including vocal tract configurations and associated rate of change variables in proprioceptive and auditory inputs. At first, the pitch achieved may be far from the desired pitch. During reorganization, the transition control system must change parameters until the right pitch is sensed. Any error reduction would result in differential amplification of a particular variant, mediated by the adaptive gain function of DA.

In Chapter 7, we reviewed the effect of selective stimulation of direct and indirect pathways when movement velocity is either above or below some threshold (Yttri & Dudman, 2016). Direct pathway activation promoted velocity similar to the recent action: if a fast action is followed by stimulation, then fast actions are more likely in the future, but if a slow action is followed by stimulation, then slow actions are more likely. On the other hand, indirect pathway stimulation had the opposite effect. These experiments were in fact inspired by bird song experiments using pitch-contingent manipulations. Results from the pitch-contingent DA stimulation experiments are also in agreement with findings from Yttri and Dudman. Stimulating DA terminals in area X can produce a similar effect when some pitch threshold is used (Xiao et al., 2018). This result can also be explained by the adaptive gain mechanism and its effect on repetition control, suggesting that DA signaling via D1 receptors promotes repetition of the last action parameter. The precision of DA signaling allows it to "tag" the recently activated spiny projection neurons (SPNs) by further increasing their excitability. Thus it promotes the re-activation of these SPNs when new commands arrive within a short time window, roughly a few hundred milliseconds based on the bird song results. This observation is consistent with the known estimate of how long DA can increase direct pathway spiny projection neuron excitability through D1 receptor activation (Lahiri & Bevan, 2020). It also suggests that, when the delay between action and outcome is much longer, proper credit assignment would require mechanisms that are different from adaptive gain using DA signaling.

The basic unit of action that can be reinforced is a simple transition like head turning or a song syllable. By promoting the reselection of the recent velocity/pitch command, DA signaling plays a key role in trial-and-error learning. When the right action variant is performed, the immediate reduction in error must be accurately attributed to the action. This is known as the credit assignment problem in RL. Adaptive gain control solves this problem by using the efference copy to reactivate or sustain the recent reference signal. Thus repetition immediately stops parameter variation to repeat the error reduction. Often the error in the higher-level controllers (e.g. hunger) cannot be reduced at once. It takes not only time but also repetition of the appropriate action allowing progressive error reduction. As we shall see below, this is an important precondition for learning and long-term plasticity.

13.8 NEURAL PLASTICITY AND REORGANIZATION

Reorganization requires long-lasting changes in the connections between components of a control system. Such changes can be implemented by synaptic plasticity. Although much progress has been made in understanding the molecular signaling mechanisms underlying synaptic plasticity, so far the connection between synaptic changes and behavior remains elusive (Gallistel, 1990; Kandel, 2001).

Concepts from traditional theories of learning like reflex strength, habit strength, and associative strength all posit some connection between state and action. This connection is implemented by synaptic strength and modified by the effective reinforcement signal (e.g. RPE). Synaptic strength is assumed to be stable at rest and that some induction event, mimicked by the stimulation protocol, results in an increased synaptic strength.

In typical *in vitro* experiments on synaptic plasticity, the response of a neuron (typically excitatory postsynaptic currents in voltage–clamp experiments) to afferent stimulation is measured repeatedly until a stable baseline is established. Then an induction protocol is used, typically involving bursts of high-frequency stimulation. Following induction, neural activity is then measured again using the baseline stimulation protocol. As a measure of synaptic strength, the amplitude of the postsynaptic response before and after the induction protocol is compared. In this design, the induction protocol is implicitly assumed to mimic the learning experience. Synaptic strength is assumed to be low and stable at first, but the induction of long-term potentiation (LTP) increases it. The equivalent *in vivo* scenario might be the following: a naïve rat in an operant chamber starts with a stable set of corticostriatal connection weights. Once it presses the lever by chance, the reinforcement signal generated by DA somehow strengthens the corticostriatal synapse. In the future, the state representation is more likely to activate the striatal units, thereby generating the lever press. This account attempts to explain how the state–action association, being low initially, could lead to the right action variant in the first place. But there is no explanation of what starts the exploration so that chance action can be produced, no clear definition of states or actions, and no model of behavior that would actually generate the changes described.

In contrast, in reorganization the parameter changes must occur first. Effective exploration requires a transient change in control system parameters to generate the requisite action variants before the weights can be finalized. Error reduction slows down the rate of variation in system parameters and saves the latest set of parameters, including the weights in the activation function.

13.8.1 ADAPTIVE GAIN AND INDUCTION OF LONG-TERM PLASTICITY

A transient change in the activation function of a neuron can occur with neuromodulation or short-term synaptic plasticity. For example, DA or acetylcholine (ACh) can also alter the activation function of SPNs. As adaptive gain, DA can repeat the recently activated action parameters (e.g. reference for relationship, serial order, or velocity), which stops or slows the parameter variation (Lahiri & Bevan, 2020). Note that this function by itself is just an example of online adaptation, which does not necessarily lead to learning but can promote learning under specific conditions. Because error reduction takes time and repetition of the action variant, the correlation between presynaptic corticostriatal inputs and postsynaptic SPN output can promote long-term plasticity at the activated synapses. This correlation is needed for standard Hebbian plasticity, similar to what is found in a spike-timing–dependent plasticity induction protocol (Bi & Poo, 2001; Shen, Flajolet, Greengard, & Surmeier, 2008).

The process of saving the right parameters through synaptic plasticity is not instantaneous but continues as long as the error is above some threshold. The parameter values are not fully saved until the error is sufficiently reduced. When the higher error is fully reduced, there have been sufficient pairings between presynaptic input and postsynaptic activation to activate intracellular pathways that ultimately lead to long-term plasticity. At the same time, activity of those neurons that were not boosted by adaptive gain might be less correlated with presynaptic input, resulting in no potentiation and possibly even weakening of the connections.

The transition control model does not rule out a role for DA in learning, but this role is indirect and different from that of a teaching signal as postulated by the RPE hypothesis. It is through its role in modulating performance online that DA could contribute to learning. Adaptive gain control makes it possible to alter the activation function transiently without committing to long-term reorganization or learning. The functional connectivity can change rapidly without permanent changes in synaptic strength.

If DA neurons are activated not by self-initiated commands but by salient stimuli, then instead of repeating some action, they will serve a 'priming' function so that their target SPNs may be more easily depolarized. Likewise, during exploration, whenever an action results in some salient sensory input (reafference), the combined effect of salient sensory inputs and adaptive gain from online estimation of action command can promote the repetition of that action (Redgrave & Gurney, 2006).

Repetition of the last action variant using adaptive gain is useful for determining agency and learning about the action–outcome contingency. As discussed in Chapter 10, corticostriatal projections from the intratelencephalic (IT) pathway from the prefrontal cortex to the posterior dorsomedial striatum are critical for such learning (Hart, Bradfield, Fok, Chieng, & Balleine, 2018). DA signaling in the associative striatum near these corticostriatal terminals may be critical in establishing the connection between a reference signal for a desired sensory state and a particular action during initial instrumental learning.

Another prediction of the present account is that learning is enabled by high error in essential variable control combined with sufficient repetition of the action variant. For example, in instrumental conditioning significant food deprivation combined with any schedule of reinforcement that promotes high rate of lever pressing would promote long-term plasticity and differential amplification of the appropriate action. Research on instrumental learning broadly supports this prediction.

The adaptive gain mechanism explains the classic reinforcement effect, but it is different from the traditional concept of reinforcement. The system involved is not one in which input determines output, but a closed-loop controller in a hierarchy. The descending reference signals that are repeated do not generate consistent outputs from the final common path, but variable outputs to ensure consistent results, as sensed by controlled inputs. In addition, while adaptive gain transiently alters the activation function, it does not necessarily lead to long-term changes in system parameters.

To understand the behavioral implications of this account, let us consider the example of a rat learning to press a lever for the first time. Food deprivation ensured a high homeostatic error that translated into a high reward rate reference as well as a high gain in the reward rate controller. But there is currently no negative feedback that can reduce this error. Persistent homeostatic error triggers the reorganization process by promoting exploratory behavior and increasing behavioral variability. This could be done, for example, by introducing random variations in reference signals into the striatum via corticostriatal projections. For example, one may randomly vary the location of the visual field that is selected as a target for the proximity control system, as described in Chapter 9.

When the rat touches the lever and obtains the reward for the first time, such feedback starts to reduce the error and stop the random variation in system parameters.

Copies of motor commands from both cortical and subcortical sensorimotor structures to the brainstem are relayed to the striatum via inputs from the PT pathway, superior colliculus, and other regions (Bickford & Hall, 1989; Shepherd, 2013). If the sensory input is a consequence of the action, then repeating the action command would recreate the salient input (Redgrave & Gurney, 2006). This produces a correlation between presynaptic input (sensory and reference inputs to striatum) and postsynaptic activity (striatal output that generates the action), the key condition for Hebbian plasticity. If there is sufficient reward rate error, we would expect a high number of repetitions and thus a condition favoring plasticity, since synaptic potentiation depends on the number of repetitions with presynaptic and postsynaptic correlation. In this case, corticostriatal plasticity does not alter the reference signal per se, but how a higher error alters the reference signal. This would allow specific goal representations to recruit SPN modules.

Reorganization, then, likely involves a two-stage process. In the first stage, there is exploration, either due to intrinsic homeostatic reference for novelty or to persistent errors in essential variables. This results in parameter variation, allowing a set of weights to be tried out before making them more permanent. As multiplicative gain, modulatory inputs like DA can determine the neuronal output at any time. A neuron with weak synaptic weights for a given set of inputs can still reach the threshold for spiking transiently. Once the parameters have been tested, during the second stage, it is possible to save the parameter permanently. The more enduring changes, which typically require significant structural changes, take some time to implement. For example, a well-established mechanism for LTP is the insertion of α-amino-3-hydroxy-5-methyl-4-isoxazolepropionic acid (AMPA) glutamate receptors on the postsynaptic neuron (Malinow & Malenka, 2002). This would contribute to a more permanent change in the activation function of the neuron, but receptors cannot be inserted instantaneously as significant time is needed for protein synthesis and receptor trafficking (Anggono & Huganir, 2012). There could also be offline processes, including those during sleep, that consolidate the changes.

One prediction from this account is that long-term plasticity depends on the repetition of the right parameters while the error remains high. For example, if a hungry rat discovers that a lever press leads to reward, learning is promoted if the homeostatic error is reduced slowly by lever pressing, and many presses are needed to reach satiety. This is similar to a plasticity induction protocol with many repetitions. Such a protocol is more likely to result in long-term plasticity. If the rat presses a lever and immediately earns all the food needed to become sated, then there will be less long-term synaptic plasticity. The number of repetitions that result in exposure to the contingency is a key determinant of learning. When the intertrial interval is too short or when the error reduction is too rapid, the rate of learning could also be reduced because there is not enough time to save the effective parameters.

13.8.2 Changes in DA Dependence with Learning

DA contributes to both short-term and long-term plasticity (Gerfen & Surmeier, 2011; Greengard, Allen, & Nairn, 1999). It can increase excitability in the short-term and promote the repetition of a particular action variant (transition error or position reference signal). This creates the necessary conditions for the induction of long-term Hebbian synaptic plasticity. If we assume that DA acts as a multiplicative gain signal, the output of dPSNs would be proportional to the product of net excitatory drive and DA acting on D1 receptors. Once corticostriatal synaptic strength is increased through LTP, there will be less dependence on DA to generate the same amount of striatal output.

This account generates an interesting prediction: behaviors that are at first highly dependent on DA for expression can become less dependent with extensive training. This prediction is supported by previous work on corticostriatal plasticity *in vivo*, as mice learned to stay on an accelerating rotarod by running continuously. On this task, time spent running on the rotarod gradually increased with training, and this improvement was also reflected in sensorimotor corticostriatal plasticity *in vivo* (Costa, Cohen, & Nicolelis, 2004). Early in learning, variability in the activity of motor cortical neurons and sensorimotor SPNs was high, but such variability decreased dramatically with training. Performance was also significantly impaired by either D1 or D2 blockade. With extensive training, however, D1 blockade no longer affected performance, yet D2 blockade continued to impair performance (Yin et al., 2009). These results suggest that the D1-dependent process plays a more transient role in learning. Once the synaptic strength is increased, the cortical input will be sufficient to elicit SPN output even with reduced D1 signaling.

The above discussion is only concerned with the role of DA in the BG, specifically in the associative and sensorimotor striatal regions. Since DA is extremely common in the brain, it may have additional roles in other areas. Nor is DA the only relevant neuromodulator in the BG. Other neuromodulators may also contribute to synaptic plasticity and learning in a similar way. Although ACh, adenosine, serotonin, and other neuromodulators in the BG have been extensively studied, at present it remains unclear how they contribute to BG circuit function and behavior.

13.9 SUMMARY

The transition control model suggests a different definition of learning. Learning is only needed when the existing organization is inadequate. Reorganization requires random changes in relevant system parameters in transition controllers, such as relationship, velocity, and serial order, and can be manifested in exploratory behavior.

In conventional RL models, the synaptic weights remain unchanged until reinforcement occurs. In reorganization, an increase in internal error beyond a threshold leads to random parameter variation first, as manifested in behavioral variability. The reduction of error stops the variation and preserves the latest parameters.

Research on bird song provides evidence for both the transition control model in performance and a reorganization-like process in learning. To begin with, birds must first establish an internal template consisting of the memory of the tutor song and an internal comparison function that compares the remembered song template with the actual performance of the song. The updating of various representations associated with the action hierarchy requires a specialized cortico-BG network. As an adaptive gain, DA signaling repeats the last syllable parameter. Thus, the BG not only contribute to an online correction that results in an improved performance but are also critical for discovering and preserving the appropriate system parameters.

Parameter variation at the transition level is characterized by transient changes in parameters first, which can be viewed as a search process in parameter space and manifested in exploratory behavior. Error reduction also reduces parameter variation or exploration. The actual plasticity takes place slowly, as parameters are gradually saved through repetition. The repetition of the transition reference is the *in vivo* equivalent of the synaptic plasticity induction protocol.

REFERENCES

Ali, F., Otchy, T. M., Pehlevan, C., Fantana, A. L., Burak, Y., & Ölveczky, B. P. (2013). The basal ganglia is necessary for learning spectral, but not temporal, features of birdsong. *Neuron*, 80(2), 494–506.

Andalman, A. S., & Fee, M. S. (2009). A basal ganglia-forebrain circuit in the songbird biases motor output to avoid vocal errors. *Proceedings of the National Academy of Sciences*, 106(30), 12518–12523. doi: 10.1073/pnas.0903214106.

Anggono, V., & Huganir, R. L. (2012). Regulation of ampa receptor trafficking and synaptic plasticity. *Current Opinion in Neurobiology*, 22(3), 461–469. doi: 10.1016/j.conb.2011.12.006.

Bardo, M. T., Donohew, R., & Harrington, N. G. (1996). Psychobiology of novelty seeking and drug seeking behavior. *Behavioural Brain Research*, 77(1–2), 23–43.

Berridge, K. C., Aldridge, J. W., Houchard, K. R., & Zhuang, X. (2005). Sequential super-stereotypy of an instinctive fixed action pattern in hyper-dopaminergic mutant mice: A model of obsessive compulsive disorder and tourette's. *BMC Biology, 3,* 4. doi: 10.1186/1741-7007-3-4.

Bi, G., & Poo, M. (2001). Synaptic modification by correlated activity: Hebb's postulate revisited. *Annual Reviews Neuroscience, 24,* 139–166. doi: 10.1146/annurev.neuro.24.1.139.

Bickford, M. E., & Hall, W. C. (1989). Collateral projections of predorsal bundle cells of the superior colliculus in the rat. *Journal of Comparative Neurology, 283*(1), 86–106.

Bottjer, S. W., Miesner, E. A., & Arnold, A. P. (1984). Forebrain lesions disrupt development but not maintenance of song in passerine birds. *Science, 224*(4651), 901–903.

Brainard, M. S., & Doupe, A. J. (2002). What songbirds teach us about learning. *Nature, 417*(6886), 351–358. doi: 10.1038/417351a 417351a.

Cador, M., Taylor, J., & Robbins, T. (1991). Potentiation of the effects of reward-related stimuli by dopaminergic-dependent mechanisms in the nucleus accumbens. *Psychopharmacology (Berl), 104*(3), 377–385.

Carr, G. D., & White, N. M. (1987). Effects of systemic and intracranial amphetamine injections on behavior in the open field: A detailed analysis. *Pharmacology Biochemistry and Behavior, 27*(1), 113–122.

Chen, R., & Goldberg, J. H. (2020). Actor-critic reinforcement learning in the songbird. *Current Opinion in Neurobiology, 65,* 1–9.

Costa, R. M., Cohen, D., & Nicolelis, M. A. (2004). Differential corticostriatal plasticity during fast and slow motor skill learning in mice. *Current Biology, 14*(13), 1124–1134.

Derégnaucourt, S., Mitra, P. P., Fehér, O., Pytte, C., & Tchernichovski, O. (2005). How sleep affects the developmental learning of bird song. *Nature, 433*(7027), 710–716.

Dhawale, A. K., Miyamoto, Y. R., Smith, M. A., & Ölveczky, B. P. (2019). Adaptive regulation of motor variability. *Current Biology, 29*(21), 3551–3562. e3557.

Domjan, M. (2005). Pavlovian conditioning: A functional perspective. *Annual Review of Psychology, 56,* 179–206. doi: 10.1146/annurev.psych.55.090902.141409.

Dworkin, B. (1993). *Learning and Physiological Regulation*: Chicago, IL: University of Chicago Press.

Fee, M. S., & Goldberg, J. H. (2011). A hypothesis for basal ganglia-dependent reinforcement learning in the songbird. *Neuroscience, 198,* 152–170.

Frank, M. J., Doll, B. B., Oas-Terpstra, J., & Moreno, F. (2009). Prefrontal and striatal dopaminergic genes predict individual differences in exploration and exploitation. *Nature Neuroscience, 12*(8), 1062.

Gale, S. D., Person, A. L., & Perkel, D. J. (2008). A novel basal ganglia pathway forms a loop linking a vocal learning circuit with its dopaminergic input. *Journal of Comparative Neurology, 508*(5), 824–839.

Gallistel, C. R. (1990). *The Organization of Learning*: Cambridge, MA: MIT Press.

Gerfen, C. R., & Surmeier, D. J. (2011). Modulation of striatal projection systems by dopamine. *Annual Reviews Neuroscience, 34,* 441–466. doi: 10.1146/annurev-neuro-061010-113641.

Goldberg, J. H., Farries, M. A., & Fee, M. S. (2013). Basal ganglia output to the thalamus: Still a paradox. *Trends in Neurosciences, 36*(12), 695–705.

Greengard, P., Allen, P. B., & Nairn, A. C. (1999). Beyond the dopamine receptor: The darpp-32/protein phosphatase-1 cascade. *Neuron, 23*(3), 435–447.

Hart, G., Bradfield, L. A., Fok, S. Y., Chieng, B., & Balleine, B. W. (2018). The bilateral prefronto-striatal pathway is necessary for learning new goal-directed actions. *Current Biology, 28*(14), 2218–2229 e2217. doi: 10.1016/j.cub.2018.05.028.

Hessler, N. A., & Doupe, A. J. (1999). Social context modulates singing-related neural activity in the songbird forebrain. *Nature Neuroscience, 2*(3), 209–211.

Hisey, E., Kearney, M. G., & Mooney, R. (2018). A common neural circuit mechanism for internally guided and externally reinforced forms of motor learning. *Nature Neuroscience, 21*(4), 589–597.

Hoffmann, L. A., Saravanan, V., Wood, A. N., He, L., & Sober, S. J. (2016). Dopaminergic contributions to vocal learning. *Journal of Neuroscience, 36*(7), 2176–2189.

Humphries, M. D., Khamassi, M., & Gurney, K. (2012). Dopaminergic control of the exploration-exploitation trade-off via the basal ganglia. *Frontiers in Neuroscience, 6,* 9.

Ikemoto, S., & Panksepp, J. (1999). The role of nucleus accumbens dopamine in motivated behavior: A unifying interpretation with special reference to reward-seeking. *Brain Research Reviews, 31*(1), 6–41.

Johnson, F., Sablan, M. M., & Bottjer, S. W. (1995). Topographic organization of a forebrain pathway involved with vocal learning in zebra finches. *Journal of Comparative Neurology, 358*(2), 260–278.

Kandel, E. R. (2001). The molecular biology of memory storage: A dialogue between genes and synapses. *Science, 294*(5544), 1030–1038.

Kao, M. H., & Brainard, M. S. (2006). Lesions of an avian basal ganglia circuit prevent context-dependent changes to song variability. *Journal of Neurophysiology, 96*(3), 1441–1455. doi: 10.1152/jn.01138.2005.

Kao, M. H., Doupe, A. J., & Brainard, M. S. (2005). Contributions of an avian basal ganglia-forebrain circuit to real-time modulation of song. *Nature, 433*(7026), 638–643. doi: 10.1038/nature03127.

Kelley, A. E., & Delfs, J. M. (1991). Dopamine and conditioned reinforcement. I. Differential effects of amphetamine microinjections into striatal subregions. *Psychopharmacology (Berl), 103*(2), 187–196.

Kimble, G. A. (1961). *Hilgard and Marquis' Conditioning and Learning* (2nd edition). New York: Appleton-Century-Crofts.

Kojima, S., Kao, M. H., Doupe, A. J., & Brainard, M. S. (2018). The avian basal ganglia are a source of rapid behavioral variation that enables vocal motor exploration. *Journal of Neuroscience, 38*(45), 9635–9647.

Konishi, M. (1965). The role of auditory feedback in the control of vocalization in the white-crowned sparrow 1. *Zeitschrift für Tierpsychologie, 22*(7), 770–783.

Lahiri, A. K., & Bevan, M. D. (2020). Dopaminergic transmission rapidly and persistently enhances excitability of d1 receptor-expressing striatal projection neurons. *Neuron, 106*(2), 277–290.

Luo, M., Ding, L., & Perkel, D. J. (2001). An avian basal ganglia pathway essential for vocal learning forms a closed topographic loop. *Journal of Neuroscience, 21*(17), 6836–6845.

Malinow, R., & Malenka, R. C. (2002). Ampa receptor trafficking and synaptic plasticity. *Annual Reviews Neuroscience, 25*(1), 103–126.

Marler, P. R., & Hamilton, W. J. (1966). *Mechanisms of Animal Behavior*: New York: John Wiley.

Mooney, R. (2009). Neural mechanisms for learned birdsong. *Learning & Memory, 16*(11), 655–669.

Nixdorf-Bergweiler, B. E., Lips, M. B., & Heinemann, U. (1995). Electrophysiological and morphological evidence for a new projection of LMAN-neurones towards area x. *Neuroreport*, 1729–1732.

Olveczby, B. P., Andalman, A. S., & Fee, M. S. (2005). Vocal experimentation in the juvenile songbird requires a basal ganglia circuit. *PLoS Biology, 3*(5), e153. doi: 10.1371/journal.pbio.0030153.

Pavlov, I. (1927). *Conditioned Reflexes*: Oxford: Oxford University Press.

Perkel, D. J. (2004). Origin of the anterior forebrain pathway. *Annals of the New York Academy of Sciences, 1016*, 736–748.

Powers, W. T. (1973). *Behavior: Control of Perception*: New Canaan, CT: Benchmark Publications.

Redgrave, P., & Gurney, K. (2006). The short-latency dopamine signal: A role in discovering novel actions? *Nature Reviews Neuroscience, 7*(12), 967–975.

Riede, T., Suthers, R. A., Fletcher, N. H., & Blevins, W. E. (2006). Songbirds tune their vocal tract to the fundamental frequency of their song. *Proceedings of the National Academy of Sciences, 103*(14), 5543–5548.

Saka, E., Goodrich, C., Harlan, P., Madras, B. K., & Graybiel, A. M. (2004). Repetitive behaviors in monkeys are linked to specific striatal activation patterns. *Journal of Neuroscience, 24*(34), 7557–7565.

Sakata, J. T., & Brainard, M. S. (2006). Real-time contributions of auditory feedback to avian vocal motor control. *Journal of Neuroscience, 26*(38), 9619–9628. doi: 10.1523/JNEUROSCI.2027-06.2006.

Sakata, J. T., & Brainard, M. S. (2008). Online contributions of auditory feedback to neural activity in avian song control circuitry. *Journal of Neuroscience, 28*(44), 11378–11390. doi: 10.1523/JNEUROSCI.3254-08.2008.

Scharff, C., & Nottebohm, F. (1991). A comparative study of the behavioral deficits following lesions of various parts of the zebra finch song system: Implications for vocal learning. *Journal of Neuroscience, 11*(9), 2896–2913.

Shen, W., Flajolet, M., Greengard, P., & Surmeier, D. J. (2008). Dichotomous dopaminergic control of striatal synaptic plasticity. *Science, 321*(5890), 848–851.

Shepherd, G. M. (2013). Corticostriatal connectivity and its role in disease. *Nature Reviews Neuroscience, 14*(4), 278–291.

Singh Alvarado, J., Goffinet, J., Michael, V., Liberti, W., Hatfield, J., Gardner, T., ... Mooney, R. (2021). Neural dynamics underlying birdsong practice and performance. *Nature, 599*(7886), 635–639.

Skinner, B. F. (1963). Operant behavior. *American Psychologist, 18*(8), 503.

Sober, S. J., & Brainard, M. S. (2009). Adult birdsong is actively maintained by error correction. *Nature Neuroscience, 12*(7), 927–931. doi: 10.1038/nn.2336.

Sohrabji, F., Nordeen, E. J., & Nordeen, K. W. (1990). Selective impairment of song learning following lesions of a forebrain nucleus in the juvenile zebra finch. *Behavioral and Neural Biology, 53*(1), 51–63.

Sossinka, R., & Böhner, J. (1980). Song types in the zebra finch poephila guttata castanotis 1. *Zeitschrift für Tierpsychologie, 53*(2), 123–132.

Sutton, R. S., & Barto, A. G. (2018). *Reinforcement Learning: An Introduction*: Cambridge, MA: MIT Press.

Swanson, C., Heath, S., Stratford, T., & Kelley, A. (1997). Differential behavioral responses to dopaminergic stimulation of nucleus accumbens subregions in the rat. *Pharmacology Biochemistry and Behavior, 58*(4), 933–945.

Tchernichovski, O., Mitra, P. P., Lints, T., & Nottebohm, F. (2001). Dynamics of the vocal imitation process: How a zebra finch learns its song. *Science*, *291*(5513), 2564–2569.

Troyer, T. W., & Doupe, A. J. (2000a). An associational model of birdsong sensorimotor learning i. Efference copy and the learning of song syllables. *Journal of Neurophysiology*, *84*(3), 1204–1223.

Troyer, T. W., & Doupe, A. J. (2000b). An associational model of birdsong sensorimotor learning ii. Temporal hierarchies and the learning of song sequence. *Journal of Neurophysiology*, *84*(3), 1224–1239.

Tumer, E. C., & Brainard, M. S. (2007). Performance variability enables adaptive plasticity of 'crystallized' adult birdsong. *Nature*, *450*(7173), 1240–1244.

Xiao, L., Chattree, G., Oscos, F. G., Cao, M., Wanat, M. J., & Roberts, T. F. (2018). A basal ganglia circuit sufficient to guide birdsong learning. *Neuron*, *98*(1), 208–221. e205.

Yin, H. H. (2009). The role of the murine motor cortex in action duration and order. *Frontiers in Integrative Neuroscience*, *3*, 23. doi: 10.3389/neuro.07.023.2009.

Yin, H. H., Mulcare, S. P., Hilario, M. R., Clouse, E., Holloway, T., Davis, M. I., ... Costa, R. M. (2009). Dynamic reorganization of striatal circuits during the acquisition and consolidation of a skill. *Nature Neuroscience*, *12*(3), 333–341. doi: 10.1038/nn.2261.

Yttri, E. A., & Dudman, J. T. (2016). Opponent and bidirectional control of movement velocity in the basal ganglia. *Nature*, *533*(7603), 402–406. doi: 10.1038/nature17639.

14 Interpretation of Clinical Symptoms

It is perhaps no coincidence that two pioneers in the study of the basal ganglia (BG), Willis and Reil, also coined "neurology" and "psychiatry," respectively. BG pathology is observed in many disorders, such as Parkinson's disease, schizophrenia, and Tourette's syndrome. Although in previous chapters we have considered the relevant symptoms whenever they are germane to the discussion at hand, we have not explored the clinical implications of the transition control model.

This chapter focuses on an analysis of clinical symptoms rather than disorders. We are not concerned with *how* the system became impaired, but with *what* is impaired and how this could impact behavior. Multiple disorders can share the same symptoms due to overlapping pathology. The same symptoms can also be generated in different ways. For example, neurons may die due to degeneration, hypoxia, or trauma, with similar functional consequences. As Denny-Brown pointed out (1962, p. 52):

> *Although fixed expression, rigid posture, and difficulty in initiating movement are prominent features of parkinsonism, these signs also accompany types of disorders such as Double Athetosis and Wilson's Disease. Conversely, the characteristic tremor of parkinsonism can appear in the course of diseases such as Wilson's Disease, Huntington's Chorea, and the Fahr syndrome.*
>
> (Denny-Brown, 1962)

We shall start with an analysis of symptoms of movement disorders. We shall then consider symptoms related to cognitive functions, including disorders of attention and working memory, as well as obsessions, hallucinations, and delusions, which are more common in psychiatric disorders.

14.1 ANALYSIS OF SYMPTOMS

To appreciate how the transition control model can elucidate clinical symptoms, it is useful to consider the different ways in which any control system can fail.

14.1.1 IMPAIRED INPUT FUNCTION

A disruption in the input function of a controller alters how sensory inputs affect the comparison function. If the damage is severe enough to abolish feedback, then the loop is opened. Without negative feedback, in a system with positive reference signals (Chapter 5), the reference signal would generate sustained output as the error could no longer be reduced. At higher levels in the action hierarchy, even a slight increase in the reference signal can produce a large error, which can be converted to large changes in the reference signal sent to the lower system. In a system with negative reference signals, loss of feedback would result in no output. Depending on the level of the hierarchy where the feedback is lost, this increased output could be manifested in different ways. In a velocity controller, for example, even a slight increase in action command could result in large changes in posture and movements. These changes will appear to be "uncontrollable."

Damage could change the definition of the controlled variable. In this case, there is still feedback, but the system does not control the variable it is supposed to control. Instead it is controlling an altered and distorted version. This situation is similar to the example of distorted pitch feedback in bird song experiments discussed in Chapter 13, except that the distortion occurs within the sensory system of the organism.

14.1.2 IMPAIRED OUTPUT FUNCTION

The output function transforms errors into output. The behavioral consequences of output function damage depend on the level of the hierarchy where the damage is found. At the lowest level, the final common path could be disrupted by motor neuron disease or muscle atrophy, resulting in paralysis. The higher level could still send commands, but to no avail because descending signals can no longer be converted into muscle contraction. There are no alternative means to compensate for damage to the final common path. On the other hand, compensation is possible if damage occurs at higher levels. For example, at the velocity control level, impaired output function could alter how much and how rapidly the position reference changes, but the lower levels can still function and generate output when their controlled variables are disturbed.

In some control systems, some combination of proportional (P), integral (I), and derivative (d) terms may be used in the output function. Each term can be computed separately using the error signal. In principle, changes in these terms could have a major impact on system stability. PID (proportional-integral-derivative) control is common in engineered control systems, but it is unclear the extent to which neural controllers use this strategy. As reviewed in Chapter 6, there is evidence for proportional (muscle length) and derivative (rate of change in muscle length) terms in lower-level controllers.

14.1.3 CHANGE IN GAIN

The loop gain determines the quality of control. Although sufficient gain is needed for a control system to function, the consequences of gain reduction might not be obvious. Halving the loop gain may only produce a slight reduction in the capacity for control. With reduced gain, the amount of change produced by the output during each time step is smaller, so it takes longer to reach a steady state. In general, as the gain is lowered, the control becomes sluggish, whereas an excessively high gain could result in oscillations. Gain can be implemented by neuromodulation. In principle, any input that changes the responsiveness of neurons to excitatory or inhibitory synaptic inputs could alter the gain. Although often multiplicative gain is used in the output function of a controller, gain changes do not have to be localized to a single component. Changes in loop gain could be due to changes at multiple components in the loop, even multiple level in the control hierarchy.

14.1.4 OSCILLATIONS

Negative feedback control systems could show oscillations, which usually indicate problems with feedback timing. Physical systems cannot change their state instantaneously. Within the control system, there are also lags in signal transmission due to synaptic transmission or axonal conduction. Around any loop, there is some delay for which the output can further reinforce later output instead of reducing it, thus turning negative feedback into positive feedback. Oscillations, both neural and behavioral, are common in neurological symptoms (Wiener, 1948). The frequency of the oscillations may also provide insight into the properties of the controllers. In a control hierarchy, higher levels typically have slowly changing inputs and slower oscillations. The lower levels, however, can show oscillations with higher frequencies.

14.2 POSTURAL CONTROL DEFICITS

When considering all the symptoms associated with the BG, it is helpful to start with postural control deficits, which standard models of the BG fail to address adequately. These deficits, traditionally labeled extrapyramidal, include rigidity and deviant postures that patients may maintain for extended periods of time. There are axial postural deformities involving the head and torso, such as torticollis, Pisa syndrome, scoliosis, camptocormia, and antecollis, and less frequently, deformities

Interpretation of Clinical Symptoms

in the extremities, such as striatal foot and striatal hand. Torticollis, for example, is characterized by a head tilt with tonic flexion of neck muscles on one side (Shaikh, Zee, Crawford, & Jinnah, 2016). Figure 14.1 shows some examples of these symptoms.

Asymmetric posture is common following unilateral BG damage (Oberlander, Dumont, & Boissier, 1977; Scheel-Krüger, Arnt, & Magelund, 1977). According to the transition control model, posture is maintained by a variety of position controllers, mostly found at the brainstem level, that receive direct projections from the BG. Unilateral manipulations of these brainstem nuclei involved in position control can also produce postural asymmetry. For example, unilateral manipulation of the interstitial nucleus of Cajal also produces a tonic head tilt (Fukushima, Takahashi, Kudo, & Kato, 1985; Klier, Wang, Constantin, & Crawford, 2002). As reviewed in Chapter 6, brainstem position controllers receive continuous reference signals from the BG during volitional behavior (Barter et al., 2015). BG output neurons, such as substantia nigra pars reticulata (SNr) gamma-Aminobytyric acid (GABA) neurons, provide tonic bias signals. Tonic firing rates of multiple classes of neurons represent a neutral position vector. Increases and decreases in their firing rates generate turning movements in different directions. During movement, the BG send reference signals to

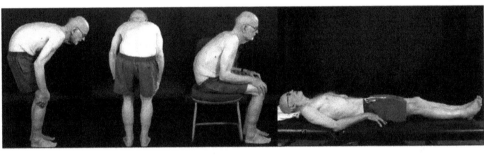

FIGURE 14.1 Postural deformities in disorders associated with the basal ganglia. Camptocormia: extreme stooped posture with thoracolumbar flexion when standing, but resolves when lying supine. Antecollis: forward flexion in the head and neck. Pisa syndrome: lateral flexion of the trunk, which also resolves when lying down. Scoliosis: lateral flexion of the torso that is not relieved by passive manipulation or a supine position. The examples here are from Parkinson's patients, but similar symptoms are found in other disorders as well. From Doherty et al. (2011) with permission.

areas that are critical for orienting and body configuration control, like the superior colliculus and interstitial nucleus of Cajal (Crawford, Martinez-Trujillo, & Klier, 2003; Onodera & Hicks, 1998; Redgrave, Marrow, & Dean, 1992).

Cervical dystonia (torticollis) and other deviant postures have been associated with asymmetric BG outputs (Lee & Kiss, 2014; Moll et al., 2014; Sedov, Semenova et al., 2019; Sedov, Usova et al., 2019). For example, tonic output from the globus pallidus internus (GPi) or SNr on the left side may be lower than that from the right side, generating asymmetric position reference vectors. Behaviorally, this is manifested in a tonic-asymmetric posture that is resistant to manipulation. The specific posture observed is determined by specific channels of BG output affected. A bent spine, for example, can be attributed to asymmetric reference signals to the controllers responsible for torso position control.

Torticollis can also be evoked by unilateral SNr inhibition, which presumably has the same effects as sustained inhibition by striatal GABA release in the SNr (Burbaud, Bonnet, Guehl, Lagueny, & Bioulac, 1998; Dybdal et al., 2013). There is a contraversive head tilt and rotation (away from the side of inhibition), often with the head pressed against the shoulder lateral flexion (Holmes et al., 2012; Oberlander et al., 1977; Scheel-Krüger et al., 1977). Similarly, unilateral striatal manipulation or asymmetric DA depletion can also produce torticollis (Crossman, Mitchell, Sambrook, & Jackson, 1988; Yamada, Fujimoto, & Yoshida, 1995). When unilateral SNr inhibition produces a head tilt, the tilt can be reduced by inactivating the dorsolateral superior colliculus. This observation suggests that the head tilt is due to asymmetric tonic position reference signals from the SNr to the superior colliculus (Holmes et al., 2012).

Stiffness and resistance to passive movement, also known as rigidity, are commonly observed in disorders like Parkinson's disease (Rushworth, 1960). Rigidity can affect nearly all parts of the body; it is an expression of position control with fixed reference signals. The increased muscle tone represents the output of position controllers in response to disturbances in the absence of descending commands from the BG. An indication that rigidity is related to position control is that it varies depending on the posture, as different postures have different levels of disturbance from gravity and other forces. If the body is well supported, there is lower muscle output and less rigidity because less output is needed to counter disturbances.

According to the focused selection model, fixed postures and rigidity result from excessive suppression of actions and failure to turn off postural reflexes (Mink, 1996). But this account fails to explain why BG output represents position vectors or why there is postural asymmetry and active resistance to disturbance that varies as a function of support. Rigidity varies with the amount of support not because the position input has changed, but because the disturbance is altered by support. To understand this phenomenon, we must recall that disturbance is not a property of the environment but emerges through the interaction between the environment and controller. The controller output (rigidity and stiffness) that is generated from the error signal mirrors the disturbance.

The transition control model suggests that volitional movements are initiated when the BG outputs adjust the reference signals to the postural control systems, rather than by suppressing postural control. Using the same mechanism, it is also possible to generate rigidity volitionally by increasing the descending reference for stiffness (Chapter 6). This effectively increases the loop gain of position control, which could be useful in anticipation of potential disturbances to position control. For example, to better resist an upcoming push to the arm, one could volitionally increase stiffness by increasing tension in antagonistic muscle groups like the biceps and triceps.

14.3 BRADYKINESIA, AKINESIA, AND PARADOXICAL KINESIA

Bradykinesia and akinesia are commonly found in Parkinson's disease. In bradykinesia, movements are slower and smaller in amplitude, and in akinesia, there is an inability to perform any volitional movement (Viviani, Burkhard, Chiuvé, dell'Acqua, & Vindras, 2009). These observations support the predictions of the velocity control model. According to this model, DA depletion reduces the multiplicative gain, so the magnitude of the striatal output (from velocity-related direct pathway spiny projection neurons [dSPNs]) in response to excitatory input is also reduced. Slower accumulation in the striatonigral integrator manifests in slower movements. The time it takes to initiate movement is also increased, as it takes longer for the accumulated signal to reach the threshold. Bradykinesia often begins in a specific body part and then gradually spreads throughout the body as the disease progresses, indicating that the reduced gain only affects a small region in the striatum at first, but as more DA neurons die, larger regions are affected. If the gain becomes too low, volitional movements become impossible (akinesia).

Since DA depletion reduces gain in transition control, reference signals entering the striatum from the corticostriatal projections do not produce sufficient output. Suppose normally the corticostriatal input at the axon terminal has a signal magnitude of 2 and the multiplicative gain value from DA is 3, then the final product in the form of spiny projection neuron (SPN) activation is 6. If the threshold for generating SPN output is 6, then the signal can only activate the SPN when the gain is at least 3. However, when the gain is reduced to 1 as a result of DA depletion, the minimum cortical input needed to reach the threshold for SPN firing is 6 rather than 2. With reduced gain, then, greater excitatory drive is required to generate striatal output. This could explain the phenomenon of paradoxical kinesia—although Parkinson's patients struggle to initiate actions, under some conditions they can exhibit surprisingly normal movements. For example, when thrown a ball, they could still catch it. Visual cues, especially those with large sizes and high contrast, can significantly improve movements. For example, transverse lines that are clearly marked on the floor can be crossed (Martin, 1963). Patients can also increase their step length when there are specific visual targets on the floor (Bagley, Kelly, Tunnicliffe, Turnbull, & Walker, 1991).

The exteroceptive guidance system uses a representation of the distance to the target to activate the necessary SPNs for movement (Chapter 10). This spatial relationship controller can be engaged by salient signs, which increase the excitatory corticostriatal input representing the distance to the target (Kim et al., 2019). In turn, the distance error is converted to an increased velocity reference, compensating for the loss of gain due to dopamine (DA) depletion. Likewise, a state of great emotional excitement may also increase excitatory drive to the striatum to reach the firing threshold even when the gain is low.

14.4 DEFICITS IN LOCOMOTION

Parkinson's patients often struggle with the initiation of locomotion (Nutt et al., 2011). But when given a push, the patient rushes forward but finds it difficult to stop once movement is initiated. This striking phenomenon is known as festination (from Latin *festinare*, to rush), which was first noted by Parkinson in his initial report:

> *The propensity to lean forward becomes invincible, and the patient is thereby forced to step on the toes and fore part of the feet, whilst the upper part of the body is thrown so far forward as to render it difficult to avoid falling on the face. In some cases, when this state of malady is attained, the patient can no longer exercise himself by walking in his unusual manner, but is thrown on the toes and forepart of the feet; being, at the same time, irresistibly impelled to make much quicker and short steps, and thereby to adopt unwillingly a running pace.*
>
> (Parkinson, 1817)

In humans, walking requires not only stepping but also maintaining balance and a standing posture throughout. As the weight is transferred from one foot to the other and back again, the torso also sways from side to side. A forward lean can promote the initiation of locomotion, and a lean in the opposite direction can contribute to stopping (Figure 14.2). Stepping can be triggered by a shove or some other externally imposed disturbance to posture, or it can be activated volitionally using top–down commands.

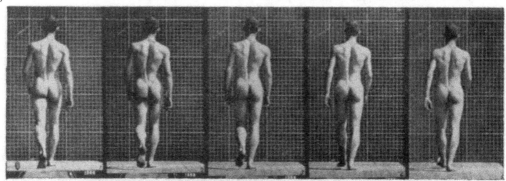

FIGURE 14.2 Control of body tilt. (a) In normal walking, the upper body sways from side to side. For example, when the left leg is swinging, the body sways to the right. In Parkinson's disease, often the patients are unable to sway the torso, which makes it difficult to trigger the stepping mechanism, which is generated by pattern generators in the spinal cord and is independent of the basal ganglia. (b) Impaired volitional control of torso tilt in Parkinson's patients. The patient is unable to propel herself forward, but if she carries a chair in front of her, the resulting forward tilt allows her to walk normally. From Martin (1963) with permission.

Martin attributes the Parkinsonian deficits in locomotion primarily to impaired postural control, in particular the control of torso swaying, rather than to impaired stepping (Martin, 1963, 1967). According to the present model, due to their impaired ability to change position reference signals, patients cannot volitionally tilt the torso to initiate or stop walking. Although they cannot walk volitionally, they can be made to walk by carrying a chair, which creates sufficient forward lean to initiate walking. Apparently, the weight of the chair creates enough disturbance to activate the leg pattern generators and promote walking to counter the disturbance (Figure 14.2). Likewise, patients with a festinating gait often have rigidly bowed bodies; when they walk, they tend to fall forward, as if they were chasing their center of gravity. If velocity is increased passively with a push from behind, the patient may walk forward, but cannot slow down volitionally.

Based on these observations, Martin argues that the BG are not needed for stepping per se but are needed primarily for top–down regulation of the torso tilt. This includes the lateral alternating sway or rocking of the torso and the control of the antero-posterior displacement of the center of gravity that enables forward propulsion (Martin, 1967; Martin & Hurwitz, 1962). According to the present model, the impaired ability to change the torso tilt volitionally can be explained by the lack of change (transition) in the position reference signals.

14.5 DEEP BRAIN STIMULATION (DBS)

Perhaps the most popular surgical intervention for Parkinson's disease and other movement disorders is deep brain stimulation (DBS), which uses chronically implanted electrodes to deliver high-frequency electrical stimulation in specific deep brain locations, most commonly the subthalamic nucleus (STN). Because the STN sends glutamatergic projections to the globus pallidus externus (GPe) and GPi, it is usually assumed to provide excitatory drive to the BG output. It is believed to suppress movement, as unilateral STN lesions can produce hemiballismus, wild uncontrollable movements on the contralateral side of the lesion (Aron & Poldrack, 2006; Denny-Brown, 1962; Matsumura, Kojima, Gardiner, & Hikosaka, 1992).

According to the rate model, the excessive behavioral suppression seen in Parkinson's disease is attributed to STN activation, increased iSPN activity, and reduced GPe activity. The net result is excessive BG output and movement suppression (Hassani, Mouroux, & Feger, 1996). This suggests that reducing STN output would reduce the excessive BG output and enable action selection (Bergman, Wichmann, & DeLong, 1990). Indeed, STN lesions were found to be effective at alleviating some of the motor symptoms, but STN DBS is superior to lesions as it is not only reversible but also allows flexible adjustment of stimulation parameters (DeLong & Wichmann, 2007).

Since STN lesions and DBS both alleviate Parkinsonian symptoms, it was originally assumed that DBS acts as a reversible lesion. The stimulation frequencies used are extremely high (> 100 Hz), which could produce depolarization block, in which sustained excitatory input eventually stops spiking in the postsynaptic neuron. The net effect is similar to STN inactivation. However, in primate models of Parkinsonism, STN DBS increased activity in GPi and GPe, suggesting that stimulation may increase STN output (Hashimoto, Elder, Okun, Patrick, & Vitek, 2003).

What happens to STN activity during DBS remains under debate (Lozano, Dostrovsky, Chen, & Ashby, 2002). Electrical stimulation is not selective, as it can activate fibers of passage as well as axon terminals of excitatory and inhibitory afferents. For example, activating GABAergic afferent terminals from the GPe could inhibit STN projection neurons, and the spread of electrical current to passing pallidothalamic fibers could suppress thalamic activity (DeLong & Wichmann, 2007).

According to one hypothesis, the primary consequence of DBS is to abolish abnormal oscillations (Bergman et al., 1998). Changes in neural oscillations and synchrony are observed in patients with movement disorders. DA depletion generates widespread abnormal patterns of neural activity in the BG, including increased synchrony and oscillations in the theta (4–9 Hz) and beta (10–30 Hz) frequency bands (Brown et al., 2001; DeLong & Wichmann, 2007). Some argue that movement deficits can be explained by abnormal patterns of neural activity in the BG and the spread of such activity to the thalamocortical networks (Hammond, Bergman, & Brown, 2007). DBS is predicted to halt the spread of abnormal oscillations, but it is unclear what the correct pattern of activity should be and how it enables action generation. Nor is it clear whether the change in firing patterns is a by-product of some other change. The fact that neural oscillations often reflect tremor frequency and its higher harmonics suggests some malfunctioning control system. But altered firing rates or increased oscillations are just observations rather than explanations.

Recent work has shown that activation of intralaminar thalamic projections to the STN can alleviate akinesia following DA depletion (Watson et al., 2021). The parafascicular (Pf) nucleus of the thalamus provides distinct pathways to the striatum and to the STN, and selective activation of the Pf–STN projection can generate movements in normal mice (Feger, Bevan, & Crossman, 1994). In mice with bilateral DA depletion and complete akinesia, activation of this pathway can restore natural movements (Watson et al., 2021). On the other hand, stimulation of the Pf–striatum pathway does not have any effect. Stimulation of STN projection neurons that receive Pf projections is also effective, even at more physiologically realistic frequencies (e.g. 25 Hz) far below what is typically used in DBS (e.g. 130 Hz). But when selectively stimulating the Pf–STN pathway, high frequency is not necessary for effective reversal of motor deficits.

STN outputs to the brainstem also contribute to turning behavior (Friedman & Yin, 2022). Stimulation of STN output neurons generates movements with a short latency (~20 ms). Unilateral stimulation quantitatively determines orientation, head yaw, and head roll consistently in the ipsiversive direction, resembling the effects of iSPN stimulation, which also produces ipsiversive turning. On the other hand, head-raising (pitch) was observed after stimulation on either side. Even a single pulse of 10 ms light could generate movement. These results question the common assumption that STN simply suppresses movement.

The importance of selective activation of specific neuronal populations is also highlighted by recent work on GPe stimulation. Gittis and colleagues showed that global activation of GPe did not rescue movement in DA-depleted mice (Mastro et al., 2017; Spix et al., 2021). However, activation of parvalbumin (PV)+ neurons in the GPe robustly restored movement. Activation of another population of GPe neurons, Lim homeobox 6-positive (Lhx6+) neurons, had no clear effect on movement, but inhibition of Lhx6+ neurons also restored movement. These results suggest that prototypical GPe neurons contain opponent populations with opposite effects on behavior. To restore movement, it is necessary to increase the output of PV–GPe neurons relative to that of Lhx6–GPe neurons. One possibility suggested by the transition model is that GPe stimulation can reduce the leak or damping effect (opposite of iSPN stimulation), but it remains to be determined whether the PV+ GPe neurons are specifically responsible for introducing the leak in the integrator.

Together, the results just reviewed suggest that the requirement for high stimulation frequency in traditional DBS is probably due to nonselective stimulation of cell populations. Selective targeting of specific populations in the BG is more effective at restoring movement in Parkinsonian animals. At present, however, the mechanisms by which stimulation of the GPe and STN can restore movement remain poorly understood.

14.6 HYPERKINETIC SYMPTOMS

BG damage can also produce excessive and involuntary movements, include dyskinesia, chorea, dystonia, tics, ballismus, and athetosis (Mettler & Mettler, 1942; Richter, 1945; Richter & Klüver, 1944). These are classified as "hyperkinetic," being the opposite of "hypokinetic" or "akinetic." But the symptoms are quite diverse, ranging from small repetitive movements in athetosis to large-amplitude flinging movements in ballism. Just like rigidity, hyperkinetic symptoms can be found in specific body parts. For example, local dystonia can be classified according to the body part affected, such as the neck (cervical), eyes (blepharospasm), hand (writer's cramp), or even the larynx (spasmodic dysphonia). Although precise quantifications of hyperkinetic symptoms are rare, available results suggest that they can be distinguished on the basis of kinematic features. For example, athetosis and chorea can be distinguished based on movement velocity (Liu, Oubre, Duval, Lee, & Daneault, 2022).

14.6.1 Chorea

Huntington's disease is often considered the prototypical hyperkinetic disorder. It is caused by an autosomal-dominant polyglutamine mutation of Huntingtin (HTT) due to an expansion of CAG (Cytosine, Adenine, and Guanine) triplets located near the beginning of the coding region of the gene (MacDonald et al., 1993). The HTT protein is critical for cell health and survival of SPNs, and deletion of HTT can disrupt synaptic connectivity in the striatum (McKinstry et al., 2014). The most well-known symptom is chorea, defined as jerking or flinching movements that are unpredictable and involuntary, yet in late stages the patient becomes primarily akinetic.

In Huntington's disease, there is significant degeneration in the striatum. Studies have shown that iSPNs degenerate early in disease progression, usually before degeneration in dSPNs and other parts of the BG and cortex (Beal et al., 1988; Reiner et al., 1988). Chorea is due to selective loss of striatopallidal neurons early on, whereas akinesia is a result of widespread neurodegeneration in the striatum. Increased DA signaling has been associated with chorea, as DA agonists exacerbate hyperkinetic symptoms, whereas D2 receptor antagonists can alleviate such symptoms (Bird, 1980). As D2 antagonists are expected to increase indirect pathway output, they may increase the damping of the transition controller, thus reducing hyperkinesia.

14.6.2 L-DOPA–Induced Dyskinesia (LID)

Nearly all treatments for hypokinetic disorders are DA agonists, whereas treatments for hyperkinetic symptoms are often DA antagonists. Increased DA signaling may therefore contribute to hyperkinetic symptoms. Indeed, although DA replacement with levodopa is effective for treating Parkinson's disease, when used chronically it could produce levodopa-induced dyskinesias (LIDs), characterized by uncontrollable movements that often resemble chorea (Schrag & Quinn, 2000).

When rats with unilateral DA depletion are treated with DA agonists, they show contraversive rotations (away from the side of the lesion), suggesting that the depleted side produces a greater response when given DA. DA depletion seems to result in super-sensitivity of D1 receptors and increased dSPN output (Gerfen, 2003). Doses of D1 agonists that result in little or no activation in the intact striatum, as measured by immediate early gene expression, produce greater activation in the DA-depleted striatum. This effect is not due to increased receptor expression but rather altered intracellular signaling (Gerfen, Miyachi, Paletzki, & Brown, 2002). Consequently, D1 signaling is expected to be enhanced following DA depletion.

In mouse models, LIDs preferentially activate dSPNs. They are also reduced by selective inhibition of the activated neurons, suggesting that dSPNs contribute to the generation of dyskinetic movements (Girasole et al., 2018). The transition control model predicts that increased D1 signaling will increase output from movement velocity-related SPNs in the sensorimotor striatum, resulting in larger velocity commands and faster movements. In support of this prediction, movement velocity was significantly increased during axial dyskinesias induced by L-DOPA (Alberico, Kim, Lence, & Narayanan, 2017). During such movements, sensorimotor striatal activity was also correlated with velocity. These results suggest that LIDs could involve increased output from velocity-related dSPNs due to a higher multiplicative gain from increased D1 signaling.

14.6.3 Reduced Damping

How can the present model shed light on hyperkinetic symptoms? There are three major possibilities: reduced damping, reduced reference selectivity, and impaired feedback. These are not mutually exclusive.

Rapid, uncontrollable movements like chorea and ballismus can be explained by inadequate damping. Damping occurs at the same time as integrator accumulation, whereas discharging occurs at the end of the accumulation process. Damping can be modeled by a leak in the bucket that slows down the accumulation. In a motion velocity controller, it can reduce velocity.

In LIDS, due to compensatory changes in the depleted striatum (e.g. receptor super-sensitivity), the effect of DA is much enhanced. Without sufficient damping, dSPN output could result in rapid accumulation in the integrator. If this occurs in the velocity control system, movement is expected to become more ballistic. Indeed, dyskinetic movements are also characterized by higher velocity (Alberico et al., 2017).

Another consequence of reduced damping is instability and oscillations, e.g. overshooting and undershooting. According to the transition control model, the oscillations reflect altered feedback timing in a closed-loop transition controller. The oscillations can be detected using kinematic measures with high temporal and spatial resolution. Local field potential recordings in the mouse striatum showed increased delta band (4–7 Hz) oscillations during LIDs (Alberico et al., 2017). Dyskinetic movements can also occur in this frequency range (Alonso-Frech et al., 2006). Increased delta oscillations may simply reflect the frequency component of dyskinetic movements, and many pathological neural oscillations have frequencies that are higher harmonics of these frequencies.

14.6.4 Loss of Selectivity

In the normal striatum, DA ensures selective activation of specific SPNs, not only by enhancing excitability but also by facilitating GABA release from dSPNs and thus recurrent inhibition from axonal collaterals (Wei, Ding, & Zhou, 2017). After DA depletion, there are well-documented compensatory changes, such as DA receptor super-sensitivity (Gerfen, 2003), loss of glutamatergic corticostriatal synapses (Day et al., 2006), reduced collateral inhibition (Taverna, Ilijic, & Surmeier, 2008), and compensatory increases in neuronal excitability in SPNs (Azdad et al., 2009; Fino, Glowinski, & Venance, 2007). As a result of these changes, SPNs can become more responsive to both glutamatergic and dopaminergic inputs, but their activation will be less selective. It would be more difficult for them to respond to one set of inputs than another, since even inputs with weak synaptic strength may be sufficient to elicit firing in SPNs. As a result, SPNs tend to respond indiscriminately to excitatory inputs.

Uncontrollable movements can occur at rest, but more often they occur when there is some intention to move. These movements do not follow the intended course of the action. They are often inappropriate and difficult to suppress. According to the present model, the corticostriatal projections allow errors from higher control systems to recruit specific action modules. If these projections lose their selectivity, a given goal can no longer activate the appropriate comparators to generate action. Instead, it could produce simultaneous activation of different and even conflicting actions.

With DA replacement treatment, then, striatal activation in response to cortical input could be restored, but the selectivity of such activation is reduced, contributing to uncontrollable movements. Moreover, the loss of selectivity in corticostriatal projections may also distort sensory feedback to transition controllers. With inappropriate feedback affecting the comparator, the production of the action could be influenced by irrelevant sensory inputs.

14.6.5 Loss of Feedback

Some hyperkinetic symptoms can also be explained by reduced feedback. Indeed, there are reports that damage to the parietal cortex can cause deficits in proprioceptive feedback and large involuntary movements (Sharp, Rando, Greenberg, Brown, & Sagar, 1994). Corticostriatal and possibly thalamostriatal projections allow sensory representations to reach the comparators, providing ongoing feedback to the transition controllers. When feedback is reduced, the error accumulates and generates persistent output because there is no negative feedback to offset it. The output now reveals the amplification in the output function, or the "open loop" gain.

Moreover, changes in BG circuitry can alter the pattern and sign of feedback to transition controllers. This appears to be the case in Tourette's syndrome, another common hyperkinetic disorder characterized by tics, which are rapid, stereotyped, and repetitive movements. Tics can range from abrupt single movements (twitches or sounds) to complex sequences of behaviors (Leckman & Riddle, 2000). In the brains of patients with Tourette's syndrome, loss of fast-spiking interneurons (FSIs) has been reported (Kalanithi et al., 2005; Kataoka et al., 2010). These GABAergic neurons receive sensory cortical inputs (e.g. from the parietal regions) and project to SPNs. They may invert the sign of the excitatory inputs to generate negative feedback and allow a comparison between reference and perceptual inputs. Loss of FSIs and possibly other GABAergic interneurons could therefore impair negative feedback. As discussed in Chapter 10, FSIs can also transform distance errors to generate velocity commands. In this circuit, the loss of FSIs could increase the instantaneous velocity command. If the polarity of feedback is changed from negative to positive, there could be explosive and regenerative output.

14.7 PERSEVERATION, STEREOTYPY, AND COMPULSION

Perseveration and stereotypy are also common symptoms in many disorders, ranging from autism spectrum disorders to obsessive–compulsive disorder (OCD) and Tourette's syndrome. For example, OCD symptoms include repetitive, intrusive thoughts, and compulsive washing, counting, and checking (Ernst & Smelik, 1966). Human neuroimaging studies of OCD patients reveal a circuit including the anterior cingulate cortex and the posterior orbitofrontal cortex and their targets in the limbic and associative striatum (Baxter et al., 1992; Graybiel & Rauch, 2000; Swedo et al., 1992).

It has long been known that injections of DA agonists into the striatum can produce stereotypy, including compulsive gnawing (Ernst & Smelik, 1966; Randrup & Munkvad, 1967). Robbins argues that psychostimulants simply increase the repetition of behaviors (Robbins, 1976). In human amphetamine addicts, repetitions of seemingly purposeless acts and behavioral stereotypy are often observed.

These observations can be explained by a malfunctioning repetition control system (Chapter 8). Increased gain from DA signaling is expected to prolong a bout of ongoing behavior. D1 signaling can increase dSPN output from the repetition control system and repeat the same reference signal from the cortex using reentrant projections to the thalamocortical network. Depending on the striatal regions affected, simple transitions or more complex behavioral sequences could be affected. In most cases, behavioral perseveration and stereotypy are not limited to simple movements but involve higher-order transitions at the event level or above, where behavioral sequences are constructed by concatenating simple transitions.

In OCD, thoughts and needs (obsessions) are responsible for repetitive behaviors (compulsions). Obsessions and compulsions cannot be explained by the increased persistence of ongoing behavioral sequences via repetition control. There appears to be impaired feedback at higher levels. Higher-level feedback is usually slower and updated less frequently compared to lower-level feedback. One example of such feedback is sensory input reporting the final outcome of some action, which indicates that the action has been completed and terminates the behavioral sequence. In OCD patients, compulsive cleaning behaviors are common. They are often accompanied by obsessions about contamination, e.g. the feeling that the hands are dirty even if they have been washed recently (Graybiel, 1998). The persistent needs or obsessions they experience are reference states and expressed as compulsions, which are persistent though futile. There is always residual error at the higher levels, no matter how many times something has been cleaned. Negative feedback related to the final outcome appears to have been disrupted. As discussed in Chapter 10, representation of such outcomes is critical for instrumental behavior and involves the associative cortico-BG network, in particular the circuit originating from the orbitofrontal cortex. Human imaging studies have implicated this circuit in OCD (Chamberlain et al., 2008; Graybiel & Rauch, 2000).

Mouse models of compulsive grooming have also been developed. For example, mice with genetic deletion of SAP90/PSD95-associated protein 3 (SAPAP3) show compulsive grooming, often resulting in the loss of facial hair and skin lesions (Welch et al., 2007). SAPAP3 is a postsynaptic scaffolding protein at corticostriatal synapses. When deleted, there is a reduction in corticostriatal excitatory transmission. Interestingly, in *Sapap3* knockout mice, the baseline firing rates of SPNs are elevated, and the number of FSIs is also reduced. These findings suggest that compulsive grooming may be due to impaired feedback signaling in the FSI–SPN circuit that suppresses overall striatal output. In support of this interpretation, optogenetic stimulation of the lateral orbitofrontal cortex, which projects to the associative striatum, alleviates compulsive grooming in *Sapap3* knockout mice (Burguière, Monteiro, Feng, & Graybiel, 2013). Such stimulation specifically activates striatal FSIs and suppresses SPNs. Compulsive grooming behavior could be reduced by enhancing transmission in the FSI–SPN circuit.

14.8 ATTENTIONAL DEFICITS

Sensory neglect, broadly defined as the inability to perceive or pay attention to some sensory input, can result from damage to cortical or striatal areas. A patient with neglect might fail to eat the food on half of their plate or ignore the side of the body contralateral to the lesioned side. Neglect can result from the inability to orient toward the contralateral side. It is not restricted to sensory processing and can include dramatic attentional and cognitive-spatial deficits (Heilman, Watson, & Valenstein, 1993).

Neglect is observed not only after cortical lesions but also following striatal lesions or DA depletion in the striatum (Reep et al., 2004). As Reep and colleagues have shown, sensory neglect could be produced in rats by unilateral lesions of the prefrontal medial agranular cortex (the homolog of Brodmann area 8 in primates) or posterior parietal cortex (area 7 in primates). Lesions of the medial agranular cortex produce severe deficits in orienting, whereas lesions of the posterior parietal result in deficits in allocentric spatial awareness. These two regions are not only interconnected via direct corticocortical projections but also send convergent projections to the dorsocentral striatum (Reep, Cheatwood, & Corwin, 2003).

According to the present model, these regions correspond to the anterior (prefrontal) and posterior (parietal) divisions that provide the reference and sensory inputs to the comparison function. In other words, the prefrontal division sends reference signals that are compared with parietal inputs in generating volitional orienting behavior. When this orienting circuit is damaged, the symptoms include a lack of responsiveness to stimuli in the space contralateral to the lesioned side and impaired representation of anything presented in the contralesional hemispace.

ADHD is characterized by frequent switching, inattention, restlessness, and impulsivity. ADHD is often associated with reduced striatal activity or striatal volume, especially in the caudate nucleus, but neuroimaging studies have not quantified the striatal output during specific behaviors (Teicher et al., 2000).

The most common treatment for ADHD is methylphenidate (Ritalin) or dextroamphetamine, a nonselective DA agonist that activates D1 receptors. The inability to sustain attention could be due to inadequate activation of the direct pathway. It is possible that D1 activation, by enhancing the gain for repetition control, can promote persisting in one activity for a long time. Methylphenidate may increase the gain of the repetition control system by potentiating the direct pathway. This results in sustained attention and persistence in performing one behavior sequence. Given the important role of the dorsocentral striatum (comparable to parts of the caudate) in attention, it may be a key component of an associative cortico-BG circuit that is affected in ADHD.

ADHD is also associated with deficits in working memory, which is often needed to maintain the goal representation during a task (Kofler, Rapport, Bolden, Sarver, & Raiker, 2010). The manipulation of working memory representations can also be achieved using the same mechanism for repetition control. Direct pathway activation can sustain the working memory representation through reentrant projections to the prefrontal cortex. In contrast, indirect pathway activation may terminate the ongoing representation by discharging the integrator. In individuals with ADHD, the inability to maintain goals in working memory using the reentrant circuit (promoted by the direct pathway) may be responsible for many of the cognitive symptoms.

14.9 PSYCHOSIS AND SCHIZOPHRENIA

Schizophrenia, the best-known psychiatric disorder, is often characterized by hallucinations and delusions, as well as catatonia and reduced motivation. Similarities between these symptoms and those resulting from striatal lesions were noted early (Mettler, 1955; Mettler & Mettler, 1942). For decades, schizophrenia has been associated with increased DA signaling, though DA is not the only transmitter affected by the disease (Howes & Kapur, 2009). Amphetamine, which blocks DA reuptake and increases synaptic DA levels, can induce psychotic symptoms similar to those observed in schizophrenia (Connell, 1957). DA replacement treatment given to PD patients can also sometimes result in psychosis. However, it remains unclear how hyperdopaminergia can account for the diverse symptoms observed in schizophrenia (Baumeister & Francis, 2002).

14.9.1 Positive and Negative Symptoms

A striking feature of schizophrenia is the presence of two classes of symptoms, known as positive and negative symptoms. Positive symptoms include hallucinations, delusions, paranoia, and compulsive behaviors; negative symptoms include apathy, catatonia, and catalepsy. Positive symptoms are often attributed to abnormally high DA in the striatum, whereas negative symptoms are attributed to reduced DA signaling in the prefrontal cortex (Davis, Kahn, Ko, & Davidson, 1991).

Classic antipsychotics (neuroleptics) are D2 antagonists. When used to treat schizophrenia, these drugs are more effective at treating positive symptoms than negative symptoms (Creese, Burt, & Snyder, 1976). Chronic use of antipsychotics can produce Parkinsonian symptoms like rigidity, catalepsy, and bradykinesia, often known as "extrapyramidal side effects" (Albin, Young, & Penney, 1989). Such side effects are less common after treatment with atypical antipsychotics. For example, haloperidol, a typical antipsychotic, induces catalepsy (postural rigidity), while clozapine, an atypical antipsychotic, does not (Vauquelin, Bostoen, Vanderheyden, & Seeman, 2012). This difference can be explained by the fact that clozapine binds more loosely to D2 receptors and dissociates rapidly, resulting in less sustained blockade of D2 receptors (Seeman, 2002). The present model predicts that D2 blockade increases iSPN activation, resulting in more damping that may contribute to catalepsy.

14.9.2 Hallucinations

Hallucinations, perhaps the most common positive symptom, are found in many psychiatric disorders. Schizophrenics experiencing hallucinations show increased activation of the inferotemporal cortex, an area known for being critical for object recognition. In normal humans, electrical stimulation of this area can produce visual hallucinations (Middleton & Strick, 1996; Penfield & Perot, 1963). The inferotemporal cortex also projects to the striatum and receives reentrant projections from the thalamus, thus forming a closed cortico-BG loop. Similarly, auditory cortical regions project to the BG and also receive reentrant projections. Auditory hallucinations are also associated with activations in the cortico-BG circuits (Silbersweig et al., 1995).

Middleton and Strick argue that, in auditory and visual hallucinations, BG outputs can activate perceptual representations in both visual and auditory cortices. Reduced SNr output could result in increased thalamocortical drive to areas like the inferotemporal cortex and activate visual hallucinations (Middleton & Strick, 2000). In support of this view, visual hallucinations have been reported following damage to BG output nuclei like the SNr or GPi (Lauterbach et al., 1994; McKee, Levine, Kowall, & Richardson Jr., 1990). In Chapter 8, we discussed how a subset of the BG outputs that project to the thalamus may implement the imagination mode. The BG outputs are in a position to regulate the tempo and sequencing of hallucinations. In the imagination mode, the BG output may compete with ongoing perceptual inputs at the level of the thalamocortical network, which gates which inputs (imaginary or perceptual) can reenter the cortex.

In support of this account, visual hallucinations are sometimes produced by levodopa treatment (Damásio, Lobo-Antunes, & Macedo, 1971). In the striatal regions receiving visual cortical inputs, increased gain from DA signaling could result in excessive activation (disinhibition) of the visual thalamocortical network. Kepecs and colleagues used a mouse model to study the role of DA signaling in auditory hallucinations (Schmack, Bosc, Ott, Sturgill, & Kepecs, 2021). In their study, mice were presented with tones and reported their perception by poking into one choice port if they perceived an auditory signal and another port if they did not. Increased tonic DA in this region before stimulus onset predicted hallucination-like perception, as indicated by false alarms in the choice behavior. Optogenetic stimulation of DA terminals in the tail of striatum did not impair performance in general but increased false alarms without affecting overall misses. This effect was reversed by the D2 antagonist haloperidol. These results suggest that increased DA signaling in the tail of the striatum, which receives auditory cortical inputs, can contribute to auditory hallucinations.

14.9.3 Delusions

In delusions, patients often feel that their actions and thoughts are generated by some external agent. It is instructive to compare this condition with a related symptom known as utilization behavior. Patients with utilization behavior show mental inertia and apathy; they fail to initiate actions volitionally. Yet given salient external stimulation, they often show inappropriate, context-driven behavior, without being aware of their deficit (Frith, Blakemore, & Wolpert, 2000; Pacherie, 2007). The major difference between delusions and utilization behavior appears to be the experience of agency. In delusions, the action is wrongly attributed to an external agent, whereas in utilization behavior, there is no awareness of agency at all. Interestingly, both conditions are associated with damage in the limbic and associative cortico-BG circuits, but it is difficult to determine the differences that are responsible for the agency attribution (Archibald, Mateer, & Kerns, 2001). The damaged circuits overlap with those that are critical for action–outcome learning and monitoring (Chapter 10). Indeed, the context-driven utilization behavior also resembles habitual behavior observed following damage to the associative cortico-BG network (Chapter 11).

The sense of agency is known to be related to the monitoring of the causal efficacy of actions. Recall that to determine the action–outcome contingency, one must compare outcomes due to self-generated actions and outcomes due to external causes (Chapter 10). The result of this comparison is directly correlated with subjective awareness of causal efficacy. According to the present model, agency attribution requires monitoring action rates using efference copies of BG output from the associative network. This explains why damage to this network, which is critical for action–outcome learning, could either reduce or distort the experience of agency. Indeed, studies by Balleine and colleagues revealed impaired action–outcome learning in schizophrenia patients (Morris, Cyrzon, Green, Le Pelley, & Balleine, 2018; Morris, Quail, Griffiths, Green, & Balleine, 2015). Participants learned to perform specific actions to obtain food rewards. While both schizophrenics and healthy adults learned to perform the actions, patients showed reduced sensitivity to changes in outcome value or in action–outcome contingency. Moreover, in habitual and well-learned skills, the representation of action–outcome contingency is not needed to guide behavior. It is possible to perform such actions without a sense of agency, as in utilization behavior. Damage to the associative network, especially the prefrontal–posterior dorsomedial striatum (pDMS) circuit, can also render performance less sensitive to changes in outcome value or changes in the action–outcome contingency.

Malenka et al. tested schizophrenics on a tracking task in which the participant controlled a cursor position by moving a joystick (Malenka, Angel, Hampton, & Berger, 1982). The relationship between the movement of the joystick and the movement of the cursor (feedback function) could be manipulated. Usually, the cursor moved in the same direction as the joystick, but sometimes the polarity of the feedback function was reversed, so the joystick and cursor moved in opposite directions. This design tested the ability to recognize and adapt to changes in the feedback polarity without exteroceptive feedback or instructions. Schizophrenics were significantly impaired in adapting to changes in the feedback function between joystick movement and cursor movement. These results also suggest impaired monitoring of the instrumental contingency in schizophrenics.

In another study, schizophrenics and normal subjects were tested with distorted sensory feedback: they heard their own voice at a different pitch (Cahill, 1996). Whereas normal subjects could easily recognize their own voice, schizophrenics with delusions reported that they heard someone else speaking. For example, when attributing the distorted sound of their own voice to an external source, they would say: "I think it's an evil spirit speaking when I speak." The frequency of making such attributions was correlated with the degree of pitch distortion and the severity of delusions. Failure to disambiguate self-generated sensory input (reafference) from externally caused sensory input (exafference) may contribute to symptoms like delusions (Pynn & DeSouza, 2013).

The associative cortico-BG network (dorsolateral prefrontal cortex and its target in the caudate nucleus) appears to be the key network affected in schizophrenia (McCutcheon, Abi-Dargham, & Howes, 2019). This circuit is analogous to the prelimbic-pDMS circuit responsible for action–outcome contingency learning, as reviewed in Chapter 10. The sense of agency in voluntary behavior requires the forward propagation of signals from the limbic system to associative networks, through which specific desires recruit actions in the motivational hierarchy.

The efference copy signal from action commands can also activate sensory representations associated with the action. This signaling is hypothesized to be mediated by BG signaling to the thalamocortical network. One candidate thalamic target region is the mediodorsal thalamus, which receives projections from the limbic and associative BG and projects to the frontal cortex (Crail-Melendez, Atriano-Mendieta, Carrillo-Meza, & Ramirez-Bermudez, 2013; Mukherjee & Halassa, 2022). When efference copy signaling is impaired, sensory inputs to the thalamocortical network may be interpreted as externally caused (exafference). When the patient is unable to anticipate the sensory outcomes of his actions, he may feel as though he is under the control of external agents. Likewise, if he does not recognize his own voice or inner speech as self-initiated due to impaired efference copy signaling, he may interpret it as hearing voices from some external source.

14.10 SUMMARY

The transition control model offers a unified account of diverse clinical symptoms. At the core of this model is a neural integrator with accumulation, damping, and discharging functions. Damage to input functions can reduce or distort feedback, potentially resulting in oscillations and uncontrollable outputs. Reduced transition gain is responsible for reduced movement velocity and tempo, as observed in akinesia and bradykinesia. In contrast, excessive gain can also produce oscillations and instability in control.

Postural control deficits are mainly due to impaired descending commands to position control systems, whereas bradykinesia and akinesia are due to impaired transition control. Asymmetric position reference signals could be responsible for deviant postures in many disorders associated with BG damage. In Parkinson's disease, the loss of DA neurons reduces the gain in various transition controllers, resulting in impaired control in movement velocity as well as higher-order transition variables depending on the circuits affected by DA depletion.

Perseveration and stereotypy are observed in various disorders such as autism, OCD, and Tourette's syndrome. Malfunctioning repetition control systems and impaired higher-level feedback may contribute to these symptoms. In OCD, compulsive behaviors result from persistent needs or obsessions and disrupted feedback about the final outcome of actions. Mouse models suggest that impaired feedback signaling in the FSI–SPN circuit may cause compulsive grooming, which can be alleviated by stimulation of the orbitofrontal cortex.

Hallucinations are a common symptom of psychiatric disorders. They could be due to excessive activation of perceptual representations in the visual and auditory cortices, leading to visual and auditory hallucinations. In the imagination mode, a subset of BG outputs may compete with ongoing perceptual inputs in the thalamus. Impaired gating at this level may contribute to the generation of hallucinations.

Schizophrenia patients have impaired action–outcome learning and monitoring. They show reduced sensitivity to changes in outcome value or action–outcome contingency and difficulty adapting to changes in feedback function. This appears to be due to impairments in the associative cortico-BG network, including the dorsolateral prefrontal cortex and the associative striatum. The output from this network not only sends reference signals to lower-level systems for action generation but also sends axon collaterals to target the thalamocortical network. Disruptions in the efference copy signaling may contribute to delusions by impairing the detection of agency.

REFERENCES

Alberico, S. L., Kim, Y.-C., Lence, T., & Narayanan, N. S. (2017). Axial levodopa-induced dyskinesias and neuronal activity in the dorsal striatum. *Neuroscience, 343*, 240–249.

Albin, R. L., Young, A. B., & Penney, J. B. (1989). The functional anatomy of basal ganglia disorders. *Trends in Neurosciences, 12*(10), 366–375.

Alonso-Frech, F., Zamarbide, I., Alegre, M., Rodriguez-Oroz, M. C., Guridi, J., Manrique, M., ... Obeso, J. A. (2006). Slow oscillatory activity and levodopa-induced dyskinesias in parkinson's disease. *Brain, 129*(7), 1748–1757.

Archibald, S. J., Mateer, C. A., & Kerns, K. A. (2001). Utilization behavior: Clinical manifestations and neurological mechanisms. *Neuropsychology Review, 11*(3), 117–130.

Aron, A. R., & Poldrack, R. A. (2006). Cortical and subcortical contributions to stop signal response inhibition: Role of the subthalamic nucleus. *Journal of Neuroscience, 26*(9), 2424–2433. doi: 10.1523/JNEUROSCI.4682-05.2006.

Azdad, K., Chàvez, M., Bischop, P. D., Wetzelaer, P., Marescau, B., De Deyn, P. P., ... Schiffmann, S. N. (2009). Homeostatic plasticity of striatal neurons intrinsic excitability following dopamine depletion. *PLoS One, 4*(9), e6908.

Bagley, S., Kelly, B., Tunnicliffe, N., Turnbull, G. I., & Walker, J. M. (1991). The effect of visual cues on the gait of independently mobile parkinson's disease patients. *Physiotherapy, 77*(6), 415–420.

Barter, J. W., Li, S., Sukharnikova, T., Rossi, M. A., Bartholomew, R. A., & Yin, H. H. (2015). Basal ganglia outputs map instantaneous position coordinates during behavior. *Journal of Neuroscience, 35*(6), 2703–2716.

Baumeister, A. A., & Francis, J. L. (2002). Historical development of the dopamine hypothesis of schizophrenia. *Journal of the History of the Neurosciences, 11*(3), 265–277.

Baxter, L. R., Schwartz, J. M., Bergman, K. S., Szuba, M. P., Guze, B. H., Mazziotta, J. C., ... Munford, P. (1992). Caudate glucose metabolic rate changes with both drug and behavior therapy for obsessive-compulsive disorder. *Archives of General Psychiatry, 49*(9), 681–689.

Beal, M. F., Ellison, D. W., Mazurek, M. F., Swartz, K. J., Malloy, J. R., Bird, E. D., & Martin, J. B. (1988). A detailed examination of substance p in pathologically graded cases of huntington's disease. *Journal of the Neurological Sciences, 84*(1), 51–61.

Bergman, H., Feingold, A., Nini, A., Raz, A., Slovin, H., Abeles, M., & Vaadia, E. (1998). Physiological aspects of information processing in the basal ganglia of normal and parkinsonian primates. *Trends in Neurosciences, 21*(1), 32–38.

Bergman, H., Wichmann, T., & DeLong, M. R. (1990). Reversal of experimental parkinsonism by lesions of the subthalamic nucleus. *Science, 249*(4975), 1436–1438.

Bird, E. D. (1980). Chemical pathology of huntington's disease. *Annual Review of Pharmacology and Toxicology, 20*(1), 533–551.

Brown, P., Oliviero, A., Mazzone, P., Insola, A., Tonali, P., & Di Lazzaro, V. (2001). Dopamine dependency of oscillations between subthalamic nucleus and pallidum in parkinson's disease. *Journal of Neuroscience, 21*(3), 1033–1038.

Burbaud, P., Bonnet, B., Guehl, D., Lagueny, A., & Bioulac, B. (1998). Movement disorders induced by gamma-aminobutyric agonist and antagonist injections into the internal globus pallidus and substantia nigra pars reticulata of the monkey. *Brain Research, 780*(1), 102–107.

Burguière, E., Monteiro, P., Feng, G., & Graybiel, A. M. (2013). Optogenetic stimulation of lateral orbito-fronto-striatal pathway suppresses compulsive behaviors. *Science, 340*(6137), 1243–1246.

Cahill, C. (1996). Psychotic experiences induced in deluded patients using distorted auditory feedback. *Cognitive Neuropsychiatry*, *1*(3), 201–211.

Chamberlain, S. R., Menzies, L., Hampshire, A., Suckling, J., Fineberg, N. A., del Campo, N., ... Bullmore, E. T. (2008). Orbitofrontal dysfunction in patients with obsessive-compulsive disorder and their unaffected relatives. *Science*, *321*(5887), 421–422.

Connell, P. H. (1957). Amphetamine psychosis. *British Medical Journal*, *1*(5018), 582.

Crail-Melendez, D., Atriano-Mendieta, C., Carrillo-Meza, R., & Ramirez-Bermudez, J. (2013). Schizophrenia-like psychosis associated with right lacunar thalamic infarct. *Neurocase*, *19*(1), 22–26.

Crawford, J., Martinez-Trujillo, J., & Klier, E. (2003). Neural control of three-dimensional eye and head movements. *Current Opinion in Neurobiology*, *13*(6), 655–662.

Creese, I., Burt, D. R., & Snyder, S. H. (1976). Dopamine receptor binding predicts clinical and pharmacological potencies of antischizophrenic drugs. *Science*, *192*(4238), 481–483.

Crossman, A., Mitchell, I., Sambrook, M., & Jackson, A. (1988). Chorea and myoclonus in the monkey induced by gamma-aminobutyric acid antagonism in the lentiform complex: The site of drug action and a hypothesis for the neural mechanisms of chorea. *Brain*, *111*(5), 1211–1233.

Damásio, A. R., Lobo-Antunes, J., & Macedo, C. (1971). Psychiatric aspects in parkinsonism treated with l-dopa. *Journal of Neurology, Neurosurgery & Psychiatry*, *34*(5), 502–507.

Davis, K. L., Kahn, R. S., Ko, G., & Davidson, M. (1991). Dopamine in schizophrenia: A review and reconceptualization. *The American Journal of Psychiatry*, *148*(11), 1474–1486.

Day, M., Wang, Z., Ding, J., An, X., Ingham, C. A., Shering, A. F., ... Surmeier, D. J. (2006). Selective elimination of glutamatergic synapses on striatopallidal neurons in parkinson disease models. *Nature Neuroscience*, *9*(2), 251–259. doi: 10.1038/nn1632.

DeLong, M. R., & Wichmann, T. (2007). Circuits and circuit disorders of the basal ganglia. *Archives of Neurology*, *64*(1), 20–24.

Denny-Brown, D. (1962). *The Basal Ganglia and their Relation to Disorders of Movement*: Oxford: Oxford University Press.

Doherty, K. M., van de Warrenburg, B. P., Peralta, M. C., Silveira-Moriyama, L., Azulay, J.-P., Gershanik, O. S., & Bloem, B. R. (2011). Postural deformities in parkinson's disease. *The Lancet Neurology*, *10*(6), 538–549.

Dybdal, D., Forcelli, P. A., Dubach, M., Oppedisano, M., Holmes, A., Malkova, L., & Gale, K. (2013). Topography of dyskinesias and torticollis evoked by inhibition of substantia nigra pars reticulata. *Movement Disorders*, *28*(4), 460–468.

Ernst, A., & Smelik, P. (1966). Site of action of dopamine and apomorphine on compulsive gnawing behaviour in rats. *Experientia*, *22*(12), 837–838.

Feger, J., Bevan, M., & Crossman, A. (1994). The projections from the parafascicular thalamic nucleus to the subthalamic nucleus and the striatum arise from separate neuronal populations: A comparison with the corticostriatal and corticosubthalamic efferents in a retrograde fluorescent double-labelling study. *Neuroscience*, *60*(1), 125–132.

Fino, E., Glowinski, J., & Venance, L. (2007). Effects of acute dopamine depletion on the electrophysiological properties of striatal neurons. *Neuroscience Research*, *58*(3), 305–316.

Friedman, A. D., & Yin, H. H. (2022). Selective activation of subthalamic nucleus output quantitatively scales movements. *bioRxiv*. 2022.01.19.477002.

Frith, C. D., Blakemore, S.-J., & Wolpert, D. M. (2000). Explaining the symptoms of schizophrenia: Abnormalities in the awareness of action. *Brain Research Reviews*, *31*(2–3), 357–363.

Fukushima, K., Takahashi, K., Kudo, J., & Kato, M. (1985). Interstitial-vestibular interaction in the control of head posture. *Experimental Brain Research*, *57*, 264–270.

Gerfen, C. R. (2003). D1 dopamine receptor supersensitivity in the dopamine-depleted striatum animal model of parkinson's disease. *The Neuroscientist*, *9*(6), 455–462.

Gerfen, C. R., Miyachi, S., Paletzki, R., & Brown, P. (2002). D1 dopamine receptor supersensitivity in the dopamine-depleted striatum results from a switch in the regulation of erk1/2/map kinase. *Journal of Neuroscience*, *22*(12), 5042–5054.

Girasole, A. E., Lum, M. Y., Nathaniel, D., Bair-Marshall, C. J., Guenthner, C. J., Luo, L., ... Nelson, A. B. (2018). A subpopulation of striatal neurons mediates levodopa-induced dyskinesia. *Neuron*, *97*(4), 787–795. e786.

Graybiel, A. M. (1998). The basal ganglia and chunking of action repertoires. *Neurobiology of Learning and Memory*, *70*(1–2), 119–136.

Graybiel, A. M., & Rauch, S. L. (2000). Toward a neurobiology of obsessive-compulsive disorder. *Neuron*, *28*(2), 343–347.

Hammond, C., Bergman, H., & Brown, P. (2007). Pathological synchronization in parkinson's disease: Networks, models and treatments. *Trends in Neurosciences*, *30*(7), 357–364.

Hashimoto, T., Elder, C. M., Okun, M. S., Patrick, S. K., & Vitek, J. L. (2003). Stimulation of the subthalamic nucleus changes the firing pattern of pallidal neurons. *Journal of Neuroscience*, *23*(5), 1916–1923.

Hassani, O.-K., Mouroux, M., & Feger, J. (1996). Increased subthalamic neuronal activity after nigral dopaminergic lesion independent of disinhibition via the globus pallidus. *Neuroscience*, *72*(1), 105–115.

Heilman, K. M., Watson, R. T., & Valenstein, E. (1993). Neglect and related disorders. *Clinical Neuropsychology*, *3*, 279–336.

Holmes, A. L., Forcelli, P. A., DesJardin, J. T., Decker, A. L., Teferra, M., West, E. A., ... Gale, K. (2012). Superior colliculus mediates cervical dystonia evoked by inhibition of the substantia nigra pars reticulata. *Journal of Neuroscience*, *32*(38), 13326–13332.

Howes, O. D., & Kapur, S. (2009). The dopamine hypothesis of schizophrenia: Version iii—the final common pathway. *Schizophrenia Bulletin*, *35*(3), 549–562.

Kalanithi, P. S., Zheng, W., Kataoka, Y., DiFiglia, M., Grantz, H., Saper, C. B., ... Vaccarino, F. M. (2005). Altered parvalbumin-positive neuron distribution in basal ganglia of individuals with tourette syndrome. *Proceedings of the National Academy of Sciences*, *102*(37), 13307–13312. doi: 10.1073/pnas.0502624102.

Kataoka, Y., Kalanithi, P. S., Grantz, H., Schwartz, M. L., Saper, C., Leckman, J. F., & Vaccarino, F. M. (2010). Decreased number of parvalbumin and cholinergic interneurons in the striatum of individuals with tourette syndrome. *Journal of Comparative Neurology*, *518*(3), 277–291.

Kim, N., Li, H. E., Hughes, R. N., Watson, G. D. R., Gallegos, D., West, A. E., ... Yin, H. H. (2019). A striatal interneuron circuit for continuous target pursuit. *Nature Communications*, *10*(1), 2715. doi: 10.1038/s41467-019-10716-w.

Klier, E. M., Wang, H., Constantin, A. G., & Crawford, J. D. (2002). Midbrain control of three-dimensional head orientation. *Science*, *295*(5558), 1314–1316.

Kofler, M. J., Rapport, M. D., Bolden, J., Sarver, D. E., & Raiker, J. S. (2010). Adhd and working memory: The impact of central executive deficits and exceeding storage/rehearsal capacity on observed inattentive behavior. *Journal of Abnormal Child Psychology*, *38*, 149–161.

Lauterbach, E. C., Spears, T. E., Prewett, M. J., Price, S. T., Jackson, J. G., & Kirsh, A. D. (1994). Neuropsychiatric disorders, myoclonus, and dystonia in calcification of basal ganglia pathways. *Biological Psychiatry*, *35*(5), 345–351.

Leckman, J. F., & Riddle, M. A. (2000). Tourette's syndrome: When habit-forming systems form habits of their own? *Neuron*, *28*(2), 349–354.

Lee, J. R., & Kiss, Z. H. (2014). Interhemispheric difference of pallidal local field potential activity in cervical dystonia. *Journal of Neurology, Neurosurgery & Psychiatry*, *85*(3), 306–310.

Liu, Y., Oubre, B., Duval, C., Lee, S. I., & Daneault, J.-F. (2022). A kinematic data-driven approach to differentiate involuntary choreic movements in individuals with neurological conditions. *IEEE Transactions on Biomedical Engineering*, *69*(12), 3784–3791.

Lozano, A. M., Dostrovsky, J., Chen, R., & Ashby, P. (2002). Deep brain stimulation for parkinson's disease: Disrupting the disruption. *The Lancet Neurology*, *1*(4), 225–231.

MacDonald, M. E., Ambrose, C. M., Duyao, M. P., Myers, R. H., Lin, C., Srinidhi, L., ... Groot, N. (1993). A novel gene containing a trinucleotide repeat that is expanded and unstable on huntington's disease chromosomes. *Cell*, *72*(6), 971–983.

Malenka, R. C., Angel, R. W., Hampton, B., & Berger, P. A. (1982). Impaired central error-correcting behavior in schizophrenia. *Archives of General Psychiatry*, *39*(1), 101–107.

Martin, J. P. (1963). The basal ganglia and locomotion: Arris and gale lecture delivered at the royal college of surgeons of england on 3rd january 1963. *Annals of the Royal College of Surgeons of England*, *32*(4), 219.

Martin, J. P. (1967). *The Basal Ganglia and Posture*: Philadelphia: Lippincott.

Martin, J. P., & Hurwitz, L. (1962). Locomotion and the basal ganglia. *Brain*, *85*(2), 261–276.

Mastro, K. J., Zitelli, K. T., Willard, A. M., Leblanc, K. H., Kravitz, A. V., & Gittis, A. H. (2017). Cell-specific pallidal intervention induces long-lasting motor recovery in dopamine-depleted mice. *Nature Neuroscience*, *20*(6), 815–823. doi: 10.1038/nn.4559.

Matsumura, M., Kojima, J., Gardiner, T. W., & Hikosaka, O. (1992). Visual and oculomotor functions of monkey subthalamic nucleus. *Journal of Neurophysiology*, *67*(6), 1615–1632.

McCutcheon, R. A., Abi-Dargham, A., & Howes, O. D. (2019). Schizophrenia, dopamine and the striatum: From biology to symptoms. *Trends in Neurosciences*, *42*(3), 205–220.

McKee, A., Levine, D., Kowall, N., & Richardson Jr, E. (1990). Peduncular hallucinosis associated with isolated infarction of the substantia nigra pars reticulata. *Annals of Neurology: Official Journal of the American Neurological Association and the Child Neurology Society*, *27*(5), 500–504.

McKinstry, S. U., Karadeniz, Y. B., Worthington, A. K., Hayrapetyan, V. Y., Ozlu, M. I., Serafin-Molina, K., ... Eroglu, C. (2014). Huntingtin is required for normal excitatory synapse development in cortical and striatal circuits. *Journal of Neuroscience*, *34*(28), 9455–9472. doi: 10.1523/JNEUROSCI.4699-13.2014.

Mettler, F. A. (1955). Perceptual capacity, functions of the corpus striatum and schizophrenia. *Psychiatric Quarterly*, *29*(1), 89–111.

Mettler, F. A., & Mettler, C. C. (1942). The effects of striatal injury. *Brain: A Journal of Neurology*, *65*, 242–255.

Middleton, F. A., & Strick, P. L. (1996). The temporal lobe is a target of output from the basal ganglia. *Proceedings of the National Academy of Sciences*, *93*(16), 8683–8687.

Middleton, F. A., & Strick, P. L. (2000). Basal ganglia and cerebellar loops: Motor and cognitive circuits. *Brain Research Reviews*, *31*(2–3), 236–250.

Mink, J. W. (1996). The basal ganglia: Focused selection and inhibition of competing motor programs. *Progress in Neurobiology*, *50*(4), 381–425.

Moll, C. K., Galindo-Leon, E., Sharott, A., Gulberti, A., Buhmann, C., Koeppen, J. A., ... Westphal, M. (2014). Asymmetric pallidal neuronal activity in patients with cervical dystonia. *Frontiers in Systems Neuroscience*, *8*, 15.

Morris, R. W., Cyrzon, C., Green, M. J., Le Pelley, M. E., & Balleine, B. W. (2018). Impairments in action–outcome learning in schizophrenia. *Translational Psychiatry*, *8*(1), 1–12.

Morris, R. W., Quail, S., Griffiths, K. R., Green, M. J., & Balleine, B. W. (2015). Corticostriatal control of goal-directed action is impaired in schizophrenia. *Biological Psychiatry*, *77*(2), 187–195.

Mukherjee, A., & Halassa, M. M. (2022). The associative thalamus: A switchboard for cortical operations and a promising target for schizophrenia. *The Neuroscientist*, 10738584221112861.

Nutt, J. G., Bloem, B. R., Giladi, N., Hallett, M., Horak, F. B., & Nieuwboer, A. (2011). Freezing of gait: Moving forward on a mysterious clinical phenomenon. *The Lancet Neurology*, *10*(8), 734–744.

Oberlander, C., Dumont, C., & Boissier, J. R. (1977). Rotational behaviour after unilateral intranigral injection of muscimol in rats. *European Journal of Pharmacology*, *43*(4), 389–390.

Onodera, S., & Hicks, T. P. (1998). Projections from substantia nigra and zona incerta to the cat's nucleus of darkschewitsch. *Journal of Comparative Neurology*, *396*(4), 461–482.

Pacherie, E. (2007). The anarchic hand syndrome and utilization behavior: A window onto agentive self-awareness. *Functional Neurology*, *22*(4), 211–217.

Parkinson, J. (1817). *An Essay on the Shaking Palsy*: London: Sherwood, Neely, and Jones.

Penfield, W., & Perot, P. (1963). The brain's record of auditory and visual experience: A final summary and discussion. *Brain*, *86*(4), 595–696.

Pynn, L. K., & DeSouza, J. F. (2013). The function of efference copy signals: Implications for symptoms of schizophrenia. *Vision Research*, *76*, 124–133.

Randrup, A., & Munkvad, I. (1967). Stereotyped activities produced by amphetamine in several animal species and man. *Psychopharmacologia*, *11*(4), 300–310.

Redgrave, P., Marrow, L., & Dean, P. (1992). Topographical organization of the nigrotectal projection in rat: Evidence for segregated channels. *Neuroscience*, *50*(3), 571–595.

Reep, R. L., Cheatwood, J. L., & Corwin, J. V. (2003). The associative striatum: Organization of cortical projections to the dorsocentral striatum in rats. *Journal of Comparative Neurology*, *467*(3), 271–292.

Reep, R. L., Corwin, J. V., Cheatwood, J. L., Van Vleet, T. M., Heilman, K. M., & Watson, R. T. (2004). A rodent model for investigating the neurobiology of contralateral neglect. *Cognitive and Behavioral Neurology*, *17*(4), 191–194.

Reiner, A., Albin, R. L., Anderson, K. D., D'Amato, C. J., Penney, J. B., & Young, A. B. (1988). Differential loss of striatal projection neurons in huntington disease. *Proceedings of the National Academy of Sciences*, *85*(15), 5733–5737.

Richter, R. (1945). Degeneration of the basal ganglia in monkeys from chronic carbon disulfide poisoning 1. *Journal of Neuropathology & Experimental Neurology*, *4*(4), 324–353.

Richter, R., & Klüver, H. (1944). Spontaneous striatal degeneration in a monkey. *Journal of Neuropathology & Experimental Neurology*, *3*(1), 49–62.

Robbins, T. (1976). Relationship between reward-enhancing and stereotypical effects of psychomotor stimulant drugs. *Nature*, *264*(5581), 57–59.

Rushworth, G. (1960). Spasticity and rigidity: An experimental study and review. *Journal of Neurology, Neurosurgery, and Psychiatry*, *23*(2), 99.

Scheel-Krüger, J., Arnt, J., & Magelund, G. (1977). Behavioural stimulation induced by muscimol and other gaba agonists injected into the substantia nigra. *Neuroscience Letters*, *4*(6), 351–356.

Schmack, K., Bosc, M., Ott, T., Sturgill, J., & Kepecs, A. (2021). Striatal dopamine mediates hallucination-like perception in mice. *Science*, *372*(6537), eabf4740.

Schrag, A., & Quinn, N. (2000). Dyskinesias and motor fluctuations in parkinson's disease: A community-based study. *Brain, 123*(11), 2297–2305.

Sedov, A., Semenova, U., Usova, S., Tomskiy, A., Crawford, J. D., Jinnah, H. A., & Shaikh, A. G. (2019). Implications of asymmetric neural activity patterns in the basal ganglia outflow in the integrative neural network model for cervical dystonia. *Progress in Brain Research, 249*, 261–268.

Sedov, A., Usova, S., Semenova, U., Gamaleya, A., Tomskiy, A., Crawford, J. D., ... Shaikh, A. G. (2019). The role of pallidum in the neural integrator model of cervical dystonia. *Neurobiology of Disease, 125*, 45–54.

Seeman, P. (2002). Atypical antipsychotics: Mechanism of action. *The Canadian Journal of Psychiatry, 47*(1), 29–40.

Shaikh, A. G., Zee, D. S., Crawford, J. D., & Jinnah, H. A. (2016). Cervical dystonia: A neural integrator disorder. *Brain, 139*(10), 2590–2599.

Sharp, F. R., Rando, T. A., Greenberg, S. A., Brown, L., & Sagar, S. M. (1994). Pseudochoreoathetosis: Movements associated with loss of proprioception. *Archives of Neurology, 51*(11), 1103–1109.

Silbersweig, D. A., Stern, E., Frith, C., Cahill, C., Holmes, A., Grootoonk, S., ... Schnorr, L. (1995). A functional neuroanatomy of hallucinations in schizophrenia. *Nature, 378*(6553), 176–179.

Spix, T. A., Nanivadekar, S., Toong, N., Kaplow, I. M., Isett, B. R., Goksen, Y., ... Gittis, A. H. (2021). Population-specific neuromodulation prolongs therapeutic benefits of deep brain stimulation. *Science, 374*(6564), 201–206. doi: 10.1126/science.abi7852.

Swedo, S. E., Pietrini, P., Leonard, H. L., Schapiro, M. B., Rettew, D. C., Goldberger, E. L., ... Grady, C. L. (1992). Cerebral glucose metabolism in childhood-onset obsessive-compulsive disorder: Revisualization during pharmacotherapy. *Archives of General Psychiatry, 49*(9), 690–694.

Taverna, S., Ilijic, E., & Surmeier, D. J. (2008). Recurrent collateral connections of striatal medium spiny neurons are disrupted in models of parkinson's disease. *Journal of Neuroscience, 28*(21), 5504–5512.

Teicher, M. H., Anderson, C. M., Polcari, A., Glod, C. A., Maas, L. C., & Renshaw, P. F. (2000). Functional deficits in basal ganglia of children with attention-deficit/hyperactivity disorder shown with functional magnetic resonance imaging relaxometry. *Nature Medicine, 6*(4), 470–473.

Vauquelin, G., Bostoen, S., Vanderheyden, P., & Seeman, P. (2012). Clozapine, atypical antipsychotics, and the benefits of fast-off d 2 dopamine receptor antagonism. *Naunyn-Schmiedeberg's Archives of Pharmacology, 385*(4), 337–372.

Viviani, P., Burkhard, P. R., Chiuvé, S. C., dell'Acqua, C. C., & Vindras, P. (2009). Velocity control in parkinson's disease: A quantitative analysis of isochrony in scribbling movements. *Experimental Brain Research, 194*(2), 259–283.

Watson, G. D., Hughes, R. N., Petter, E. A., Fallon, I. P., Kim, N., Severino, F. P. U., & Yin, H. H. (2021). Thalamic projections to the subthalamic nucleus contribute to movement initiation and rescue of parkinsonian symptoms. *Science Advances, 7*(6), eabe9192.

Wei, W., Ding, S., & Zhou, F.-M. (2017). Dopaminergic treatment weakens medium spiny neuron collateral inhibition in the parkinsonian striatum. *Journal of Neurophysiology, 117*(3), 987–999.

Welch, J. M., Lu, J., Rodriguiz, R. M., Trotta, N. C., Peca, J., Ding, J.-D., ... Luo, J. (2007). Cortico-striatal synaptic defects and ocd-like behaviours in sapap3-mutant mice. *Nature, 448*(7156), 894–900.

Wiener, N. (1948). *Cybernetics*: Paris: Hermann & Cie Editeurs.

Yamada, H., Fujimoto, K.-i., & Yoshida, M. (1995). Neuronal mechanism underlying dystonia induced by bicuculline injection into the putamen of the cat. *Brain Research, 677*(2), 333–336.

15 Synthesis

> Nature uses only the longest threads to weave her patterns, so that each small piece of her fabric reveals the organization of the entire tapestry.
>
> Richard Feynman

Since Willis' discovery of the corpus striatum nearly four centuries ago, much has been learned about the basal ganglia (BG). In attempting to explain the integrative function of the BG in the preceding chapters, it was necessary to take a long and circuitous route, often in unfamiliar territories. We are now ready to survey the sites visited and describe the new vistas before us.

15.1 BEHAVIOR AND CONTROL

Traditional models assume that the BG output suppresses behavior by tonically inhibiting target structures in the brainstem and thalamus and transiently allowing an action to be selected (Alexander, DeLong, & Strick, 1986; Mink, 1996). The action is defined at a level of abstraction, devoid of space or time, where there is nothing left to explain other than its occurrence. This "event-based" assumption has dominated BG research and indeed all of neuroscience for many decades. It is reflected in conventional experimental designs and analysis used to understand the role of the BG in behavior. In a typical experiment, behavior is labeled as a series of events, with time stamps for cues, actions, and rewards, and neural activity is examined relative to these time stamps. But this approach has produced ambiguous results and conflicting interpretations. Recent work has questioned the assumption that actions are all-or-none events. Instead, as discussed in earlier chapters, studies have revealed a striking relationship between continuous behavioral measures like kinematics and BG activity (Yin, 2017). These results can be explained by a new model of BG function based on hierarchical negative feedback control.

The assumption in traditional models of BG function is based on the linear causation paradigm, according to which organisms receive sensory inputs and convert these inputs to motor outputs using the brain. But contrary to this assumption, observable behavior is not the output of the organism. Although all behaviors require output from the final common path, muscle contraction cannot be equated with behavior. For example, when keeping a hand raised, the muscle output fluctuates as a function of environmental disturbances (e.g. gravity, muscle fatigue, and wind). Muscle tension varies to offset the impact of these disturbances, so that the position-related input is maintained at a relatively constant level. To generate behavior successfully, one must produce variable outputs from the final common path to counter the effect of disturbances without computing or modeling detailed properties of the environment. The only way to achieve this in biological organisms is to use negative feedback control.

To control, in the sense used here, is to specify the value of some input variable and to reach that value by acting on the environment. A control system compares perceptual inputs with internal reference signals and uses the difference or error signal to generate output. Through an external feedback function, the controller can reduce error. In doing so, it automatically mirrors the disturbance since its output is continuously adjusted to offset the impact of disturbance on its inputs (Chapter 5).

When properly understood, negative feedback control has revolutionary implications for understanding behavior and brain function, as it contradicts the dominant paradigm of input–output linear causation. Due to widespread misunderstanding of feedback control, however, these implications remain virtually unknown. In particular, misleading engineering conventions have prevented successful application of control theory to neuroscience. These conventions treat the control system as a servo that converts user input to desired output. Control engineering typically attempts to regulate some required output from a single loop from the perspective of an ideal observer. This approach poses the control problem from the perspective of an external observer rather than the controller itself. It focuses on modeling the dynamics of the system or plant being controlled and relies on the accuracy of the internal models of the environment. But just because engineers can perform matrix calculations with powerful computers does not mean that the brain also performs similar computations in real time.

In biological organisms, reference signals are intrinsic to the control system rather than introduced by the designer. What engineering conventions neglect is the system's autonomy. A negative feedback control system does not perform sensorimotor transformations. Instead of linear causation, it is characterized by circular causation in a closed loop, in which output affects input at the same time that the input affects output. It generates variable outputs in order to achieve consistent inputs. Conventional analysis of cause/effect systems, designed to determine how outputs are generated as a function of input, is futile and misleading when applied to closed-loop systems.

The central premise in this book is that the nervous system consists of a collection of control systems, which differ widely in the variables they sense and control. They are also organized hierarchically. Each controller is therefore a basic building block in the hierarchy, in which higher levels use descending commands to prescribe the reference signals of lower levels. This organization embodies the key computations through its interaction with the environment. What is normally called behavior is the outward manifestation of the underlying process of control. Most behaviors are generated by a complex hierarchy of control systems. At the lowest level, muscle tension, length, and joint angles are controlled. These control systems are implemented by spinal circuits. Specifying the reference signals for these lower-level controllers is the means by which higher variables can be controlled.

To situate the BG in a neural control hierarchy and to relate their contribution to those from the other levels of the hierarchy, we must understand what the BG output represents and how it influences output from the final common path. BG outputs do not usually send direct projections to motor neurons but instead influence midbrain and brainstem regions that in turn project to motor neurons. They are therefore in a position to send top–down commands to alter the reference signals of lower-level controllers. The hypothesis advanced here is that the BG implement transition control. A transition is the rate of change in a higher-level perceptual representation. Transitions can be controlled by generating variable behaviors, including postural adjustments, orienting, steering, and locomotion. Often the behaviors must be learned, as arbitrary sequences of simple transitions must be generated in a particular order depending on environmental contingencies.

The transition control hypothesis explains the wide variety of higher-level sensory inputs from the cerebral cortex to the striatum, as these inputs represent potentially controllable input variables at the level of transitions. It could also explain the diverse regions targeted by the BG outputs, which can ultimately access all effectors, including skeletomotor, autonomic, and neuroendocrine.

15.2 KINEMATICS AS A GATEWAY TO UNDERSTANDING BG FUNCTION

To understand the basic operations performed by the BG circuits, we can start by considering the control of simple proprioceptive transitions, as reflected in movement kinematics. Recent work has shown that, when kinematic variables like velocity and position are measured continuously, there is a striking relationship between these parameters and BG activity. The high correlation between the firing rates of BG neurons and kinematics is unprecedented.

There are distinct classes of neurons representing different vector components of kinematic variables. The components correspond to directions of motion (e.g. left, right, up, down). For example, spiny projection neurons (SPNs) can represent the velocity vector. In the sensorimotor striatum, some SPNs are correlated with leftward velocity; they are silent during rightward movement (Kim, Barter, Sukharnikova, & Yin, 2014; Kim et al., 2019). Other SPNs show the opposite pattern and show correlations with rightward velocity. On the other hand, in the output nuclei of the BG, for example the substantia nigra pars reticulata (SNr), there are different classes of neurons that represent different components of the position vector (Barter et al., 2015). SNr output neurons that increase firing when moving to the right decrease firing when moving to the left, and the firing rate of each neuron reflects the instantaneous position coordinates. The striatal representation of velocity and nigral representation of position suggest the presence of a neural integrator in the striatonigral circuit. The rate of change in the BG output is therefore proportional to the magnitude of striatal output.

Neural representations of kinematic variables in the BG are analog and vectorial. BG output neurons in the SNr and VTA show activity that is correlated with position variables but slightly leads such variables (Barter et al., 2015; Hughes et al., 2019). In the VTA, selective stimulation of the GABAergic projection neurons quantitatively determines the direction and amplitude of head-turning movements (Hughes et al., 2019). These observations suggest that the BG output can signal reference position vectors, which specify the displacement and direction of volitional movements. Different classes of BG output neurons representing different vector components are expected to have distinct projection patterns to lower-level position controllers. In agreement with this hypothesis, in brainstem nuclei that receive direct projections from the BG, there appears to be a topographical organization of nuclei corresponding to different position vector components for horizontal, vertical, and rotational movements (Masino, 1992; McElvain et al., 2021). Stimulation at different locations can evoke the isolated components, such as purely horizontal or purely vertical movements.

The hypothesis advanced here is that the BG project to the comparators of lower-level position controllers, and these projections can adjust reference signals in body orientation and configuration (Yin, 2014a, b, 2016). The discovery of precise kinematic representations suggest that BG output does not merely open a gate for action, as traditionally believed, but quantitatively determines action parameters. Inhibitory output from the BG does not simply suppress behavior, but rather provides a bias signal in a push–pull circuit for bidirectional control. For a continuous variable like position, the effective zero corresponds to a neutral position, like orienting straight ahead. The tonic bias allows bidirectional signaling, even though the firing rate can never be negative. Even in a simple movement like turning one's head, multiple classes of BG output neurons change their firing rates systematically, some increasing while others decreasing. When the firing rates of BG output neurons are constant, there is no descending command for movement, and the current posture is maintained.

It is important to emphasize that kinematic variables like position and velocity are independent of muscle contraction, the output of the final common path. In the control hierarchy, muscle activity can vary to reach a specific position or velocity. The final common path output is therefore the means by which higher reference states can be achieved.

Velocity is an example of a transition variable. More generally, the BG can control different types of transition variables in different sensory modalities. To control any transition variable, there must be some environmental feedback function linking the behavioral output to transition input. For example, walking forward can generate optic flow, and typing can generate changes in tactile sensations.

A simple transition is a movement like a head turn or a saccade. It is a basic unit of volitional behavior. Such simple transitions can be concatenated to form more complex transitions. The BG also mediate repetition control as in knocking, in which the tempo and bout length can be varied, or serial order control as in a dance routine, in which the order of the elements can be varied.

Movement is a change in posture, yet in conventional models, movement and posture are considered mutually exclusive (Mink, 1996). These models fail to explain why disorders of the BG can produce abnormal postures and rigidity. The control hierarchy sheds light on the posture/movement problem. Initiation of volitional behavior does not require turning off postural control. Rather, by changing the reference position vector using descending projections, it is possible to produce volitional movements.

A key feature of volitional actions is that their velocity can be arbitrarily controlled, i.e. one can move as quickly or as slowly as needed. According to the present model, the velocity controller is situated just above the position controller. The rate of change in position is reflected in the rate of change in the activity of output neurons.

It is hypothesized that the sensorimotor BG network contains a neural integrator that converts velocity commands into the rate of change in position reference signals. The descending position reference vector represents a request for a specific change in posture. The rate of this change can be controlled upstream using the velocity controller, which in turn reflects errors in higher-level controllers. When the BG output stays the same, there is no volitional behavior, but the lower-level controllers generate outputs continuously. Such outputs can be produced independently of the BG as they mirror disturbances in lower-level controlled variables like muscle length or joint angle, and automatically generating corrective outputs. Although the BG are not necessary for such "reflexive" behavior, they can command the same lower-level controllers to generate voluntary behavior.

15.3 CORTEX VERSUS BG

The cerebral cortex is the major source of reference and input functions in transition control, whereas the BG supply the comparison functions and output functions. There are two major cortical divisions. The anterior cortical division dictates what "should be," or reference representations, while the posterior division describes "what is," or perceptual representations. The frontal cortical areas contain a library of goal representations, mostly acquired through learning. Goals are perceptual memories that are used as reference signals for higher-level transition control systems. These goal states can be activated by predictors. For example, the smell of strawberries activates the strawberry representation as a potential goal, which may then send a reference signal to the transition controllers to generate specific actions. On the other hand, the posterior cortical areas carry signals that monitor ongoing sensory states of the variables being controlled. From lower-level sensory inputs, they can generate invariant object representations, which are often used for transition control.

There are corresponding anterior and posterior cortical areas, which are typically reciprocally connected, that converge in specific striatal areas. For example, related areas in S1 and M1 can both reach the same sensorimotor striatal region. This convergent pattern of corticostriatal connectivity routes input and reference signals to the comparison function in the striatum. Because cortex and BG function together, it is not appropriate to contrast their functions at the level of behavior, though they perform distinct operations within a given transition controller.

The BG are not required for perception per se or for representing goals of actions. Rather, they serve to compare inputs representing input and reference and contain a neural integrator that converts error signals to a sequence of reference signals for position controllers and to generate efference copy signals. The BG are critical for detecting and controlling the rate of change by modulating descending motor commands, whether to change the pitch, to run toward a reward, or to avoid an obstacle.

15.4 BG OUTPUTS AND THE COORDINATION PROBLEM

According to conventional models of BG function, BG projections to the thalamocortical network are necessary for action selection by activating the corticospinal projections from motor cortical areas. But this account neglects the projections from the BG to the brainstem and midbrain in movement or the patterns of axonal collateralization in BG output neurons that send similar BG signals to multiple destinations (Beckstead, 1983; Beckstead, Domesick, & Nauta, 1979; McElvain et al., 2021).

According to the present model, the BG projections represent descending reference signals to position controllers or initiate central pattern generators for rhythmic behaviors like licking, chewing, and locomotion. The projections to the thalamocortical network are often collaterals of descending projections. They are not necessary for movement per se, but they can send efference copies of ongoing action commands to the thalamocortical network (Chapter 8). Such signaling is critical for coordinating the activation of different cortico-BG circuits in generating learned behavioral sequences. Even simple behaviors involve a complex movement sequence in which multiple effector systems must be coordinated in space and time, much as different members of an orchestra must coordinate their efforts while performing a symphony. The parallel and interactive BG loop organization makes it possible for different transitions to be coupled to each other. For example, knocking on the door once is a simple transition in which the velocity of the movement can be controlled. Repeating this knock three times requires repetition control, in which each knock is treated as a distinct event and the tempo can be controlled. According to the present model, repetition control uses a reentrant circuit to reactivate the original knock reference. An efference copy from the BG output can be used to reactivate a reference signal using the reentrant projection back to the frontal cortical area where the reference signal originates.

Via projections to the thalamocortical network, the BG output can promote the activation of a different reference representation. This function is useful for combining simple transitions in a specific order by coordinating different controllers. If the knocks are then followed by turning the knob, it is then necessary to activate another transition reference for knob turning, which requires a different spatial target and a different pattern of effector activation. This priming of the reference representation by the efference copy from the previous component is hypothesized to be the mechanism underlying serial order control.

Each simple transition, lasting perhaps a few hundred milliseconds, can be treated as a single "pulse" that is accumulated in some other integrator circuit. This is similar to the relationship between the second hand and the minute hand on a mechanical clock. The larger integrators require reentrant projections to the frontal cortex. The two transitions can be coupled, so that the rate of change in one controlled variable can be related to that of another.

An emergent property of transition control is the sense of time and the capacity to time behaviors. There is no dedicated sensory receptor for time, as it emerges from higher-level monitoring of transitions in any sensory modality as well as efference copies of ongoing actions. Any transition controller can be used as a flexible timer, and a similar mechanism can be used for counting (Meck & Church, 1983). Not surprisingly, the BG are also critical for interval timing (Yin, 2014a).

15.5 DIRECT AND INDIRECT PATHWAYS

A key feature of BG anatomy is the existence of direct (striatonigral) and indirect (striatopallidal) pathways. These pathways have opposite effects on the BG output nuclei: the direct pathway reduces BG output, while the indirect pathway increases BG output. But how these pathways contribute to behavior has remained controversial, in part due to the assumption that actions are all-or-none. Influenced by this assumption, the direct pathway is often considered the GO pathway that generates action, whereas the indirect pathway is considered the NO–GO pathway that suppresses action.

According to the present model, the direct pathway provides the "inflow" to a neural integrator (Chapter 7). The striatonigral projection provides the input, and the nigral output is proportional to the time integral of the input. Accumulation is enabled by two successive inhibitory synapses, so that the net integrator output is reflected in the firing rates of neurons that receive nigral projections. On the other hand, the indirect pathway is hypothesized to have a damping or discharging function. It is analogous to a leak in the integrator.

How the integrator dynamics are manifested in behavior depends on the specific BG circuit involved and the controlled variable. In the sensorimotor striatonigral circuit for velocity control, direct and indirect activation could produce movement acceleration and deceleration, respectively. Stimulation of two pathways can produce movements in opposite directions: direct pathway activation generates contraversive movement, while indirect pathway activation generates ipsiversive movement. This can be explained by their opposite effects on the process of integration. An increase or decrease in the BG output reflects a change in position in a specific direction (Chapter 7). The relative timing of direct pathway activation and indirect pathway activation and the lag between the two determine the velocity profile. The indirect pathway could also be used to return to the original position by discharging the integrator.

At the event level, the activation of the direct pathway can result in action repetition, whereas the activation of the indirect pathway can result in switching to the next element in a behavioral sequence. At the relationship level, direct pathway activation could result in the selection of a particular target and indirect pathway activation in switching to a different target. Finally, for cognitive representations like mental imagery, direct pathway activation could promote the maintenance of the representation online, while indirect pathway activation could promote switching to a different representation.

15.6 DA AND ADAPTIVE GAIN

According to the present model, dopamine (DA) in the BG represents the gain for transition control (Chapter 7). DA can enhance dSPN output by activating D1-like receptors and reduce iSPN output by activating D2-like receptors (Gerfen & Surmeier, 2011; Lahiri & Bevan, 2020). Its net effect is to promote the accumulation in the integrator while pausing or preventing the leak. This effect is similar to multiplicative gain, so that the magnitude of the DA signal multiplied by the net excitatory input is proportional to the magnitude of the striatal output.

Moreover, the gain is not fixed but adaptive, changing according to the current demands. Adaptive gain is needed because the gain for transition controllers can vary greatly depending on the task requirement. For example, in the velocity controller, a low gain that allows stability when moving slowly may not be sufficient for fast movements, and the peak gain used for fast movements would result in instability if it is used as the default value at rest. Because DA neurons receive projections from BG output neurons that send descending action commands, DA signaling is also modulated by current behavioral states. This is an online estimator method for achieving adaptive gain.

The adaptive gain hypothesis explains why phasic DA activity, like SPN activity, can represent velocity vectors or force vectors (Barter et al., 2015; Hughes et al., 2020). The DA activity reflects an efference copy of striatal output, and as a direct online estimate of the velocity command, DA can also activate those striatal regions giving rise to the command in the first place, provided the descending excitatory drive from the corticostriatal projections is present.

Moreover, since DA neurons receive a wide variety of inputs from many brain regions (Watabe-Uchida, Zhu, Ogawa, Vamanrao, & Uchida, 2012), it is also possible to increase the gain in anticipation given predictors like discriminative or conditioned stimuli (Hughes et al., 2020; Schultz, Dayan, & Montague, 1997). For example, a dog may appear lazy and lethargic when resting alone in the living room, but as soon as it detects the return of the owner, the anticipation can rapidly increase the gain in all transition controllers, producing repetitive movements in a state of great excitement.

As adaptive gain, the contribution of DA to behavior depends on the target being modulated by DA neurons. The mesolimbic DA pathway from the VTA mainly modulates the limbic cortico-BG network, while the nigrostriatal pathway from the substantia nigra pars compacta mainly modulates associative and sensorimotor networks. In a velocity controller, adaptive gain increases the instantaneous velocity error and the rate of change in the reference position vector. DA depletion is then expected to reduce the peak velocity command that is attainable, as shown in bradykinesia or akinesia in Parkinson's patients. In the repetition control system, the adaptive gain function explains the well-known role of DA in self-stimulation. By increasing adaptive gain, phasic DA signaling can prolong and repeat ongoing behavioral selection by re-boosting the corticostriatal reference signal. The immediate activation of DA signaling following some action will therefore promote the repetition of that action—an effect often called reinforcement. As this is only a transient effect on performance, it should be distinguished from long-term learning. DA can promote learning only under specific conditions (Chapter 12).

15.7 MOTIVATIONAL HIERARCHY

Disorders involving the BG are often associated with symptoms in motivation, such as abulia and apathy. Such deficits are mostly attributed to damage to the limbic and associative BG networks. Experimental evidence in rodents suggests that the nucleus accumbens shell is critical for gustatory or consummatory behavior, whereas the nucleus accumbens core is critical for controlling approach behavior and effort regulation.

The parallel organization of cortico-BG networks can implement a labile motivational hierarchy. At the top of the hierarchy are a set of controllers that maintain the values of variables that are essential for survival. For example, in feeding, the homeostatic error can be reduced by activating innately coupled control systems responsible for consummatory behaviors. Higher-level error signals represent what is typically called desire.

The ends and means in the motivational hierarchy are transition variables, like rate of reward and rate of action. The lower level is the means, and the higher level is the end. The higher end can be defined as error reduction in the control of the higher variable, but it can also be the means to yet another higher end. For example, lever pressing is the means by which one may reach the end of collecting a food reward, but collecting a reward is the means to a still higher end, namely reducing hunger (error in energy homeostatic control).

The relationship between ends and means is reflected in a key feature of BG anatomy: the complex patterns of convergence and divergence in corticostriatal projections. Many cortical regions send projections to a small striatal region, and a small cortical region can project to multiple striatal regions (Chapter 4). That is to say, in principle multiple reference signals corresponding to multiple goal representations can be sent to the same comparator. For example, the striatal modules for turning left may be recruited for different purposes. Alternatively, often the same goal can also be reached using multiple actions. This is reflected in the divergent pattern in corticostriatal projections. Multiple actions are equivalent in reducing the error in a higher-level controller, but which one is recruited first depends on the corticostriatal synaptic strength as well as concurrent modulatory inputs.

Limbic cortico-BG networks can regulate interoceptive inputs, including chemical sensing like taste and smell, often directly coupled to consummatory behaviors, such as chewing and licking. These behavioral repertoires are only sufficient when the food is readily available for consumption, i.e. chewing may be initiated upon contact with hard food in the mouth. However, such behaviors cannot be used to find food. On the other hand, the behaviors used to obtain food cannot be innately specified due to the complexity and unpredictability of the environment. All that can be specified are general-purpose control systems—e.g. those for locomotor pattern generators, spatial proximity control, etc.—that generate preparatory behaviors to seek any number of goals. Through learning, these behaviors may be modified and combined in order to reach desired foods before the consummatory behaviors can occur. Homeostatic errors must be transformed into specific desires and channeled in the action hierarchy to generate adaptive behavior.

The connectivity of different cortico-BG networks also allows them to interact, in particular in the forward direction from limbic to associative and sensorimotor networks (Aoki et al., 2019; Haber, Fudge, & McFarland, 2000; Redgrave, Prescott, & Gurney, 1999). For example, as adaptive gain in the limbic BG, DA can determine the gain in reward rate control to recruit any number of effective actions (Chapter 9). Using the feedforward striato-nigro-striatal projections, it is possible for the limbic BG to influence specific circuits in the associative and sensorimotor BG. This influence reflects the channeling of adaptive gain signals from one controller to a relative controller. The higher controller can bias specific sets of inputs to the lower controller by adjusting gain, resulting in preferential activation of some reference and perceptual inputs so that a particular set of actions may be generated. For example, outcome representation activated by predictive cues can also activate the instrumental action that previously earned the specific outcome. This could explain the Pavlovian-to-instrumental transfer (PIT) effect, in which predictors of a reward can potentiate the performance of actions previously acquired to earn that reward. PIT depends on amygdalo-ventral striatal circuits as well as mesolimbic DA projections (Corbit & Balleine, 2011; Corbit, Janak, & Balleine, 2007).

15.8 GOAL SEEKING AND CONTROL OF RELATIONSHIP

To reach goals like food, prey, or mates, actions must be guided by distance from the goal. Spatial distance control is the foundation of relationship control. The limbic cortico-BG networks are critical for a conditioned approach to predictors of reward as well as consummatory behavior, but they are not needed for goal-directed instrumental actions. Rather, the associative cortico-BG network, including the prefrontal cortex and its projections to the posterior dorsomedial striatum (pDMS), is critical for learning and performance of goal-directed actions, as determined by behavioral assays like outcome devaluation and instrumental contingency degradation (Hart, Bradfield, Fok, Chieng, & Balleine, 2018; Yin, Ostlund, Knowlton, & Balleine, 2005). The pDMS is a key hub in this network, as it is necessary for the learning of action–outcome contingencies and the performance based on such contingencies.

Homeostatic errors are usually coupled to innate pattern generators that are available at birth, but the behaviors they generate are limited when facing the extensive and unpredictable disturbances in any environment. Through learning, the controllers for essential variables can recruit specific actions to achieve control. Homeostatic needs can be transformed into specific desires and incentives in the limbic network, which can then activate intentions and instrumental contingencies in the associative network. In the limbic network, the generation of desires can activate or deactivate a pattern generator or regulate the tempo, but it has limited control of action parameters and lacks global monitoring of action–outcome contingencies. To monitor such contingencies, efference copies of specific actions are integrated over time to generate a representation of action rate. This rate signal can then be correlated with a representation of rate of final outcomes (e.g. reward). The associative cortico-BG network, in particular the medial prefrontal cortical projections to the pDMS, appears to be critical for this function (Chapter 10).

On the other hand, the outputs of the sensorimotor network can regulate specific action parameters. It can control tactile and proprioceptive inputs, which provide critical feedback for skill learning and refinement of behavioral repertoires. It can also generate more arbitrary behaviors by varying kinematic variables.

Habit formation and skill learning have overlapping underlying neural mechanisms, including plasticity in the corticostriatal projections to the DLS and a shift in the locus of control from the associative to the sensorimotor BG network. This shift is accompanied by reduced attentional demand and increased effector specificity. It reflects a shift from a higher level in the motivational hierarchy to a lower level, with corresponding changes in the type of feedback that governs behavioral performance (Chapter 11). Extensive training, lean reinforcement schedules, and weak action–outcome contingencies promote habit formation.

Habitual behaviors can be generated in response to discriminative stimuli. They appear to be based on local feedback from each component of the sequence rather than feedback related to the final outcome. Skill learning additionally requires fine tuning at multiple levels of the control hierarchy, precision of online control, and detailed specification of action parameters. With learning, performance of each element within the sequence can be refined, for example, by specifying the exact proprioceptive or tactile feedback.

15.9 HIGHER FUNCTIONS

The transition control model also explains higher functions such as working memory, attention, and mental imagery. Perceptual representations from the thalamocortical network are recombined and ordered by the BG.

In the imagination mode, the output of a transition controller enters its own input function (Powers, 1973). It is hypothesized that this mode is enabled by the BG circuit, through which the cortex can activate the thalamocortical representations responsible for perception in the absence of overt behavior. This could be implemented by the BG outputs that mainly target higher-order thalamic nuclei, which reenter sensory cortical regions (Middleton & Strick, 2000). By using imagination alone, it is possible to activate the reference signals internally and retrieve the corresponding sensory states previously experienced with these reference signals. This makes it possible to rehearse different actions internally by simulating the expected outcomes without activating the downstream effectors. Through thalamocortical projections, the BG can affect perceptual representations in sensory cortical regions, enabling mental rehearsal and imagery for action planning. Malfunctions in this mechanism can contribute to hallucinations.

The BG also play an important role in gating and manipulating working memory. Lesions to components of the prefrontal cortex-associative BG circuit can lead to significant working memory deficits. The direct and indirect pathways in the BG can play opposite roles in this process, with the direct pathway responsible for repeating the same routine and the indirect pathway contributing to updating or switching to the next element. Thus, the manipulation of working memory representations uses the same mechanisms as control of serial order.

15.10 LEARNING AND REORGANIZATION

According to a popular view, the BG implement reinforcement learning (RL), which allows the organism to maximize rewards through trial and error in the absence of explicit instructions (Houk, Adams, & Barto, 1995; Miller, 1981). In evaluating this idea, we must distinguish between RL as a set of observations and RL as a computational model that attempts to explain these observations. There is extensive evidence implicating the BG in unsupervised learning based on feedback. These observations are valid, but they cannot be explained by the standard RL model. As a formal model that uses prediction errors to update state–action connection weights, the RL model is inadequate because it makes questionable assumptions about behavior. For example, in the actor-critic model, the most common RL model thought to be implemented by the BG, the behavioral output is assumed to be a discrete policy. There is no adequate model of how behavior is generated, and learning is conflated with changes in performance.

The actor-critic model relies on the reward prediction error (RPE) as a teaching signal to change the weights. This signal is often assumed to be implemented by phasic DA (Barto, 1995; Schultz et al., 1997). However, as shown in Chapter 12, results that purportedly support the RPE hypothesis can be explained by changes in performance rather than learning. In head-restrained mice, recent work using precise measurements of force exerted found opponent populations of VTA DA neurons that predict forward or backward movement (Bakhurin et al., 2023). During learning, there was a shift in the timing of forward force exertion, which coincided with the shift in phasic DA activity from the unconditioned stimulus to the conditioned stimulus. The shift in DA activity can be explained by the shift in force exertion during learning. In addition, the force tuning in DA neurons was also independent of learning or outcome valence. Manipulation of VTA DA neurons generated force on the task without affecting learning.

In the actor-critic model, learning promotes the repetition (reinforcement) of a particular policy, but repeating an output from a single control system cannot generate the desired behavior, given environmental disturbances. To achieve consistent results, the outputs repeated must represent high-level reference signals that request specific sensory inputs using negative feedback control.

In the present model, learning involves long-term changes in control system parameters, which can be achieved through long-term synaptic plasticity. Reorganization is initiated when the existing control hierarchy fails to function successfully. Persistent error generates systematic parameter variations in the relevant control systems, manifested in behavioral exploration and generation of behavioral variability.

Traditional RL models assume that a reinforcement or teaching signal strengthens the state-policy connection, thereby making the action more likely in the future. The present model, however, proposes that the parameter changes in the control system must occur transiently first, to promote effective exploration of the action space, before they can be saved in a more permanent form. Transient changes in the activation function of a neuron can be implemented by neuromodulation or short-term synaptic plasticity, allowing variations in system parameters during exploration. The reduction of error stops reorganization and saves the latest set of parameters. This process could be gradual if the accumulated error is large. When error reduction takes sufficient time and repetition, the parameters repeated are more likely to be saved via long-term synaptic plasticity. As adaptive gain, DA can promote learning through the repetition of the most recent set of parameters and the activation of intracellular pathways that ultimately lead to long-term plasticity.

15.11 GAPS IN UNDERSTANDING

The model of BG function advanced here is far from complete. Beyond the basic model for the control of simple transitions, the mechanisms for controlling higher-order transitions remain poorly understood. We still lack quantitative data on the relationship between BG activity and behavior. The studies discussed in this book are limited in the type of behavior measured, the temporal and spatial resolution of the behavioral measures, and the type of neural activity recorded.

Although significant progress has been made in understanding the broad functional contributions of specific cortico-BG networks, especially those involving the nucleus accumbens, pDMS, and DLS, our ignorance at the neural circuit level remains profound. For example, little is known about the contributions of the striatal chemical compartments, of intratelencephalic pathway and pyramidal tract corticostriatal pathways, of the various types of striatal interneurons, and of the GPe and STN.

Likewise, there are still major gaps in our knowledge of the cerebral cortex and thalamus, the two major sources of input to the BG. The relationship between neural activity and behavior has long appeared obscure and ambiguous, leading to many efforts to read the neural code using information theory. But this is primarily due to inadequate behavioral measures and failure to appreciate how neural signaling contributes to analog computing with neural circuits. As we have seen, the study of BG function has benefited from experimental approaches that simultaneously record neural activity and continuous behavioral measures. Whenever behavior is measured adequately and continuously in the BG, simple monotonic relationships between neural activity and behavioral variables are found, revealing analog signaling using firing rates. Although this process-based approach has shed light on BG representation of behavioral variables, it has rarely been used in the study of other brain areas. The experimental designs used in most studies on other brain regions are similar to those used in traditional studies of BG function and suffer from similar limitations. Consequently, despite decades of research, little is known about the representations of behaviorally relevant variables in the cerebral cortex and thalamus.

More globally, the cortico-BG network is only one of the major neural circuits with a loop organization. There are similar loop organizations connecting the thalamocortical network with areas like the midbrain and the cerebellum (McHaffie, Stanford, Stein, Coizet, & Redgrave, 2005; Middleton & Strick, 2000). Why are such loop organizations so common in the nervous system? How do the other loops differ from the cortico-BG loops in function? How do they interact with the cortico-BG loops? These questions remain to be answered.

15.12 A NEW VISTA AND THE WAY FORWARD

Despite the numerous gaps in our knowledge, the concepts and discoveries discussed in this book provide a new perspective on the BG. We have identified transition control as a unified function of the BG working in concert with the thalamocortical network. We have also outlined a model using basic building blocks and a limited set of analog computing operations to achieve transition control. The present model explains how, instead of selecting categorical actions, the BG contribute to the specification of action parameters in real time, and to the long-term modification of these parameters through learning. The transition control model makes predictions not only about the types of signaling to be found in different components of the BG but also about the behavioral consequences of manipulating these components. It can explain many empirical observations, including rigidity, postural deformity, locomotion deficits, abulia, bradykinesia, dyskinesia, working memory, compulsion, hallucination, and mental imagery.

As we have seen in Chapter 1, in the earliest speculations on BG function, Willis proposed that these nuclei are critical for sensation, movement, and imagination, though he did not explain what these terms mean. In the present model, movement is achieved by controlling sensory inputs. Imagination is a similar control process in simulation mode at the highest level of the hierarchy. Sensation, movement, and imagination are thus united by the control loop.

The basic cortico-BG circuit, which implements transition control, is a basic motif that is repeated in the brains of all vertebrates. Such control is only possible because the BG can monitor transition feedback from all sensory modalities online and generate variable outputs to achieve desired transitions. The common function of control explains the evolutionary conservation of the basic organization of the brain over the past 500 million years (Grillner & Robertson, 2016). Living organisms contain collections of control systems, which allow them to maintain homeostasis and achieve consistent behavioral results in the face of environmental disturbances.

The conceptual framework presented in this book contradicts some of the most influential assumptions in neuroscience, including linear causation, sensorimotor transformation, and information encoding and decoding. It suggests a new approach to studying the relationship between brain and behavior, using continuous behavioral measures and the test for the controlled variable. It may also stimulate the development of working models that can be implemented physically (Barter & Yin, 2021). The combination of a new experimental approach and the method of modeling provides a new roadmap for future research on how the brain works.

REFERENCES

Alexander, G. E., DeLong, M. R., & Strick, P. L. (1986). Parallel organization of functionally segregated circuits linking basal ganglia and cortex. *Annual Review of Neuroscience*, 9, 357–381.

Aoki, S., Smith, J. B., Li, H., Yan, X., Igarashi, M., Coulon, P., ... Jin, X. (2019). An open cortico-basal ganglia loop allows limbic control over motor output via the nigrothalamic pathway. *Elife*, 8, e49995.

Bakhurin, K. I., Hughes, R. N., Jiang, Q., Fallon, I. P., & Yin, H. (2023). Force tuning explains changes in phasic dopamine signaling during stimulus-reward learning. bioRxiv, 2023-04.

Barter, J. W., Li, S., Sukharnikova, T., Rossi, M. A., Bartholomew, R. A., & Yin, H. H. (2015). Basal ganglia outputs map instantaneous position coordinates during behavior. *Journal of Neuroscience*, 35(6), 2703–2716.

Barter, J. W., & Yin, H. H. (2021). Achieving natural behavior in a robot using neurally inspired hierarchical perceptual control. *Iscience*, 24(9), 102948.

Barto, A. G. (1995). Adaptive critics and the basal ganglia. *Models of Information Processing in the Basal Ganglia*, 215–232.

Beckstead, R. (1983). Long collateral branches of substantia nigra pars reticulata axons to thalamus, superior colliculus and reticular formation in monkey and cat. Multiple retrograde neuronal labeling with fluorescent dyes. *Neuroscience*, 10(3), 767–779.

Beckstead, R. M., Domesick, V. B., & Nauta, W. J. (1979). Efferent connections of the substantia nigra and ventral tegmental area in the rat. *Brain Research*, 175(2), 191–217.

Corbit, L. H., & Balleine, B. W. (2011). The general and outcome-specific forms of pavlovian-instrumental transfer are differentially mediated by the nucleus accumbens core and shell. *Journal of Neuroscience*, 31(33), 11786–11794. doi: 10.1523/JNEUROSCI.2711-11.2011.

Corbit, L. H., Janak, P. H., & Balleine, B. W. (2007). General and outcome-specific forms of pavlovian-instrumental transfer: The effect of shifts in motivational state and inactivation of the ventral tegmental area. *European Journal of Neuroscience*, 26(11), 3141–3149.

Gerfen, C. R., & Surmeier, D. J. (2011). Modulation of striatal projection systems by dopamine. *Annual Review of Neuroscience, 34*, 441–466. doi: 10.1146/annurev-neuro-061010-113641.

Grillner, S., & Robertson, B. (2016). The basal ganglia over 500 million years. *Current Biology, 26*(20), R1088–R1100.

Haber, S. N., Fudge, J. L., & McFarland, N. R. (2000). Striatonigrostriatal pathways in primates form an ascending spiral from the shell to the dorsolateral striatum. *Journal of Neuroscience, 20*(6), 2369–2382.

Hart, G., Bradfield, L. A., Fok, S. Y., Chieng, B., & Balleine, B. W. (2018). The bilateral prefronto-striatal pathway is necessary for learning new goal-directed actions. *Current Biology, 28*(14), 2218–2229 e2217. doi: 10.1016/j.cub.2018.05.028.

Houk, J. C., Adams, J. L., & Barto, A. G. (1995). A model of how the basal ganglia generates and uses neural signals that predict reinforcement. In: Houk, J. C., Davis, J. L. & Beiser, D. G. (Eds.), *Models of Information Processing in the Basal Ganglia* (pp. 249–270). Cambridge, MA: MIT Press.

Hughes, R. N., Bakhurin, K. I., Petter, E. A., Watson, G. D., Kim, N., Friedman, A. D., & Yin, H. H. (2020). Ventral tegmental dopamine neurons control the impulse vector during motivated behavior. *Current Biology, 30*, 1–14.

Hughes, R. N., Watson, G. D., Petter, E. A., Kim, N., Bakhurin, K. I., & Yin, H. H. (2019). Precise coordination of three-dimensional rotational kinematics by ventral tegmental area gabaergic neurons. *Current Biology, 29*(19), 3244–3255. e3244.

Kim, N., Barter, J. W., Sukharnikova, T., & Yin, H. H. (2014). Striatal firing rate reflects head movement velocity. *European Journal of Neuroscience, 40*(10), 3481–3490. doi: 10.1111/ejn.12722.

Kim, N., Li, H. E., Hughes, R. N., Watson, G. D. R., Gallegos, D., West, A. E., ... Yin, H. H. (2019). A striatal interneuron circuit for continuous target pursuit. *Nature Communications, 10*(1), 2715. doi: 10.1038/s41467-019-10716-w.

Lahiri, A. K., & Bevan, M. D. (2020). Dopaminergic transmission rapidly and persistently enhances excitability of d1 receptor-expressing striatal projection neurons. *Neuron, 106*(2), 277–290.

Masino, T. (1992). Brainstem control of orienting movements: Intrinsic coordinate systems and underlying circuitry. *Brain, Behavior and Evolution, 40*(2–3), 98–111.

McElvain, L. E., Chen, Y., Moore, J. D., Brigidi, G. S., Bloodgood, B. L., Lim, B. K., ... Kleinfeld, D. (2021). Specific populations of basal ganglia output neurons target distinct brain stem areas while collateralizing throughout the diencephalon. *Neuron, 109*, 1–18.

McHaffie, J. G., Stanford, T. R., Stein, B. E., Coizet, V., & Redgrave, P. (2005). Subcortical loops through the basal ganglia. *Trends in Neurosciences, 28*(8), 401–407.

Meck, W. H., & Church, R. M. (1983). A mode control model of counting and timing processes. *Journal of Experimental Psychology: Animal Behavior Processes, 9*(3), 320.

Middleton, F. A., & Strick, P. L. (2000). Basal ganglia and cerebellar loops: Motor and cognitive circuits. *Brain Research Reviews, 31*(2–3), 236–250.

Miller, R. (1981). *Meaning and Purpose in the Intact Brain*. New York: Oxford University Press.

Mink, J. W. (1996). The basal ganglia: Focused selection and inhibition of competing motor programs. *Progress in Neurobiology, 50*(4), 381–425.

Powers, W. T. (1973). *Behavior: Control of Perception*. New Canaan, CT: Benchmark Publications.

Redgrave, P., Prescott, T. J., & Gurney, K. (1999). The basal ganglia: A vertebrate solution to the selection problem? *Neuroscience, 89*(4), 1009–1023.

Schultz, W., Dayan, P., & Montague, P. R. (1997). A neural substrate of prediction and reward. *Science, 275*(5306), 1593–1599.

Watabe-Uchida, M., Zhu, L., Ogawa, S. K., Vamanrao, A., & Uchida, N. (2012). Whole-brain mapping of direct inputs to midbrain dopamine neurons. *Neuron, 74*(5), 858–873. doi: 10.1016/j.neuron.2012.03.017.

Yin, H. H. (2014a). Action, time and the basal ganglia. *Philosophical Transactions of the Royal Society B: Biological Sciences, 369*(1637), 20120473.

Yin, H. H. (2014b). How basal ganglia outputs generate behavior. *Advances in Neuroscience, 2014*, 768313.

Yin, H. H. (2016). The role of opponent basal ganglia outputs in behavior. *Future Neurology, 11*(2), 149–169.

Yin, H. H. (2017). The basal ganglia in action. *Neuroscientist, 23*(3), 299–313. doi: 10.1177/1073858416654115.

Yin, H. H., Ostlund, S. B., Knowlton, B. J., & Balleine, B. W. (2005). The role of the dorsomedial striatum in instrumental conditioning. *European Journal of Neuroscience, 22*, 513–523.

Index

Note: *Italic* page numbers refer to figures.

acetylcholine (ACh) 46–47, 55–56, 61, 273, 276
action selection models *86*
    central selection 86–87
    focused selection 85–86
    problems with 87
actor-critic models 88, 244–245, 248, 257–258, 309
    bird song learning 272
    feedback control and architecture *258*
adaptive gain
    and action parameters reinforcement 147–148
    bird song learning 273
    DA and 306
    described 146–147
    long-term plasticity, induction of 273–275
    and reinforcement 255–257
adenosine
    GPCRs 58
    signaling 58, *59*
akinesia 146, 284–285
anatomical organization of BG 33–34
    associative striatum 18
    cerebral cortex 15–16
    corticostriatal projections *see* corticostriatal projections to striatum
    development of *16*
    direct and indirect pathway neurons 23–25
    globus pallidus external segment (GPe) 27–28
    globus pallidus internal segment (GPi) 28
    limbic striatum 18
    names 15–17
    output nuclei 25–27
    putamen and caudate nucleus 15
    sensorimotor striatum 18–19
    striatum chemical compartments *19*, 19–20
    substantia nigra 30–32
    subthalamic nucleus (STN) 32–33
    thalamostriatal projections *22*, 22–23
    topographical inputs 17–19
    ventral pallidum (VP) 28–29
    ventral tegmental area (VTA) 29–30
anterior forebrain pathway (AFP) 270
A–O learning 213
approach behavior
    and feedback 205, *206*
    learning 206–208
arkypallidal axons 28
associationist models 209
associative cortico-BG network 213
associative striatum 18
attentional deficits 292
attention-deficit hyperactivity disorder (ADHD) 292
auditory feedback 269–270
auditory hallucinations 293
automaticity development 232–233

basal ganglia (BG) 310–311
    action selection 6–7
    behavioral analysis 8–9
    cortex *vs.* 304
    cortical function 4
    damage to 11
    described 1
    discovering 1–2
    exploration and 267–268
    extrapyramidal system 3
    fortunes 2–3
    function *see* BG function
    homology and homoplasy 6
    locomotion and 123, 130
    motif of cerebral organization 4–6, *5*
    output nuclei *see* BG output nuclei
    outputs and coordination problem 304–305
    outputs, functional significance of 127–129
    proto-BG circuit in invertebrates 6
    pyramidal/extrapyramidal distinction 3
    reinforcement learning (RL) 7–8
    steering and orienting, regulation of 120–121, 130
bed nuclei of stria terminalis (BNST) 18
behavior
    calculation problem 98
    challenge of analysis 89–90
    compulsive 236–238
    and control 301–302
    flexibility 234
    insufficiency principle 97–98
    licking behavior *see* licking behavior
    preparatory 184
    reparatory 185–186
    turning 120–121, *121*, 130
    variability 267
BG function 9–10, 302–304
    action selection models 85–87
    challenge of behavioral analysis 89–90
    concept of reentrant BG loops 75
    convergence and divergence patterns 78–79
    disinhibition 81–83
    interaction between loops 79–81
    parallel loops model 75–78
    rate model 83–85
    reinforcement learning 87–89
BG output nuclei 34
    pallidal axons 27
    pallidum 25
    striatopallidal projections 26
    striatum *vs.* pallid 26
    summary *26*
bidirectional control 115, 184
    common-mode signal 115
    stiffness control 115

bird song learning
  actor-critic models 272
  BG contributions to 270–271
  credit assignment problem in RL 273
  DA and 271–272
  direct and indirect pathways 272–273
  feedback and control 268–270, 272
  lessons from 272–273
  stages 268
blocking 252–253
body configuration 112
bradykinesia 84, 146, 284–285
brain
  behavior *see* behavior
  DBS *see* deep brain stimulation (DBS)
  midbrain contributions to orienting *see* midbrain contributions to orienting
brain stem 2–5, 18, 20, 25, 31–33, 79, 89, 257, 275, 283, 301–304
  BG *see* basal ganglia (BG)
  motor neuron 158–159
  STN outputs 288
  transition control *see* transition control
bridging collaterals 24
Bush–Mosteller model 248

caffeine 58
caudate nucleus 15
central pattern generators (CPG) 158
central selection model 86–87
cerebral cortex 4, 310
Cerebri Anatome (Thomas Willis) 1
cervical dystonia 120, 284
cholinergic interneurons (CINs) 46–47, 55
chorea 288–289
clinical symptoms
  ADHD 292
  analysis of 281–282
  bradykinesia, akinesia, and paradoxical kinesia 284–285
  control of body tilt *286*
  DBS *see* deep brain stimulation (DBS)
  gain, change in 282
  hyperkinetic symptoms 288–291
  impaired input function 281
  impaired output function 282
  locomotion, deficits in 285–287
  oscillations 282
  perseveration, stereotypy, and compulsion 291–292, 295
  postural control deficits 282–284, 295
  psychosis and schizophrenia 293–295
  rigidity 284
coding 107–108
compulsive behavior 236–238
compulsive grooming 291–292
compulsory approach 201–202
computing in control system
  bidirectional system control 100–101
  delays and oscillations 101
  perceptual and reference input 100
conditioned response (CR) 206
consummatory behavior 184–186
contingency 208–210

contributions of different BG networks to learning
  from actions to habits 229–230
  automaticity development 232–233
  compulsive behavior 236–237
  feedback, nature of 235–236
  multiple memory systems *see* multiple memory systems
  place and response 225–227, *226*
  procedural learning, BG and 227–228
  sensorimotor striatum and habit formation 230–232, *231*
control hierarchy
  behavioral flexibility 234
  computing system 100–101
  content and operation 108
  defined 99
  hierarchical 103
  misunderstanding 101–103
  modern control engineering 109
  negative feedback, components of closed-loop 99–100
  neural signaling and control systems 105–107
  reinforcement concept 104–105
  sensorimotor transformations, beyond 104
  signals 98
  teleology 104–105
control of relationship 308
conventional RL models 267
corpus Luysii *see* subthalamic nucleus (STN)
cortex *vs.* BG 304
cortico-BG networks 307–308, 310, 311
corticostriatal circuit 148–150
corticostriatal LTP 60
corticostriatal projections to glutamate 50
corticostriatal projections to striatum
  convergence and divergence 21–22
  input patterns *21*
  pyramidal type (PT) and intratelencephalic type (IT) neurons *20*, 20–21

DA *see* dopamine (DA)
DA and RL
  learning and performance 253–255
  necessary and sufficient for learning, DA 251–253
  phasic activity DA and RPE 246–247
  phasic DA and performance 248–251
  prediction error 243–244
  principles of RL 244–245
  reinforcement learning *vs.* control 257–258
  results that challenge RPE hypothesis 247–248
  value, performance, and RPE 248
damping, reduced 289–290
DA replacement treatment with levodopa (L-DOPA) 147
DA transporter (DAT) 161, 252
deep brain stimulation (DBS) 287–288
delusions 294–295
direct and indirect pathways 142–143, 305–306
  interval timing 166–167
  licking behavior 160
  neural basis of action–outcome learning 215–216
direct pathway spiny projection neurons (dSPNs) 23–25, *25*, 137, 232
disinhibition in BG circuits 81–83
  animal behavior, studies of 82

Index

cortical inputs 81–82
and direct excitation 82
striatonigral and nigrocollicular pathways in monkey eye movements 82–83
dopamine (DA)
　activity of SNc dopamine (DA) neurons *137*
　adaptive gain and 306
　axons 53
　and clock speed *164*
　contributes to effort 190–191
　dependence with learning, changes in 275–276
　dopaminergic modulation of GABAergic transmission 55
　and force generation 191–193, *192*
　and gain control 145–147
　influencing striatal outputs 54–55
　and kinematics 137–138
　midbrain neurons 53
　modulation of timing 165–166
　receptors 54
　RL and *see* DA and RL
　role in goal-directed actions 217–218
　sign tracking, and 206–207, *208*
dopaminergic projections to striatum 23, *24*
dorsolateral striatum (DLS) 168
dorsomedial striatum (DMS) 169
dyskinesia 84
dyskinetic movements 289–290

effector specificity 234
effort exertion
　conflating reward rate and performance 191
　and control hierarchy *189*
　DA contributes to effort 190–191
　effort defined 188
　feedback function 188
　limbic BG and effort regulation 190
　reward rate 188–189
electromyography (EMG) 135
endocannabinoid (eCB) 57
enkephalin 29
entopeduncular nucleus 27
exploration
　BG and 267–268
　reorganization and 267
explosive running 120
exteroceptive guidance system 285
extracellular signal-regulated kinase (ERK) 217
extrapyramidal side effects 293
extrapyramidal system 3–4
eye movements 119–120

factorization of length and tension (FLETE) 139
fast-spiking interneurons (FSIs) 45, 232
fast-spiking interneuron–spiny projection neuron (FSI–SPN) circuit *203*, 204–205
feedback
　and control in bird song *see* feedback and control in bird song
　function defined 188
　negative feedback control systems 282
　polarity in approach behavior 205, *206*, *210*
feedback and control in bird song
　auditory feedback 269–270

fundamental frequency 268–269
imaginary auditory inputs 270
sensorimotor learning stage 269
festination 285
final common path 97
firing rate 129
focused selection model 85–86, 150–151
FSIs *see* fast-spiking interneurons (FSIs)
function, BG *see* BG function

GABAergic projections
　VP 29
　VTA 30
GABAergic synapses 144
gain hypothesis 151
gamma-aminobutyric acid (GABA) 6, 43
　feedforward inhibition in striatum 51
　GABA-A receptors 50–51
　GABA-B receptors 50–51
　GABAergic transmission in pallidum 52
　lateral (recurrent) inhibition 52–53
gear coupling mechanism 161–162, *162*
globus pallidus external segment (GPe)
　described 27
　GPi and 27
　neurons 47, *48*
　pallidostriatal projections 28
globus pallidus internal segment (GPi) 27
　described 28
　neurons 48
glutamate 15
　corticostriatal transmission 50
　NMDARs 49
　thalamostriatal transmission 50
goal-directed actions 308
　approach behavior and feedback 205, *206*
　approaching goal 201
　BG contributions to neuroprosthetic control 218–219
　compulsory approach 201–202
　contingency, associative structures, and analysis of conditioning experiments 208–210
　FSI–SPN circuit and control of distance to target *203*
　learning to approach 206–208
　neural basis of action–outcome learning 211–218
　striatal circuit for relationship control and continuous pursuit 202–205
"go" pathway 83
G-protein–coupled inwardly rectifying potassium (GIRK) channels 50
G-protein–coupled receptors (GPCRs) 43
grooming 168

habit formation 308
habits
　action 229–230
　effector specificity 234
　plasticity mechanisms underlying 232
　reduced attentional demand 234
　sensorimotor striatum and habit formation 230–232, *231*
　and skills 234
habit strength (sHr) 7
habitual behaviors 308

hallucinations 295–296
   auditory 293
   visual 294
Hebbian plasticity 60, 207, 275
hedonic hotspots 184
hierarchical control 103
higher-order transitions
   dorsolateral striatum and grooming 168
   event repetition and control of tempo 157–158, *158*
   imagination 171–173, 175
   interval timing 163–167
   related rates and gear coupling 161–162
   rhythmic behavior regulation 158–161
   sequence learning 168–171
   serial order 167–168, *169*, 175
   transitions defined 157
   working memory 173–174
Histology of the Nervous System (Cajal) 3
homeostatic errors 308
Hull's model 7, 225
Huntington's disease 288–289
hyperkinetic symptoms
   chorea 288–289
   feedback, loss of 290–291
   LID 289
   model presentation 289
   reduced damping 289–290
   selectivity, lack of 290

imaginary auditory inputs 270
imagination *171*, 175
   action planning and simulation 172
   DA depletion 172–173
   described 171
   mental rotation 173
indirect pathway *see* direct and indirect pathways
indirect pathway spiny projection neuron (iSPNs) 232
ingestive behaviors 184
instrumental learning 266
insufficiency principle 97–98
integrator dynamics
   integrators described 140–141
   movement velocity control 142
   neural integration and behavior *141*
   phase splitter circuit 141–142
   STN and SPNs 142
interval timing *163*
   DA and clock speed *164*
   DA modulation of timing 165–166
   described 163–164
   direct and indirect pathways 166–167
   higher-order transitions 163–167
   pacemaker-accumulator model 164
   scalar expectancy model 164
intracranial self-stimulation 186–188
intrinsic homeostatic control systems 193

kinematics 302–304

labile motivational hierarchy 193–194, 196
laterodorsal tegmental nucleus (LDT) 55
L-DOPA–induced dyskinesia (LID) 289
learning
   associative structures 266

bird song learning *see* bird song learning
   blocking and 252–253
   from control perspective 265–266, *266*
   and performance 253–255
licking behavior 158–161
   direct and indirect pathways 160
   orofacial movements 158–159
   velocity control 160
   ventrolateral striatal dopamine (DA) 160–161
liking expression 182–184
limbic BG circuits 182–184
limbic cortico-BG networks 307
limbic striatum 18
LMAN projects 270
locomotion 123
   deficits in 285–287
   described 123
long-term depression (LTD) 62, 63
long-term potentiation (LTP) 62, 63
low-threshold spiking interneurons (LTSIs) 45

machine learning *see* reinforcement learning (RL)
medium spiny neurons *see* spiny projection neurons (SPNs)
mental rotation 173
mesencephalic locomotor region (MLR) 116, 123
mesolimbic DA and incentive salience 182–184, 196
methylphenidate 292
midbrain contributions to orienting
   fixation neurons 118
   gaze shift 118–119
   superior colliculus 118
misunderstanding control 101–103
motivation
   accumbens outputs regulating reparatory and consummatory behaviors 185–186
   aspects of 181
   DA and force generation 191–193, *192*
   effort exertion 188–191
   labile motivational hierarchy 193–194, 196
   limbic BG and reward 182–184
   parallel BG networks and *194*, 194–196
   reinforcement 186
   self-stimulation 186–188
   valence and bidirectional control 184, *185*
motivational hierarchy 193–194, 307–308
motor control in calculation problem 98
motor cortex 4
motor neuron
   alpha 97, 112–115
   gamma 114
movement disorders
   described 303–304
   rate model 83–85
multiple memory systems 237
   limitations of 228–229
   place/response task 225, *226*
muscle length control and fusimotor system
   descending gamma activation 114
   intrafusal muscle fibers 113–114
   knee jerk reflex 114–115
   muscle spindles, feature of 114
muscle tension and length, control of *113*
   fusimotor system, and 113–115
   Golgi tendon organs 112

Index

N-BACK working memory task 190
negative feedback control system *100*, 108
    calculation problem 99–100
    components 99
    misunderstanding *see* misunderstanding control
neural activity
    and behavioral variables 107
    information and coding 107–108
neural basis of action–outcome learning 211–212
    associative cortico-BG network and A–O learning 213
    DA role in goal-directed actions 217–218
    direct and indirect pathways 215–216
    pDMS 213–215
    striatal activity modulated by reward expectancy 212–213
    synaptic plasticity in pDMS 217
neural integrator in BG
    control system 140
    integrator dynamics 140–142
neural signaling
    analog circuits *106*, 106–107
    analog computations 106, 109
    and control systems 105–107
    digital code 105
    digital computer 106
    electric circuits 107
    firing rate 106
neurobiological implementation of integration 143–144, *145*
neuroprosthetic control 218–219
nicotinic ACh receptors (nAChRs) 55–56
nigral neurons 48–49
nigrocollicular axons 52
nigrocollicular pathway and eye movements 119–120
N-methyl-D-aspartate (NMDA) activation 207
N-methyl-D-aspartate receptors (NMDARs) 49, 62, 207, 215, 218
"no-go" pathway 83
"not enough" error signals 184
nucleus of the medial longitudinal fasciculus (nMLF) 117

obsessive–compulsive disorder (OCD) 291
olfacto-striatum *see* limbic striatum
orofacial movements 158–159

pacemaker–accumulator model 164, *165*
pallidal neurons
    GPe 47, *48*
    GPi 48
    nigral 48–49
pallidothalamic projections 52
pallidum 15–16, 33
    GABAergic transmission in 52
    striatum *vs.* 26
    ventral 28–29
paradoxical kinesia 284–285
parallel BG networks and motivated behaviors 194–196
parallel loops model 75–78
    BG damage 76
    concept of reentrant BG loops 75
    convergence 76–78
    limitations of 79–81
    loops 75–76, *76*
    recurrent or reverberating projections 75

    signals 76–77
Parkinson's disease
    bradykinesia and akinesia 284–285
    dopamine (DA) depletion 4
    rigidity 112, 116, 284
    symptoms of 135
pathway spiny projection neurons (dSPNs) 195
Pavlovian-instrumental transfer (PIT) 195–196
Pavlovian stimulus-reward task 136
pedunculopontine nucleus (PPN) 116
performance
    phasic DA and 248–251
    value and RPE 248
perseveration 291–292, 295
phasic dopamine (DA) activity
    patterns of *249*
    and performance 248–251
    and RPE 246–247
    at time of reward omission *250*
pimozide 255
place/response task 225, *226*
plasticity mechanisms underlying habits 232
ponto-medullary reticular formation (PMRF) 116
position controller 139, 144
position control system
    bidirectional control 115–116
    conventional model 111
    functional significance of BG outputs 127–129
    locomotion, BG and 123, 130
    midbrain contributions to orienting 118–119
    muscle tension and length, control of 112–115
    nigrocollicular pathway and eye movements 119–120
    Parkinsonian rigidity 112
    position controllers for orientation 116–118
    posture and movement 111–112
    reticulospinal pathway 116
    SNr and *122*, 122–123
    steering and orienting, BG regulation of 120–121, 130
    velocity control *vs.* 138–139
    volitional action 112
    VTA output and head position 126–127, *127*
positron emission tomography (PET) imaging in humans 135
posterior dorsomedial striatum (pDMS) 256
    A–O contingencies 215
    prelimbic–pDMS pathway 215
    pre-training or post-training lesions 213
    rats with 213, *214*
    rewards 213–214
    synaptic plasticity in *216*, 217
postural control deficits 282–283, *283*, 295
    asymmetric posture 283–284
    cervical dystonia and other deviant postures 284
posture 303–304; *see also* position control system
prediction errors
    described 243
    Rescorla–Wagner model 243–244
    TD algorithm 244
preparatory behavior 184
primate oculomotor systems 118
psychiatry 2
psychosis 293–295
putamen 15

Q-value 244–245

ramp-to-threshold mechanism 167
rate coding 107
rate model *83*
   described 83–84
   limitations of 84–85
reduced attentional demand 234
reduced damping 289–290
reinforcement 186
reinforcement learning (RL) 87–89, *89*, 309–310
   action selection and *8*
   adaptive gain and 255–257
   BG function, impact on 7–8
   control *vs.* 257–258
   credit assignment problem in 273
   DA and *see* DA and RL
   described 7
   law of effect 7
   principles of 244–245
reorganization
   exploration and 267
   neural plasticity and 273–276
Rescorla–Wagner model 243–244, 248, 252
response substitution 266
reticulospinal pathway 116
reward expectancy, striatal activity modulated by 212–213
reward prediction errors (RPE) 138, 309
   computing 247
   hypothesis, results challenging 247–248
   phasic DA and 246–247
   value, performance and 248
reward rate
   conflated with performance 191
   effort exertion 188–189
rhythmic behavior regulation 158–161
rigidity in Parkinson's disease 112, 116, 284
robotics and calculation problem 98

salivation 266
satiety 256
scalar expectancy model 164, 165
schizophrenia 296
   delusions 294–295
   described 293
   hallucinations 293–294
   positive and negative symptoms 293
secondary reinforcement 235
self-stimulation 186–188
sensorimotor learning 268
sensorimotor striatum 18–19
sequence learning 168–171
serial order 167–168, *169*, 175
sign tracking 206–207, *208*
skill learning 238
spasmodic torticollis 120
spike-timing–dependent plasticity (STDP) protocols 59–60, *60*
spiny projection neurons (SPNs) 16, 44, *45*, 78–79, 204, 302
   integrator dynamics 142
   velocity control *136*, 136–137
spiraling striatonigral projections 195
stereotypy 291–292, 295
stiffness control 115
stimulus-response (S-R) association 7

striatal activity modulated by reward expectancy 212–213
striatal long-term depression (LTD) 60, *62*
   D2 activation 61
   GABAergic synapses 61
striatal long-term potentiation (LTP) 60, *61*
striatal neurons 43–44
   CINs 46–47
   FSIs 45
   LTSIs 45
   SPNs 44, *45*
striatal synaptic plasticity, functional implications of 62–63
striatonigral synapses 143–144
striatonigral transmission 144
striatum 33
   cortical projections *149*
   corticostriatal projections *20*, 20–21
   described 3, 15–16
   DLS 168
   dopaminergic projections to 23, *24*
   feedforward inhibition in 51
   vs. pallidum 26
   striosome and matrix *19*, 19–20
striosome 19
substantia nigra
   described 30–31
   nigral output 31–32
   SNl 32
substantia nigra pars lateralis (SNl) 32
substantia nigra pars reticulata (SNr)
   instantaneous position coordinates using Cartesian coordinates *125*
   and position coordinates 123–126, *124*
   position vectors, representation of 124–126
   and postural control *122*, 122–123
   unilateral lesions 120–121
subthalamic nucleus (STN) 32–33
   anatomical organization of BG 32–33
   integrator dynamics 142
   and SPNs 142
synaptic depression 144
synaptic facilitation 143–144, *145*
synaptic plasticity in BG
   induction 59–60
   mechanisms 59
   plasticity described 59
synaptic plasticity in pDMS *216*, 217
synaptic transmission and plasticity 43

target velocity 202
teleology 104–105
temporal difference (TD) algorithm 244
thalamostriatal projections to glutamate 50
thalamostriatal projections to striatum
   cortical and intralaminar thalamic axon terminals 23
   intralaminar nuclei project 22
   ventral motor nuclei 22–23
thalamus 310
   rate model 84–85
Thorndike's law of effect 7
timing
   DA modulation 165–166
   role in BG 164

Index

tonically active neurons (TANs) *see* cholinergic interneurons (CINs)
"too much" error signals 184
topographical inputs to BG 17
    associative striatum 18
    limbic striatum 18
    sensorimotor striatum 18–19
torticollis 284
Tourette's syndrome 84, 290–292
transition control 165, 259, 276, 303
    adaptive gain and reinforcement of action parameters 147–148
    BG function models, compared to other 150–151
    clinical symptoms *see* clinical symptoms
    corticostriatal circuit and 148–150
    DA and gain control 145–147
    DA and kinematics 137–138
    direct and indirect pathway pathways 142–143
    focused selection model 150–151
    higher functions 308–309
    history-dependent gain hypothesis 151
    limitations in experimental designs 151–152
    neural integrator in BG 140–142
    neurobiological implementation of integration 143–145
    position *vs.* velocity control 138–139
    sensorimotor BG network 152
    transitions defined 157
    velocity control 135–137, 152
    VITE model 139–140

transport lags in controller 98
turning behavior 120–121, *121*, 130
two-way position control mechanism 129

unconditioned response (UR) 266
unilateral SNr lesions 120–121

valence or value
    described 184
    performance and RPE 248
vector integration to endpoint (VITE) model 139–140
velocity control *139*, 152, 303
    licking behavior 160
    Pavlovian stimulus-reward task 136
    PET imaging in humans 135–136
    position control *vs.* 138–139
    SPNs *136*, 136–137
velocity controller 138
ventral pallidum (VP) 18, 28–29
ventral tegmental area (VTA) 29–30
ventromedial thalamus 161
visual hallucinations 293
volitional action 112
VTA DA neurons 191–192

wanting expression 183–184
working memory 173–174, 292

zero-reference signal 129

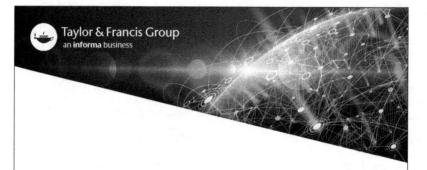

Printed in the USA
CPSIA information can be obtained
at www.ICGtesting.com
LVHW081143221223
766782LV00084B/145